Managing Microorganisms

Managing Microorganisms

Authors

David Smith

Director Biological Resources, CABI

Matthew J. Ryan

Research Lead Biological Resources, CABI

Alan G. Buddie

Microbiology Research Leader, CABI

CABI

CABI is a trading name of CAB International

CABI
Nosworthy Way
Wallingford
Oxfordshire OX10 8DE
UK
Tel: +44 (0)1491 832111
E-mail: info@cabi.org
Website: www.cabi.org

CABI
200 Portland Street
Boston
MA 02114
USA
Tel: +1 (617)682-9015
E-mail: cabi-nao@cabi.org

A catalogue record for this book is available from the British Library, London, UK.

ISBN-13: 9781800622111 (hardback)
9781800622128 (ePDF)
9781800622135 (ePub)

DOI: 10.1079/9781800622135.0000

Commissioning Editor: Rebecca Stubbs
Editorial Assistant: Lauren Davies
Production Editor: James Bishop

Typeset by Straive, Pondicherry, India
Printed and bound in the USA by Integrated Books International, Dulles, Virginia

Contents

Contributors

Anthony Kermode, CABI Genetic Resources Collection Curator, Contributed to Chapters 4, 5 and 6 and Appendix D and F
Georgina Burdon, ex CABI, Contributed to Chapter 2
Emma Cobb, CABI, Contributed to Appendix E
Lisa Offord, CABI, revised Chapter 2 and contributed to Appendix B
Helen Stewart, CABI, Contributed to Chapter 2 and Appendix B

Preface

This book provides the background required to understand key aspects of isolation, handling, characterization and storage of fungi and bacteria, the key aspects of "managing microorganisms". It also provides a sound basis for what culture collections do and what they can offer.

It provides methods for preserving microorganisms and shows how they can be characterized to facilitate their use in research and industry. Importantly it provides guidance in isolating and growing bacteria and fungi, particularly plant pathogens that are identified on a daily basis in CABI through the Diagnostic Advisory Service (DAS). It also describes how the CABI living collection, with its role to support the organization's global role in agriculture and the environment and specifically as the UK National Collection of Fungus Cultures, plays a key role in these processes. The chapters of the book are also structured to support CABI's training course providing the theory behind the operations of the collection that leads into the laboratory-based training offered by CABI. The reader will learn how to utilize the services and products of the global network of culture collections but importantly they will be able to preserve and characterize strains and maintain their own collections.

The book builds on two key publications: Smith and Onions' *Preservation and Maintenance of Living Fungi* (Smith and Onions, 1994) and *The UK National Culture Collection Biological Resource: Properties, maintenance and management* which was first published by CABI for the UK National Culture Collection (UKNCC) in 2001. These have been merged, updated and added to, providing an ideal resource book for all working with microorganisms.

The UKNCC book was originally edited by **David Smith** and **Matthew J. Ryan** at CABI with **John G. Day** then Curator, Culture Collection of Algae and Protozoa (freshwater), Centre for Ecology & Hydrology, Windermere. Individual chapters and text were also contributed by **Sarah Clayton**, CABI, **Paul D. Bridge**, then Professor in Biosciences, School of Biological and Chemical Sciences, Birkbeck, University of London, **Peter Green**, Curator, National Collections of Industrial, Food and Marine Bacteria, Aberdeen, with the preface written by **Professor Mike Goodfellow**, Dept of Agricultural and Environmental Science, University of Newcastle Upon Tyne.

It is important that lessons learned in the handling, preservation, characterization and use of microorganisms is passed on to others in order that reference, production and application strains are preserved intact and remain reproducible for future study and use. This book builds upon and updates a wealth of knowledge amassed by past and present specialists in the field. It provides reference material for all working with microorganisms and is further supported by CABI training and information resources. Although the focus is on bacteria and fungi, reference is made to the preservation and maintenance of other microorganisms covering best practices and compliance with law impacting on microbiology. It provides a resource to enable a business case to be constructed for establishing and sustaining a collection if this is what the reader wishes to do but also provides information on the services and support already available as an alternative approach to ensure their science and investment in generating resources is protected.

The CABI Bioscience team amass more than 120 years of work with the CABI living collection and the associated identification, research and consultancy services. The book provides the theory that supports practical training in microbiology and leads the reader towards the e-learning resources CABI is establishing through CABI Academy where an initial version of *Managing Microbes* with interesting learning exercises and quizzes can test knowledge in managing microorganisms.

Guidance is provided on how to set up and manage a culture collection to ensure cultures used in research are fit for purpose. The references and links to outside sources of information and training provide a wealth of knowledge and can lead the reader to key information needed in the management and use of microorganisms.

Acknowledgements

The authors wish to thank all the contributors and CABI Bioscience staff who have contributed to this book and have made working in the CABI Genetic Resource Collection such a wonderful and rewarding experience. CABI is an international intergovernmental organization, and we gratefully acknowledge the core financial support from our member countries (and lead agencies) including the United Kingdom (Department for International Development), China (Chinese Ministry of Agriculture), Australia (Australian Centre for International Agricultural Research), Canada (Agriculture and Agri-Food Canada), Netherlands (Directorate-General for International Cooperation), and Switzerland (Swiss Agency for Development and Cooperation). See https://www.cabi.org/what-we-do/how-we-work/cabi-donors-and-partners/ for full details.

Notes

1) Mentions of specific commercial brands should not be taken as endorsements; 2) Before undertaking any of the protocols/techniques mentioned, the user must undertake appropriate H&S risk assessments and take due precautions in accordance with the findings and/or any statutory/local H&S regulations

Introduction: An Overview of *Managing Microorganisms*

Microorganisms require careful handling. If you work with microorganisms in research, health, academia or industry or manage microbial collections, you'll know that it's a complex field. There are a broad range of different cell types to work with requiring different approaches to their isolation, purification, characterization, preservation, storage and use. There are many texts and papers describing how these operations can be managed and be successful but where do you start? What are the key principles in selecting the best approach for your needs? Impacting on all of this is a broad range of legislative and regulatory guidance which everyone in a laboratory needs to know and work within to ensure that everyone is compliant and protected at every stage of their work. This volume builds on the information provided in Smith and Onions (1994) *The Preservation and Maintenance of Living Fungi* and the UKNCC *Biological Resource: Properties, maintenance and management* manual, updating and expanding the content.

As curators of fungi and bacteria since CABI's origins in 1920 and managing the UK National Collection of Fungus Cultures since 1947, CABI has built up a wealth of knowledge about working with microorganisms, particularly fungi and bacteria. Our team has written, taught and consulted on all aspects of culture collection operation and management to a wide range of international audiences. They understand the complexity of the technical and legislative framework and have applied their insight to present this in *Managing Microorganisms* to provide a comprehensive, easy to use reference source. Packed with critical information and useful references, users will be able to cover the broad scope of issues from effective microbial identification and safe laboratory practice to international, regional and industry specific law. The background and reference materials are presented in the hardcopy version of this book with the e-book version facilitating access to references, protocols, guidance and further text in all the areas covered.

Managing Microorganisms covers all aspects of collecting, storing and handling microorganisms and all the associated regulatory guidance. To provide a deeper knowledge of handling and using microorganisms the cited literature gives examples of approaches and protocols. Links to international, regional and local regulations are given where available and appropriate to cover all aspects of collecting, developing and managing a microbial culture collection; the latter includes:

- Understanding their characteristics, properties and identification
- Sampling organisms and naming them
- Maintaining organisms – growing them, storing them, keeping them stable
- Delivering and transporting microorganisms to other laboratories.

CABI offers laboratory-based courses supported by the *Managing Microorganisms* book and is building an extensive set of e-learning experiences through CABI Academy (https://www.cabi.org/product-training/cabi-academy/). The courses are aimed at anyone who is involved in storing, working with and handling microorganisms in a range of sectors – from agriculture, food, cosmetics and pharmaceuticals through to research and higher education including laboratory managers and technicians, culture collection managers and students studying microbiology. All would benefit by:

- Integrating *Managing Microorganisms* into existing induction and in-house training, underpinning individual professional development goals
- Ensuring consistency of knowledge development
- Meeting the laboratory certification requirements for the sector
- Gaining the benefits of CABI's experts derived from their work in setting up microbial collections worldwide.

The combination of the *Managing Microorganisms* book with e-learning resources enables participants to:

- Learn at their own pace, at their convenience at home or in the office
- Improve their ability to identify and treat different microbial infections
- Apply their knowledge practically on a daily basis, which builds retention and ensures consistent development over the course of their work
- Access an extensive collection of literature and live weblinks and online documents from which to build their own reference library
- Improve the quality of their research and their ability to produce high quality strains
- Share their progress with colleagues and build their confidence as they reflect on their knowledge development.

Managing Microorganisms presents a practical overview of best practice in working with microorganisms. Information and reference sources are provided which will assist in implementing professional practice in a microbiological laboratory to enable the reader to learn the core skills required to establish and manage a collection of microorganisms. The book covers topics on a wide range of theoretical considerations, practical skills and legislative guidance which are essential for professionals working with microorganisms.

The book does not replace practical training courses or "on the job" training but serves to provide the theory and direction towards implementation of best practices. CABI will provide further assistance where necessary to clarify content or provide some direction in implementation of best practice and assist with the laboratory-based training so that the knowledge gained can be implemented in practice.

The goal is to provide:

- An understanding of the different aspects of culture collection operations and their quality management
- The methods available for microorganism characterization and use
- The best practices in the isolation, growth, identification, properties and storage of microorganisms
- An understanding of the methodologies for long-term maintenance and preservation of microorganisms
- An understanding of the health and safety requirements of microbiology as required by law.

This will enable the reader to:

- Design a sustainable business plan for a culture collection
- Determine whether using a culture collection for preservation and supply of strains would be more appropriate for their research purposes than creating their own collection
- Categorize the methods of preservation and characterization for different groups of microorganisms
- Demonstrate the key principles of culture collection management
- Determine the most appropriate course of action for protecting investment in research with microorganisms
- Describe where to source microorganisms
- Explain how to store, handle and distribute organisms effectively and safely
- Identify the crucial elements of handling and storage of microorganisms.

Resource books and references

The book cites many sources of information and protocols but has specifically incorporated the following resources with updates and recommendations.

The UK National Culture Collection (UKNCC) Biological Resource: Properties, maintenance and management

This book updates the original content covering bacteria and fungi of Smith, D., Ryan, M.J. and Day, J.G. (eds) (2001) *The UK National Culture Collection Biological Resource: Properties, maintenance and management*. pp 382. UK National Culture Collection, Egham. The book was published by CABI on behalf of the UKNCC and the text relevant to CABI input has been used extensively throughout the chapters of this book, updated to consider developments over the past 20 years since the original was published.

Biological Resource Centre (BRC) Generic Business Plan

This has been presented as Chapter 7 and provides an outline for a business case or plan for a culture collection. It is written on the principle that the long-term sustainability of a BRC (culture collection) will require the generation of income lines that are traditionally a mix of culture sales, contracted services, contracted research, sponsorship and more. It would be necessary to attain a level of income that will allow expansion and development in the medium term requiring a portfolio of new products and services, mechanisms for funding a conservation role and increased competitiveness in securing research funding.

CABI Genetic Resources Collection Compliance with the Law

There is extensive legislation concerning the safe handling and distribution of microorganisms at the national, regional and international levels. This document was made public online, the content formed much of the EU European Biological Resource Centres Network (EBRCN) project output available at https://ebrcn.org/docs.php. The EBRCN project ran from 2004 – 2006 (http://www.cabri.org/FAQ/faq_ebrcn.html); it set common policies and maintained high standards of operation, linking BRC catalogues and other data sets through a single common website and co-ordinated responses to new initiatives relevant to living resources. This content is integrated into Chapter 5.

European Consortium of Microbial Resources Centres BRC standard

This is an EU project output based on the Organisation for Economic Co-Operation and Development (OECD) Best Practice Guidelines for Biological Resource Centres for the public domain. It was produced as Deliverable D.NA1.2.1: as a working draft for an ISO Standard with its delivery led by CABI. It can be found online on the European Consortium of Microbial Resources Centres (EMbaRC) website which was an EU project funded under the Seventh Framework Programme Research Infrastructures (INFRA-2008-1.1.2.9: Biological Resources Centres) for microorganisms (http://www.embarc.eu/). The standard provides mandatory guidance and provides the basis for best practice in the management of microbial domain Biological Resource Centres and all laboratories maintaining replicable microbiological materials. It draws together the key principles and best practice of quality management systems and operational guidelines prescribed by individual BRCs, national public service collections, and national, regional and world culture collection organizations. It has been designed to meet the needs of the user community through discussion in the OECD Biological Resource Centre Task Force, EBRCN, GBRCN and EMbaRC projects (http://www.embarc.eu/deliverables/EMbaRC_D.NA1.2.1_2.28_BRC_standard.pdf). This document is referenced and forms the basis of Chapter 6.

Open access references cited throughout the modules

References are listed at the end of each chapter in which they are listed and all those with online links can be connected to via their Digital Object Identifier (doi) or to online pdfs where they are open access. Unfortunately, some are only available for purchase or subscription to Journals. We have tried to focus on those that are open access but this is not always possible. The electronic version of the book enables linking out to these resources.

CABI training courses

Formal courses are available in:

Preservation, storage, maintenance of microorganisms

CABI expert scientists provide training in molecular identification of microorganisms, morphological identification of fungi, microbial techniques, isolation methods for fungi and bacteria, plant pathology, biological control and preservation of microorganisms. Visit the CABI website to see the current courses being offered at: https://www.cabi.org/products-and-services/bioscience-services/training/.

Study visits

Contact CABI to discuss the possibility of study visits in the CABI laboratories.

CABI products that will provide additional support

This section presents short descriptions of CABI products and the relevance to the course and what the user can get from them.

Description sheets

Diagnostic descriptions, illustrations and information on over 1700 fungal and bacterial taxa which cover:

- Fungal and bacterial pathogens of animals and plants
- Fungi of benefit to crops and other plants
- Fungi of importance in applied fields such as biocontrol, biodegradation, biodeterioration, biotechnology, food spoilage and industrial mycology
- Fungi pathogenic to man, and other fungi of medical significance
- Fungi of importance in biodiversity and conservation
- Common fungal saprobes and fungi important in forest nutrient recycling
- Microfungi, and some larger basidiomycetes
- Fungi significant in education
- Fungi significant in ethnomycology, folk traditions and traditional uses

This CABI product https://cabidigitallibrary.org/journal/dfb will help the user recognize significant beneficial and pathogenic fungi and bacteria. The sheets are designed for use particularly by plant pathologists and veterinary and medical mycologists but can also help recognize fungi important in assays, quality assurance testing protocols and screening programmes. Many species covered by the series are therefore important economically (for example in food spoilage or as crop diseases). Organisms of significance in biodiversity, conservation and other applied fields such as biological control of insects and weeds, biodeterioration, biotechnology and industrial mycology are also covered. Some more recent sheets deal with commonly encountered saprobes or facultative pathogens. Almost all fungi covered are microscopic, although a few sheets describe larger species, mostly basidiomycete pathogens.

Dictionary of the Fungi

The *Dictionary of the Fungi*, 10th Edition is an acclaimed reference work, has more than 21,000 entries, and provides the most complete listing available of generic names of fungi, their families and orders, their attributes and descriptive terms (as valid for the date of publication in 2008). For each genus, the authority, the date of publication, status, systematic position, number of accepted species, distribution, and key references are given. Diagnoses of families and details of orders and higher categories are included for all groups of fungi. In addition, there are biographic notes, information on well-known metabolites and mycotoxins, and concise accounts of almost all pure and applied aspects of the subject (including citations of important literature). This is the one essential handbook for all who work with or are interested in fungi (including lichens, slime moulds, yeasts and fungal analogues). It provides definitions of all fungal terms and enables even the least mycological minded to understand publications and keys on fungi. The definitions in the *Dictionary* are used to form the glossary of *Managing Microorganisms* giving added value to the e-book making it more of a learning tool.

Index of Fungi

The *Index of Fungi* lists recently published names of fungi, living and fossil, at all ranks including those organisms now considered to be better classified in the Protozoa and Chromista, compiled from world literature. The resource is a bi-annual listing providing full bibliographic and nomenclatural details of some 2000 names of fungi per annum. It includes names of new taxa, new and revised names at all ranks, arranged alphabetically and with host/substratum and epithet/name indexes, cumulating into a volume over 10 years. Available as a print only journal.

Books – including *Bibliography of Systematic Mycology*; the Mycology series includes over 70 items that are available for purchase online: http://bookshop.cabi.org/.

The current Books Catalogue can be downloaded and the user can create their own personal library to support their studies, research and work activities. CABI makes over 560 e-books available that cover Agriculture, Animal and Veterinary Sciences, Human, Food and Nutrition Sciences, Plant Sciences, Leisure and Tourism, Environmental Sciences and several of the Mycology series.

Main databases

CAB Direct https://www.cabdirect.org/

CAB Direct is the most thorough and extensive source of reference in the applied life sciences, incorporating the leading bibliographic databases CAB Abstracts and Global Health. CAB Direct provides a convenient, single point of access to all CABI database subscriptions.

CAB Abstracts https://www.cabi.org/publishing-products/cab-abstracts/

CAB Abstracts is a bibliographic information service providing access to the world's applied life sciences literature. It comes with **CABI Full Text**, giving users automatic access to over 570,000 journal articles, conference papers and reports (80% of which are not available electronically anywhere else).

CAB Abstracts Archive https://www.cabi.org/publishing-products/cab-abstracts-archive/

Covering the literature from 1913 to 1972, CAB Abstracts Archive makes over 1.8 million records available electronically. It brings the science of the early 20th century to the desktops of modern researchers, adding links between CAB Abstracts and any older full text articles that other publishers have now digitized.

Global Health https://www.cabi.org/publishing-products/global-health/

Global Health gives researchers and students access to over 4.3 million records covering all the world's relevant public health research and practice – providing knowledge without borders.

Global Health Archive https://www.cabi.org/publishing-products/global-health-archive/

The archive is a fully searchable modern database of public health and biomedical research dating from 1910 to 1983, containing 800,000 records derived from six printed abstract journals.

Internet resources

AgBiotechNet https://www.cabi.org/publishing-products/agbiotechnet/

AgBiotechNet is the complete online agricultural biotechnology information service that covers transgenics and tissue culture of plants and animals.

Environmental Impact https://www.cabi.org/publishing-products/environmental-impact/

Environmental Impact is an internet resource created in response to demand from the scientific community, policy makers and information specialists for a single comprehensive bibliographic information resource on climate change and other influences of humans on the biosphere.

Nutrition and Food Sciences https://www.cabi.org/publishing-products/nutrition-and-food-sciences-database/

Nutrition and Food Sciences is a specialist internet resource that gathers together research, reviews and news on nutrition and the food sciences in an easily searchable database. It covers the whole food chain from the raw ingredients to the physiological and health effects of nutrients. Key subject areas include public health, clinical nutrition, nutrition physiology, food safety, food security, consumer behaviour, food science and technology and food product groups such as functional foods.

VetMed Resource https://www.cabi.org/publishing-products/vetmed-resource/

VetMed Resource is the most comprehensive tool for veterinarians available. With its specific veterinary focus and access to over 2.4 million relevant records from the acclaimed CAB Abstracts database, VetMed Resource is the tool of choice when it comes to keeping up-to-date in your practice.

Specialized databases

Abstracts on Hygiene and Communicable Diseases https://www.cabi.org/publishing-products/abstracts-on-hygiene-and-communicable-diseases/

Communicable diseases and public health including epidemiology and risk, diagnosis, therapy, health promotion and disease prevention.

AgBiotech News and Information https://www.cabi.org/publishing-products/agbiotech-news-and-information/

AgBiotech News and Information brings you all the latest information on agricultural biotechnology, covering genetic engineering, molecular genetics and in vitro culture of plants and animals.

Biocontrol News and Information https://www.cabi.org/publishing-products/biocontrol-news-and-information/

This fully searchable abstract database provides the latest information on important developments in all aspects of the use of living organisms to control pest animals, weeds and plant diseases.

Bioprotection Portal https://www.cabi.org/publishing-products/bioprotection-portal/

The CABI BioProtection Portal is a free, web-based tool that enables users to discover information about registered biocontrol and biopesticide products around the world.

Biofuels Abstracts https://www.cabi.org/publishing-products/biofuels-abstracts/

Biofuels Abstracts provides you with the latest information on biofuels and bioenergy, including production of energy crops, energy forestry, short-rotation coppice, processing of crops and agricultural wastes into biodiesel or fuel ethanol, utilization of fuelwood, and all CAB Abstracts records relating to key biofuel species such as *Jatropha*.

Environmental Science Database https://www.cabi.org/publishing-products/environmental-science-database/

The Environmental Science Database provides access to international scientific literature relating to all aspects of environmental quality, monitoring, resource management and conservation.

Review of Medical and Veterinary Mycology https://www.cabi.org/publishing-products/review-of-medical-and-veterinary-mycology/

Review of Medical and Veterinary Mycology is a fully searchable abstracts database of internationally published research on medical and veterinary mycology.

Review of Plant Pathology https://www.cabi.org/publishing-products/review-of-plant-pathology/

Review of Plant Pathology provides the latest information on plant pathogenic fungi, viruses, bacteria, phytoplasmas and spiroplasmas, including pathogen biology from genetics to geographical distributions.

Rice Abstracts https://www.cabi.org/publishing-products/rice-abstracts/

Rice Abstracts is a fully searchable abstracts database of internationally published research, bringing you the latest information on rice production, from breeding and agronomy to storage, consumption and marketing.

Soil Science Database https://www.cabi.org/publishing-products/soil-science-database/

The Soil Science Database contains over 30 years of international research, providing an ideal source of information on soils.

Soils and Fertilizers Abstracts https://www.cabi.org/publishing-products/soils-and-fertilizers-abstracts/

Soils and Fertilizers Abstracts brings you the latest information on soil science, soil management, fertilizers, and environmental aspects of agriculture.

Sugar Industry Abstracts https://www.cabi.org/publishing-products/sugar-industry-abstracts/

Sugar Industry Abstracts is a fully searchable abstracts database of internationally published research on the sugar industry.

Veterinary Science Database https://www.cabi.org/publishing-products/veterinary-science-database/

The Veterinary Science Database covers all aspects of veterinary medicine and arthropod, helminth, protozoal and fungal diseases of domestic and wild animals.

Weed Abstracts https://www.cabi.org/publishing-products/weed-abstracts/

Weed Abstracts brings you the latest information on weed biology, from genetics to ecology, including parasitic, poisonous, allergenic and aquatic weeds.

Overview of *Managing Microorganisms*

The book supports an online learning programme built around 6 modules; an early version is available on CABI Academy. The modules for this course are given below and the information to deliver them is presented in this book.

Module 1: Culture Collections

The module covers the different aspects of culture collections and their quality management systems including:

- Why Retain Microorganisms
- Roles of Culture Collections
- Issues Affecting Culture Collections
- Culture Collection Accession Policy and Criteria
- Culture Collection Quality Management
- Managing a Small Working Collection
- Managing CABI's Public Service Collection

Public service culture collections have provided a service to the scientific community for over 100 years. The traditional role of such collections is to provide the scientific community with access to authenticated cultures and specialist advice on their cultivation and preservation. In recent years, other services have been added to the curatorial role of service culture collections, such as patent deposit facilities and the supply of cultures for quality control research to add value to strains. Indeed, service culture collections have evolved into Biological Resource Centres (BRCs), thereby responding to revolutionary developments in areas such as molecular biology and bioinformatics.

This module covers a detailed assessment of the importance of conserving microorganisms as in culture collections and the role they play in research. It also deals with the various factors involved in the maintenance and management of a good working collection. Even if the course participant is not working in a public service culture collection, the principles if not the scale of the task can be transferred to even the smallest laboratory-based collection.

Module Objective
After undertaking the module, the user should be able to:

- Describe the different aspects of culture collections and their quality management systems
- Explain the need to retain microorganisms
- Identify issues affecting culture collections
- Describe how to manage a small working collection

As an example, recognize how CABI's public service collection is managed.

Module 2: Characterization and Use of Microorganisms
This module covers the methods available for culture/strain characterization and describes the uses of microorganisms in science, industry and research including:

- Culture Based Tests
- Enzymatic Activity of Extracts and Broths
- Analysis of Proteins
- Immunological and DNA Molecular Methods
- Chromatographic Methods
- Microscopy Techniques
- Harnessing Microbial Resources

Microorganisms make up a major portion of the living biomass on this planet. They are essential in many aspects of life and hence it is important to properly identify and characterize them. This is where the various characterizing tests described in this module will be useful.

Characterization from a preservation prospective:

- Ensures the correct taxonomy to allow selection of the correct preservation procedure (e.g. anatomy, morphology)
- Allows the assessment of the suitability of the preservation regime applied and, if necessary, the need to modify and improve the preservation procedure
- Facilitates the monitoring of long-term stability of the organism
- Provides information to help identify the scope of use

Screening organisms or extracts from them helps in detection of novel products (metabolites, vitamins, enzymes) and the potential for other economic uses.

Module Objectives
After undertaking the module, the user should be able to:

- Identify methods available for culture characterization
- Describe the uses of microorganisms

Module 3: Isolation and Growth of Microorganisms
This module covers the best practices in isolation, growth, identification, properties and storage of microorganisms including:

- Basic Techniques for Isolation
- Collection of Plant Specimens
- Incubation in a Moist Chamber
- Tissue Clearing
- Aseptic Isolation of Plant Material

Microorganisms are useful to man in various ways. Their use in research and biotechnology is vast hence it is essential to know the latest methods of identification and isolation. This module deals with each of the microorganisms in detail, namely Bacteria, Algae, Protozoa, Yeasts, Fungi and Animal cells. The techniques used vary according to the organism as each one is specific in its growth requirements.

Module Objectives

After undertaking the module, the user should be able to:

- Describe best practices in isolation, growth, identification, properties, storage of microorganisms
- Identify the correct methods for collecting and dispatching samples
- Identify techniques to handle the specimen and isolate the microbe

Module 4: Preserving Microorganisms

This module describes the methods used for longer-term maintenance and preservation of specific microorganisms including:

- Standard Preservation Regimes
- Selection and Development of Preservation Techniques
- Preservation Methods for Bacteria
- Preservation Methods for Fungi

Microorganisms, because of their importance in different aspects of research and development, need to be preserved for future reference. There are a number of preservation methods which have been developed according to the type of organism to be stored and the purpose for which it needs to be stored. These methods maintain them in a viable and stable form without affecting their morphology or physiology. This module highlights the different preservation and storage techniques available and provides guidelines for the selection of these techniques based on which microbial culture it is most suitable for.

Module Objective

After undertaking the module, the user should be able to:

- Describe the methods used for longer-term maintenance and preservation of specific microorganisms
- Describe the standard preservation regimes
- Describe the selection and development of preservation techniques for different types of collections
- Describe preservation methods for bacteria and fungi and the advantages and disadvantages of each method

Module 5: Safe Handling and Distribution of Microorganisms

This module covers health and safety requirements for microorganisms as stated under the law including:

- Health and Safety in Handling Microorganisms
- Classification of Microorganisms on the Basis of Hazard
- Safety in the Laboratory
- Regulations Governing Distribution of Cultures

Microbial culture collections are maintained in many organizations all over the world to facilitate research and development. Some of the organisms stored are potential pathogens and many present health hazards and therefore need to be handled with care. The necessary precautions must be taken while storing, handling and distributing samples. Best practice has been defined offering guidance to the microbiologist and specifically to those maintaining collections and by which the collection organizations abide. This module covers in detail the health and safety requirements that are taken into consideration while handling microorganisms.

Module Objective

After undertaking the module, the user should be able to:

- Explain the importance of health and safety requirements when working with microorganisms and the impact of regulations and the law
- Describe hazard group classification of organisms based on their risks and the different containment levels
- Identify safety measures regarding equipment, culture media, stains and chemicals
- Describe safety criteria and compliance
- Describe the controls that various countries have in place for distribution of dangerous organisms and the impact of these measures on microorganism distribution

Module 6: Understanding and Compliance with Best Practice and International Standards in Microbiology

This module covers best practices in other areas of microbiology including:

- Best Practices and Standards Applied in Microbiology
- Details of OECD Biological Resource Centre Best Practice
- ISO International Quality Management Standards
- Implementing the CWA Biorisk Standard
- Introducing Best Practice to Preservation – Freeze Drying
- Mechanisms to Demonstrate Compliance

Best practices are derived from many sources, for example, from simple protocols published in peer reviewed journals to standards developed by practitioners in the field. Examples of the latter are the International Standards Organization (ISO) standards that are implemented by National Accreditation Bodies. The work of microbiology laboratories can be subject to such standards or best practices. This module covers in detail some of the best practices that are available and the differences in their use and application. It will help prepare participants to introduce best practices in their laboratory and comply with applicable standards.

Module Objectives

After undertaking the module, the user should be able to:

- Identify Best Practices and standards applied in microbiology
- Describe general Best Practices for BRCs
- Describe the implementation and operation of CWA biorisk standard
- Describe mechanisms to demonstrate compliance

Summary

Managing Microorganisms provides guidance on handling, characterization and storage of research, production and reference strains to facilitate their use to help the microbiologist make the right choices in their work. The goal is to ensure that the organisms remain stable, their properties are reproducible and the investment made in research is protected for the future.

References

Smith, D. and Onions, A.H.S. (1994) *The Preservation and Maintenance of Living Fungi.* 2nd edn. IMI Technical Handbooks No. 2, pp 122. CAB International, Wallingford, UK. https://www.cabi.org/bookshop/book/9780851989020/

Smith, D., Ryan, M.J. and Day, J.G. (eds) (2001) *The UK National Culture Collection Biological Resource: Properties, maintenance and management.* pp 382. UK National Culture Collection, Egham. https://www.ukbrcn.org/media/1025/ukncc-biological-resource.pdf

Abstracts

Chapter 1

The form, function and activities of culture collections are described, demonstrating the support they provide to research and development. Optimal preservation in culture collections protects investment in time and resources for isolating and characterizing strains. This approach ensures continuity of use, enabling confirmation of published scientific findings, and retention of strain properties. The information provided highlights the range of collections, along with their supportive networks, organizations and societies. Importantly, key considerations for establishing and managing a small in-house collection are covered. Invariably, before creating a collection, a robust business case must be made incorporating operational and financial plans for sustainability. Guidance is provided for new users, whilst also answering questions such as: why do you need a collection? What accommodation, technology and staffing will be needed? What data storage will be needed? Are there specific regulatory requirements that need to be met? What is the funding model and how much will it cost?

Chapter 2

CABI has been working with microorganisms, mainly bacteria and fungi, for over 100 years. This chapter presents the techniques and guidance that have been finely tuned for use in the isolation and growth of such organisms, over this time, particularly, with emphasis on organisms causing disease in plants. The processes undertaken by CABI's diagnostic advisory service are described from the arrival of samples in the laboratory, including how to package and ship the samples; making isolations of presumptive pathogens; identifying them and producing pure cultures on the most appropriate growth media. A checklist of activities in plant disease diagnostics is provided. Description of techniques such as single spore isolations, staining for microscopy, incubation in moist chambers, use of black light to induce fungus sporulation, prevention of mite infestation and many others used in the diagnostic laboratory are described. The chapter focus is on bacteria, filamentous fungi and yeasts.

Chapter 3

This chapter describes methods and tests for characterizing microbes that range from the simplest morphological observations, through biochemical and physiological tests, to recent nucleic acid sequencing advances. Approaches are presented that facilitate selection for tailored uses, also useful from a preservation perspective where stability during storage can be monitored. The chapter provides examples of strains with particular suitability for, amongst others, biodegradation, bioremediation, and food spoilage linking through to appendices listing CABI collections strains that demonstrate such properties. This chapter also presents a range of methods for screening organisms, or extracts derived from them, to find useful products (metabolites, vitamins, enzymes) and open the way to other eventual uses. Approaches for the identification of strains as well as revealing new uses, the likely benefits of microbial resources for commercial use and how to identify potential new ones for exploitation, are presented.

Chapter 4

This chapter provides information to guide selection of the most appropriate preservation method for microorganisms and the protocols to achieve this. The methods include serial subculture, the long-term drying technologies of freeze drying and the more widely applicable cryopreservation. Information is provided on how to select the preservation method optimally suited to the organisms that are being stored and the resources and capacity available. The focus is mainly on bacteria and fungi but preservation of other microorganism taxa is touched on, along with microbiomes and microbial consortia. Cryopreservation and freeze drying are the preferred choice for a public service collection such as CABI but there may be circumstances where this is not possible. Criteria for selection of the most appropriate preservation technique covered, include relevance to type of collection and the use of its contents, to logistical criteria and taxonomy; a decision based key is provided to guide the selection.

Chapter 5

This chapter covers the regulatory environment that governs access to, handling, use and distribution of microorganisms. In particular, there are health and safety requirements that must be taken into consideration while handling microorganisms. The impact of regulations and the law is explained and specifically covers the importance of health and safety and the effect of law on safety with regard to microorganisms (from the perspective of CABI in the UK). It includes the classification of microorganisms on the basis of hazard and the risks involved, in addition to the different containment levels needed. Information about safety measures regarding equipment, culture media and stains, and chemicals and plant material are presented along with the regulations governing distribution of cultures. The chapter also provides guidance on establishing a regulatorily compliant and safe environment in which to operate a culture collection and handle microorganisms in the laboratory.

Chapter 6

The operation of microbiology activities to appropriate standards is essential to ensure quality, reproducibility and continuity of work. To achieve this, collections must apply quality control and assurance measures to maintain these standards, consider the needs of users and of the facilities and resources available. There are numerous standards available, both informal best practice and international industry standards that are externally audited. These are described and discussed in regard to managing microorganisms. The salient features of a quality management system are described based upon the World Federation for Culture Collections and the OECD Best Practice Guidelines for Biological Resource Centres. The chapter goes on to describe approaches to national accreditation or certification and mechanisms to demonstrate compliance. The applicable International Organization for Standardization (ISO) International quality management standards are described and an approach to self-auditing presented. The impact of quality management on freeze drying is demonstrated.

Chapter 7

This chapter outlines the medium-term mechanisms for achieving a sustainable business strategy for any modest sized culture collection.

Collections need to provide authentic stable strains with reproducible properties that have been preserved by the best available techniques, whilst also providing confirmed and validated information to facilitate their access and use. There is not one single correct solution but the many commonalities are described that can help to create the most appropriate financial plan for sustainability. The chapter demonstrates how this might be achieved in a stepwise manner, as available investment or revenue lines allow. Finally, information is presented as to how collections can generate revenue or other sources of funding. Traditionally, these have been through culture sales, institutional funding, contracted services, contracted research and sponsorship.

Acronyms and abbreviations

ABS – Access and Benefit Sharing
ABSCH – Access and Benefit Sharing Clearing House
ACDP – Advisory Committee on Dangerous Pathogens
ACGM – Advisory Committee on Genetic Manipulation
ACM – Asian Consortium for the Conservation and Sustainable Use of Microbial Resources
AFLP – Arbitrary fragment length polymorphism
AFNOR – Association Francais de Normalisation
AHTEG – Ad Hoc technical expert group
AnaEE – Analysis and experimentation on ecosystems
APIZYM – A simple rapid system for the detection of bacterial enzymes
ASTM – American Society for Testing Materials now ASTM International
BBMRI – Biobanking and BioMolecular Resources Research Infrastructure
BBR – Bioinformatics and Biological Resources
BBSRC – Biotechnological and Biological Sciences Research Council
BCA – Biological control agent
BCCM – Belgian Coordinated Collections of Microorganisms
BIOLOG – Biolog phenotype assays Microbial Identification System
BOLD – Barcode of Life Data System
BRC – Biological Resource Centre
BSL – Biosafety level
BWC – Biological Weapons Convention
CABI – CAB International
CABRI – Common Access to Biological Resources and Information
CBD – Convention on Biological Diversity
CCAG – Culture Collection Advisory Group
CCAP – Culture Collection of Algae and Protozoa
CDC – Centre for Disease Control
CGIAR – Consultative Group on International Agricultural Research
CGRFA – Commission on Genetic Resources for Food and Agriculture
CHEF – Contour-clamped homogeneous electric field
COI – Cytochrome oxidase I
COP – Conference of the Parties (CBD)

COP/MOP – a Conference of the Parties (CBD) serving as the meeting of the parties to the Kyoto Protocol
COSHH – Control of Substances Hazardous to Health
CRISPR – Clustered regularly interspaced short palindromic repeats
CTAB – Cetryltrimethyl ammonium bromide
cDNA – Complementary DNA
DAD – Diode array detection
DGGE – Denaturing gradient gel electrophoresis
DGR – Dangerous Goods Regulations
DMSO – Dimethyl sulphoxide
DNA – Deoxyribonucleic acid
DOI – Digital object identifier
DSI – Digital sequence information
DSD – Digital sequence data
DSMZ – German Collection of Microorganisms and Cell Cultures
EATRIS – European Advanced Translational Research Infrastructure in Medicine
EBRCN – European Biological Resource Centre Network
EC – European Commission
ECACC – European Collection of Authenticated Cell Cultures
ECCO – European Culture Collections' Organisation
EDTA – Ethylenediaminetetraacetic acid
EFB – European Federation of Biotechnology
EM – Electron microscopy
EMbaRC – European Consortium of Microbial Resource Centres
EMBL – European Molecular Biology Laboratory
EMBRIC – European Marine Biological Research Infrastructure Cluster
EMPHASIS – European Infrastructure for Multi-scale Plant Phenomics and Simulation
ELISA – Enzyme-linked immunosorbent assay
EOLSS – Encyclopaedia of Life Support Systems
EPA – European Patent Office
ERIC – European Research Infrastructure Consortium
ERINHA – European Research Infrastructure on Highly Pathogenic Agents
ESFRI – European Strategy Forum on Research Infrastructures
EU – European Union
EU-OPENSCEEN – European Infrastructure of Open Screening Platforms for Chemical Biology
Euro-Bioimaging – European Research Infrastructure for Imaging Technologies in Biological and Biomedical Sciences
FAIR – Findability, accessibility, interoperability, and reusability
FAME – Fatty acid methyl ester
FAO – Food and Agriculture Organization
FAT – Fluorescent antibody technique
FDS – Full data set
FISH – Fluorescence *in situ* hybridization
GBRCN – Global Biological Resource Centre Network
GC – Gas chromatography
GEM – Genetically engineered microorganisms
GLP – Good Laboratory Practice
GMO – Genetically modified organism
GMP – Good Manufacturing Practice
GR – Genetic resource
GRC – Genetic Resource Collection
GRFA – Genetic Resources for Food and Agriculture

GRSD – Genetic resource sequence data
HPLC – High performance liquid chromatography
IATA – International Air Transport Association
ICAO – International Civil Aviation Organization
ICRO – International Cell Research Organization
ID – Identification
IDA – International Depositary Authority
IEL – Individual export licence
IMI – International Mycological Institute formerly Imperial Mycological Institute
IMS – Industrial methylated spirit
INFRAFRONTIER – European Research Infrastructure for the generation, phenotyping, archiving and distribution of mouse disease models
INSDC – International Nucleotide Sequence Database Collaboration
INSTRUCT – Integrated Structural Biology Infrastructure
IP – Intellectual property
IPBS – Infectious perishable biological substance
IPR – Intellectual Property Rights
IRCC – Internationally Recognized Certificate of Compliance
ISBE – Infrastructure for Systems Biology Europe
ISBER – International Society for Biological and Environmental Repositories
ISO – International Standards Organization
ITPGRFA – International Treaty on Plant Genetic Resources for Food and Agriculture
ITS – Internal transcribed spacers
IUBS – International Union of Biological Societies
IUCN – International Union for Conservation of Nature
IUMS – International Union of Microbiological Societies
JFCC – Japanese Federation for Culture Collections
LEV – Local exhaust ventilation
LGR – Ligase chain reaction
LPSN – List of Prokaryotic Names of Standing in Nomenclature
MALDI-TOF – Matrix-assisted laser desorption/ionization-time of flight
MAT – Mutually agreed terms
MDS – Minimum data set
MEL – Maximum exposure limit
METROFOOD-RI – Infrastructure for promising Metrology in Food and Nutrition
MGR – Marine genetic resources
MIDI – Microbial identification using cellular fatty acid composition
MIL – Military (standard)
MINE – Microbial Information Network Europe
MIRCEN – Microbiological Resource Centre
MIRRI – Microbial Resources Research Infrastructure
MLST – Multi-locus sequence typing
MOSAICC – Microorganisms Sustainable Use and Access Regulation International Code of Conduct
MSDN – Microbial Strain Data Network
MSDS – Minimum safety data sheets
MTA – Material transfer agreement
NAMAS – National Measurement Service
NASBA – Nucleic acid sequence based amplification
NASD – Nucleotide and amino acid sequence data
NASSI – Nucleotide and amino acid sequence and structural information
NASSFI – Nucleotide and amino acid sequence, structural and functional information
NCFC – National Collection of Fungus Cultures

NCIMB – National Collection of Industrial and Marine Bacteria Ltd
NCPPB – National Collection of Plant Pathogenic Bacteria
NCTC – National Collection of Type Cultures
NCYC – National Collection of Yeast Cultures
NFP – National focal point
NHM – Natural History Museum
NIH – National Institute of Health
NMR – Nuclear magnetic resonance
NP – Nagoya Protocol
NPBS – Non-infectious perishable biological substance
NSD – Nucleotide sequence data
NSF – National Science Foundation
NSI – Nucleotide sequence information
OECD – Organisation for Economic Co-Operation and Development
OES – Occupational exposure standards
OEWG – Open-ended working group
OFAGE – Orthogonal field-alternating field electrophoresis
OGEL – Open general export licence
OST – Office of Science and Technology
PCR – Polymerase chain reaction
PFGE – Pulse field gel electrophoresis
PGRFA – Plant Genetic Resources for Food and Agriculture
PI – Principle investigator
PIC – Prior informed consent
PIP – Pandemic influenza preparedness
QA – Quality Assurance
QMS – Quality Management System
RAPD – Random amplified polymorphic DNA
R&D – Research and Development
RBG – Royal Botanical Garden
rDNA – Recombinant DNA
RDS – Recommended data set
REP – Repetitive extragenic palindromic
RFLP – Restriction fragment length polymorphism
RIKEN – Kokuritsu Kenkyū Kaihatsu Hōjin Rikagaku Kenkyūsho (Institute of Physical and Chemical Research)
RNA – Ribonucleic acid
ROFE – Rotating field gel electrophoresis
SBSTTA – Subsidiary Body on Scientific, Technical and Technological Advice
SCAR – Sequence characterized amplified region
SDA – Strand displacement amplification
SEM – Scanning electron microscope
SNPs – Single nucleotide polymorphisms
SOP – Standard Operating Procedure
SSCP – Simple strand conformation polymorphism
TEM – Transmission electron microscope
TLC – Thin layer chromatography
TRIPs – Trade-related aspects of Intellectual Property Rights
UKAS – United Kingdom Accreditation Service
UKBRCN – United Kingdom Biological Resource Centre Network
UKNCC – United Kingdom National Culture Collections
UN – United Nations

UNCLOS – United Nations Convention on the Law of the Sea
UNEP – United Nations Environment Programme
UNESCO – United Nations Educational, Scientific and Cultural Organization
UPU – Universal Postal Union
USCCN – United States Culture Collection Network
USPHS – United States Public Health service
UV – Ultra violet light
VAT – Value added tax
VNTR – Variable number tandem repeat
WCO – World Customs Organization
WDCM – World Data Centre for Microorganisms
WFCC – World Federation for Culture Collections
WHO – World Health Organization
WIPO – World Intellectual Property Organization
XFEL – X-ray free-electron lasers

1

Culture Collections

Introduction

Living cultures of microorganisms are needed for study, reference or use in industry and as a consequence, they need to be made accessible and stored without change. Culture collections are established to store strains, as it is not always possible to re-isolate an organism with the same strain properties. The primary aim is to keep the organisms viable and stable to enable reproducibility and continuity of work. These collections have different forms and function; they may be created for the use of a private individual scientist or an institution and be small and focused on a specific organism or they may be large and offer a public service and many things in between. However, they all will have common functions and operations; these are described here to provide the basis for establishing a collection that meets the specific needs of its intended users. The chapter will also provide the fundamentals for the design of a robust business plan to facilitate a collection's sustainability built on the lessons learned from the operations of the public service culture collections now more commonly termed 'Biological Resource Centres (BRC)'.

Living microorganisms form the basic elements of microbiology and cell biology, they are the research tools, the producer of compounds, fuel and food and the basis for teaching and education in microbiology (Overmann, 2015; Overmann and Smith, 2017; NSF, 2020). Globally, these cells are grown and utilized in huge numbers in laboratories and are the key to many research programmes, industrial processes and training courses. They act as references in taxonomy and provide the living references that underpin sequence databases. Cultures are often placed in the back of a refrigerator, or kept in the incubator until they are needed again. If research is carried out using deteriorated, contaminated or incorrect strains large investments in time and resources can be lost. Microbiologists often establish their own laboratory collections to ensure their key strains are maintained for future use and for confirmation of results. There are many such collections maintaining organisms for use in their laboratories and, on occasion, sharing with other researchers in the field. The organisms in such collections are stored by many methods of preservation and the selection of the right method to suit the purpose can often be difficult. Often equipment and facilities are not available and are expensive. Collections must be efficient, yet cost effective, strains must be retained without change yet often the resources are not always to hand.

Microbiological resource collections range from small private collections through to large service collections, and have widely differing policies and holdings. There are relatively few collections that attempt to maintain organisms solely for the scientific community in general, or have a remit for *ex situ* conservation of biodiversity.

DOI: 10.1079/9781800622135.0001

Collections of organisms are normally linked to their use in operations related to those of the parental organization activities. For example, screening for exploitable metabolites or enzymes, direct use as food, or food modification, as biocontrol agents, waste bioconversion, or waste detoxification.

The need for total stability of an organism's abilities stems from the continuing discoveries of new uses and natural products (Hüttner *et al.*, 2020). We are not blessed with a crystal ball where we can foresee what technologies and methods for analysis will be developed in the future. A collection may be asked to provide representative strains that may express a property not considered when the strain was first deposited. If strains are not maintained appropriately then these as yet unknown properties could be lost. Methods, as described in Chapter 4, that are normally used for long-term stability of organisms are freeze-drying and cryopreservation, either in, or above, liquid nitrogen, or in a low temperature freezer (Smith and Onions, 1994; Smith and Kolkowski, 1996; Ryan and Smith, 2007). Preservation methods should be optimized for the different cell types and understanding the science of low temperature chemistry and physics and the ability to observe what happens to cells is essential for technique design (Smith, 1992, 1993, 2012; Ryan *et al.*, 2012; Ryan *et al.*, 2019). A BRC's role includes the provision of access to authenticated cultures and specialist advice such as cultivation, to undertake preservation of strains used in public research and to carry out research that adds value to the strains held. They also offer safe, confidential and patent deposit services to help organizations protect their IP and key production strains.

Microbiologists often establish their own laboratory collections or deposit strains in public service collections to ensure that their key strains are maintained for future use and for confirmation of results. Public Service Culture Collections or BRCs provide the authenticated biological material upon which high quality research is based, they serve as repositories for strains as part of patent deposits, and provide safe and confidential storage for key organisms in research and industrial production (Smith *et al.*, 2013). Importantly they provide access to microorganisms cited in scientific papers enabling the follow up discoveries and enabling innovation.

Culture collections are uniquely positioned as custodians of *ex situ* genetic resources and have a key role in their conservation (Kirsop and Hawksworth, 1994; Smith, 2012). Biologists, who publish information on organisms, should make their most important strains available for confirmation of results and future use by depositing them in public service collections (Stackebrandt *et al.*, 2014). Such collections mainly serve as:

- Repositories for strains; as part of patent deposits and *ex situ* conservation of organisms
- Providers of safe and confidential services; to store key organisms for research and industry
- Sources of organisms cited in scientific papers that can be used in the confirmation of results and for further study
- Custodians of key resources of national importance
- Providers of living resources; to underpin life sciences

The OECD recognize these roles as essential for biotechnology and recommend their networking. Details on the OECD Biological Resource Centre Initiative can be found in the OECD report *Biological Resource Centres: Underpinning the future of life sciences and biotechnology* at https://www.oecd.org/sti/emerging-tech/2487422.pdf.

The *in situ* conservation of microorganisms is difficult to assess as we have so little information on the extent of microbial diversity. Microorganisms are subject to relatively rapid evolution and adapt and change according to environmental factors. It is estimated that there are up to 5.1 million species of fungi on earth (Blackwell, 2011) yet only 120,000 species have been described (2.4%) as not all can yet be cultured. To date most culture collections store isolated pure strains but only 10-15% of the described species of fungi are available from public service collections. The World Data Centre for Microorganisms' (WDCM) statistics on 19 January 2022 show holdings of 885,898 strains of fungi representing around 40,000 fungal names representing 15-20,000 species (16%). WDCM data (http://www.wdcm.org/) when analysed show that 50% of *ex situ* holdings are represented by one isolate in one collection. If a culture is lost it is not always possible to go back to a location and isolate the same species with exactly the same properties.

There are many threats to biodiversity; these include habitat loss, conversion of forests to farmland and the over-exploitation of natural resources such as harvesting lichens from rainforest. However, today it is more likely that climate change and industrial pollution present the greatest threats to a number of species. The sequencing of microbial DNA and RNA in environmental samples allows the assessment of the hidden (yet to be cultured) microbial communities, the microbiome: a new role is emerging to preserve these samples with the data to facilitate their access and use (Ryan *et al.*, 2019).

It is recognized that microorganisms can provide solutions to world problems pertaining to public health, food, environment, poverty alleviation and food security. They are vital components of the world's biodiversity, greatly contributing to ecological and economic issues. Thus, safeguarding of microbial diversity for future use is of vital importance for wealth creation. There is a clear need for culture collections, particularly those with the remit of a BRC in the conservation of microorganisms.

Role of Culture Collections

BRCs are the next generation culture collections, they are a key element of the scientific and technological infrastructure for life sciences and biotechnology. To operate such functions, collections must follow standard procedures to offer products and services of a consistent nature and in a way that is compliant with legislation and that provides legal clarity for the user. Organizations have been established to co-ordinate collection activities and aid the sharing of knowledge and experience in maintaining collections of organisms, their safe handling and distribution.

Culture Collection Operations

The key operations to deliver the culture collection role outlined above are processes to acquire, access and store microorganisms with the associated data on collection, properties and use and mechanisms to ensure their safe and legitimate use. These are presented in the following two flow charts (Figs 1.1 and 1.2) with references provided to the details of each stage in the process.

Culture Collection Organizations

There are several levels at which co-ordination, collaboration and discussion on approaches to microbial resource collection establishment and organization is carried out. Organizations exist for the support of collection activities on national, regional and international levels. These include national federations such as the United Kingdom Biological Resource Centre Network (UKBRCN), United States Culture Collections Network (USCCN), and the Japanese Federation for Culture Collections (JFCC). At the regional level the European Culture Collections' Organisation (ECCO), the Microbial Resources Research Infrastructure (MIRRI) and at the international level, the World Federation for Culture Collections (WFCC), and Microbial Resource Centres (MIRCENs) operate. Collections have also been drawn together to operate at a more intimate level through national and international affiliations such as the Belgium Co-ordinated Collection of Micro-organisms (BCCM).

International Federations

The World Federation of Culture Collections (WFCC)

The WFCC was founded in 1968 and is a multidisciplinary commission of the International Union of Biological Sciences (IUBS) and since the separation of the International Union of Microbiological Societies (IUMS) from IUBS in 1979 it has operated as an inter-union commission. It seeks to promote and foster activities that support the interests of culture collections and their users.

The WFCC has published guidelines for the establishment and operation of collections. These Guidelines are updated regularly and can be viewed on the WFCC website via the link provided. In order that a collection's user can rely upon the organisms supplied, and the services provided, it is imperative that the collection

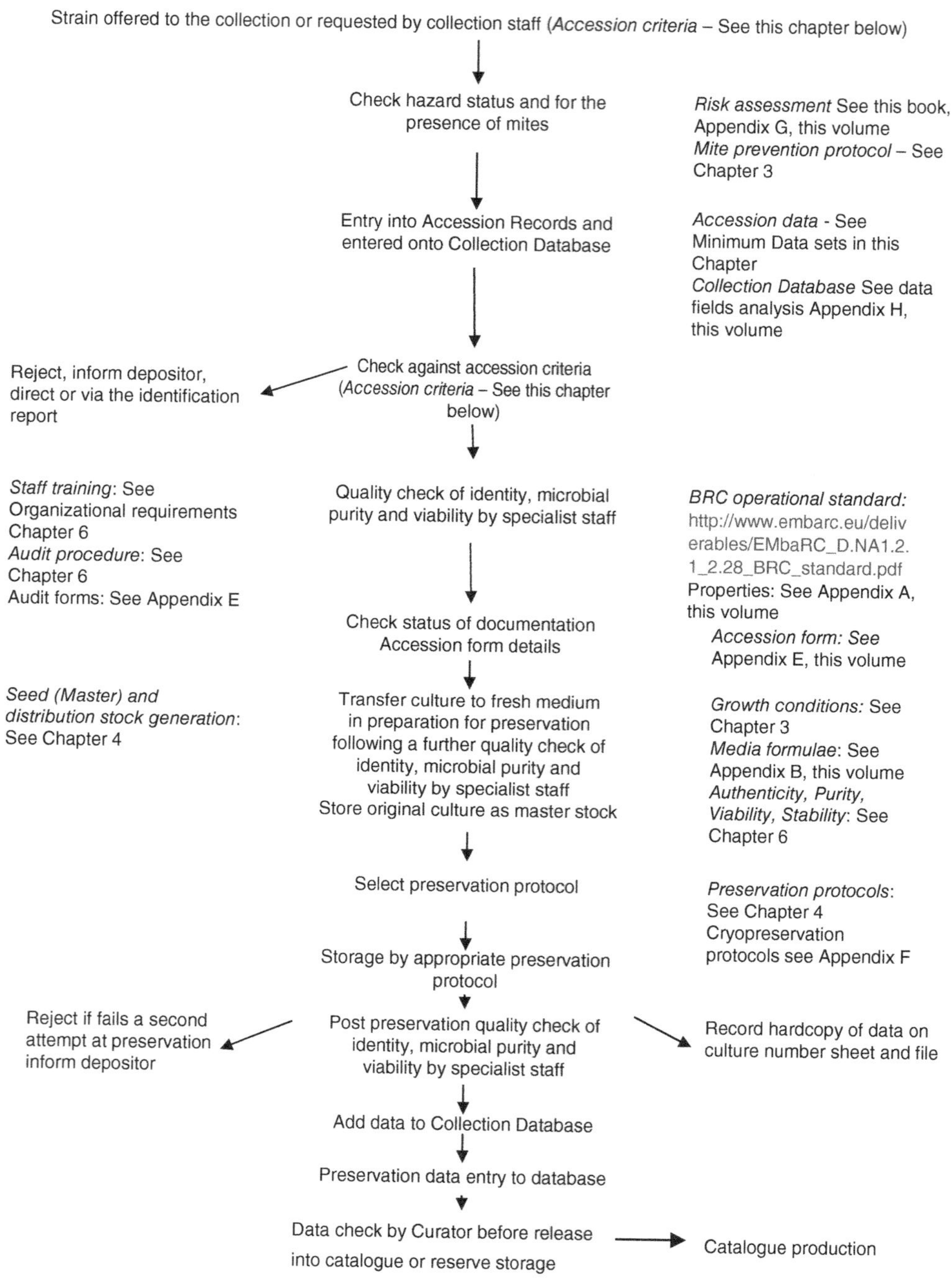

Fig. 1.1. Accessions into the culture collection flow diagram

follows good practices. Acceptance of a collection as a member of WFCC offers a limited form of accreditation, but in the future a more formal accreditation scheme may be desirable (Stevenson and Jong, 1992).

Member collections of the WFCC register with the World Data Centre for Micro-organisms and there are currently c. 124 affiliated collections amongst the 800 plus collections registered from over 70 countries (http://www.wdcm.org/).

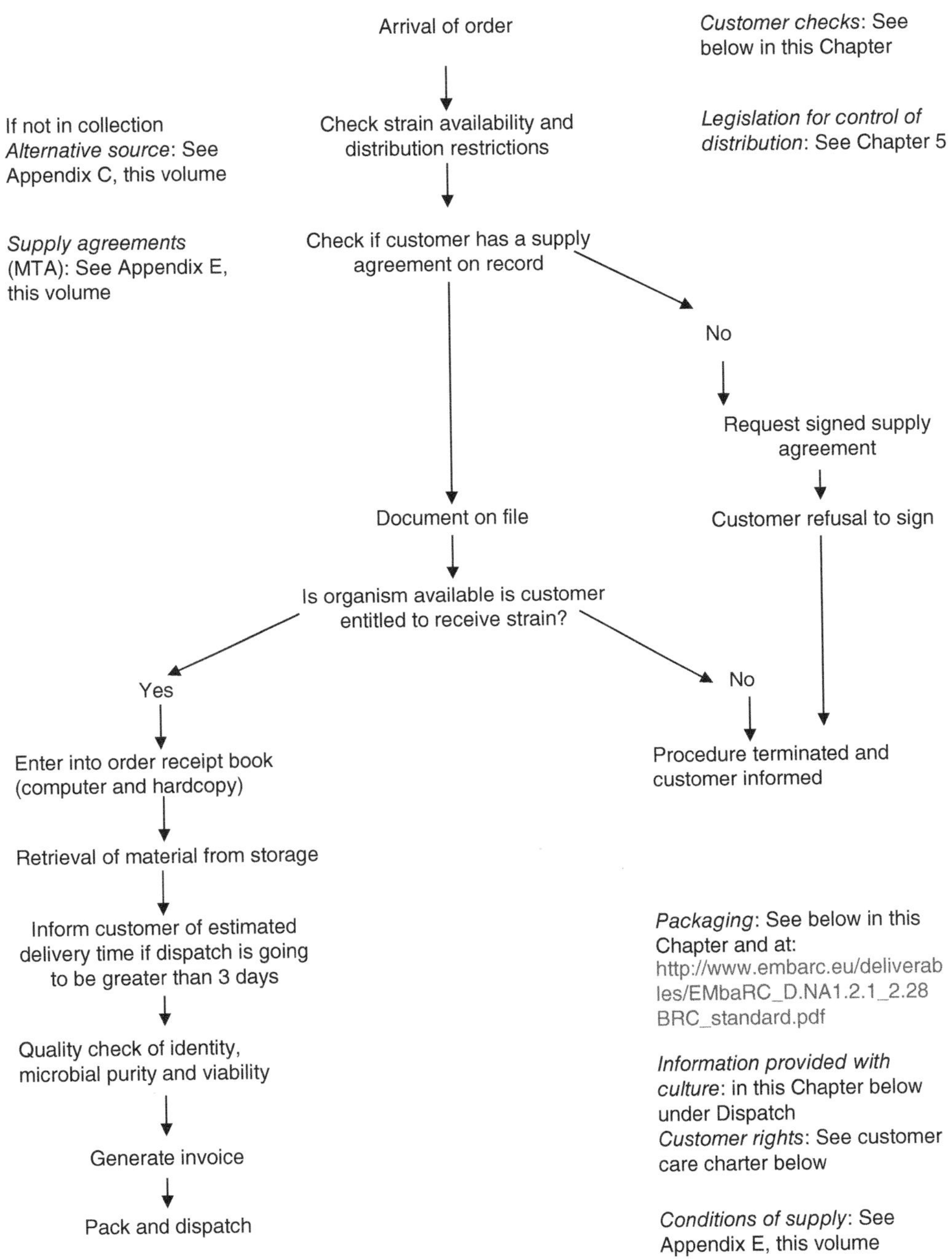

Fig. 1.2. Culture supply flow diagram

A congress is held every four years to discuss advances in technology and common policies with regard to biodiversity and the role of culture collections. The WFCC keeps its members informed on matters relevant to collections via its newsletter and has standing committees reporting on patent depositions, postal, quarantine and safety regulations, safeguarding of endangered collections, education, publicity, standards and biodiversity.

The World Data Centre for Micro-organisms (WDCM)

Since 1986, the WFCC has overseen the activities of the WDCM and it is now the data centre for the WFCC and Microbial Resource Centres Network. The WDCM is supported by UNEP and UNESCO and the database holds information on collections, the species they hold and details on their specialization. It was established in 1968 and produced the first hard copy volume of the World Directory of Collections of Cultures of Micro-organisms in 1972, whilst based at the University of Queensland, Australia. The WDCM relocated in 1986 to RIKEN, Saitama, Japan, to the National Institute of Genetics, Japan in 1999 and now it is housed and run by Institute of Microbiology at the Chinese Academy of Sciences, Beijing (Wu *et al.*, 2017). The World Directory (Sugawara *et al.*, 1993, 1999) illustrates some of the data held; it has indexes by country, main subjects studied, cultures held, the culture availability, their staff, and services offered. These data can also be accessed at the WDCM website (http://www.wdcm.org/).

The WDCM collections hold in excess of 2 million strains, 26% are fungi, 44% bacteria, 2% viruses, 1% live cells, and 10% others (including plasmids, plant, animal cells and algae). Of the total, 35% of the holdings are maintained by only 10 collections.

Microbial Resource Centres (MIRCENs)

In 1974 UNEP, UNESCO and ICRO established the MIRCEN network and in 1978 the first evaluation of MIRCEN activities was carried out. This led to the establishment of MIRCENs in industrialized countries and the production of *MIRCEN News* in 1980 to help publicize the activities of the network. In 1985 Oxford University Press published the *MIRCEN Journal of Applied Microbiology and Biotechnology* and UNESCO has recommended consolidation and expansion of the MIRCEN network (Da Silva, 1988).

The objectives of the worldwide network of MIRCENs (https://uia.org/s/or/en/1100013806) are to preserve and exploit microbial gene pools, make them accessible to developing countries and to carry out research and development in environmental microbiology and biotechnology. They support and contribute to UN Sustainable Development Goals 4: Quality education, 5: Achieve gender equality and empower all women and girls and 7: Ensure access to affordable, reliable, sustainable and modern energy.

Sixteen centres are listed by Kirsop and Da Silva (1988). The MIRCEN network carries out various activities to meet these ends including training and the provision of information. The role of microbial resources centres and UNESCO in the development of biotechnology is described by Horst W. Doelle in the *Encyclopaedia of Life Support Systems* (EOLSS) https://www.eolss.net/sample-chapters/c17/E6-58-13.pdf.

Regional Federations

European Culture Collections' Organisation (ECCO)

As they describe, ECCO was established in 1981 with the aim to promote collaboration and exchange of ideas and information about all aspects of culture collection activity. They bring together collections from 25 countries with over 60 member collections together holding over 350,000 strains representing yeasts, filamentous fungi, bacteria and archaea, phages, plasmids including plasmid bearing strains and other recombinant DNA constructs, animal cells including human and hybridoma cell lines, animal and plant viruses, plant cells, algae and protozoa. They specifically provide a coordinated approach to compliance with the Convention on Biological Diversity Nagoya Protocol on Access and Benefit Sharing. ECCO meetings are held annually and are a valuable forum for discussion and innovation on the future development of member collection activities. The ECCO website provides background to the activities and the broader services provided by its members (https://www.eccosite.org/).

Asian Consortium for the Conservation and Sustainable Use of Microbial Resources (ACM)

The ACM was established in 2004 with participation of 12 Asian countries and now 28 organizations in 15 countries/areas are members (https://www.nite.go.jp/en/nbrc/global/acm/asia_acm.html). The consortium holds an annual meeting for information exchange between researchers in the field of microbiology and representatives of governmental organizations concerning the utilization of biological resources in Asia. The ACM aims to strengthen the relationship among Asian countries/areas as well as to encourage and facilitate international cooperation on biotechnology in which biological resources in Asia are utilized.

National Networks

In this section we give examples of two national networks – the UK and US. Organizations can also be found in Belgium, Brazil, China, Czech Republic, France, Indonesia, Japan, Portugal, Russia and Thailand (see http://www.wfcc.info/networks/).

The United Kingdom Biological Resource Centre Network (UKBRCN)

The United Kingdom Biological Resource Centre Network formerly the UK National Culture Collection (UKNCC) brings together the UK national public service collections providing a platform for access to their work, products and services. The UKNCC was established through the implementation of the UK government's strategy for UK microbial collections (OST, 1996). It brought together nine national collections (one on two sites) following the Government's response to an independent review of UK microbial culture collections. The UKNCC addressed three specific initiatives: to improve the profile and marketing of the collections, to fund a molecular characterization programme and to establish an animal virus collection. The Biotechnology and Biological Sciences Research Council (BBSRC) administered the first phase through the Culture Collection Advisory Group (CCAG), a steering group selected to advise and support these activities. Disbanded in 2006, the UKNCC formed the nucleus of the current UKBRCN which was established to coordinate the activities of the UK national service collections of microbial organisms and natural resource biobanks to better address the challenges of today.

Currently, in 2020 there are 14 member collections with holdings that number more than 100,000 (*c.* 2300 algae and protozoa, more than 20,000 animal cell lines, over 30,000 bacteria, 30,000 fungi including yeasts, plus actinomycetes, cyanobacteria, nematodes, mycoplasma and viruses). The supply of these organisms, associated services and expertise of the national culture collections that make up the UKBRCN are available to support microbiological and cell culture needs. The services and products provided by the member collections are detailed on the UKBRCN website where access to the individual collections is given. The collections' primary interests are generally in identification and preservation of their holdings but they and their parental organizations offer many more services.

The key objectives of the UKBRCN are to:

- co-ordinate engagement with users and common stakeholders (editorial boards, grant funding authorities, UK government)
- share best practice regarding quality and compliance, ensuring that users of living genetic resources and collections are within the law (Nagoya Protocol)
- provide expert responses and solutions to the problems faced by those working with and harnessing biological resources
- raise awareness of the value of culture collections to the scientific community by providing information and training
- co-ordinate grant applications in key areas of research to enhance scientific procedures and services of the UK's biological resource centres

Such activities have made UK biological resources more accessible but to facilitate biosciences further such activities must develop globally. There are similar initiatives such as the Belgian Co-ordinated Collection of Micro-organisms (BCCM) and on a European scale the Microbial Resources Research Infrastructure (MIRRI); the latter gives access to 300,000 microbial isolates (including archaea, bacteria, yeast, fungi, algae and viruses).

United States Culture Collection Network (USCCN)

The USCCN brings together scientists working with laboratory-based collections of microbes which seeks to optimize the quality and availability of microbial resources and become a central resource for US microbe culture collections and their users. Its mission is to facilitate the safe and responsible utilization of microbial resources for research, education, industry, medicine and agriculture for the betterment of humankind by providing opportunities for US culture collection workers to engage with each other and with the broader culture collection community (https://usccn.org/). The USCCN has an online collection registry of its members who provide users with cultures, guidance and information. It consists of a searchable database of microbial culture collections from universities and government agencies that currently distribute materials to the research community (publicly available), collections that aspire to become a publicly available resource (aspiring), and scientists with research collections (research).

Research Infrastructures

There are many national and regional infrastructures being built to facilitate access to new technologies to facilitate innovation allowing characterization and research that are most often outside the budgets of research organizations. In Europe, the European Strategy Forum on Research Infrastructures (ESFRI) provides a range of platforms that develop scientific integration giving competitive and open access to high quality research infrastructures (https://www.esfri.eu/). Examples that are relevant to culture collections and microbiology are given here.

AnaEE Infrastructure for Analysis and Experimentation on Ecosystems https://www.anaee.eu/: paving the way for understanding the complex impact of today's multiple, interacting global change drivers on terrestrial and aquatic continental ecosystems across Europe. It will forge evidence-based adaptation and mitigation strategies that assure plant, soil, water, biodiversity and ecosystem health today and in the future.

EMPHASIS European Infrastructure for Multi-scale Plant Phenomics and Simulation https://emphasis.plant-phenotyping.eu/: enabling researchers to use facilities, resources and services for plant phenotyping across Europe; with the vision to help scientists better understand plant performance and translate this knowledge into application.

ISBE Infrastructure for System Biology Europe http://www.sysbio.it/isbe-service/: This ESFRI Research Infrastructure (RI) is currently under construction. It will provide open access to state-of-the-art facilities, data, models, tools and training in systems biology.

METROFOOD-RI Infrastructure for promoting Metrology in Food and Nutrition http://roadmap2018.esfri.eu/projects-and-landmarks/browse-the-catalogue/metrofood-ri/; https://www.metrofood.eu/: is a new distributed RI aimed to promote scientific excellence in the field of food quality and safety. It provides high-quality metrology services in food and nutrition, comprising an important cross-section of highly interdisciplinary and interconnected fields throughout the food value chain, including agrifood, sustainable development, food safety, quality, traceability and authenticity, environmental safety and human health.

MIRRI Microbial Resource Research Infrastructure https://www.mirri.org/: This distributed ESFRI is currently under construction and consists of 11 country members, notably excluding the UK, Germany, Ireland, Austria and the Nordic countries. It is pan-European for the preservation, systematic investigation, provision and valorization of microbial resources and biodiversity. It serves the bioscience and the bioindustry communities by facilitating the access, through a single point, to the broadest range of high-quality microorganisms, their derivatives, associated data and services.

BBMRI-ERIC Biobanking and BioMolecular Resources Research Infrastructure https://www.bbmri-eric.eu/: is a European research infrastructure for biobanking, bringing together researchers, biobankers, industry and patients to boost biomedical research. They offer quality management services, support with ethical, legal and societal issues, and a number of online tools and software solutions.

EATRIS ERIC European Advanced Translational Research Infrastructure in Medicine https://eatris.eu/: Providing access to a vast array of pre-clinical and clinical expertise and facilities that are available across Europe. They focus on improving and optimizing preclinical and early clinical development of drugs, vaccines and diagnostics, and overcome barriers to health innovation.

ECRIN ERIC European Clinical Research Infrastructure Network https://ecrin.org/: A public, non-profit organization that links scientific partners and networks across Europe to facilitate multinational clinical research. Providing advice, management services and tools to overcome hurdles to multinational trials and enhance collaboration.

ELIXIR A distributed infrastructure for life-science information https://elixir-europe.org/: Unites Europe's leading life science organizations in managing and safeguarding the increasing volume of data being generated by publicly funded research. It coordinates, integrates and sustains bioinformatics resources across its member states and enables users in academia and industry to access services that are vital for their research.

EMBRC ERIC European Marine Biological Resource Centre https://www.embrc.eu/: A European research infrastructure that provides researchers and companies with access to marine organisms and the facilities to study them, including experimental facilities and technological platforms. It enables researchers to better understand our impact on the oceans and ensure the sustainable use of marine resources.

ERINHA European Research Infrastructure on Highly Pathogenic Agents https://www.erinha.eu/: A pan-European distributed RI dedicated to the study of high-consequence emerging and re-emerging pathogens. It brings together leading European Bio-Safety Level 4 and complementary (e.g. BSL3) facilities with the over-arching goal to provide capacities to conduct projects which are broad in scope, ambition and require a range of capabilities that no single facility or even country can provide on its own.

EU-OPENSCREEN ERIC European Infrastructure of Open Screening Platforms for Chemical Biology https://www.eu-openscreen.eu/: integrates high-capacity screening platforms throughout Europe. They use specially selected EU-OPENSCREEN compound collections; all compound structures and primary screening data is published in the open-access. Access to the EU-OPENSCREEN resources is open to researchers from academia and industry from countries inside and outside of the European Union.

Euro-BioImaging European Research Infrastructure for Imaging Technologies in Biological and Biomedical Sciences https://www.eurobioimaging.eu/: Provides imaging services that bridge biological and biomedical imaging and facilitates innovative research. Euro-BioImaging provides the tools and support to explore and answer your research questions.

INFRAFRONTIER European Research Infrastructure for the generation, phenotyping, archiving and distribution of mouse disease models https://www.infrafrontier.eu/: Carrying out the generation, phenotyping, archiving and distribution of model mammalian genomes; it provides access to first-class tools and data for biomedical research, and thereby contributes to improving the understanding of gene function in human health and disease using the mouse model.

INSTRUCT ERIC Integrated Structural Biology Infrastructure https://instruct-eric.eu/: making high-end technologies and methods available to all European researchers for the interpretation of molecular and cellular functions. Providing access to the significant advances in the necessary tools (direct electron detectors in EM, advances in synchrotron sources and detectors, XFELs, ultra-high field NMR, super-resolution cryo-light microscopy).

Further information is accessible on the individual RI websites given above and coverage is provided on access and use of European Research Infrastructures on the following video:

https://www.youtube.com/watch?v=ISOrDWOiwuw&feature=youtu.be

Ancillary Services Provided by Culture Collections

The main interests of culture collections are generally the identification and preservation of microbes. In certain cases, they and their parental organizations, often offer many more services.

Research, Consultancy, Contract and Investigation Services

In addition to their culture collection activities, collections can offer consultancies and contract research expertise in areas of biodeterioration, biodegradation, biological control, biodiversity, ecology, parasitology, pest management, biochemistry, genomics, bioinformatics, microbial ecology, taxonomy and identifications, metabolite and enzyme screening, analytical chemistry, food and beverage microbiology, process development and large-scale culture, stored product microbiology, plant pathology and evaluation of microbial identification kits.

Identification of Cultures

Individual collections offer identification, often to species level, using a range of morphological, physiological and genomic techniques for many organisms. These include: Actinomycetes, algae, animal cells, bacteria, cyanobacteria, filamentous fungi, nematodes, protozoa, mycoplasmas and yeasts.

Preservation Protocols

Collections work to optimize existing preservation protocols and development of new ones. Most research is aimed at improving cryopreservation protocols with emphasis on species specific criteria (see Appendix F, this volume).

Depositing Cultures

Culture collections welcome deposits relevant to their objectives, with the hope that scientists will deposit interesting strains in these and other principal collections. For strains that are rare, storing it in more than one collection is desirable as it insures against accidental loss. Ideally, all strains cited in published research should be deposited in a public service collection as reference material. Strains of new taxa, type cultures and strains with interesting properties should always be deposited. Depositors are requested to provide as much information and meta-data relating to the strain as possible. Collections offer a range of benefits for depositors. These include:

- Safe preservation of an isolate
- Worldwide access to deposited strains
- Removal of the burden of distribution
- Validation of type strains
- Exchange of cultures
- Safe deposits

A confidential safe deposit service is offered by some collections. Cultures are stored as freeze-dried ampoules and/or in liquid nitrogen to ensure long-term viability, and are reserved for the use of the depositor.

Depositors often find this service useful for important strains during research and development prior to application for a patent. Viability checks can be carried out during storage if requested. Cultures known to be unstable can be reprocessed at regular intervals specified by the depositor. The deposit of microbial strains in public service collections as part of the publication process to underpin good practice in science is covered by Stackebrandt *et al.*, (2014).

Patent Deposits

Over 40 collections worldwide are designated as International Depositary Authorities under the Budapest Treaty of 1977 and together accept a wide range of microorganisms deposited as part of the patent disclosure process. These are listed on the WIPO website (https://www.wipo.int/treaties/en/registration/budapest/).

Patent deposits are given a unique collection number upon receipt. This is classed as the date of deposit. However, the strain cannot be formally accepted until a successful viability test has been carried out (in some instances this follows preservation). In most cases, the depositor cannot even be notified of the number until the viability test is completed. Application and accession forms must be correctly completed and received. Some collections even require full payment before acceptance. After successful preservation, an ampoule or culture is returned to the depositor for confirmation of identity and activities. Viability checks are not normally carried out during storage unless requested by the depositor. The Collection will suggest intervals depending on the organism and preservation method. A fee is levied for each viability check. Samples of the organisms are made available at any time to the depositor, so that its retention of activities can be checked (Kelley and Smith, 1997).

Strains previously deposited may be converted into patent deposits. Depositors must provide the relevant (completed) forms and a viability check is carried out on the deposited material. A sample of this will be returned to the depositor for confirmation of identity and activities. A new accession number and patent deposit date is then issued. Cultures are released only on the written authorization of the depositor or on receipt of a certificate of release from the relevant Patent Office; depending on the type of deposit. The depositor is informed of the name and address of the relevant Patent Office and of the requesting party.

Information Services and Training

Some culture collections, such as CABI, have built up a wealth of information in the process of their activities and can provide answers to specific questions and solutions to many problems concerning their particular activities. There are also several training opportunities and programmes available.

Information services

CABI online resources can be found at https://www.cabi.org/products-and-services/publishing-product/online-resources?section=2&order=text-asc. The key resource is CAB Direct (https://www.cabdirect.org/ which offers a most thorough and extensive source of reference in the applied life sciences, incorporating the leading bibliographic databases CAB Abstracts and Global Health. CAB Direct provides a convenient, single point of access to over 14,416,000 records. Examples of resources of relevance to the handling and use of microorganisms are:

Aquaculture Compendium: Aquatic animal and plant production, natural resources and environment, biodiversity, trade, food security and safety, poverty alleviation, livelihoods. An encyclopaedic, mixed media resource covering all aspects of aquaculture (https://www.cabi.org/publishing-products/aquaculture-compendium/).

CAB Abstracts: CAB Abstracts is the leading English language bibliographic information service providing access to the world's applied life sciences literature. It comes with CABI Full Text, giving users automatic access to over 570,000 journal articles, conference papers and reports (80% of which are not available electronically anywhere else - https://www.cabi.org/publishing-products/cab-abstracts/).

Crop Protection Compendium: Extensive global coverage of pests, diseases, weeds and their natural enemies, the crops that are their hosts, and the countries in which they occur (https://www.cabi.org/publishing-products/crop-protection-compendium/).

Descriptions of Fungi and Bacteria: Diagnostic descriptions, illustrations and information on over 2000 fungal and bacterial taxa (https://www.cabi.org/publishing-products/descriptions-of-fungi-and-bacteria/).

Distribution Maps of Plant Diseases: A leading resource for accurate data on the worldwide distribution of plant diseases (https://www.cabi.org/publishing-products/distribution-maps-of-plant-diseases/).

Distribution Maps of Plant Pests: A leading resource for accurate data on the worldwide distribution of plant pests (https://www.cabi.org/publishing-products/distribution-maps-of-plant-pests/).

Invasive Species Compendium: A repository of available science on the invasive species that have the worst impacts in the world. The focus is global. The Compendium covers all taxa in both natural and managed ecosystems (https://www.cabi.org/publishing-products/invasive-species-compendium/).

Nutrition and Food Sciences: A specialist internet resource that gathers together research, reviews and news on nutrition and the food sciences in an easily searchable database. It covers the whole food chain from the raw ingredients to the physiological and health effects of nutrients. Key subject areas include public health, clinical nutrition, nutrition physiology (https://www.cabi.org/publishing-products/nutrition-and-food-sciences-database/).

Plantwise Knowledge Bank: An open access gateway to actionable plant health information and services – from diagnostic and management advice to maps of pest locations and customized alerts on pest news (https://www.cabi.org/publishing-products/plantwise-knowledge-bank/).

VetMed Resource: A comprehensive tool for veterinarians with a specific veterinary focus and access to over 2.4 million relevant records from the CAB Abstracts database(https://www.cabi.org/publishing-products/vetmed-resource/).

The World Federation for Culture Collections provides many resources on the operation and management of culture collections via its website (http://www.wfcc.info/). There it provides a list of world collections, a library of resources that includes the WFCC Guidelines for Culture Collections.

Training

Many WFCC member collections offer a wide range of training opportunities including tailor-made training programmes, short courses, extended visiting scientist programmes and postgraduate studies arranged via local Universities (see Table 1.1). Topics most often covered include:

- Preservation of microorganisms
- Identification of microorganisms
- Methodologies for characterization of strains
- Genomics and Bioinformatics

Visit the World Data Centre for Microorganisms CCINFO database to select collections offering training.

Culture Collection Quality Management

It is crucial that the organisms used in biotechnology are maintained in a way that will retain their properties. Biological resource collections must ensure delivery of a quality product providing standard reference material that will give reproducible results. To achieve this, collections must apply quality control and assurance measures to maintain these standards considering the needs of users and of the facilities and resources available.

The need for common standards for reproducibility is evident, as the task of maintaining representative samples of microbial diversity cannot be achieved by one collection alone. It is therefore essential that a world-wide network of collections exists to provide the coverage the user requires. There are various estimates for the number of species of fungi, 1.5 million (Hawksworth, 1991), 2.2 to 3.8 million (Hawksworth and Lücking, 2017) and 5.1 million (Blackwell, 2011). However, there is not one strain of the majority of these species that can represent the full spectrum of morphology and physiology expressed within that species and therefore there is a requirement to retain a range of representative strains. To make any impression on such an enormous task with the fungi, and other microorganisms, there must be a focus for each collection and a sharing of tasks between them. In order that a customer of such a network would get a consistent level of service and quality, it is necessary to set agreed international standards for all collections to attain. These standards would also provide a useful target for new collections to aspire to. Although there are guidelines set for the establishment and operation of collections, they do not cover all protocols or procedures, nor do they set minimum requirements. Standards are necessary to increase the quality in collections and to provide a sound footing to ensure they offer the service to science and industry that is required today and provide stable reference material for the future (see Chapter 6).

There are critical elements to standard procedures in the handling, storage, characterization and distribution of microorganisms and the handling of associated information. Mechanisms for ensuring reproducibility and monitoring such procedures lead to the establishment of common standards. The industry expresses the need for quality control and standards within collections. Although publications on collection management and methodology give information on protocols and procedures (Hawksworth and Kirsop, 1988; Kirsop and Kurtzman, 1988; Kirsop and Doyle, 1991; Smith and Onions, 1994), a quality management system must go further setting operational standards.

Whatever the authoritative document, standard or certification or accreditation process that is selected, there are several underlying principles of a quality management system. A culture collection or BRC quality management system must address several specific areas:

- Organizational requirements
- Equipment use, calibration, testing and maintenance records
- Documentation management
- Data management, processing and publication
- Preparation of media and reagents
- Accession of deposits to the BRC
- Preservation and maintenance
- Supply
- Quality audit and quality review

Table 1.1. List of educational tools in microbiology and related topics

Tool	URL/availability
Managing microbes an e-learning tool from CABI	Available from CABI Academy: https://www.cabi.org/products-and-services/academy/
ICT innovations for plant doctors	To access the Plantwise Knowledge Bank on your tablet, download the free Plantwise Factsheet Library app from Google Play. https://play.google.com/store/apps/details?id=org.cabi.pfff&hl=en_GB
The WWW Virtual Library: Mycology	http://mycology.cornell.edu/
International Society for Human and Animal Mycology (ISHAM): Mycological Links	https://www.isham.org/mycology-resources/mycological-links
North American Mycological Association (NAMA), Education Resources	http://www.namyco.org/education.php
EMbaRC Technical videos	http://www.embarc.eu/Gram Staining; Fungi observation under Optical Microscope; Preservation Techniques: http://www.embarc.eu/information.html
Cybertruffle	http://www.cybertruffle.org.uk/eng/index.htm
Course Manuals	
CABI Course manual	CABI pdf available for course participants
Classwork with Fungi	CABI out of print – pdf available
Atlas of Clinical Fungi	CBS - http://www.clinicalfungi.org/
CBS Laboratory Manual Series 1 and 2	https://wi.knaw.nl/news/category/Laboratory%2520Manual%2520Series
Identification of common Aspergillus species	https://wi.knaw.nl/news/category/Other%2520publications
Plantwise Diagnostic Field Guide	https://www.cabi.org/cabi-publications/plantwise-diagnostic-field-guide/
Mycological Techniques	CABI pdf available for course participants
Courses	
CBS courses: Food and Indoor Mycology; Introduction to novel identification methods	https://wi.knaw.nl/news/category/courses
Fungi for schools - BMS	http://www.davidmoore.org.uk/Assets/fungi4schools/index.htm
Level 1 Mushroom Mycelium. Learn about Fungi Kingdom, Mycology	https://study.com/academy/lesson/what-is-mycelium-definition-function.html
Hundreds of Free Online College Courses	https://mycotopia.net/topic/91038-hundreds-of-free-online-college-courses/
Masters portal	http://www.mastersportal.eu/e.g. The University of Helsinki offers the MSc in Microbiology and Microbial Biotechnology course http://www.mastersportal.eu/studies/154236/microbiology-and-microbial-biotechnology.html
Cranfield University Applied Food Mycology Course	https://www.cranfield.ac.uk/courses/short/agrifood/applied-food-mycology
MYCOLOGOS: World's First Online Mycology (Mushroom) School	https://www.kickstarter.com/projects/mycologos/mycologos-the-worlds-first-mycology-school

Amongst the microorganism domain criteria set down by the OECD Best Practice Guidelines are recommendations on the following:

- Staff qualifications and training
- Hygiene and biosafety
- Equipment use, calibration, testing and maintenance records
- Preparation of samples
- Information provided with the biological material supplied

There are several standards set for microbiology laboratories such as Good Laboratory Practice (GLP), British Standard 5750, UK Accreditation Service (UKAS) - ISO Guide 25 and ISO 9000

(Martin *et al.*, 2015). Industry is expressing the need for quality control and standards within collections and although publications on collection management and methodology give information on protocols and procedures (Hawksworth and Kirsop, 1988; Kirsop and Kurtzman, 1988; Kirsop and Doyle, 1991; Smith and Onions, 1994) the OECD BRC best practice (https://www.oecd.org/sti/emerging-tech/oecdbestpracticeguidelinesforbiologicalresourcecentres.htm) goes further toward setting minimum standards. A Quality Assurance scheme exists for plant genetic resources. In 1993 the Commission on Plant Genetic Resources endorsed *The Genebank Standards* for use as the international reference in national, regional and international genebanks (Anon, 1994). The international network of genebanks have the responsibility to hold, for the benefit of the international community, plant genetic resources and make them available without restriction. The genebank standards are solely concerned with the storage of seeds of orthodox species that can survive very considerable desiccation and for which longevity is improved by reducing seed storage moisture content and/or temperature. It is evident that microbial resource collections could benefit from the general application of similar standards. The genebank standards provide targets for participating institutes to achieve with the ultimate objective of long-term safe and sustainable conservation. A tiered system is described providing both minimal and preferred higher standards. They acknowledge differences in the storage of original material, not usually for distribution directly to the user, and that of collections where the material is immediately available, laying down criteria including the number of generations acceptable between the original material and that supplied to the end users. Examples of quality standards and guidelines are:

- Common Access to Biological Resources and Information (CABRI) Guidelines for microbial public service collections (http://www.cabri.org/).
- The WFCC Guidelines for the establishment and operation of collections of microorganisms (http://www.wfcc.info/guideline).
- The Microbial Information Network for Europe (MINE) project standards for the member collections (Hawksworth and Schipper, 1989).
- Genebank Standards – for plant germplasm collections, produced by the Food and Agriculture Organization of the United Nations, Rome, and International Plant Genetic Resources Institute, Rome (https://www.fao.org/3/i3394e/i3394e.pdf).
- OECD Best Practice for BRCs (https://www.oecd.org/sti/emerging-tech/oecdbestpracticeguidelinesforbiologicalresourcecentres.htm)

There are critical elements to standard procedures in the handling, storage, characterization and distribution of microorganisms and cell lines and to the handling of associated information. Mechanisms for ensuring reproducibility and monitoring such procedures lead to the establishment of common standards, these are covered in Chapter 6.

Culture Collection Accession Policy and Criteria

A culture collection cannot hold representatives of all species and cover the variety within the species. It must set accession criteria (in the form of a policy) that all strains must meet that are added to the collection. Key elements of such a policy include:

Organisms for which adequate facilities are in place to handle safely and within national and international law and in particular, within the spirit of the Convention on Biological Diversity.

A definition of the type of organisms to be collected relevant to the collection's objectives e.g. strains of new species, ex-type cultures, strains with defined industrial and physiological properties, those cited in publications.

Unless for a valid scientific or industrial reason, duplication should be avoided. New accessions should have unique properties of potential long-term interest and facilities should be available to handle them. The lodging of cultures in more than one collection is desirable for rare and important strains as insurance against accidental loss. Ideally all strains referred to in published research should be deposited in a Biological Resource Collection as reference material. However, the collection reserves the right to refuse acceptance of strains if they are not required or in bad condition.

Acceptance of a strain requires careful evaluation. An accession form should be provided, consisting of detailed information about the strain. This should be completed and submitted for review before acceptance of the deposit. Individuals depositing strains for inclusion in the 'open' publicly available collection may be provided with a culture for checking after preservation, and may typically request one free culture of their own deposit at any one time, although more may be negotiated through the Collection Curator. On occasion, they may also be entitled to select one isolate of any culture in the open collection in exchange, providing this is done at the time of deposit.

A unique accession number should be allocated to the strain which should never be reassigned if the organism is later discarded. The viability, purity, identity, growth requirements and methods of maintenance and/or preservation of the strain must be determined and the information recorded against the accession number. These records can be kept by several means but should enable cross-referencing between the collection number and culture name. Initial hard copy data can be placed on a computerized database.

CABI Culture Collection Accession Policy

The CABI culture collection has a global remit and deposits should be strains and biological material that are relevant to the scientific activities and future requirements of CABI. CABI staff are encouraged to deposit all important strains into the collection, specifically those that are used routinely or as part of project work. Strains accepted may include:

- Living fungi, bacteria, nematodes or archive material of insects or plant material where inclusion is beneficial to CABI or our member countries.
- New species and type cultures that are of interest to CABI
- Strains isolated from agricultural, biotechnological or environmental systems, specifically soil health, invasive species, pests and diseases, trade and biodiversity
- Reference strains relevant to plant disease, quarantine and biological control
- All plant disease reference strains mentioned in new disease reports
- Strains with defined biological, industrial and physiological properties likely to be of potential future value to the CABI business
- Strains cited in publications
- Strains with potential to be required in CABI areas of operation to meet business needs
- Strains relevant to the National Biodiversity Action Plans of member countries.

On occasion, it may be necessary to pursue a targeted accession policy for notable gaps in our holdings in order to meet CABI's business needs. If such accession needs fall outside the scope of this policy, then approval must be sought from the culture collection curator so that available resources are allocated according to business priorities.

Strains must be obtained from reputable scientists, documented with the history of their isolation, maintenance and preservation. Wherever possible, strains should be accompanied by a full data set, 'the strain passport' that may include sequence, pathogenicity, biochemistry or genetic information such as Barcodes. Total numbers accessed per year is dependent on the needs of the CABI principal investigators (researchers), business stakeholders and available resources. To ensure fair and equitable sharing of the benefits arising from the utilization of genetic resources under the terms of the Nagoya protocol, and adherence to the Convention of Biological Diversity, strains accessed must have a documented history to include Prior Informed Consent (PIC) and the Mutually Agreed Terms (MAT) which must include the authority to deposit in a culture collection and further distribute. *Depositors should contact collection staff for more information.*

Summary acceptance criteria considered for each strain are:

- That its inclusion in the collection meets the business needs of CABI and/or that it has known unique properties of value to CABI. This is a similar requirement in other collections.
- That it is not a duplication in the collection or that of another readily available source. Some CABI projects often have repeat surveys to collect material and it is useful to deposit strains from the same locality as there may be differences in virulence, biological properties and cryptic strains. Similarly, sub-species, special

forms and different races may often exhibit different and unique properties such as virulence and pathogenicity and inclusion of such strains is encouraged.

- If genetically modified, it falls within licensed parameters.
- It is representative of a CABI interest systematic, pathogenic or ecological group not currently in the collection.
- There is a potential interest in the strain (long-term).
- There is documented evidence of collection and authority to deposit and utilize.
- The collection will not access Human or Animal Pathogens of Hazard Group 3 or above, strains where a genetic modification will have a perceived risk to the environment or human health, strains collected without permission or strains where holding them may compromise CABI's profile.
- Important and correctly identified strains isolated from CABI member countries should also be stored in national repositories where available.

Accession Data

Accession

Details are recorded on the accession form and entered into the CABI culture collection database (see data field content and analysis in Appendix H, this volume). A unique accession number is allocated to the strain which is never reassigned if the organism is later discarded. The viability, purity, identity, growth requirements, and methods of maintenance and/or preservation of the strain must be determined and the information recorded. These records are kept on the GRC strain database and the accession form is kept to provide a hard copy of the data. The preservation and growth data after receipt in the collection are recorded allowing cross-referencing between the collection number and culture name on the accession form.

Strain data records

Deposits: The normal minimum data recorded for each deposit is:

a) Name (where one can be applied)
b) Depositor's name and address
c) Substrate or host from which the organism was isolated
d) Geographical location of isolation (the country of origin/source is required for all deposits)
e) Depositors strain number or other collection number(s)
f) Assigned strain number

Supply:

a) Name, address and contact numbers of the requester
b) Organisms requested
c) Collection order number
d) Date of receipt of order and supply

Record management

Availability for inspection: Records should be readily accessible for inspection by authorized personnel from regulatory agencies and Quality Assurance (QA) personnel. Access to confidential information must be restricted to those with suitable security clearance.

Record retention: Collection, processing, storage, distribution, and QA records should be maintained for a minimum of 10 years after the last expiration of the specimens involved. When there is no expiration date, records should be maintained for 10 years after the date of distribution.

Security: Electronic records should be backed up daily on a network or remote server and weekly on a CD, tape or other appropriate media. Consideration should be given to establishing an arrangement with an offsite data security company that retrieves and stores all critical data at a remote location.

Archival system: A culture collection may develop a system for archiving records that are older than one year and fewer than 10 years. This system should be accessible for audits and inspection.

Corrections and/or changes: Corrections or changes in a hand-written record should be made with a single line drawn through the altered text. Corrections should be initiated and dated by the individual making the correction or change. Changes in electronic records should be noted and tracked.

Minimum Data Sets and Recommended Data Sets for Microbial Accessions to BRCs

To make best use of a microbial sample it needs a minimum set of data associated with it. The **E**uropean Consortium of **Microbial R**esources **C**entres (EMbaRC) (http://www.embarc.eu/) brought together European culture collections to improve microbial resources for research and as part of the

Table 1.2 Minimum Data Sets (MDS) and Recommended Data Sets (RDS) for microbial accessions to BRCs (See OECD, 2007)

Filamentous fungi	**Filamentous fungi**
Minimum Data Set (MDS)	**Recommended Data Set (RDS)**
Accession number	Misapplied names
Other collection numbers	Isolated from
Name	Mutant
Organism type	Literature
Restrictions	Sexual state
Status	Race
History of deposit	Genomic meta-data
Conditions for growth	
Form of supply	
Geographic origin	
Yeasts	**Yeasts**
Minimum Data Set (MDS)	**Recommended Data Set (RDS)**
Accession number	Isolated from
Other collection numbers	Mutant
Name	Sexual state
Organism type	Literature
Restrictions on distribution	Misapplied names
Status	Race
History of deposit	Genomic meta-data
Geographic origin	
Conditions for growth	
Form of supply	
Microalgae	**Microalgae**
Minimum Data Set (MDS)	**Recommended Data Set (RDS)**
Accession number	Literature
Other collection number	Conditions for storage
Name and taxonomy	Isolate history
History of deposit	Genomic meta-data
Isolate history	
Form of supply	
Geographic origin	
Conditions for growth	
Bacteria	**Bacteria**
Minimum Data Set (MDS)	**Recommended Data Set (RDS)**
Accession number	Serovar
Other collection numbers	Other names
Name	Isolated from
Infrasubspecific names	Mutant
Organism type	Genotype
Restrictions on distribution	Literature
Status	Genomic meta-data
History of deposit	
Geographic origin	
Conditions for growth	
Form of supply	
Cyanobacteria	**Cyanobacteria**
Minimum Data Set (MDS)	**Recommended Data Set (RDS)**
Accession number	Other names
Other collection numbers	Isolated from
Name and taxonomy	Mutant
Infrasubspecific names	Genotype
Organism type	Literature
Restrictions on distribution	Genomic meta-data

Continued

Table 1.2 Continued.

Status	
History of deposit	
Conditions for growth	
Form of supply	
Geographic origin	
Archaea	**Archaea**
Minimum Data Set (MDS)	**Recommended Data Set (RDS)**
Accession number	Other names
Other collection numbers	Isolated from
Name	Mutant
Infrasubspecific names	Genotype
Organism type	Literature
Restrictions on distribution	Genomic meta-data
Status	
History of deposit	
Geographic origin	
Conditions for growth	
Form of supply	
Plasmids	**Plasmids**
Minimum Data Set (MDS)	**Full Data Set (FDS); RDS is not applicable**
Collection Accession number	Constructed from
Name	Incompatibility group
Other culture collection numbers	Transfer ability
Type	Helper
Class	Copy number
Literature	Molecular weight
History of deposit	Cloned gene
Restricted distribution	Transposable element
Host for distribution	Promoter
Medium	Ribosome binding site
Selectable phenotype	Start codon
Replicon	Terminator
Host range	Further information (Remarks on propagation and/or on properties and/or on history, other name(s), etc)
	Restriction sites
	Sequence detail
	Price code
	Properties and application
Protozoa	**Protozoa**
Minimum Data Set (MDS)	**Recommended Data Set (RDS)**
Accession number	Biochemical or molecular characteristics
Other collection numbers	Other name
Name	Substrate or host
Organism type	Year of isolation
Stage	Literature
History of deposit	Genomic meta-data
Status	
Restriction on distribution	
Conditions for growth	
Form of supply	
Geographic origin	
Phages	**Phages**
Minimum Data Set (MDS)	**Recommended Data Set (RDS)**
Accession number	Cell surface receptor

Continued

Table 1.2 Continued.

Element name	Genomic meta-data
Element type	
Other culture collection numbers	
Restricted distribution	
Literature	
History of deposit	
Host for propagation	
Host used for propagation	
Lysogenicity	
Virus used for	
Viruses	**Viruses FDS = MDS**
Minimum Data Set (MDS) = Full Data Set (FDS)	Genomic meta-data
Accession number	
Virus name	
Virus name abbreviation	
Former name	
Genus	
Pathotype, serotype, strain	
Original host	
Geographic origin	
Isolate history	
Reference isolate	
Quarantine regulations	
Remarks	
cDNA and gDNA Libraries	**cDNA and gDNA Libraries, MDS = RDS**
Minimum Data Set (MDS)	
Library Name	
Organism	
Type (cDNA or gDNA)	
Vector	
Insert Size	
Library Coverage	

project output developed a European Standard at: http://www.embarc.eu/deliverables/EMbaRC_D.NA1.2.1_2.28_BRC_standard.pdf for operation of BRCs based on the OECD BRC Best Practice (https://www.oecd.org/sti/emerging-tech/oecd-bestpracticeguidelinesforbiologicalresourcecentres.htm). Recommendations were made for minimum data sets. Appendix H provides a detailed coverage of the CABI living collection data fields.

Collection Operations: Ordering, Charges, Payments, Quarantine and Safety

Culture collections must endeavour to provide their clients, on every occasion, with the products and services they require. The products and services supplied must be of marketable quality and fulfil product claims as defined in the culture collection catalogues. At all times good techniques and procedures as defined in the individual collection's laboratory procedures manual or standard operating procedures must be in operation and regular audits must be carried out to ensure that these procedures are followed and are effective. Many national culture collections either employ a recognized accreditation scheme, e.g., ISO 9000 (Simões *et al.*, 2016) or follow the other quality management systems (Martin *et al.*, 2015).

Amongst the specialist services that a collection can offer is a culture/cell supply service. The strains required can be selected by accessing the culture collection databases normally available via the culture collection website (see Appendix C, this volume) or through organizational links such as the UKBRCN website (https://www.ukbrcn.org/) or for example coordinated databases such as the Global Catalogue of Microorganisms (http://gcm.wdcm.org/). Strains with particular properties held in the CABI living collection are listed in Appendix A, this volume, in the list of microbial properties or through CABI's online catalogue at: https://www.cabi.org/services/microbial-services/culture-collection-microorganism-supply/grc/.

Ordering Cultures

Order forms, increasingly via an online process, should be sent directly to the relevant collection (example order form, material transfer agreement are provided in Appendix E, this volume). Most collections accept telephone, written company orders, e-mail, internet and mail orders when accompanied by an official customer order number and/or official forms. An official department order form is preferable. Clients not providing an official order number and orders from overseas may be sent a proforma invoice requesting payment in advance. In the case of CABI a signed undertaking, declaring that the client will operate within the spirit of the Convention on Biological Diversity (CBD) (see Chapter 5, this volume) must be provided and will cover all subsequent orders unless otherwise agreed. Postal quarantine and other regulations control the distribution of strains (see Chapter 5, this volume for details of what is required to comply).

Customer Undertaking

The customer receiving material protected by the CBD must agree not to claim ownership over organisms provided, nor to seek intellectual property rights over them or related information without permissions to do so. Collections should provide such strains under an MTA or at least provide the terms and conditions on which the strain was accessed from the provider country. If the recipient wishes to utilize or exploit such organisms commercially, suitable and adequate recompense to the country or origin, in the spirit of the Convention of Biological Diversity and the Nagoya protocol on access and benefit sharing must first be discussed with the appropriate authority in the provider country or with the relevant culture collection in the first instance (Smith *et al.*, 2017a). Benefit sharing commitments agreed in Mutual Agreed Terms (MAT) must be delivered where provider countries have Access and Benefit Sharing law in place (Verkley *et al.*, 2019). The customer must also have the appropriate facilities to receive and handle the microorganisms supplied.

Information Required from the Customer

Customers are required to provide: strain details (name and accessions number if applicable); a customer order number; contact name; an invoice/delivery address; telephone number; VAT registration number (if the customer is within Europe); details of how they would like the strain delivered (i.e. as a freeze-dried ampoule, living culture, etc.). Supply and use of certain organisms require a permit or authorization by local or international legislation, where this is required a copy must accompany the request (see Chapter 5, this volume). Details can be obtained from the relevant collection if in doubt.

Strain Availability

When ordering cultures, clients are advised not to quote accession numbers unless they are sure that a particular strain is most suitable for their purpose. It is frequently more satisfactory to state the purpose for which the cultures are required and to leave the choice of strain to collection experts. For educational purposes it is often sufficient to order by the generic name only, or to allow the collection to select the best strain of a species available. Many isolates are now supplied immediately as freeze-dried ampoules, but if stocks are old they are checked for viability before despatch which may cause delay.

On receipt of the order, the culture collection databases are checked to ensure strain availability. If the required culture is not available, and if the customer so desires, the database will be searched for alternatives. Users will be advised as far as possible where to obtain the strains required if they are unavailable and alternative source collections can be suggested. Freeze-dried cultures are often sent within 24 hours, but if the customer requires a live culture, or one that is stored in oil or liquid nitrogen, then delivery time is normally in the order of two to three weeks or longer for cryopreserved slow-growing strains. If the culture cannot be supplied as a freeze-dried sample, or the customer requires the culture in another form, the customer will be contacted with an estimated supply date. If any unexpected delays arise (e.g., because of failure to recover from preservation), customers are informed immediately by telephone or email to ensure new deadlines are satisfactory. Despite rigorous quality control and standard procedures being followed it may be possible that the strain may fail to grow, may be (or become) contaminated or may not exhibit the property stipulated in the order or that is reasonably expected of the organism. It is normal policy to replace the strain free of charge with a growing culture on agar where possible. If this is not possible a refund will be given often in the form of a credit note. Cultures from cryopreservation or other methods of storage are grown on agar and viability, purity and identity are checked before dispatch; again if refunds are considered appropriate they will normally be given in the form of a credit note. If the situation changes during recovery of the strain(s) the customer will be informed.

Restrictions

Once the culture collection has confirmed that the strain is available, information about the containment level of the organism, the format it is preserved in and any restrictions on the sale or distribution of the strain are discussed. All culture parcels must be opened in a laboratory and all hazard group 2/3 (Anon, 2021) organisms require the appropriate containment level facilities. The forms in which the organism is preserved affect delivery time. If the strain is a non-indigenous plant pathogen, the customer will have to apply for permission to receive the strain and the customer will be advised. In the UK a Department for Environment, Food and Rural Affairs (Defra) licence will have to be obtained before the strain can be supplied (https://www.gov.uk/guidance/moving-prohibited-plants-plant-pests-pathogens-and-soil). Overseas customers may require an import permit available from National Authorities (see the FAO advice on the requirements for phytosanitary certificates, https://www.fao.org/3/y3241e/y3241e06.htm). The customer will be informed of any special restrictions, and the relevant paperwork that is required from them prior to provision of requested cultures e.g., with plant pathogens (see below). Restrictions also apply to the supply of dangerous organisms that could be mis-used. BRCs take special precautions to prevent such occurrences and have special procedures in place (Smith *et al.*, 2017b). The handling and distribution of some human and animal pathogens and toxin producers is restricted by many countries, e.g. see the UK Biological Security Strategy (https://assets.publishing.service.gov.uk/government/uploads/system/uploads/attachment_data/file/730213/2018_UK_Biological_Security_Strategy.pdf). The Biological Weapons Convention (BWC) effectively prohibits the development, production, acquisition, transfer, stockpiling and use of biological and toxin weapons (https://www.un.org/disarmament/biological-weapons/).

Price of Strains and Special Considerations

When ordering by telephone or online, customers will be informed of the current price.

Several organizations and collections exchange organisms on a free basis. Organizations with a Memorandum of Understanding, collections belonging to the European Culture Collections' Organisation (ECCO) and depositors of strains may be entitled to one free culture in exchange (consult relevant collection). The fee paid by customers is for the supply of the culture and is not a charge for the culture itself.

Invoices will normally be dispatched with the strains from most collections unless otherwise instructed or where proforma invoices have been paid in advance, however, some collections supply the strain and invoice within a specified period of time.

Quarantine and Postal Regulations/restrictions

Adequate precautions must be taken to ensure the safety of all those persons involved in the packaging, transport and receipt of strains (Smith and Ryan, 2019). Steps must also be taken to make sure that material transported presents no hazard to plants, people, animals or the environment. In order to reduce risk in these areas, national and international regulations have been formulated to govern the shipment of biological materials. All those who are involved in the shipment of cultures must follow the rules and regulations that apply (see Chapter 5, this volume).

Human and Animal Pathogens

International Postal Regulations regarding the postage of human and animal pathogens are very strict because of the safety hazard they present (see Chapter 5, this volume). Culture collections must hold a special permit for the despatch of such material. The packaging and other regulations are strictly observed. Some countries will not accept such potential human infective material through the post including the UK. Further information is provided in the WHO Guidance on regulations for the Transport of Infectious Substances 2022: https://www.who.int/publications/i/item/9789240019720.

Dispatch

Cultures are dispatched to customers in good condition after checks (if applicable). They require periods from 5 days to several weeks to grow satisfactorily. They are packed in a way to prevent damage during transit. On receipt, if the culture is provided as an active culture, bottle caps should be loosened. Cultures on agar have a limited shelf-life and should be sub-cultured after periods of 2-26 weeks (see Chapter 3, this volume or contact the relevant collection). However, freeze-dried fungi and bacteria in ampoules or vials can be kept for much longer periods, tens of years in some cases.

The information a culture collection should provide with the biological material supplied should at least cover the following:

a) Name or identifier and batch number of the biological material, listed in a strain information sheet
b) Stability, storage instructions and if appropriate, conditions of growth
c) Instructions for opening ampoules or vials (when appropriate and in all cases where materials are being provided to new recipients)
d) Safety data sheet in the case of a hazardous organism or derivative including the containment level required for handling the biological material, disposal measures and measures to take in case of spillage
e) A Material Transfer Agreement to meet the CBD.

Quality Assurance

The identities of cultures are confirmed (wherever possible). In several of the UKBRCN member collections this is done before acceptance by a team of expert biosystematists and strains are checked again by these experts before and after preservation and during maintenance procedures to ensure quality is maintained. Many culture collections operate to conform with ISO Standards or to guidance provided to culture collection organizations e.g. the WFCC Guidelines (http://www.wfcc.info/guideline).

Managing a Small Working Collection

Establishing a Culture Collection: What are the key questions to ask

Why do you need a collection?

This may be to provide cultures to support your research or product production or perhaps to store the strains you are isolating. An estimation of the size of collection and its predicted rate of growth along with knowledge of the type of organisms you will store is required. Do not duplicate what is already available elsewhere if this can be avoided. Often reciprocal arrangements can be made between collections, capacity to store biodiversity is limited and investment in discovery, isolation and storage is significant and best used wisely. There are several local, national and international organizations or infrastructures that can facilitate distributed access to research strains (see this chapter, above).

What accommodation, technology and staffing will be needed?

Ample and appropriate space will be needed for example if the organisms need special facilities for isolation and handling (containment level), staff with specific expertise and the appropriate preservation technology established. Suitable facilities for identification and characterization need to be provided. Storage for the preserved organisms must be considered. Maintenance programmes will be needed, monitoring of equipment, quality management processes selected and implemented (see Chapter 4, this volume on preservation methodologies).

What data storage will be needed?

The research carried out will dictate the establishment of databases. These may well be associated with specific pieces of equipment. However, consideration of how you wish to present this information alongside the collection metadata, how the data are linked and analysed will dictate the type of databases and software needed.

Are there specific regulatory requirements that need to be met?
See Chapter 5, this volume on the safe handling and distribution of microorganisms and compliance with Health and Safety (Smith *et al.*, 1999), classification of microorganisms on the basis of hazard, quarantine regulations, ownership of Intellectual Property Rights (IPR), the Convention on Biological Diversity (CBD), the safety information that must be provided to the recipient of microorganisms, regulations governing shipping of cultures and the control of distribution of dangerous organisms (Macreadie *et al.*, 2019).
What is the funding model and how much will it cost
No matter the funding source a business model and case for establishment will be needed, an efficient and effective plan will consider key revenue lines for products and services that might be offered. See chapter 7, this volume.

Culture collection space and equipment

An environment, suitable for handling authenticated materials appropriate to the organism type should be provided. The environment should also be suitable for the acquisition, maintenance and provision of biological materials and associated services. Hence it is essential that:

- The member of staff allocated to a task should check that the accommodation is clean and well-lit and that usual aseptic techniques are followed.
- Appropriate protective clothing is worn and safety procedures followed.
- Appropriate arrangements, in accordance with national and international regulations, for site security are made to ensure hazardous organisms cannot be released to unauthorized users.
- The BRC facilities, as well as the environment and equipment in the premises, must conform to all relevant national and international standards and regulations.
- The containment level for the biosafety level (or risk group) of the biological materials are documented and in place and that the BRC does not operate beyond these limits.

Equipment use, calibration, testing and maintenance records
It is essential that the use of equipment, its calibration, testing and its maintenance are properly managed. This must be documented and recorded and retained for future reference. One must ensure that:

- Equipment management procedures including use, control of performance, maintenance and calibration are detailed in a predefined schedule. Instructions for these activities are contained in the manufacturer's handbooks/manuals or detailed in the BRC procedure.
- Service records are maintained and copies of key documents should be held in the BRC Equipment Maintenance and Calibration Log books, in the care of the Quality Manager.

Information management

Databases
Collection databases should contain information relating to strains held by a BRC or other data needed for collection management (inventory control for example). The data should be retained as long as a strain remains viable. If the strain is lost, then the database record should either be printed and stored on file or copied to a digital archive before the entry is removed from the working database. It should be placed in reserve or annotated to indicate that it is no longer available as living material. If genomic metadata is stored in association with the strain this must be retained with the digital archive.

Data Back-up
The informatics system should ensure regular data back-up. BRCs should introduce appropriate measures (protocols, tools and standards) in their own informatics systems to assure reasonable security of information.

Data Availability
The BRC should make data describing the biological material and its origin available and provide electronic catalogues to digital users (e.g. on its website) or through focused, national, regional or global networks. Data should also be retained for traceability in compliance with relevant national laws and regulations. Exchange of information should be in line with the OECD Guidelines on the Protection of Privacy and Transborder Flows of Personal Data. Links should be reported

to genomic accessions on the European Nucleotide archive or GENBANK wherever possible.

BRCs should restrict access to the electronic catalogues where appropriate. Users should be authenticated and be provided with specific identities and passwords to access different categories of information and services. The validity of identifiers and passwords should be checked.

Accession

There should be a minimum amount of information available for each accession in the Minimum Data Set (MDS) collection. Additional data may be included in the Recommended Data Set (RDS) and Full Data Set (FDS). The MDS comprises essential information to identify a unique item in the BRC. This should always be recorded and made available. The RDS includes useful information for an improved description of the material. The FDS provides all remaining information that is available at the BRC for any given biological material.

Receipt and handling of biological materials

Receipt of biological material

On receipt of biological material quality control procedures should be carried out to confirm its purity, identity and viability. Before accepting a deposit, the BRC should check the material against risk group lists and other lists to ensure that the biological material does not exceed the laboratory's biological safety containment level. The following practices are recommended:

The BRC should document and implement safe procedures for receipt and storage appropriate to the type of biological materials handled. All incoming parcels that contain known or unknown microorganisms should be opened in a suitable containment laboratory or appropriate microbiological safety cabinet. Facilities for the safe handling and disposal of biological materials should be available in the immediate vicinity. The depositor should provide assurance that biological materials were obtained legitimately with confirmation of mutually agreed terms or prior informed consent under the CBD. Conditions of deposit should be determined and agreed (e.g. laid down in a material transfer agreement [MTA]), for example, to protect assigned intellectual property rights (IPR).

Where deposits are outside the expertise of the BRC, alternative suitable BRCs should be recommended.

Packaging and transport

The packaging of biological material and its transport by postal and other transport services is controlled by international and regional agreements and national laws. To ensure safe and secure packaging and transportation of biological material, BRCs should follow the WHO Guidance on regulations for the Transport of Infectious Substances 2020 (https://www.who.int/publications/i/item/97892) and the document Guidance on regulations for the transport of infectious substances 2021-2022 (https://www.who.int/publications/i/item/9789240019720).

These best practice guidelines provide practical guidance to facilitate compliance with current international regulations for the transport of infectious substances by all modes of transport, both nationally and internationally. More information on how they are adopted in European Culture collections are described in the EMbaRC project European BRC Standard at: http://www.embarc.eu/deliverables/EMbaRC_D.NA1.2.1_2.28_BRC_standard.pdf.

Funding

A BRC requires adequate and reliable sources of funding for long-term sustainability. Most BRCs require some degree of commitment to core funding which could include support from their respective national governments.

Other kinds of funding sources include support from industry, grants from agencies that support research, cost-recovery through fees for service or supply of products, consultancies including development of databases and other tools and even funding from charitable sources, especially those associated with public health or sustainable development.

BRCs undertake a variety of activities directly related to quality control, collection development and operation that may include opportunities for some additional cost recovery activities. The generation of genomics and proteomics data, that complement and add value to biological materials, can be a potential new source of revenue. However, the degree to which such activities may actually provide support, sufficient to ensure financial sustainability of a BRC, is unproven (Smith *et al.*, 2014).

Operating a Collection

The guiding principles of a collection remain the same whether they are large or small, public or private. The samples must be uniquely identified to avoid selecting the wrong sample, have associated data to understand what the sample is and where it comes from and information on its properties or applications if available. They must be reproducible in property and function and maintained viable and pure, fit for purpose.

This section describes:

- The general organizational requirements of a culture collection
- Operations within a culture collection
- Procedures for biological material tracking
- Why quality assurance and control are essential
- How to design a plan to operate a small in-house culture collection

General organizational requirements for a culture collection

The purpose of the culture collection should be clearly formulated and documented. It should be defined as to whether it is a free-standing entity or part of an institution. The culture collection should have sufficient professional staff and a commitment to maintain, effectively operate, and preserve records and operating procedures for future reference and continuity. The collection ensures that work can be repeated, results can be confirmed, additional data gathered and samples retained for future use beyond a project's life span and often beyond the retirement of the scientist who created it. The view must be for the long-term.

Quality Assurance Programme/Quality Management System

The Director or another designated party should ensure that a Quality Assurance (QA) Programme (also termed a Quality Management System or QMS) is in place to make certain that the entire operation conforms to the culture collection's Standard Operating Procedures (SOPs), necessary audits, and government regulations. It is good practice for the QA/QMS Programme staff to report independently. The operations should be audited both internally through regular documented internal reviews to ensure compliance with the SOPs and regulations and externally by an independent third party. A user of the samples in the collection must be able to rely on the quality and reproducibility of the material they receive.

Staff need to be appropriately trained and if possible, a multidisciplinary team created under a Director/Lead scientist/Curator who is qualified by training and has experience to fulfil the scope of activities conducted by the culture collection. Staff must possess sufficient educational background, experience, and training to assure that assigned tasks are performed in accordance with the culture collection's established procedures. The Curator or lead of the collection should have sufficient background knowledge in research, the organism groups they curate and the operation of culture collections in order to fulfil their roles. Technical staff should be responsible for implementation of policies and procedures as established by the Director/Curator. Staff must demonstrate competency in operations to which they are assigned. Authority and reporting relationships for each member of the staff should be clearly described.

Responsibilities of the Director/Curator/Lead scientist

- Implement institutional policies ensuring compliance with current and applicable regulations
- Ensure operation within legislative guidelines, having a good working knowledge of key protocols
- Ensure that all provider country and depositor's terms and conditions are made available with the material and that they do not impede institutional practices and aims
- Liaise with researchers providing information and guidance to recipients of material
- Provide details of alternative sources of materials not available from the culture collection

- Construct and maintain a current organizational chart that delineates the functional relationships within the culture collection
- Approve and maintain job descriptions and should document staff responsibilities
- Ensure that personnel responsible for performing culture collection activities are adequate in number, have adequate experience, and should be assigned responsibilities commensurate with their capabilities
- Be responsible for developing and reviewing employee training programmes
- Ensure that there is an appropriate and acceptable safety programme

Microbiologist support

The effective management of a culture collection requires extensive knowledge not only of the organisms themselves, but also their growth and preservation requirements, properties and potential applications. Taxonomic skills are essential for quality control, if such skills are not available on staff, steps have to be taken for collaborative arrangements with specialists for the different taxonomic groups kept in the culture collection. Collections cannot be operated without this knowledge.

Best practices

Different microorganisms often require special preservation methods in order to ensure optimal viability, storage and purity. For security and in order to minimize the probability of strains being lost, each strain should be maintained by at least two different procedures. At least one of the methods used has to minimize the risks of genetic change during long term storage. The best choices are cryopreservation in liquid nitrogen and freeze drying. In order to minimize the risk of loss of special and unique strains, collections should have duplicates securely housed at a separate site and preserved with some alternative method.

Record management

Each culture collection should develop a record management system that permits detailed records to be made at the same time as the performance of each step in the collection, processing, and distribution of specimens. This may include, but is not limited to informed consent for access and collection of material, procurement, processing, preservation, quarantining, testing, record review, releasing, labelling, storage, distribution and quality control of specimens. Records should be created and maintained in a manner that allows steps to be traced clearly. Record security systems should be adequate to ensure confidentiality and safety. Record management covers the following areas:

Data structuring: The data should be structured to be compatible with internationally accepted standards (see WDCM,). Special software programmes for data management of small to medium sized culture collections are available, often direct from WDCM listed collections. It is wasteful to develop independent systems. See minimum and recommended data sets (in this Chapter below).

Research Programmes: Research programmes are usually a strong part of a small culture collection activity and very often the primary reason for their establishment. The output of such research work can often make important contributions to knowledge on the morphology, taxonomy, physiology, biochemistry and genetics of the groups of organisms maintained. It is very important to keep this data well-structured and organized in a computerized form to facilitate its access and use.

Shipping and exchange of cultures: Collections should maintain detailed records of culture recipients, the material sent, method and date of shipment, and the name and address of the person to whom it is sent for traceability.

Culture Collections as Information Centres: Culture collections often serve as information centres and have to provide consultancy, advisory or investigation services and data on holdings, in different forms. The provision of online electronic catalogues of the strains available for distribution is invaluable.

Networking: For every culture collection, regardless of its size, the establishment of formal or informal links with active regional groups is very important. In order to make the culture collection's holdings widely known, to facilitate international communication and data exchange, it is a good idea to register with the World Data Centre on Microorganisms (WDCM) and to join the WFCC.

Operations within a culture collection

Each culture collection should develop documented policies and procedures in a standardized format that should be incorporated into a Standard Operating Procedures (SOP) manual. The SOPs should state policies and define and describe all procedures in detail. These SOPs should be utilized to ensure that all samples are appropriately stored so that they may be effectively disseminated for subsequent research and other uses.

Freezer and refrigerator monitoring

The operation of all freezers and refrigerators must be monitored at all times. The function and temperature of each storage unit should be checked and recorded each workday. All storage units should have a temperature-monitoring device that can be read and recorded, and alert staff in case of failure. Continuous monitoring systems should be in place for all low-temperature storage units. All storage units must have a mechanism to generate an alarm in the event established temperature ranges are exceeded. All personnel should respond to alarms in the designated time frame. Personnel with adequate training who can take corrective action should be on call 24 hours a day, 7 days a week.

Inventory verification

A random check of the specimen inventory system (database) should be conducted on a small percentage of samples on an annual basis. This verification will confirm that the appropriate specimens are in the correct freezer locations, as indicated by the computerized inventory system. This may be a requirement for high risk organisms to ensure no material has been lost or misappropriated.

Emergency preparedness

The facility should have an emergency preparedness plan that addresses a wide variety of unlikely, but possible, emergencies. This would include natural disasters such as earthquakes, hurricanes, tornados, flood, fire, terrorist activities or political demonstrations. Ideally, the whole collection should be backed up at a secure location.

Storage environments

Adequate back-up capacity for low-temperature units must be maintained in anticipation of possible equipment failure. Extra capacity equal, at a minimum, to the capacity of the largest single storage unit must be maintained at operating temperature at all times. This applies to each temperature storage condition.

Standard Operating Practices

The SOP manual should specifically include, but should not be limited to:

- Specimen handling policies and procedures including supplies, methods and equipment
- Laboratory procedures for tests performed in-house and any specimen aliquoting or other specimen processing
- Policies and procedures for shipping and receiving specimens
- Records management policies
- Quality assurance and quality control policies and procedures for supplies, equipment, instruments, reagents, labels and processes employed in sample retrieval and processing
- Policies regarding safety programmes would include pre- and post-employment medical evaluations and immunization records
- Emergency and safety policies and procedures, including reporting of staff injuries and exposure to potential infectious pathogens
- Policies and procedures for the investigation, documentation, and reporting of accidents, errors, complaints and adverse outcomes
- Policies, procedures, and schedules for equipment inspection, maintenance, repair, and calibration for the purpose of maintaining equipment
- Procedures for disposal of hazardous waste
- Policies and procedures describing requirements of training programmes for technical and QA staff

Implementation: Either the culture collection Director/Curator and/or the individual responsible for the Quality Assurance Programme should review and approve all SOPs and associated process validation studies prior to implementation. Upon implementation, all SOPs must be followed as documented.

Staff access and review: Current copies of the SOP manual should be stored in designated locations and available to the staff at all times. New and revised policies and procedures should be reviewed by the staff prior to implementation. Documentation of staff review and any associated training should be maintained in a Training Record.

All operations should be documented in Standard Operation Procedures (SOPs) and their implementation monitored through a Quality Management System (See Chapter 6, this volume).

Biological material tracking

It is necessary to have certain systems such as a Laboratory Information Management System in place that ensures that specimens can be effectively tracked. The system should track these specimens while they are being moved from their collection site to the culture collection and eventually shipped elsewhere. Such systems include:

- The use of labels that identify the samples as they are transported and stored
- Shipping logs that document specimen arrival and departure from the culture collection
- An inventory system that allows all appropriate staff to locate the specimen within the culture collection
- Use of standardized labelling techniques such as the use of barcodes and sample identification information comprises the first step toward the standardization of sample identification and tracking across the industry.
- Each specimen needs to receive a label that remains attached under all projected storage conditions. Label printing should be resistant to all common laboratory solvents. Labels should include human-readable indications of contents.

Inventory systems

A digital inventory system must be in place that enables tracking of the location and status of every sample in the culture collection. The inventory system should be capable of storing electronic signatures that meet national regulations, including a full audit trail of changes made to the database.

The computerized inventory system should ideally have the ability to generate configurable labels to provide the most complete information for the type of sample being stored, its originator, and its recipient. An inventory system should also have linkage functions with other established databases if additional data are needed.

Specimen Location: Each freezer, refrigerator or room temperature storage cabinet should have a unique identifier. A convention should be established for numbering shelves, racks, boxes, as well as each location within the container.

Other Specimen Descriptors: The inventory system should track sample identifiers such as sample ID, barcode ID (if different), date of collection and sample type. Information should be included on the availability and volume of aliquots, the history of sample movement, sample thaws (as appropriate) and shipment to and from external sites.

Validation: A system should be in place to maintain the accuracy of the inventory. The system can employ either periodic counts of inventory storage units or a cycle counting methodology.

Formal validation of computer systems and software is required for some collections. These include organizations subject to Food and Drink Authority regulations and by clients, if it is a commercial operation. The inventory system should also be subject to regular QA audits.

Armed with the requirements for the establishment and operation of a culture collection, read Chapter 7, this volume to see how a sound business case can be generated to justify its establishment and a business model can be constructed to make it sustainable.

References

Anon (1994) *Genebank standards*. Food and Agriculture Organization of the United Nations, Rome, International Plant Genetic Resources Institute, Rome. ISBN 92 9043 236 5. https://www.fao.org/publications/card/en/c/7b79ee93-0f3c-5f58-9adc-5d4ef063f9c7/

Anon (2021) *The Approved List of Biological Agents Advisory Committee on Dangerous Pathogens*, 4th edn. Health and Safety Executive, HMSO, London. https://www.hse.gov.uk/pubns/misc208.pdf

Blackwell, M. (2011) The fungi: 1, 2, 3 ... 5.1 million species? *American Journal of Botany* 98, 426.

Da Silva, E. (1988) UNESCO, Microbial Resources and the GIAMS Twenty Five Years: A Brief Review. In: Chang, S.T., Chan, K.Y. and Wood, N.Y.S. (eds) *Recent Advances in Biotechnology and Applied Biology*. The Chinese University Press, Hong Kong, pp 89-101.

Hawksworth, D.L. (1991) The fungal dimension of biodiversity; magnitude, significance and conservation. *Mycological Research* 95, 641-655. https://doi.org/10.1016/S0953-7562(09)80810-1

Hawksworth, D.L. and Kirsop, B.E. (eds) (1988) *Living Resources for Biotechnology: Filamentous Fungi*. Cambridge University Press, Cambridge, pp 209.

Hawksworth, D.L. and Lücking, R. (2017) Fungal diversity revisited: 2.2 to 3.8 million species. *Microbiology Spectrum* Jul;5(4). https://journals.asm.org/doi/10.1128/microbiolspec.FUNK-0052-2016.

Hawksworth, D.L. and Schipper, M.A.A. (1989) Criteria for consideration in the accreditation of culture collections participating in MINE, the Microbial Information Network Europe. *MIRCEN Journal* 5, 277-281.

Hüttner, S., Johansson, A., Gonçalves Teixeira, P., Achterberg, P. *et al.* (2020) Recent advances in the intellectual property landscape of filamentous fungi. *Fungal Biology and Biotechnology* 7, 16. https://doi.org/10.1186/s40694-020-00106-z

Kelley, J. and Smith, D. (1997) Depositing micro-organisms as part of the patenting process. *European BioPharmaceutical Review*. Ballantyne Ross Ltd., London.

Kirsop, B. and Da Silva, E. (1988) Organisation of resource centres. In: Hawksworth, D. and Kirsop, B. (eds), *Filamentous Fungi: Living Resources for Biotechnology*. Cambridge University Press, Cambridge, pp. 173-187. doi:10.1017/CBO9780511752278.010 https://www.cambridge.org/core/books/abs/filamentous-fungi/organisation-of-resource-centres/AB47B0967FAA92164EA9CBD0F29B25FC

Kirsop, B.E. and Doyle, A. (eds) (1991) *Maintenance of Micro-organisms and Cultured Cells: A Manual of Laboratory Methods*. Academic Press, London, pp 308.

Kirsop, B. and Hawksworth, D.L. (1994) *The Biodiversity of Micro-organisms and the Role of Microbial Resource Centres*. World Federation for Culture Collections (WFCC), Brauschweig, Germany.

Kirsop, B.E. and Kurtzman, C.P. (eds) (1988) *Living Resources for Biotechnology: Yeasts*. Cambridge University Press, Cambridge, UK, pp 234.

Macreadie, I., Macreadie, H. and Clark, R. (2019) Sustainable Use and Preservation of Biological Resources. *Microbiology Australia* 40(3), 97-152. https://www.publish.csiro.au/MA/pdf/MAv40n3.

Martin, D., Stackebrandt, E. and Smith, D. (2015) MIRRI Promoting Quality Management Systems for Microbiology. *EC Microbiology* 2.2, 278-287. https://www.ukbrcn.org/media/1031/smith-et-al-mirri-promoting-qms-for-microbiology-ec-microbiology-2015-278-287ecmi-02-000036.pdf

NSF (2020) National Academies of Sciences, Engineering, and Medicine. 2020. *Biological Collections: Ensuring Critical Research and Education for the 21st Century*. The National Academies Press, Washington, DC. https://doi.org/10.17226/25592.

OECD (2007) *OECD Best Practice Guidelines for Biological Resource Centres*. Organisation for Economic Co-operation and Development, Paris, France. https://www.oecd.org/sti/emerging-tech/38777417.pdf

OST (1996) A New Strategy for the UK Microbial Culture Collections: Government Response to the Independent Review of UK Microbial Culture Collections. Office of Science and Technology (OST), UK. https://webarchive.nationalarchives.gov.uk/ukgwa/20010315073639/http://www.dti.gov.uk:80/ost/ukmcc/ukmcc.htm

Overmann, J. (2015) Significance and future role of microbial resource centers. *Systematic and Applied Microbiology* 38, 258-265. https://doi.org/10.1016/j.syapm.2015.02.008

Overmann, J. and Smith, D. (2017) Microbial Resource Centres contribute to bioprospecting of bacteria and filamentous microfungi. In: Hawksworth, D., Paterson, R. and Lima, N. (eds), *Topics in Biodiversity and Conservation, Volume 16 Bioprospecting – Successes, Potential and Constraints*. Springer, New York. pp 51-79. ISBN: 978-3-319-47933-0 (Print) 978-3-319-47935-4 (Online). http://link.springer.com/book/10.1007/978-3-319-47935-4

Ryan, M.J. and Smith, D. (2007) Cryopreservation and freeze-drying of fungi employing centrifugal and shelf freeze-drying. In: Day, J.G. and Stacey, G.N. (eds) *Cryopreservation and Freeze-Drying Protocols. Methods in Molecular Biology™*, vol 368. Humana Press. https://doi.org/10.1007/978-1-59745-362-2_9

Ryan, M.J., Ritchie, B.J. and Smith, D. (2012) Maintenance and storage of fungal plant pathogens. In: Lane, C., Beales, P. and Hughes, K. (eds) *Fungal Plant Pathogens – Principles and Protocols*. CAB International, Wallingford, UK. ISBN 978 1 84593 668 6 (pbk) http://bookshop.cabi.org/?page=2633&pid=2265&site=191

Ryan, M.J., McCluskey, K., Verkleij, G. and Smith, D. (2019) Fungal biological resources to support international development: challenges and opportunities. *World Journal of Microbiology and Biotechnology* 35, 139. https://doi.org/10.1007/s11274-019-2709-7

Simões, M.F., Dias, N., Santos, C. and Lima, N. (2016) Establishment of a Quality Management System based on ISO 9001 Standard in a public service fungal culture collection. *Microorganisms* 4, 21. https://doi.org/10.3390/microorganisms4020021

Smith, D. (1992) Optimising preservation. *Laboratory Practice* 41, 25-28.

Smith, D. (1993) Tolerance to freezing and thawing. In: Jennings, D.H. (ed.) *Tolerance of Fungi*. Marcel Dekker Inc., New York, pp. 145-171.

Smith, D. (2012) Culture collections. *Advances in Applied Microbiology* 79, 73–118. https://doi.org/10.1016/B978-0-12-394318-7.00004-8

Smith, D. and Kolkowski, J. (1996) Fungi. In: *Preservation and Maintenance of Cultures used in Biotechnology and Industry*. Academic Press, USA, pp. 101-132.

Smith, D. and Onions, A.H.S. (1994) *The Preservation and Maintenance of Living Fungi*. 2nd edn. *IMI Technical Handbooks No. 2*. CAB International, Wallingford, UK, pp. 122. https://www.cabi.org/bookshop/book/9780851989020/

Smith, D. and Ryan, M.J. (2019) International postal, quarantine and safety regulations. *Microbiology Australia special issue: Sustainable Use and Preservation of Microbial Resources* 40, 3. 10.1071/MA19032 http://microbiology.publish.csiro.au/?paper=MA19032

Smith, D., Rhode, C. and Holmes, B. (1999) Handling and distribution of micro-organisms and the law. *Microbiology Today* 26, 14-16. https://socgenmicrobiol.org.uk/pubs/micro_today/pdf/14_16.pdf

Smith, D., Fritze, D. and Stackebrandt, E. (2013) Public service collections and Biological Resource Centers of microorganisms. In: Rosenberg, E., DeLong, E.F., Lory, S., Stackebrandt, E. and Thompson, F. (eds) *The Prokaryotes*. Springer, Berlin, Heidelberg. https://doi.org/10.1007/978-3-642-30194-0_14

Smith, D., McCluskey, K. and Stackebrandt, E. (2014) Culture collection funding models and BRC business plans. *SpringerPlus* 3, 81. http://www.springerplus.com/content/3/1/81

Smith, D., da Silva, M., Jackson, J. and Lyal, C. (2017a) Explanation of the Nagoya Protocol on Access and Benefit Sharing and its implication for microbiology. *Microbiology* 163, 289-296. doi:10.1099/mic.0.000425.http://mic.microbiologyresearch.org/content/journal/micro/10.1099/mic.0.000425

Smith, D., Martin, D. and Novossiolova, T. (2017b) Microorganisms: good or evil, MIRRI provides biosecurity awareness. *Current Microbiology* 74, 299. https://doi.org/10.1007/s00284-016-1181-y

Stackebrandt, E., Smith, D., Casaregola, S., Varese, G.C., Verkleij, G. *et al.* (2014) Deposit of microbial strains in public service collections as part of the publication process to underpin good practice in science. *SpringerPlus* 3, 208. https://springerplus.springeropen.com/articles/10.1186/2193-1801-3-208

Stevenson, R.E. and Jong, S.C. (1992) Application of good laboratory practice (GLP) to culture collections of microbial and cell cultures. *World Journal of Microbiology and Biotechnology* 8, 229-235. https://doi.org/10.1007/BF01201869

Sugawara, H., Ma, J., Miyazaki, S., Shimura, J. and Takishima, Y. (eds) (1993) *World Directory of Collections of Cultures of Micro-organisms*. 4th edn. WFCC World Data Centre on Micro-organisms, Japan, pp 1148.

Sugawara, H., Ma, J. and Miyazaki, S. (eds) (1999) *World Directory of Collections of Cultures of Micro-organisms*. 5th edn. WFCC World Data Centre on Micro-organisms, Japan, pp. 140.

Verkley, G., Perrone, G., Pina, M., Scholz, A., Zuzuarregui, A., *et al.* (2019) New ECCO tools for Core Material Deposit and Transfer Agreements to promote compliance in procedures for deposit and supply of cultures from a public microbial culture collection. *FEMS Microbiology Letters*, fnaa044, https://doi.org/10.1093/femsle/fnaa044

Wu, L., Sun, Q., Desmeth, P., Sugawara, H., Xu, Z., *et al.* (2017) World Data Centre for microorganisms: an information infrastructure to explore and utilize preserved microbial strains worldwide. *Nucleic Acids Research* 1, 8. doi:10.1093/nar/gkw903 https://academic.oup.com/nar/article/45/D1/D611/2770659

2

Isolation and Growth of Microorganisms

Introduction

The origin of CABI in the early 1900s was borne out of a necessity to study tropical diseases in overseas plantations. Since that time the work of CABI has diversified into many areas but the diagnosis of crop problems and the provision of remedies has remained a central pillar of the work for over 100 years (Butler, 1921; Aitchison and Hawksworth, 1993; Blight and Ibbotson, 2011). Now known as the Diagnostic and Advisory Service (DAS) this facility is based in Egham and is responsible for providing diagnostic advice on all manner of crop problems and offering realistic and appropriate management strategies bespoke to the individual circumstances of the grower. Although enquiries arrive from all over the world, most originate from the 49 countries that are members of CABI. CABI membership offers many benefits to a country such as CABI information and services but one that is particularly important is the free access to the diagnostic and advisory service. In addition to the members' service there is a commercial aspect and enquiries come from all over the world on all manner of crops. The DAS service offered by CABI is fundamental to its support of agriculture and is core to CABI's mission to 'improve yields and reduce losses'. In an uncertain world where climate is changing and diseases and pests are spreading, the work of DAS is more relevant than ever, provision of timely diagnosis and appropriate remedial advice is helping overcome the very real problems experienced by farmers right now.

The starting point of the DAS investigation may be a photograph but is often living plant material infected with pathogens sent from overseas. The material may well have deteriorated in transit, this can pose challenges in the diagnosis which is initially based on symptoms and microscopic examination. Where it is not possible to identify the pathogen directly from the sample, the pathogen is isolated into pure culture before being identified. The routine techniques involved in observation and isolation haven't changed for decades, despite recent advances in technology which have revolutionized the activities of most other laboratories. The equipment and instruments used are simple, those that would be found in many biological based laboratories, it is the skill and experience of the technical staff that make the DAS laboratory a world class facility.

The DAS laboratory does not operate in isolation and has strong links with many other departments within CABI and with experts in other institutions to ensure accurate diagnoses are made and our customers are provided with the best service possible. Molecular biology and the study of DNA has revolutionized classification and diagnosis. Historically, all identifications were based on the morphology of the pathogen but increasingly we are turning to molecular methods for identification. This is where

DOI: 10.1079/9781800622135.0002

the help of the Genetic Resource Collection (GRC) and Microbial Services department situated just along the corridor from DAS is employed.

CABI's Microbial Services team offer a variety of molecular methods to identify and characterize unknown organisms with particular emphasis on agricultural and biotechnological applications. The standard method relies on rDNA sequencing to generate identifications for bacteria, filamentous fungi and yeasts. This service is vital to DAS which is dependent on the services provided by Microbial Services to provide species names for the organisms isolated from plant related enquiries. In addition to sequencing, the Bioscience team are able to offer bespoke molecular services such as fingerprinting, bacterial identification using MALDI-ToF, and Next Generation Sequencing etc.

The Microbial Services team has the Genetic Resource Collection (GRC) at its disposal, a wonderful asset and an integral part of the work of CABI (see Chapter 1, this volume). The GRC is an enormous store of living fungi and bacteria from all over the globe and contains over 28,000 microorganisms. Many of these were collected by collectors over the years but a considerable proportion were originally isolated from material sent to our laboratories. The contents of our culture collection are unique and provide a vital resource that we can draw on at any time, e.g. as a reference sample to compare features with an unknown, or as a positive control for our diagnostic test methods. The GRC is literally and metaphorically a growing collection and we have the opportunity to preserve isolates of particular interest from our diagnostic enquiries on site, or collected for studies or projects (although this is dependent on obtaining the appropriate permissions according to the Nagoya Protocol and other regulations, see Chapter 5, this volume).

Historically the laboratories at CABI have centred on fungal and bacterial pathogens, nematodes and some insect groups. Other pathogens such as viruses or phytoplasmas have not been so extensively studied. Where the symptoms indicate that one of these nonculturable organisms are the likely cause, the infected plant material is sent to one of our collaborators e.g. FERA Science (https://www.fera.co.uk/) for analysis.

Until recently, insect specimens sent from overseas were routinely passed on to collaborators for morphological analysis by entomological specialists with access to reference specimens for comparison. However, molecular methods now allow extremely precise identification via the DNA extracted from the sample. These molecular identifications which negate the need for an experienced entomologist have been beneficial as the number of insect queries has rocketed in recent years and now make up a significant proportion of our enquiries, However, the identification of insects (and nematodes) through molecular methods is not always successful, even when the samples are received in good condition and long and accurate sequences are obtained. This is because the online databases are not comprehensive and reference sequences for some groups are not represented. When this situation occurs, specimens are sent to the appropriate experts for morphological examination.

The standard molecular identification method at CABI uses DNA sequence comparisons of specific regions of the rDNA of the target organisms. The region of the DNA which is taken for comparison depends on the organism in question. For bacteria it is DNA coding for the 16s region of the ribosome whereas it is the ITS region of fungi and yeasts, and the COI region of invertebrates. Primers are used to amplify the section of DNA for comparison before sequencing of the amplified DNA. If a satisfactory answer is not provided by these sequences supplementary primers which amplify alternative loci are also used, these are often bespoke for any given organism. Sequencing is a reliable method which generates adequate results for most of our microbial enquiries, although closely related species which cannot be differentiated may result in a genus or family level identification. The examination of morphological features and/or additional information accompanying samples can sometimes be used to improve the identification results.

The analysis of genetic populations is becoming much more important and CABI has a Next Generation Sequencer (NGS) and Oxford Nanopore MinION both of which are able to analyse the entire microbial community within a sample. Whilst used extensively in other aspects of CABI's work they are not often used for samples submitted to DAS and are more suited to projects or studies involving large numbers of samples. The MinION has the added benefit of being portable and provides a real-time option of analysis in the field (See Chapter 3, this volume). The complete analysis of the microbes within a sample has obvious benefits, but high running costs/expensive consumables and complex data analysis do not currently lend themselves to our routine work.

Mass spectroscopy often provides a quicker and cheaper method to identify certain components of bacterial cells; matrix-assisted laser desorption/ionization time of flight (MALDI-ToF) has revolutionized day-to-day bacterial identifications in the DAS laboratory. The highly automated process involves mixing a very small amount of bacterial colony and inserting it into the machine; laser light is shone onto the sample which causes the sample to break up and vaporize. Once in a vaporized form, the fragments of bacteria are immediately ionized and accelerated towards the detector. The time taken for the fragments to reach the detector is indicative and can be used to create a profile for the sample which can be compared against other profiles to get a match and identify the bacterium.

Single bacterial colonies can be identified within minutes, which enables rapid decisions about whether to proceed with isolates (as potential pathogens) or discard them. The technique does rely on a comprehensive database to which the results can be compared. The standard library contains over 3000 species and although it has strong medical and food contaminant bias there are nevertheless a surprisingly high number of relevant species, environmental and plant pathogenic species do feature significantly. Despite the high upfront costs, the low consumables costs and the benefits of rapid processing time makes this a good option particularly if the number of samples is significant.

In addition to the relatively routine diagnostic work, research is also taking place on new protocols and techniques. The Microbial Services team have developed PCR-based protocols to detect known plant pathogens within host material, without the need to isolate the pathogen in pure culture and we expect to use this method to a greater extent in the future. This method is increasingly important with regard to sanitary and phytosanitary (SPS) measures. SPS measures are tightening right across the globe and it is often the remit of the DAS laboratory to show the *absence* of a pathogen rather than its presence. When this is the mandate an alternative approach is required to our standard isolation/identification procedure. In these cases, we use a PCR-based method to detect whether a DNA sequence unique to the pathogen in question can be detected in the host tissue. This approach does not run the risk of not isolating the pathogen and getting it into pure culture and has the benefit of being quick and relatively easy. The drawbacks are having to create bespoke primers for each pathogen; it is essential to run both positive and negative controls within the experiment. Even if the pathogen is identified within the tissue there is no culture produced and finally if the result is negative there is no further information as to what may be causing the problem. Whilst newer techniques have their place, our standard microbial isolation methods have stood the test of time, they provide us with the opportunity to grow viable microorganisms that can be studied, identified and preserved for future investigations.

The aim of this chapter is to show how the Diagnostic and Advisory Service operates, and gives detailed information about our procedures; from the original request to producing an end report. The steps that we take in our laboratory have been represented in the flowchart below (Fig. 2.1), with a brief summary of each stage. These techniques and methods are explained in greater detail later in the chapter. Many of the procedures are simple but effective and have stood the test of time. Additional information about these techniques, and others which are no longer routinely used at CABI, are available in previous CABI publications e.g., *Plant Clinic Handbook* (Waller *et al.*, 1998). CABI has also published the *Plant Pathologist's Pocket Book* (Waller *et al.*, 2002) and key elements of that work have been drawn upon in this chapter including the elements of culturing, preservation and maintenance of fungi (Smith, 2002). Additionally, CABI has produced the *Plantwise Diagnostic Field Guide* (Taylor, 2019) produced to assist the Plantwise (https://www.plantwise.org/) project's trained plant doctors running plant clinics. It provides images and short descriptions and is a handy resource for anyone wanting to learn more about how to diagnose plant health

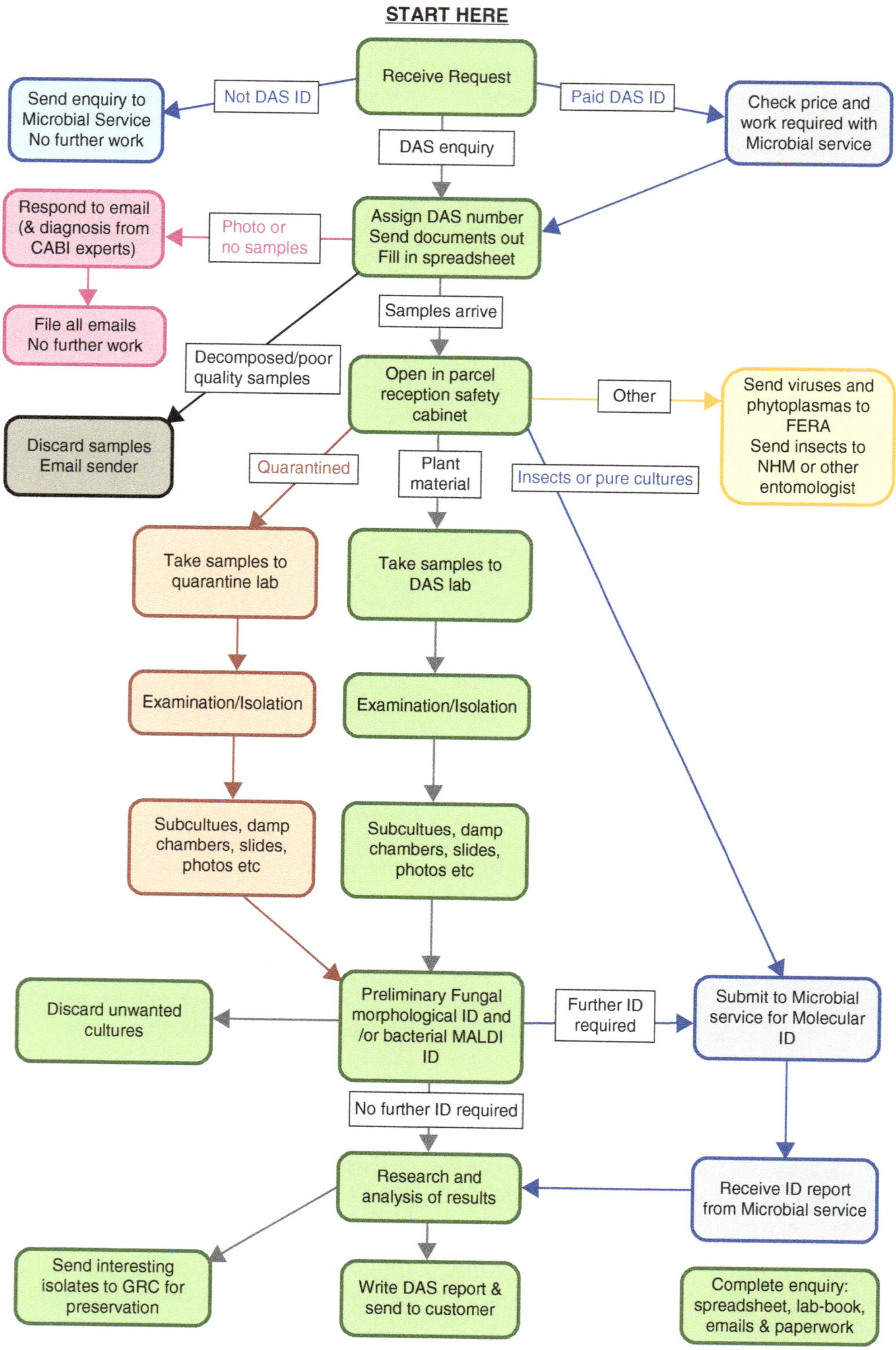

Fig. 2.1. DAS workflow chart

problems and to explore the kinds of challenges with which plant doctors commonly have to deal. This work is also supplemented by the CABI training courses (see the CABI website, https://www.cabi.org/products-and-services/bioscience-services/training/). Specifically on plant pest diagnosis is CABI's PestSmart Diagnostics eLearning course which is introduced on YouTube (https://www.youtube.com/watch?v=cP-2bqgdOzMM) and the pest diagnostic simulator can be found at https://play.google.com/store/apps/details?id=com.BondiLabs.PD&hl=en&gl=US. Based on CABI's Plantwise programme is a self-study course in Crop Pest Diagnosis (https://www.cabi.org/publishing-products/crop-pest-diagnosis/). CABI also offers a number of digital learning courses via CABI Academy (https://www.cabi.org/products-and-services/academy/courses/).

Diagnostic Advisory Service (DAS) Procedures

Receiving enquiries

Enquiries usually arrive via the diagnostic.service@cabi.org inbox, although they are sometimes forwarded from elsewhere in CABI. Initially a decision needs to be made about whether the enquiry can be processed by CABI, and if so whether DAS or a different department are best suited to complete it, e.g. Microbial Services who deal with pure cultures and some paid enquiries. At this stage it is often necessary to discuss the enquiry with members of staff in different departments, or request additional information from the sender.

Logging DAS enquiries

Each new enquiry must be given a unique DAS enquiry number, and this is used on all subsequent correspondence relating to that enquiry. The details are logged, and should include: sample host, symptoms, geographical location, client, client organization, charging code and any other relevant details.

Gathering information

It may be immediately obvious how to respond to some enquiries but others will require further information before advice can be given. It is important to know the background to the problem and have complete details which will assist in making preliminary decisions and diagnosis, including:

- Client details
- Client organization/occupation, e.g. local farmers, the Dept of Agriculture, Commercial company

Fig. 2.2. Rice grain discolouration

- Geographical location
- Enquiry type e.g. Member Country, follow on from previous work, new enquiry
- Crop type and variety
- Description of symptoms and plant parts affected
- History, severity and spread of disease symptoms
- Other contributing factors including unusual weather, application of chemicals etc.

In addition to these enquiry details, we always request several good quality photographs to assess the problem before any recommendations are made. The images and information are analysed by appropriate members of the DAS team and a plan is formulated for each enquiry.

Sending plant or insect samples

If a satisfactory diagnosis and/or identification can be made from information and photographs provided by the client, and no samples are required, the enquiry can be closed via email. Suggestions and diagnoses are made by senior members of staff using their expertise and knowledge about the most likely pathogens or other factors which may be responsible for the symptoms observed. In cases like these it is not necessary to post samples for further analysis.

CABI services incur charges which are usually paid by the client, unless they qualify for free services, which are available to our 49 Member Countries. Additional Member Country identifications are possible at a reduced members rate, and all of the details are available from Microbial Services or from the CABI website. Once adequate information has been gathered about the enquiry, and the funding/means of payment has been decided, it is often necessary to arrange for plant samples to be sent to the DAS laboratory for analysis.

The following documents are sent to clients wishing to submit samples:

- DAS sample submission form (see Form 4 Appendix E, this volume)
- Document: Collecting and sending samples (https://www.cabi.org/wp-content/uploads/PDFs/Bioscience/Preparing-and-sending-samples.pdf)
- Current Letter of Authority from DEFRA (LoA) to send samples to the UK (https://www.cabi.org/Uploads/CABI/services/microbial-services/identification/general-id-import-loa-2012-13.pdf)

The LoA is valid for one calendar year and needs to be renewed annually. CABI holds several licences including: Plant pathogenic cultures, Known restricted plants and Live invertebrates. Most infected plant samples, microbes and soil can be imported using the Microbial LoA.

Sending samples to DAS

The documents and any other relevant information are sent to the customer, and the following points are reinforced:

- Samples should be sent as fresh as possible, and be wrapped individually in newspaper
- Licensed material should be sent in 3 layers of containment, one of which should be crushproof
- Packages should be marked with zero or a nominal monetary value and samples be labelled 'No commercial value - to be destroyed after scientific testing'
- Samples should be shipped via courier and waybill numbers provided for tracking
- DAS should be advised when the samples have been sent, and what has been sent
- The relevant sections of the LoA must be completed prior to sending
- A copy of the LoA must be attached to the outside of the package for inspection if necessary.

Preparation for arrival of samples

It is important to ensure that all necessary resources are available when the samples arrive to avoid unnecessary delays, including:

- Staff
- Appropriate media for fungal or bacterial isolations
- Other specialized items, consumables or equipment as necessary

Unusual requests or difficult processing procedures should be discussed with the relevant members of CABI staff prior to accepting work. In the past this has included:

- NGS or Minion analysis of the entire microbial population of plant samples
- Specific primers to look for the presence of a pathogen (e.g. citrus greening)
- Population studies within samples
- Repeat sampling
- Other non-standardized testing

Identifying Invertebrates

The number of insect enquiries has been on the increase for several years, and the identification of insect pests now makes up a significant proportion of our work. This is due, in part, to spread of Fall Army Worm (*Spodoptera frugiperda*) across Africa and South Asia in 2016/2017. The techniques that made molecular identification of insects possible has been a game changer. We now use sequencing to identify a wide range of insects, and regularly identify both the pests and their natural enemies, i.e. potential biocontrol agents which could be used to control the pest.

Insect samples are always sent as dead specimens which have been preserved in 70-100% ethanol. Small vials containing insects must be drained of ethanol prior to sending to avoid restrictions imposed by couriers and airlines. Despite miniscule risk even small quantities of ethanol are deemed as hazardous and must be declared on paperwork; this triggers complex packaging and labelling requirements, and the potential for companies to refuse to ship the samples. An alternative option to ethanol is non-hazardous 'food grade propylene glycol' which has been tried and tested effectively at CABI as a suitable substitute for ethanol in most circumstances. Propylene glycol is considered non-hazardous, does not need to be declared and can be safely transported without any additional documentation. It appears to be an excellent substitute for ethanol and will preserve the DNA equally as well as ethanol.

Clients wishing to send insects will require:

- DAS sample submission form (see Form 4 Appendix E, this volume)
- Document: Collecting and sending insect samples (available on request)
- Letter of Authority* to send samples to the UK

*dead insect samples can be safely shipped using the General Import Authorization for animal products (IMP/GEN/2014/06), which can be obtained online. This document is used for shipping dead invertebrate specimens and products derived from them, and is not applied for annually by CABI. It is freely available online.

Samples arriving at CABI are opened in the parcel reception area, logged and photographed.

The paperwork and samples are then passed to the Microbial Service team for molecular processing. Once completed the sequencing results are analysed and the identification results reported to DAS.

We also have the expertise and facilities available to identify nematodes at CABI. Specimens may be sent live in soil/plant material for morphological examination, or as dead preserved specimens for sequencing analysis. Live specimens must be sent with the appropriate Letter of Authority for Live Invertebrates, and require careful packaging and shipping to ensure survival whilst in transit. The preparation of dead specimens can be difficult due to their extremely small size.

DAS Checklist of Activities

Parcel reception:

- Collect package from reception and take to Parcel Reception Laboratory
- Open parcel in laminar flow or microbial safety cabinet
- Photograph contents of parcel for future reference
- Ensure that all samples are adequately labelled and correspond to samples listed on attached paperwork
- Check for obvious mites or problems, and assess whether samples can be transferred to main building (mites cause a lot of problems in mycology laboratories and are not welcome)
- Dispose of packaging in yellow incineration bags

- Transport samples in 3 layers of containment, including a secure plastic box to Defra approved DAS laboratory.

Arrival in laboratory:

- Log samples. They will already have a DAS reference number
- Ensure that all samples are labelled
- Record initial observations: how many samples, general condition, immediate observations, how they were packaged etc.
- Take photographs of each sample, showing both the entire sample and closeups of the diseased areas
- Email the sender to acknowledge receipt of samples.

Insect samples, pure microbial cultures and samples for direct processing in molecular laboratory (e.g. Citrus greening):

- Pass samples immediately to the molecular team for processing
- Provide a copy of the paperwork to Microbial Services
- Record associated processing numbers generated for sequencing analysis.

Prior to working on plant material:

- Online information, reference books and prior knowledge should be used to draw up a shortlist of the likely causes of disease
- Senior plant pathologist contacted for advice or assistance as required.

Working on plant samples in DAS laboratory:

- Only work on licensed material and dirty plant material in DEFRA approved laboratory areas
- Examine plant material using a binocular microscope. Record observations
- Photograph diseased areas and/or surface fungal structures if present
- Fungal structures on the surface of the material can be picked off with a needle and mounted on a slide for examination using a compound microscope (or create a Sellotape® mount)
- Decide on work plan based on symptoms observed e.g. fungal and/or bacterial isolations, damp chambers, baiting etc.

Fungal isolations:

- Internal lesions/infected tissues within fruits, stems or large plant pieces should be aseptically removed and placed directly on appropriate media (e.g., PCA or TWA, see Appendix B, this volume) without the need for surface sterilization, or
- Surface lesions on the external surfaces of plant material should be removed, rehydrated, surface sterilized, washed and plated onto TWA (Tap Water Agar)
- Isolation plates containing infected tissues should be incubated for several days until fungal growth is 1-2 cm away from plant material
- Each colony type should be sub-cultured onto fresh media and incubated until adequate growth has developed to make slides of fungal structures/spores
- Examine slides using a compound stereo microscope, and make tentative IDs if possible
- Isolates of interest should be sent to Microbial Services for identification using rDNA sequencing analysis. Non-sporulating cultures will need UV light, alternative media or extended incubation times.

Bacterial isolations:

- Remove affected areas of plant material which are thought to contain bacteria, and look for bacterial streaming into sterile distilled water. Usually this will involve placing a piece of infected leaf material onto a glass slide and observing using a compound microscope. Occasionally, where the bacteria have entered the stem and have induced a wilt, heavily infected material will show bacterial streaming which is visible to the naked eye, e.g. *Ralstonia solanacearum* streaming from a cut stem suspended in a glass bottle of sterile distilled water, although by the time we have received the material it has usually deteriorated too much to see this effect.
- If streaming is observed, select appropriate plant material (stem/leaf spot see Fig. 2.3) and sterilize the surface with alcohol.
- Macerate plant tissue in a small quantity of sterile distilled water in a sterile container. Leave for a few minutes then plate out onto

suitable media (e.g. Nutrient agar) with a sterile loop

- Check growth daily and subculture colonies of interest onto fresh agar plates and incubate
- Process colonies of interest through MALDI for a rapid ID
- Submit samples requiring full sequencing analysis to Microbial Services.

Storing material:

- All samples should be labelled with DAS enquiry and sample numbers, and dated.
- Licensed material must be identified as such, and its arrival, location and destruction must be logged.
- Samples and cultures should be kept until all work has been completed. Fridges must be emptied regularly and plant material/ cultures discarded once work has been completed.
- Original plant material should be rewrapped in newspaper, sealed in an envelope or bag and stored in the DAS sample fridge. Isolation plates and subcultures of microbes should be taped up, bagged and stored in the DAS Cultures fridge.
- Once all of the samples have been processed, a final report is written and sent to the customer.
- Report templates are available, and all relevant processing information should be included in the final report with emphasis on results and conclusions. Relevant photographs should be included if possible.
- CABI datasheets or other references can be inserted to provide additional information.
- Conclusions and suggestions should be completed by a senior plant pathologist.

Sending samples offsite for identification:

- Plant material with suspected viruses or phytoplasmas are usually sent to FERA for analysis. Licensed material must be sent with a PHI10 from APHA which permits the transfer of material between 2 licence-holding institutes. These licences must be linked before samples are sent.
- When required dead insects are sent to the Natural History Museum (NHM) or alternative entomology expert for morphological identification. No restrictions or postal requirements are enforced within the UK.

Plant Pathology Techniques

Microorganisms can be isolated from almost all known habitats, and have adapted to live in even the most extreme environments. In order to study them effectively they usually need to be isolated and grown as a single organism in pure culture, and this is done by replicating the conditions which favour their growth. This section describes the methods used to isolate fungal and bacterial

Fig. 2.3. Orchid with spot.

pathogens from host plant material, and how to grow them to enable identification and diagnosis of disease. Detailed procedures are described below for the isolation, growth and observation of culturable microorganisms in the laboratory.

Plant Disease Diagnosis

Diagnosis aims to determine the cause of an abnormal condition, and involves more than just the identification of the pathogen. The control measures and reasons for the occurrence of the disease are also considered. These factors require an understanding of what a disease is and how it is caused. Some disease-like symptoms are caused by abiotic factors and typically do not spread, whilst pathogens do multiply and may spread to cause epidemics (either on a small or large scale). Pathogenic diseases are influenced by environmental conditions, and the combination of a susceptible host, a virulent pathogen and favourable conditions are required for disease to take hold.

Disease is characterized by abnormality in form and/or function, so it is important to know what healthy plants look like. Pathogens may affect the whole plant, or a confined area, and symptoms may vary depending on the host plant cultivar, environment and strain of pathogen (see Fig. 2.4). Identification of the causal organism is vital, and an understanding of pathogens, the symptoms that they cause and how they can be detected or grown is essential. Frequently more than one disease-causing organism can be isolated from a plant sample, and it is important to determine their significance, and which is responsible for the symptoms observed. Ideally, Koch's postulates would be used to confirm pathogenicity and replicate symptoms, although this is rarely possible.

Collection of Plant Disease Specimens

The initial selection of suitable plant material is vital to maximize the chance of successful isolation of the disease-causing organism. A basic knowledge of disease symptoms and pathways of infection are required to ensure that the plant parts collected are most likely to harbour the pathogen. This may not always be where the symptoms occur — e.g. wilt disease pathogens will be present in the roots or mature stems, although the symptoms are most apparent in the leaves and young shoots which collapse and leaves are often shed by the plant.

To maximize the chances of success, plants showing the early to middle stages of disease should be selected. During these early stages the pathogen is more likely to be actively invading the surrounding healthy plant tissues. Heavily diseased and rotten samples should be avoided as the pathogen will be in decline, and the material is more likely to contain secondary organisms or saprobes. These subsequent organisms will colonize

Fig. 2.4. Maize leaf with streaking

previously infected or necrotic tissues once the primary pathogen has weakened the plant.

Precautions

Samples must be carefully packed to avoid unnecessary damage caused by crushing or conditions which accelerate microbial deterioration. Plant material often contains high levels of moisture, and if the conditions are not controlled saprobes will rapidly colonize the plant tissues and which will expediate decomposition. Samples should be individually wrapped in absorbent material to maintain optimal conditions and prevent cross-contamination. Newspaper is ideal as it is readily available, absorbs excess moisture and does not disintegrate when damp. The use of plastic for packaging plants is generally discouraged as it creates a moist humid environment which will facilitate the growth of unwanted microorganisms. Additional paper, cardboard, or polystyrene granules can be used to cushion samples whilst in transit if necessary.

Plant samples with suspected bacterial disease are more difficult to send to a distant laboratory as plant tissue disintegration is often rapid. In some plant hosts, e.g. Gramineae, bacterial pathogens may die within 12-24 hours in drying host tissue. However, in other plant hosts, e.g. cotton, the bacteria can remain viable for much longer periods. It is important that the samples are as fresh as possible and are not allowed to desiccate prior to sending to give the best chance for the successful isolation of bacterial pathogens.

Method for Suspected Viral Infections

Plant material with a suspected viral infection can be partially dried prior to sending. Viruses cannot be grown in culture, and identification methods usually involve serological tests or molecular methods of the affected tissues.

Packaging Plant Material

Leaves

Leaves should not be collected when wet, but if this is unavoidable, the leaves must be blotted to remove excess moisture and then placed between layers of newspaper or other absorbent paper to dry the surface. Paper tissues are not recommended as they disintegrate and are difficult to remove from the surface of the sample. The leaves should be spread out so that they are flat and do not overlap, and the newspaper may need to be changed daily. When leaves are suitably dry they can be wrapped in fresh newspaper prior to sending (see Fig. 2.5).

Roots

Excess soil should be removed from root samples before each is individually wrapped in newspaper. If roots are washed they should be blotted dry before wrapping.

Stems

Stem portions showing the interface between healthy and diseased areas of tissue should be selected, and carefully wrapped in newspaper. Individual stems should be wrapped separately, as they are easily damaged if bundled together.

Sending samples for analysis

Samples should be sent to their destination as quickly as possible, using the most rapid method available as the material will rapidly deteriorate – time is vital. Couriers are usually the most effective means of delivery, although postal services and options will be location

Fig. 2.5. Grape leaf with suspected virus

specific and should be checked prior to collection of samples.

Packages should be able to withstand the rigours of the postal and courier services, and strong cardboard boxes or padded envelopes are recommended. Packaging conforming to UN3373 [specifically for category B, biological material] is desirable if available.

Licensed plant material MUST be safely packaged in at least three layers of containment (one of which must be unbreakable) to prevent accidental release.

Packages should be clearly marked with their destination and the following text:

'Perishable Biological Material Keep in Shade. Keep material cool but do not refrigerate'

All necessary paperwork including a completed Letter of Authority, invoice (stating the monetary value) and any additional requested documents must be sent with the samples. A copy of the paperwork and licence/licence number should be attached to the outside of the box for inspection by customs officials.

Important phytosanitary precautions

To safeguard plant health and minimize quarantine risks, National and International regulations have been formulated to govern the shipment of biological materials. National post offices provide information online covering local regulations, necessary permits etc. Additionally, the IATA (International Air Transport Association) regulations must also be followed (see Chapter 5, this volume).

Collection details

It is important to gather detailed collection details about the samples because an unidentifiable plant sample from an unknown geographical area will be of limited use, no matter how good the sample may be. The identity of the host is often necessary for the correct identification of many obligate pathogens, many of which are host specific.

Relevant collection details to accompany specimens should include:

- Host and part of plant
- Country, state, locality
- Map references and altitude (if known)
- Collection date (necessary if isolations are to be attempted)
- Collector's name and collection number (you need to be able to identify your collections)
- Symptoms observed
- Disease severity, e.g. number of plants affected (is it one plant on the edge of a field or is the whole area affected. This helps to identify the importance of the problem)
- Whether the symptoms have been observed previously
- Any other information such as extreme weather, use of chemicals etc.

Working with Plant Specimens

It is important to start working with plant material as soon as possible after arrival, as it will quickly deteriorate and become un-usable. Valuable time and effort can be wasted on samples which are in a poor state.

Licensed plant material should only be processed in approved laboratories with the necessary containment facilities. Further information can be obtained from DEFRA or equivalent Government department.

Initial examination should involve a visual check of symptoms and examination of diseased areas using a binocular microscope. Care should be taken to avoid overhandling the material or cross-contaminating samples. At this stage it may be possible to narrow down which type of organism is the most likely cause of the symptoms, e.g. fungus, bacteria, virus, phytoplasma, or whether the cause is more likely to be abiotic; caused by a nutrient deficiency, environmental conditions, chemicals etc.

Sellotape® mount

This is a quick and easy method which can be used to view the structures present on the surface of plant material prior to carrying out any isolation methods. It is most commonly used on leaves and fruit, and is better suited to relatively clean samples. The process involves cutting a piece of Sellotape which is slightly smaller than a glass microscope slide. The tape is then pressed onto the surface of the specimen, and any loose material on the surface will adhere to the sticky tape. The tape is then removed and stuck to a microscope slide which has had a large drop of mountant (lactic acid with cotton blue or lactofuchsin) applied to the surface. The tape should be smoothed over the coloured mountant so that all areas of the tape are stained (see Fig. 2.6).

The slide can now be viewed using a compound microscope, and fungal structures observed.

Clearing and staining

There are various methods available for tissue clearing and staining, which allow the fungal structures present in plant material to be viewed *in situ*. The process involves removing the chlorophyll, gums and tannins with chemicals, prior to staining the fungi, which can then be viewed using a microscope. This method is not routinely used in the DAS lab, although the method is available in the Plant Clinic Handbook (Waller *et al.*, 1998).

If plant material is very thin, and not too deeply coloured it is sometimes possible to view fungal structures directly in/on the plant tissue by staining and viewing using the compound microscope without clearing the tissue first (see Fig. 2.7).

Preparation of squash mounts from specimens

This method is suitable for viewing fungal structures on the surface of leaves, stems, twigs etc.

Examine the surface of the plant specimen to determine likely areas of fungal growth, and scrape a small portion of fungal material from the surface with a scalpel blade. If the material fragments easily, it can be softened with a few drops of water, prior to scraping. Transfer material to a drop of mountant on a clean slide.

The fruiting bodies of ascomycetes and coelomycetes should be carefully removed from the surface of the host material and transferred to a drop of mountant on a glass slide (see Fig. 2.8). A cover slip can then be placed over the fungal material, and tapped gently to break open the fruiting bodies and spread the material prior to viewing.

Where fruiting structures are immersed or partly-immersed within host material, the surface of the specimen may need to be cut away to expose the fruit bodies before they can be removed. The procedure outlined above can then be followed.

It is also possible to cut sections of fungi by hand if the internal structure of fruiting bodies in ascomycetes and coelomycetes are to be examined.

Place the material on the stage of a dissecting microscope and examine the material to determine the location of fruiting structures. If the material is dry, soften the specimen with a little water.

If the fruit bodies are immersed or partly-immersed within the host tissue, cut away the surface layers to expose the fruiting structures. Having located fruitbodies, carefully cut thin vertical sections with the razor blade.

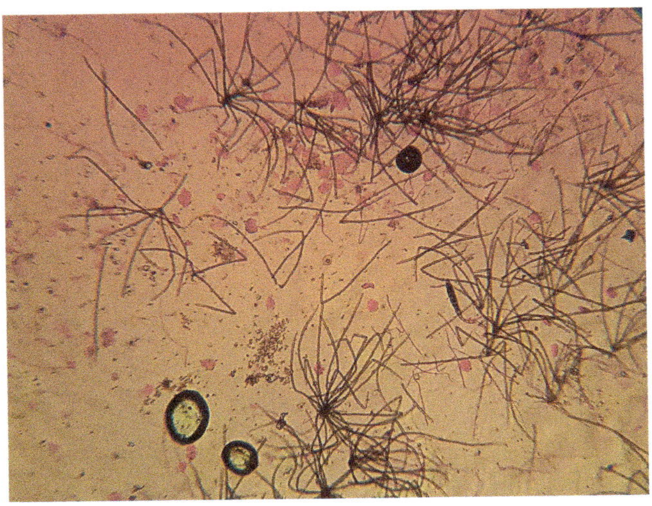

Fig. 2.6. Example of a Sellotape mount from banana

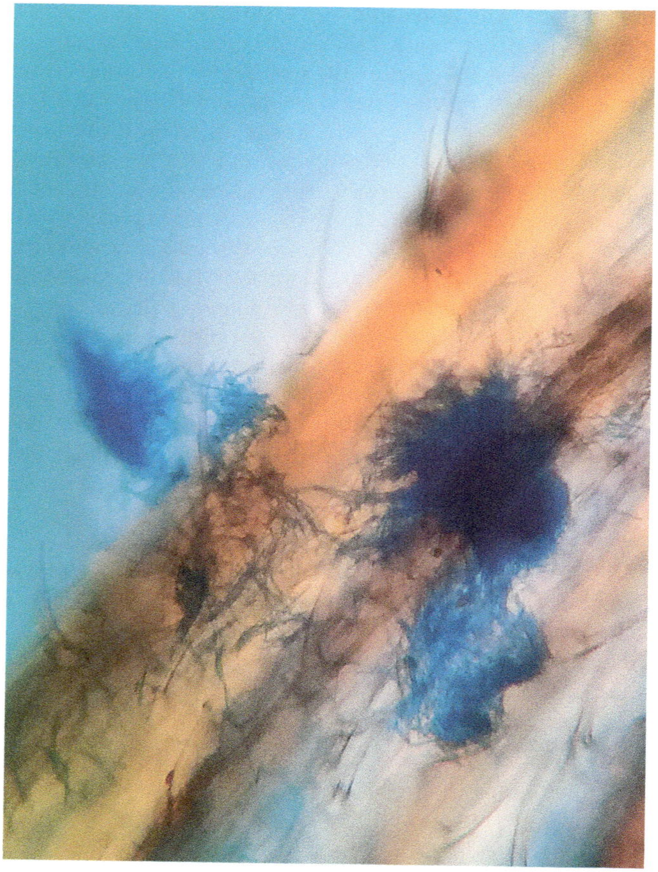

Fig. 2.7. Stained tissue: *Septoria* on the surface of wheat leaves

Remove sections carefully on a needle tip and transfer to a drop of mountant on a slide.

Lower a cover slip carefully then flame to release trapped air and seal. If hand-cut sections do not show required structures, microtome-cut sections may be required. Details of this method are available in the Plant Clinic Handbook (Waller *et al.*, 1998).

Incubation in a moist chamber

A moist/damp chamber can be used to induce fungal fruiting body production and/or sporulation on suitable plant samples. This method provides the ideal growth conditions for many microbes, including both the target organism and unwanted saprobes which will be present on the surface of non-sterile plant surfaces. A brief surface swab with industrial methylated spirit (IMS) or sodium hypochlorite may be helpful but care should be taken not to damage the existing surface structures of the pathogen. Alternatively,

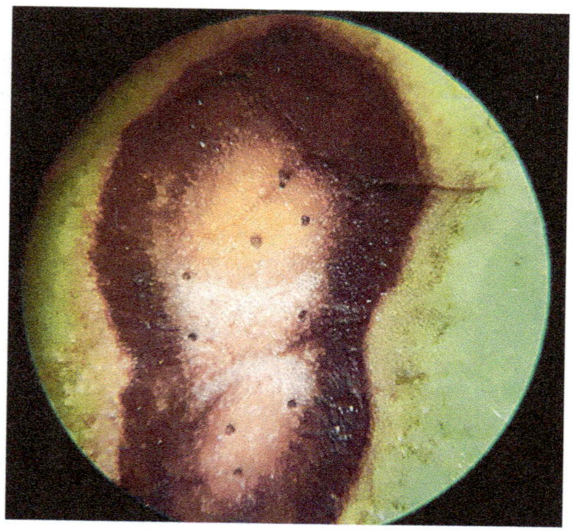

Fig. 2.8. Fungal fruiting bodies within a leafspot on yam

specimens can be washed in sterile water and allowed to dry prior to incubation.

If the target fungus is immersed beneath the surface of host tissue, the growth of surface saprophytes and bacteria can be reduced or prevented by surface sterilization of the host prior to incubation by immersing in a 1% solution of sodium hypochlorite for 30 seconds – 2 minutes, and then washing in sterile distilled water.

Moist chambers should not be incubated at high temperatures and are best kept in the light, e.g. on the laboratory bench, but out of direct sunlight and in a fairly constant temperature to avoid condensation. They need to be checked daily.

Examination under a low power stereoscopic microscope is required to observe any fungal sporulation on the surface of plant material. Sporulating structures can then be removed with a fine sterile needle and transferred to a microscope slide for closer microscopic examination, or cultured onto agar plates.

Sterile Petri dishes are useful incubation chambers for very small or flat specimens such as leaves or small twigs. Clean plastic boxes, such as sandwich boxes, are ideal for larger material and very large specimens can be incubated in appropriately sized plastic containers or clean plastic bags, e.g. unused autoclave disposal bags.

Preparation of moist chambers

- Items used to prepare moist chambers (see Fig. 2.9) should preferably be sterile, but clean, non-sterile containers can be swabbed with IMS and allowed to dry. Plastic bags should be unused.
- Sterile filter papers (we wrap in foil and autoclave) or clean tissue can be added to the Petri dishes or boxes. The addition of sterile distilled water will generate a moist environment.
- A few drops of glycerol may be added to the water to slow down evaporation of the water, or alternatively the water can be topped up as required.
- Plant specimens are best kept above the damp paper by placing them onto a suitable platform. This may be a shallow dish, grid or other platform within the chamber, which can be sterilized with IMS before use. We sometimes use the sterile lid of a small Petri dish which can be accommodated within a larger sterile Petri dish.

Isolation of fungi from plant material

Generally only pathogens capable of saprobic growth (facultative parasites or necrotrophs) can be grown as fungal cultures in the lab. Some of these organisms may be very demanding in their growth requirements, demanding specific nutrients, temperature or other conditions in order to grow. The isolation of fungi from plant material is usually achieved by placing small pieces of infected tissue onto a suitable agar growth medium in sterile Petri dishes. The resultant growth can then be removed and transferred to fresh agar plates where it can be grown on as a pure isolate.

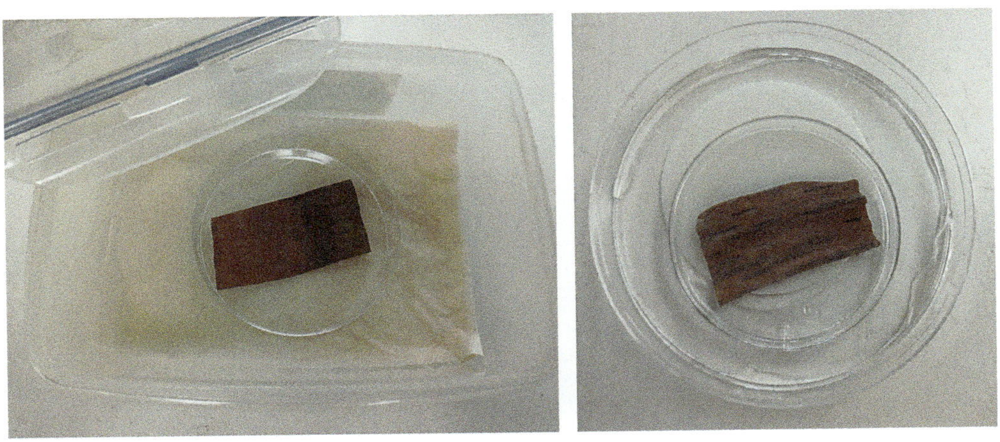

Fig. 2.9. Moist chambers

This technique is suitable for a wide variety of samples including stems, leaves, roots and fruit. Plant diseases caused by fungi growing inside the tissues, such as wilt, will also require a similar isolation method in order to determine which fungus is causing the problem.

Spores which are already visible on the plant material can be removed from fungal fruiting structures with a fine sterile needle and placed on or streaked across the surface of an agar plate.

Isolation can be undertaken in several ways, and the methods that have proved effective for many of the disease-causing fungi are listed below.

Aseptic isolation

The aim of this exercise is to isolate the plant pathogen of interest from the host, but this isn't as simple as it sounds, and it is very easy to grow unwanted organisms or end up with mixed cultures if appropriate steps aren't taken. Many saprobic fungi and bacteria are able to grow on or within contaminated plant tissue, and are common secondary colonizers of previously diseased lesions. Consideration should also be given to the presence of spores or other microscopic propagules in the air and on most non-sterile surfaces, and it is important to take simple precautions to avoid growth of these contaminants on isolation plates by using aseptic techniques. The surface sterilization of excised tissue is usually necessary to remove saprobic organisms which commonly grow over plant surfaces before plating out, and good aseptic technique is essential – even when working in a laminar flow or microbial safety cabinet (see Figs 2.10 and 2.11).

Method

- Wipe the working surface with a suitable disinfectant solution. A smooth, hard work surface is preferable, and a glazed tile or similar can be used if the work surface is porous or unsuitable.
- Instruments can be sterilized by dipping into IMS and carefully passing through the flame of a Bunsen burner or spirit lamp, or by heating until red hot. Hold the instrument horizontally; care must be taken to ensure that burning alcohol does not contact skin, clothing or combustible materials such as paper.
- As a general rule, isolations from plant material should be undertaken on tap-water agar (TWA). This minimal media favours the growth of the pathogen, which uses the host plant material as a food source, and the agar as a platform to grow across. Nutrient rich agars favour the growth of faster growing saprobes, and will hinder the isolation of plant pathogens.
- When plating out pieces of tissue onto agar, Petri dish lids should carefully be lifted and replaced, in preference to completely removing lids which will allow airborne contaminants to settle on the agar. Ideally a laminar

Fig. 2.10. Example of maize seed isolation plate

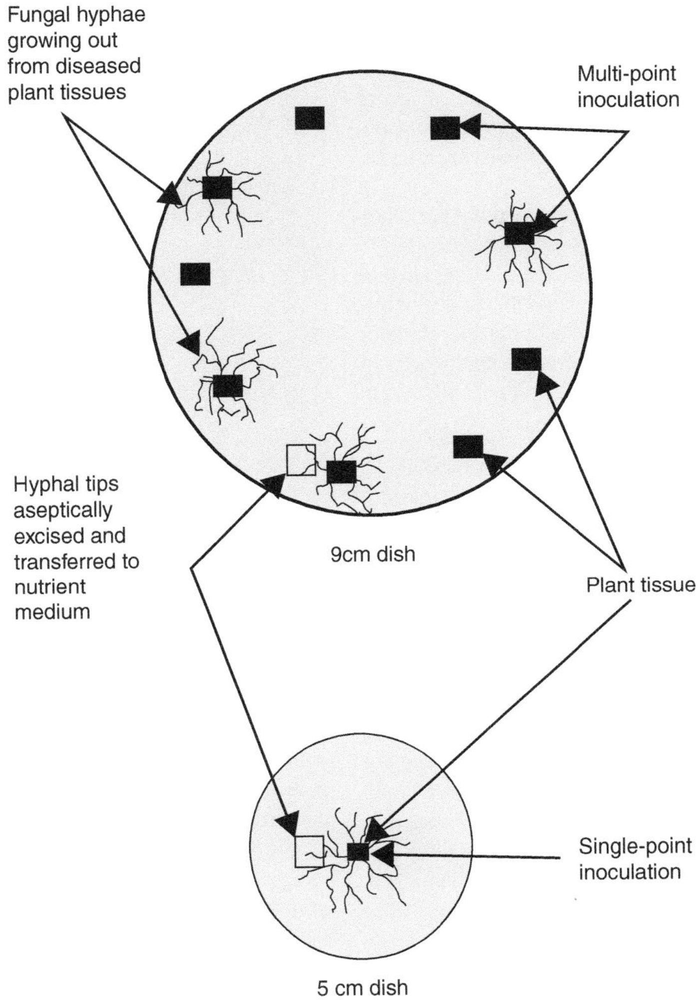

Fig. 2.11. Isolation plate transfer of fungal mycelium

flow or microbial safety cabinet should be used if available, although it is quite possible to work on an open bench.

- Petri dishes should usually be incubated upside down as this minimizes the formation of condensation on the lid of the dish. If water droplets fall onto the agar surface there is a risk that the fungal spores will be spread over the surface of the isolation plate, mixing up or contaminating otherwise pure colonies. If substantial or heavy plant tissues or seeds are plated out, care must be taken when inverting the Petri dish so that the material doesn't fall off the agar. It is sometimes necessary to incubate the plates the right way up for a day or so to allow the material to anchor itself to the agar before plates are inverted.

Surface sterilization

Surface sterilization aims to remove unwanted saprobes and other contaminants from the surface of diseased plant material, and permits the pathogen to grow unhindered when the cleaned material is plated onto agar or incubated in a moist chamber.

The simplest method is to wash the sample under running water to remove surface soil, and contaminants from the sample. However, organisms which have penetrated the surface of the material won't be removed using this technique. This method is not appropriate for licensed material, as all waste and run-off water must be collected and suitably destroyed to ensure that non-indigenous pathogens aren't accidentally released.

If the sample is visibly overgrown with saprobes, the surface may be swabbed with IMS to remove loose matter, providing a basic surface clean. A more comprehensive method involves treating small carefully selected pieces of plant material with sterilant to remove the majority of the unwanted organisms. The method used in our lab is given below:

- Cut out small sections of diseased tissue.
- Rehydrate dry plant material in sterile distilled water. Failure to do this may result in the sterilizing solution flooding into the desiccated cells, which will destroy all of the organisms present – including those of interest.
- Place the tissue pieces into a suitable sterilizing solution (choose from the list below). The amount of time needed for sterilization depends on type of material, and will be based on a number of factors, e.g. how dense or fragile it is. Sterilizing times usually range from 1-5min, and a series can be completed to ensure that the optimal time is captured i.e. where the saprobes are killed but the pathogen is maintained.
- Remove the tissue pieces from sterilizing solution and wash thoroughly in sterile distilled water. Place tissue pieces onto the surface of an agar plate. Several pieces can be placed onto the same plate if they are adequately distanced (see Fig. 2.12).

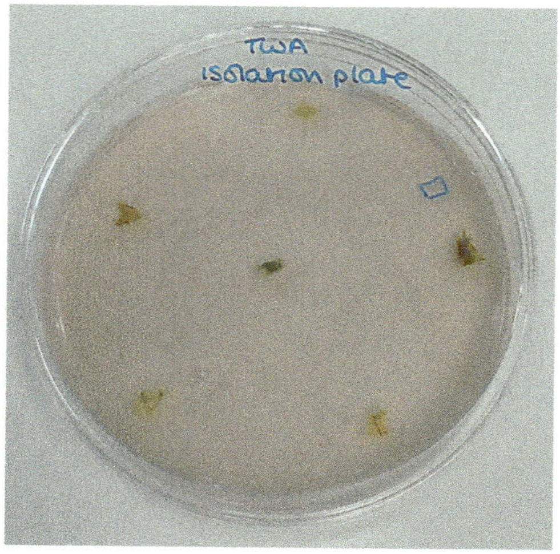

Fig. 2.12. Example of isolation plate

Surface sterilising agents

Industrial Methylated Spirit (IMS; industrial alcohol)

IMS is used as a surface swab or a dip, it is a useful cleansing and wetting agent especially for hard surfaces. IMS needs no post-treatment washing, and can be flamed off or left to evaporate.

However, IMS is not effective against many fungi.

Sodium or Calcium Hypochlorite (NaCl)O or ($Ca(ClO)_2$)

Sodium hypochlorite (or bleach) is a very effective sterilizing agent which is commonly used and widely available. Household/general purpose bleach typically contains about 5% available chlorine, whilst industrial concentrated bleach may contain over 10% available chlorine. A working solution of ~1% available chlorine should be made just before use, and dilutions will depend on the original concentration. Sterilization times usually range from 1-5 minutes depending on the type of material.

It is possible to store diluted sterilants in a refrigerator if required, but their potency will diminish with age and a fresh solution should be made at least every 2-3 weeks.

Hydrogen Peroxide (H_2O_2)

Hydrogen peroxide is used at 30% (100 volumes) strength and is ideal for robust tissues, such as seeds, wood, etc. Dilutions down to 10% can be used for more delicate tissues. It is gentler in action than hypochlorite and immersion times are longer – up to 10min. Bottles of sterilant must be kept in a refrigerator and protected against light.

Alternative sterilants

Other chemicals such as mercuric chloride (0.1% $HgCl_2$) are not recommended for general use as they require several post-sterilization washes to remove traces of the chemical, are toxic and their disposal presents an environmental hazard.

Physical methods

Heat, UV and gamma radiation are all used as sterilizing agents although their use is generally confined to materials, apparatus and instruments. They are not recommended for sterilizing living organisms due to the potential risk of causing mutations or other irreversible consequences.

Isolation techniques from different types of plant material

1. Stem/root lesions

Deep or internal lesions within vascular tissues usually benefit from removing the internal 'clean' tissues from beneath the 'dirty' outer surface layers of bark or roots. The removal of tissue from these cleaner areas means that the requirement to surface sterilize is unnecessary.

To examine the extent of disease within a stem or root, the sample can be split longitudinally from healthy to diseased area, using a flamed knife or appropriate tool. It is sometimes possible to pull the two halves of a stem or root apart manually once the initial cut has been made. A sterile scalpel can then be used to carefully remove tissues from the 'leading edge' of the lesion of the newly exposed surface, taking care not to touch the outer 'dirty' tissue (see Fig. 2.13). Small slivers of tissue (3-5mm long) can then be transferred to a Petri dish containing suitable isolation agar.

When working on thin stems or twigs (see Fig. 2.14), where the lesion is mainly confined to the outer tissues, or in other situations where it is not possible to take internal tissue samples, small pieces of diseased tissue of a few mm^2 should be excised from the lesion margin using a sterile scalpel. As it isn't possible to avoid touching the dirty outer tissues, the excised pieces should be surface sterilized (1-3min in NaOCl) and washed in sterile water, before being placed onto suitable agar plates.

2. Fine root lesions

Small and fine roots should be washed initially to remove excess soil. They can then be placed in a wide flat-based white or glass dish, containing approximately a few centimetres of water. The diseased root portions will float apart in the water, which will make them easier to examine and moved by carefully teasing apart using a scalpel or forceps. A sterile scalpel or scissors can then be used to cut out leading edges from lesions on roots (pieces approx. 5mm). A very light surface sterilization

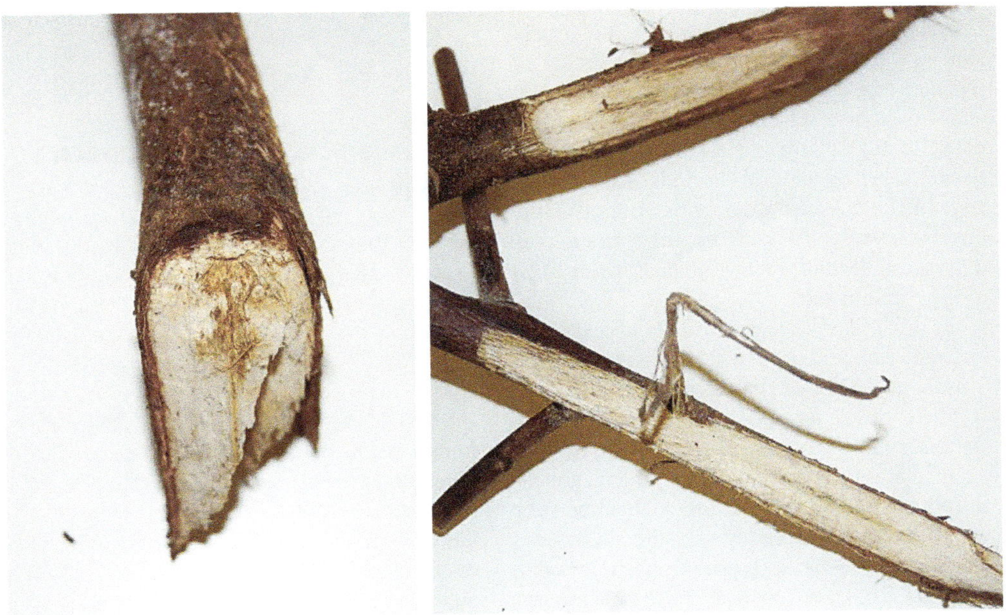

Fig. 2.13. Internal discoloured areas in hibiscus stem

Fig. 2.14. Fungal growth on cassava stem

(using IMS or NaOCl) can be undertaken, followed by a wash in sterile distilled water. Delicate material which may be infected with Oomycete pathogens is particularly sensitive to surface sterilizing chemicals, so it may be preferential to wash root pieces in this situation. A sterile scalpel or forceps can then be used to transfer the root pieces to a Petri dish containing suitable agar.

3. Leaves

Leaves should be selected displaying young lesions – where the fungus will be at its most active (see Fig. 2.15). Several 5mm pieces of tissue should be cut from the edges of the lesion using sterile scissors or a sterile scalpel. The tissue selected should include the leading edge of the diseased area – i.e. where the fungus is actively invading healthy tissue, but where the secondary organisms are unlikely to have colonized.

Surface sterilization is usually necessary for leaf material, and we usually immerse tissue for 1-3min in NaOCl, and follow with a wash in sterile water. Sterile forceps should be used to place leaf pieces onto the agar surface of a Petri dish.

Isolation plates should be incubated and assessed regularly until the fungal hyphae have grown out from the plant tissue sufficiently. Usually we would wait until growth has reached at least 10 mm to ensure that pure cultures are obtained. It is then possible to remove the hyphal tips from the isolation plate onto a new agar plate of suitable growth medium, e.g., Potato Carrot Agar (PCA) (see Fig. 2.16).

Fig. 2.15. Example of leafspots on yam

4. Fruits etc.

Internal infected tissues can be aseptically removed from beneath the outer skin, and plated directly onto suitable media (see Fig. 2.17). The removal of tissue from these cleaner areas means that it is usually only necessary to wipe the surface of the plant material with IMS before making an incision to prevent contamination by surface microbes.

If lesions are only apparent on the outer surface (see Fig. 2.18), they should be removed

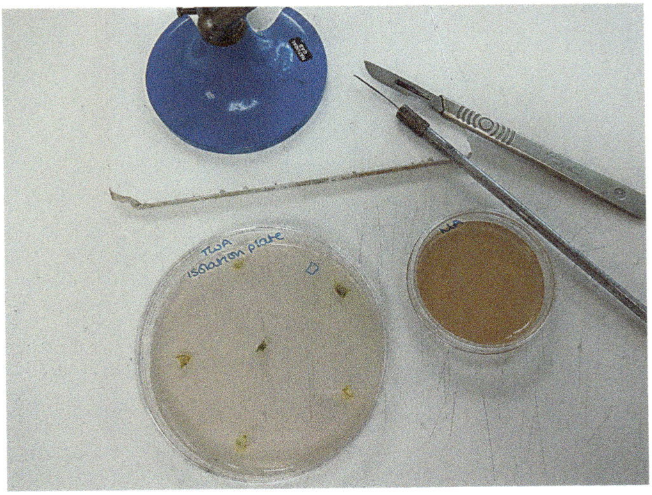

Fig. 2.16. Transfer of fungal growth to new agar plate

Fig. 2.17. Internal tissue of pomegranate

from the plant material and surface sterilized prior to plating out onto agar (using the techniques suggested for leaves). Sterilization times will depend on the thickness and type of material. Lesions often penetrate into the internal tissues beneath the outer skin.

Baiting

An alternative approach for isolating pathogens is to use non-infected plant material as bait to draw out specific pathogens from diseased plants which are contaminated with secondary invaders and saprobes. Clean seedlings, fruit and leaves etc., can all be used to pull out particular pathogens present in water or soil.

Baiting using plant material in water

This method is useful for isolating Oomycetes, such as *Pythium* or *Phytophthora* isolates from soil. Sterile seeds, germinating seedlings or other clean plant material is floated on the surface of water to which the infected test material, such as soil, has been added. Oomycetes are able to travel through the water to infect the bait material, causing lesions which can be transferred to suitable growth medium once they have developed. The resultant growth should represent a pure

Fig. 2.18. Soursop with fungal lesion

pathogen, although it may be necessary to purify the culture further by transferring it to new sterile water and fresh sterile plant material bait before this step.

Baiting using whole fruits

An alternative baiting method uses whole hard or unripe fruits, such as apples, which can be used as bait by inoculating them with a small quantity of soil or diseased plant tissue. The fruit should be surface sterilized by wiping with IMS prior to inoculation to remove any surface microbes which could be transferred to the internal tissue. A small flap should be cut into the flesh with a sterile scalpel and aseptically removed to reveal a hole into which a small portion of the test material can be placed. The flap is then replaced and the cut edges sealed with adhesive tape. When a lesion has developed, the fungus can be recovered by excising a portion of the internal tissue from the edge of the lesion some distance away from the original inoculation site and cultured onto a suitable medium.

This method is also useful for isolating Oomycetes, e.g., *Phytophthora* and *Pythium* species.

Precautions

Careful monitoring of progress is necessary. Incubation regimes should be as stable as possible and extremes of temperature, humidity and light should be avoided. Subsequent sub-culturing should be onto a suitable medium, avoiding nutrient/sugar rich media that will promote excessive vegetative growth at the expense of sporulation.

Isolation and Growth of Fungi

This section covers some of the key techniques used for the observation, isolation and growth of fungi. The focus is on plant pathogens but the methods can also be applied to fungi from other substrates.

Slide preparation: squash mounts from fungal cultures

Use of mountants

Cotton-blue plus lactic acid or lactofuchsin are the mountants of choice in our laboratories, which are used for the examination of most fungi using a compound microscope, particularly those with little or no natural pigmentation (see Fig. 2.19). Clear lactic acid is suitable for mounting fungi with darkly pigmented mycelium and spores which require no further staining.

Preparation method

- Examine the culture under a dissecting microscope to determine areas of sporulation.
- Place a drop of an appropriate mountant on a slide.

- Sterilize a mounted needle by flaming, then cool in the agar at the edge of the colony before carefully removing a small quantity of fungal material.
- Remove the fungal material from the needle using the corner of a cover slip or a second needle and transfer to the drop of mountant on the slide.
- Carefully lower a cover slip on to the mountant.

For fungi with ascocarps or pycnidia, remove several fruit-bodies from the colony and transfer to the slide. If necessary, crush the bodies gently under the cover slip to release spores by tapping gently with the handle end of mounted needle or similar item (see Fig. 2.20).

To remove trapped air from beneath the coverslip, carefully flame the slide, and remove it from the heat as soon as air is expelled from the stain. Examine the slide to check if air bubbles have been removed. If not, flame again gently to expel remaining air.

A tissue or cloth can be used to remove excess mountant from the edges of the cover slip to ensure that area around the cover slip is clean and dry.

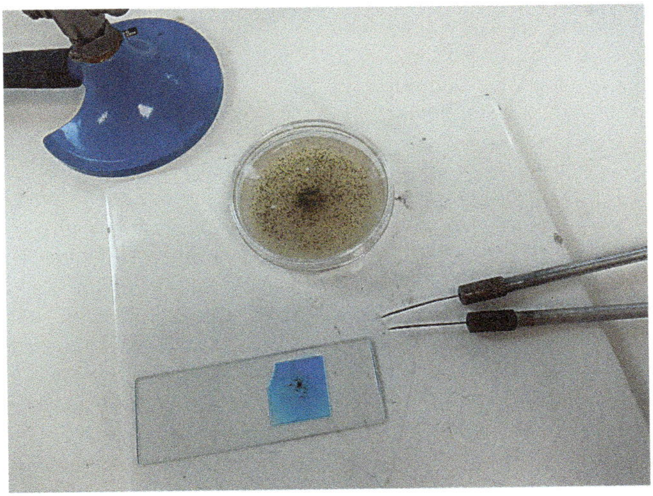

Fig. 2.19. Slide preparation

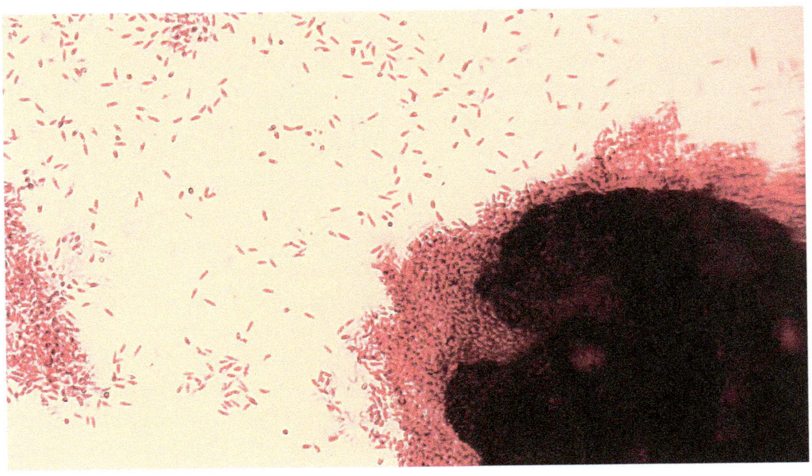

Fig. 2.20. Example of pycnidium

Sub-culturing Fungi

The simplest method of maintaining living fungi is by serial transfer from used media to fresh (solid or liquid) media, and then to incubate under appropriate conditions for individual isolates. Many fungi can be maintained in this way for several years; although this practice isn't recommended for long-term storage. The successful maintenance of fungal cultures is dependent on ensuring that contaminants or genetic variants do not replace the original strain. Such methods are labour intensive and time consuming when large collections are involved. The time period between transfers varies from fungus to fungus, some require transfer every 2-4 weeks, the majority every 2-4 months. Although Chu (1970) maintained several representatives of forest tree pathogens for 1 year at 5°C most organisms were best transferred after much shorter periods.

Disadvantages of frequent transfer:

- Danger of variation, loss of pathogenicity or other physiological or morphological characteristics.
- Danger of contamination by air-borne spores or mite carried infections.
- Requirement for constant specialist supervision to ensure that the fungus is not replaced by a contaminant or sub-cultured from an atypical sector.

Advantages of frequent transfer:

- Collections can be kept viable for many years if supervised by a specialist.
- The method requires no specialized equipment, is relatively inexpensive and, for a small collection, the time involved is not excessive.
- Retrieval is very easy.

Standard method

To transfer material, sterilize an inoculating needle by flaming to red-hot and briefly flame the base of the needle holder. Cool the needle in the agar at the side of the dish containing the existing culture. Scrape a little material from the culture and deposit it in the centre of a fresh agar plate. Unless a laminar flow or microbiological cabinet is available to work in, it is important to keep the Petri dish lid above the plate so as not to expose it completely to the air. This will help avoid aerial contamination.

If difficulty is experienced removing fungal strands from the inoculating needle, instead of removing mycelium alone, remove a block of inoculum from the existing culture, including some of the agar in which the existing colony is growing and transfer this to the centre of a fresh agar plate.

Sub-culturing Penicillium *and* Aspergillus *using three-point inoculation*

Penicillium and *Aspergillus* species produce large numbers of dry spores, making them more challenging to work with. The standard procedure for sub-culturing fungi, in which a small piece of inoculum from an existing culture is transferred to a fresh plate, is generally not satisfactory for *Penicillium* and *Aspergillus* as dry spores tend to fall from the inoculating needle during transfer resulting in the development of multiple colonies.

To prevent the formation of colonies by stray spores, a special method for preparing sub-cultures of *Penicillium* and *Aspergillus* was devised by J.I. Pitt (1973). This technique involves the use of a semi-solid gel (Pitt's medium) to prepare a spore suspension of the required fungus prior to sub-culturing. For *Penicillium* and *Aspergillus*, inoculations are usually prepared on Czapek Agar and Malt Extract Agar.

Inoculation of *Penicillium* and *Aspergillus* is best carried out in a laminar flow or microbiological cabinet to prevent spores escaping into the lab, although this is not essential if care is taken. Select an area that is sporulating well. Sterilize a loop by heating in a Bunsen flame and cool in the agar at the edge of the plate or bottle in which the colony is growing (or alternatively use a sterile plastic loop).

Using the loop, remove a small quantity of spores from the colony and transfer to a Pitt bottle containing semi-solid gel. Stir the contents of the bottle with the loop to form a spore suspension. Once the spore suspension has been prepared, re-sterilize the loop (or use another plastic loop), cool in the spore suspension and withdraw a loopful of suspension.

Touch the loop on to the agar surface at a point approximately 2cm from the edge of the dish and repeat inoculations twice more to form a triangular arrangement. Leave the plates upright for a minute or two until the inoculum has dried. Attach two pieces of tape to secure the lid and base of the plate so as to allow exchange of oxygen. Invert and place in an incubator at 25°C for 7-10 days. The plates should be left undisturbed for the entire growth period. Three distinct colonies result, and each should look identical.

Mite Infestation Prevention

It is important to be vigilant when looking for mites and signs of their presence in the lab; fungal cultures are highly susceptible to infestation with mites which occur naturally in soil and on organic material. They can be brought into the laboratory on fresh plant material, decaying mouldy products, on shoes, on the bodies of flying insects or in cultures received from other laboratories.

The damage mites cause is two-fold. Firstly, they eat the cultures; a heavy infestation can completely strip the colonies from an agar plate. Secondly, they carry fungal spores and bacteria on and in their bodies and as they move from one culture to another the cultures can become contaminated and sometimes heavily infected with unwanted fungi and bacteria.

The mites commonly associated with fungal cultures are about 0.25 mm in length. In the UK we would expect to see species belonging to the genera *Tyrophagus* and *Tarsonemus*. Mites can just about be seen with the naked eye as tiny white dots, but they are almost at the limit of our vision so infestation can easily go undetected. Mites may sometimes be observed in the condensation on the lids of Petri dishes rather than on the surface of the agar, so it is important to check cultures thoroughly if their presence is suspected. When given favourable conditions: high humidity, warm temperatures and a ready food source, they breed rapidly and spread quickly. The first indication of their presence may be that older cultures appear to have deteriorated, whilst young plates may exhibit trails of bacteria or small fungal colonies where the mites have travelled across the surface of the agar. It is worth bearing in mind that good hygiene practices and preventative precautions are preferable to trying to control an outbreak. All material coming into the lab should be examined on arrival and a separate room for checking and processing dirty material is desirable. The sealing of incoming cultures, storage in a refrigerator or some form of screening and quarantine system can be helpful, as it is possible for cultures with only a light infestation at the time of receipt to develop a heavy infestation later. There are many different methods of control, and a combination of precautions may be appropriate.

Hygiene

Hygiene coupled with robust quarantine procedures is perhaps the best protection against mites. All laboratory work surfaces must be kept clean and cultures protected from aerial and dust contamination. The workbenches, incubators, fridges, shelves and cupboards should be regularly cleaned, especially if an infestation of mites is suspected. Historically, the decontamination procedure at CABI involved the use of Actellic 25EC, a non-fungicidal acaricide. However, its use is no longer permitted, and we are currently restricted to using standard lab disinfectants and IMS to clean and sterilize surfaces. All cultures in the immediate area must be checked and isolated from the infected collection as soon as an outbreak is acknowledged, and regular examination is essential to ensure that they remain 'clean'.

If Actellic 25EC or a similar chemical is permitted for use in a particular country, care must be taken to adhere to the associated health and safety guidelines. Actellic 25EC is of moderate toxicity and irritating to skin: therefore, all contact must be avoided.

Fumigation

This method is used as a last resort or when moving into new premises and should be carried out by a licensed specialized company. Where large numbers of important cultures or specimens are involved, these items can be fumigated off site in specialized equipment. If this is necessary it is advised that specialist contractors are employed. Such fumigation generally involves the use of chemicals that are toxic to fungi, therefore unaffected cultures should be removed or protected.

Mechanical and chemical barriers

Many physical methods to prevent the infestation and spread of mites have been evaluated, and simple procedures can be very effective. Culture bottles, tubes or plates can be elevated on a platform surrounded by water or oil – like a moat, or placed on a solid surface surrounded by a barrier of petroleum jelly or other sticky material which traps any crawling insects. In our lab we place our cultures on shallow plastic trays covered with a sheet of foil with Vaseline® smeared around the outer edge to form a barrier to anything trying to enter or leave the tray.

We seal most of our culture containers; such as Petri dishes and Universal bottles, with electrical tape or Parafilm® to reduce the chance of mite infestation and contamination - although this may not provide a complete barrier if cracks and wrinkles occur. Sellotape®, masking tape and other sticky tapes may also be used.

A method useful to protect cultures in Universal bottles and cotton wool plugged tubes is the cigarette paper method first described by Snyder and Hansen (1946). This procedure involves gluing a thin cigarette paper over the opening of the bottle/tube; preventing the entry/exit of mites, but allowing free passage of air. Care must be taken to ensure that the paper is not damaged through handling and that a good seal is made. It has the advantage that it not only keeps mites out, but it also keeps them in - thus preventing spread of infestation. It is also useful when sending cultures between labs (see Figs. 2.21 and 2.22).

Method

Cut cigarette papers in half and sterilize in an oven at 180°C.

Glue a cigarette paper onto the neck of an inoculated universal bottle using copper sulphate gelatin glue (20g gelatin dissolved in 100ml water and then add 2g copper sulphate).

Burn off the excess cigarette paper protruding from the outer edge of the tube or bottle.

The lid of the universal can be replaced once the glue has dried.

Protected storage

The various methods used for the long-term storage of cultures in culture collections: silica gel, freeze drying and liquid nitrogen, actively prevent the chances of infestation and spread of mites. The use of sealed tubes, extreme temperatures and desiccation create both physical barriers and exceptionally unfavourable environments.

However, replication of these conditions is not practical for the short-term storage of lab cultures, and alternative techniques must be used to eliminate mites from working collections.

Cold storage at 4-8°C reduces the spread of mites which are almost immobile at this temperature, but refrigeration temperatures are not low enough to kill mites, and when removed from the fridge they will rapidly become active again.

Storage of infested cultures in the deep freeze (<-20°C) for at least 3 days gives better control, and whilst most fungal cultures will remain viable when exposed to this temperature, mites are unlikely to survive. Once removed from the freezer the culture will need to be re-isolated onto new agar as the original will probably be contaminated with other microbes introduced by the mites.

Covering cultures with mineral oil prevents mites escaping should they get into the culture vessel, although contamination of the culture may have already occurred through the spread of spores and bacteria carried by the mites.

Growth Conditions for Fungi

Generally, fungi grow best on media formulated from the natural materials from which they were originally isolated. In the Diagnostic Lab we routinely use potato carrot agar (PCA) for our initial isolations, although we also use a

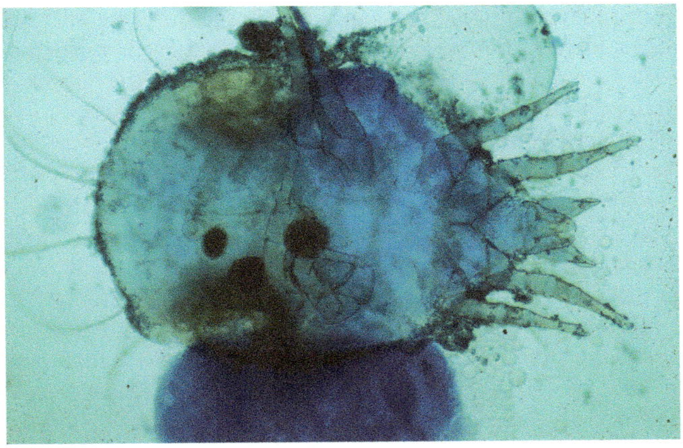

Fig. 2.21. Mite stained under the microscope.

Fig. 2.22. Mite prevention equipment for paper seal of Universal bottles

range of other media. CABI also utilizes plant materials such as leaves, stems or seeds which can be placed on the surface of solid agar to supplement the standard recipes and provide optimal growth conditions. It is important to provide the best growth conditions possible when maintaining cultures, especially those required for long-term use. The major factors affecting growth are medium, temperature, light, aeration, pH and water activity. If any of these requirements aren't met, outcomes may include the selection of variants from within the population, deterioration of strains, or complete inability to culture the target organism.

Media

The growth requirements for fungi may vary from strain to strain, although cultures of the same species and genera tend to grow best on similar media (see Appendix B, this volume). The source of isolates can give an indication of suitable growth conditions. For example, species from leaves may sporulate best on low sugar media when exposed to light, those from marine situations may require salt, those from the tropics may prefer high growth temperatures, and isolates from sugar rich sources can be expected to grow well on high-sugar media.

Lab cultures are usually maintained on agar slopes, in test-tubes or in culture bottles. However, some fungi deteriorate when kept on the same medium for prolonged periods, so the medium should be alternated from time to time. Most laboratories prefer not to keep a large stock of different media and the majority of isolates can be maintained on a relatively small range depending on the specialization of the collection.

Experience at CABI indicates that fungal cultures grow more satisfactorily on freshly prepared media, especially natural media containing fresh vegetable decoctions. These media have several advantages: they are usually easy and relatively cheap to prepare, require few facilities, small quantities can be sterilized using a domestic pressure cooker, and the pH can be adjusted using drops of hydrochloric acid or potassium hydroxide. However, variation between batches is often evident, due to differences in the constituent ingredients, water quality, weighing discrepancies, interpretation of recipes or simply may vary depending on who has made the media.

The shift towards proprietary media is increasing, even at CABI, and many of the recipes which were historically made to order using natural products have been replaced with equivalent synthetic media. A commercially-produced dehydrated product tends to be more standardized than its freshly prepared equivalent, and gives little variation between batches. This feature is important when replicating work. The use of dehydrated media is a useful option for labs with limited facilities, who want to make batches of media to order, or wish to make amendments to the standard recipes. For example, at CABI we might choose to make 50% strength media, vary the sugar or agar content, or add antibiotics to a dehydrated media product to optimize its use.

Taking this a step further, many commonly used media are now available as pre-poured plates which can be ordered directly from the supplier as required – negating the need for any in-house media preparation facilities at all. Although this has obvious benefits, the disadvantages include higher costs, the requirement to pre-plan work and predict quantities of media needed, the need to adhere to the shelf life of batches and to cover the costs of associated waste, the limitations of stock availability and restricted media choice. The decision about whether to make media from scratch, use dehydrated formula or purchase pre-poured plates will be based on the resources and requirements of the individual user.

Some media for special purposes such as assay work will require very careful preparation. A wide range of media are used by different workers and most mycologists have preferred media. For example, Raper and Thom (1949) used Czapek's Agar, Steep Agar and Malt Extract Agar for the growth of penicillia and aspergilli, while Pitt (1980) in his monograph on penicillia recommended Czapek Yeast Autolysate (CYA) and Malt Extract Agar (MEA). Preferences for growth on particular media are normally developed over many years and are the result of experience. The standardization of defined media formulae is necessary for most work. Media will affect colony morphology and colour, whether particular structures are formed and may affect the retention of properties. Most fungi can be grown on Potato Carrot Agar (PCA) or Malt Agar (MA). However, others have specified growth requirements. Some water moulds are best stored in water with the addition of plant material (Goldie-Smith, 1956), whilst other more sensitive water moulds, may require aeration (Clark and Dick, 1974; Webster and Davey, 1976). Examples of particular preferences are given below and in Smith and Onions (1994).

Mucorales grow well on Malt Agar (MA) and will not grow in Czapek Agar (CZ) as they lack the enzymes to digest sucrose.

Many fungi thrive on Potato Dextrose Agar (PDA), but this can be too rich, encouraging the growth of mycelium with ultimate loss of sporulation, so a period on Potato Carrot Agar (PCA), a starvation medium, may encourage sporulation.

Fusarium species grow well on Potato Sucrose Agar (PSA).

Wood inhabiting fungi and dematiaceous fungi often sporulate better on Cornmeal Agar (CMA) and Oat Agar (OA) both of which have less easily digestible carbohydrate.

Cellulose destroying fungi and spoilage fungi, such as *Trichoderma*, *Chaetomium* and *Stachybotrys* retain their ability to produce cellulase when grown on a weak medium such as TWA or PCA with a piece of sterile filter paper, wheat straw or lupin stem placed on the agar surface.

All sorts of vegetable decoctions are possible and apart from the advantages of standardization it is reasonable to use what is readily available, e.g., yam media might be preferable to potato media in the tropics.

The introduction of pieces of tissue, such as rice, grains, leaves, wheat straw or dung, often produces good sporulation. For formulae of the most commonly used media see Appendix B, this volume.

Temperature

The majority of filamentous fungi are mesophilic, growing at temperatures within the range of 10-35°C, with optimum temperatures between 15 and 30°C. Some species (e.g., *Aspergillus fumigatus*, *Talaromyces avellaneus*) are thermotolerant and will grow at higher temperatures, although they are still capable of growth within the 20-25°C range. A small number of species (e.g., *Chaetomium thermophilum*, *Penicillium dupontii*, *Thermoascus aurantiacus*) are thermophilic and will grow and sporulate at 45°C or higher but fail to grow below 20°C. A few fungi (e.g., *Hypocrea psychrophila*) are psychrophilic and are unable to grow above 20°C, while many others (e.g., a wide range of *Fusarium* and *Penicillium* species) are psychrotolerant and are able to grow both at freezing point and at mesophilic temperatures (Smith and Onions, 1994).

Single Spore Isolation

Single spore isolation is a useful technique for several purposes. It can be used for isolating fungi directly from host material, for separating fungi from mixed cultures and for obtaining single-spore colonies for biochemical or molecular studies. The technique involves obtaining sporulating material, either from the host or in culture, preparing a spore suspension of the required fungus in sterile water, streaking a loopful of the spore suspension across an agar plate and after a period of incubation, transferring single germinating spores to fresh agar plates.

Preparation of agar plates

It is necessary to use a clear agar such as Potato Carrot, Malt or Tap Water Agar on which to carry out this procedure. The location of single spores under the microscope in the latter stages of the technique requires the transmission of light through the agar from beneath the dish, so dense/grainy agar and thickly poured plates are unsuitable. Ideally plates should be thinly poured, no more than 3mm deep. This is because the plate may need to be examined from the reverse side under a compound microscope and it is necessary for the lens to be able to focus beyond the base of the plate to the agar surface. When isolating from host material or from cultures contaminated with bacteria, prepare plates containing an antibiotic such as Penicillin G+Streptomycin sulphate.

Mark guidelines on the base of the plates along which the spore suspension will be streaked. Four strokes are used to distribute spores and the guidelines are best drawn in the pattern of a 'W' or 'E' using a marker pen.

Preparing a spore suspension

Having obtained a sporulating colony of the required fungus on host material or in culture, a spore suspension is obtained as follows:

Place a Petri dish containing the sporulating material on the stage of a binocular dissecting microscope and select an area from which spores are to be isolated. Fungi immersed beneath the host surface can be exposed by cutting away the surface tissue with a scalpel or razor blade. Sterilize an inoculating needle by flaming to red hot and cool in a Universal bottle containing a little sterile water which also serves to moisten the needle tip.

In the case of hyphomycetes producing abundant spores, remove a small quantity of sporulating material from the specimen or culture and transfer to a Universal bottle containing 2-3 ml of sterile water. Replace the cap and shake the bottle to form a spore suspension.

If spores are not abundant, use the following technique to obtain a more concentrated spore suspension. Sterilize a glass slide by dipping in alcohol and flaming, and place a few drops of sterile water in the centre of the slide using a flamed glass rod or wire loop. Remove a small quantity of spores on a needle tip as described above, transfer to the water droplet on the slide and stir with the needle or a loop to form a spore suspension.

For coelomycetes and ascomycetes producing spores within fruit bodies, either on host tissue or in culture, remove a few bodies with a sterile needle and squash on a sterilized slide to release the contents. If spores are abundant, place a loopful in a bottle of sterile water and shake to form a spore suspension. If spores are not abundant, transfer a loopful to a drop of water on a slide and stir.

Streaking out the spore suspension

Sterilize a wire loop by flaming, cool in sterile water and remove a loopful of the spore suspension. Lift the lid of a large agar plate and starting from the beginning of the 'W' or 'E' pattern of guidelines, streak the loop across the agar surface using four strokes to distribute the spores. As the spore suspension is spread with each successive stroke, spores become more widely separated. Once prepared, incubate the plates at 25°C overnight and observe after 18-24 hours for signs of germination.

Locating and isolating large germinating spores

After incubation, place the Petri dish on the stage of a dissecting microscope, and view using light transmitted from beneath. The path taken by the loop is evident as a series of parallel striations following the guidelines along which spores are scattered. If spores are sufficiently large, use of the more powerful compound microscope is unnecessary for the purpose of locating and removing single spores.

Scan the guidelines until widely separated single spores are found. Usually, spores are deposited close together at the beginning of the first stroke, thinning out towards the end of the second stroke and becoming sparse or absent in the third and fourth strokes depending on the quantity of spores contained in the original spore suspension. With larger spores, it is often possible to observe germination under the higher power of the dissecting microscope, germ tubes being visible as tiny translucent filaments emerging from the spores.

Once a single germinating spore is found, it is best removed with a scalpel blade by cutting out a 2 mm block of agar containing the spore. A fine scalpel blade e.g,. a Swann Morton No.11 blade fitted to a No.3 blade holder is ideal for this purpose. Sterilize the blade by dipping in alcohol and flaming, cool in the agar at one side of the dish and cut the agar closely around the required spore. Remove the agar fragment bearing the spore and deposit in the centre of a fresh agar plate. This procedure is repeated to obtain further spores.

Petri dishes containing single germinating spores should be labelled, sealed with tape and placed in an incubator or under black light. The resulting single spore colonies are characterized by a uniform growth pattern.

Locating and isolating small germinating spores

In the case of fungi with very small spores that cannot be clearly seen under the dissecting microscope, it is necessary to place the agar plate on the stage of a compound microscope and scan along the guidelines until a single germinating spore is found.

It is very difficult to remove germinating spores when the plate is mounted on the stage of a compound microscope, unless a micro-manipulator is used, because the image is in reverse to reality and can cause confusion to the observer. A solution to this problem is to view the plate from the reverse side and mark the position of spores.

Place the plate upside down on the stage of a compound microscope and view the agar surface through the base of the Petri dish. Scan along the guidelines and find a single germinating spore widely separated from neighbouring spores. Manoeuvre the plate until the spore is in the centre of the field of view. The light source from the microscope produces a circle of light on the base of the Petri dish which corresponds to the field of view. (Adjust the condenser if necessary, to focus the light on to the Petri dish base). When the spore is in the centre of the field of view, it is also in the centre of the spot of light. Close the diaphragm to make the spot of light as small as possible, then using a fine marker pen, circle the spot of light. The spore will be in the centre of the marked circle. Remove the plate from the microscope stage and transfer to the stage of a dissecting microscope. Find the circle marking the position of the germinating spore. The germ tube is usually just discernible. The spore can then be removed with a scalpel blade as previously described and transferred to a fresh agar plate.

Isolation of fungi by direct plating

When isolating fungi, a translucent general-purpose media e.g., Potato Carrot Agar, Malt Agar or Difco Cornmeal Agar enables developing colonies to be clearly visible. On standard strength media, fungal growth can be rapid, so to facilitate separation of colonies, the growth rate can be reduced by preparing standard media at quarter or half-strength. Tap Water Agar is also an effective isolation medium since it restricts development of fast-growing fungi and discourages bacterial growth – although growth can be very slow. Antibiotics added to media to eliminate bacterial contamination include Penicillin G, Streptomycin sulphate and Chloramphenicol.

Isolation of surface fungi

Fungi can be isolated directly from material by carefully removing a small quantity of spores, mycelium, or individual fruit bodies using a sterile needle or scalpel blade and transferring directly to suitable media e.g., Tap Water Agar or Potato Carrot Agar containing antibiotic. When fungal growth has reached a diameter of 1-2 cm, pieces of inoculum can be transferred aseptically to fresh plates of a suitable nutrient medium.

Use of black light

Fungi that do not sporulate readily in culture can be stimulated to sporulate by exposure to short wavelength irradiation produced by near ultra-violet light (near UV or black light, wavelength 300-380 nm). Units for producing black light consist of a black light fluorescent tube fixed by brackets beneath a bench top, together with 1-2 cool white fluorescent tubes which are fixed adjacent to the black light tube. These are operated by a timing mechanism so that they can be switched on and off at set times. Beneath the lamps, a shelf is placed on which to grow the cultures. The recommended distance between the shelf and black light tube is 32 cm. A wooden shield extending 15 cm below the bench top is fitted around the unit to protect laboratory staff from near-UV light rays.

Cultures are best grown without exposure to black light for 3-4 days to allow time for the colony to develop before placing them under the black light unit. It is recommended to use an alternating cycle of 12 h black light, followed by 12 h darkness. This often results in bands of sporulation, corresponding to periods of exposure to the black light. The optimum temperature range for growth under black light is usually between 21-28°C. Temperatures in excess of 30°C should be avoided, as the effects of black light can be lost at high temperatures. Rich growth media should be avoided, as they may give rise to excessive growth of mycelium; nutritionally weak media such as potato carrot agar (PCA) are more suitable for inducing sporulation.

In order for black light to be effective, fungi should be grown in plastic Petri dishes or plastic universal bottles. Glass is not suitable, as it does not allow adequate transmission of black light.

Formulae and uses of mountants

Although many different stains and mountants are available, we currently use Cotton blue in lactic acid and Lactofuchin routinely in our lab.

Cotton blue in lactic acid

0.1% cotton blue in lactic acid is widely used as the standard mountant for examination of fungi, particularly those which have little or no natural pigment. It does not stain cell walls but rapidly stains the cytoplasm of fungal cells and has excellent lasting properties.

Formula:

Cotton blue	0.1g
Lactic acid (85%)	100.0ml

Add cotton blue powder to lactic acid. Heat in a glass beaker, stirring continuously until dissolved. Allow to cool then filter to remove any sediment. Store in a dark bottle. For a lighter stain, dilute with further lactic acid as required.

Lactic acid

Lactic acid is a suitable mountant for fungi that have sufficient natural pigment to make staining unnecessary. Mounts prepared in lactic acid can last for many years when sealed with several layers of nail varnish.
Formula:

Use neat lactic acid (ca 85%). Store in a dark bottle.

Lactofuchsin

Lactofuchsin, is produced by combining acid-fuchsin with lactic acid. This produces a red stain with properties similar to those of cotton blue in lactic acid. Although it fades more rapidly than cotton blue, it has a slightly better refractive index, giving excellent clarity. It is beneficial to use lactofuchsin rather than cotton blue when photographing fungal structures, since use of blue filters in photography does not reduce definition with this stain, as is the case with cotton blue.
Formula:

Acid-fuchsin	0.1g
Lactic acid	100.0ml

Dissolve acid-fuchsin in lactic acid. Stir thoroughly. Store in a dark bottle.

Isolation and Growth of Bacteria

Bacteria are diverse organisms which have adapted to grow and thrive in most known environments, therefore it is difficult to make generalizations about optimal growth conditions. However, the majority of organisms will grow at pH values near neutrality, at temperatures between 20 - 30°C and in a medium containing an energy source (e.g., glucose), and an organic nitrogen source (e.g., peptone.) Many UK culture collections supply information on media and growth conditions for the strains that they supply and can provide advice on an individual basis.

Bacteria can be isolated from numerous sources: air, water, soil, plants, animals, solid surfaces, fuel and even from extreme environments. The methods used for isolating bacteria from plant material can generally be applied to organisms from other sources.

Isolation of Bacteria from Plant Material

Bacterial diseases of plants are much more difficult to determine, and are often indicated by the absence of more visible pathogens and pests, such as fungi, insects etc. Bacterial disease may prove difficult to distinguish from some virus-induced local lesions. The original material should be handled as little as possible to avoid contamination with microbes from hands, etc. Sterilization is often undesirable, especially with near-dry or dry material. This is due to the rapid tissue penetration of the sterilizing agent effectively killing all the bacteria.

Dilution and streaking method

The usual procedure for the isolation of bacteria from plants is a dilution and streaking method. The bacterium is most active in the tissues at the leading edge of the lesion and small pieces of affected plant tissue should be carefully removed and placed in sterile water to allow the bacterial cells to flow out of the diseased material. The aim is to dilute the bacterial cells so that at the end of the streak those colonies growing are most likely the product of a single bacterial cell. It is then a matter of checking for purity by culturing from one of the small single colonies and suspending the bacterial cells in sterile distilled water. It should be re-streaked as before to ensure that the colonies at the end of the streak are likely to be the product of a single bacterial cell.

Isolation of Bacteria from Environmental Samples

Isolation of bacteria from other sources is usually also by dilution methods. Most plant pathogenic bacteria are not particularly fastidious and the standard dilution medium is usually sterile distilled or deionized water. Other bacteria from alternative substrates can be much more exacting, and Ringer's solution would be suggested as a dilution medium. Bacteria can also be isolated directly onto a suitable culture medium using an inoculating loop. It is important to ensure that the medium is carefully selected to ensure that it is suitable for the bacteria. Culture plates should be monitored regularly to allow the transfer of individual colonies to fresh media to establish pure cultures.

Identification of a bacterium can have various routes; traditionally the Gram Stain would be the first step in identification through diagnostic keys, but is not usually necessary when using molecular methods. However, it is essential that any culture taken forward for identification is pure and contains only a single organism.

Isolation of Bacteria from Plants

The isolation of bacteria from plants can also be achieved using similar methods used for fungi. If pieces of infected tissue are placed directly onto the agar medium, bacterial growth will occur around them close to the plant material. Whilst fungi are able to grow away from the point of inoculation, bacteria are unable to do this and growth is likely to comprise several organisms. If saprobes are present, growth is liable to consist largely of these organisms, and the slower growing pathogens will be lost. Therefore, this method should only be used when the dilution method is deemed unsuitable, for example when the presence of *Xanthomonas populi* is suspected, and a suitable isolation medium can be selected.

Determining presence of bacteria in infected plant material

If bacteria have colonized plant tissue in great numbers it is sometimes possible to view the cells to verify their presence. This is achieved by cutting a small section from the boundary between diseased and healthy tissue, mounting in a drop of water on a microscope slide, and examining under the high-power objective (40X) of a compound microscope. Thicker plant tissues should be teased apart after mounting prior to microscopic examination. Unstained bacteria can be difficult to see under the microscope and it is necessary to reduce light by closing the substage diaphragm or lowering the condenser. Toluidine Blue O may be used as a stain to gain contrast between the bacterial cells and plant material (Ghemawat, 1977). When observing bacteria great care must be taken not to confuse particulate material such as latex, plastids or starch granules with bacteria.

If bacteria have colonized the plant tissues, it may be possible to observe bacterial streaming; cells moving away from the plant material into the surrounding water. A compound microscope is usually required to observe this, but sometimes inoculation levels are so high that the cells can be seen with the naked eye – with the appearance of a milky solution falling from the host tissue. Large masses of bacteria observed streaming from the plant tissue will generally ensure a successful subsequent isolation.

Surface Sterilization of Plant Material

A quick immersion into 0.5-1% NaOCl, (sodium hypochlorite solution) is sufficient to sterilize most material if required. The material should be rinsed in sterile water after treatment.

Larger or thicker plant parts, such as buds, fruits, cankers, galls etc. can be immersed in 70% ethanol (aq., v/v) and flamed, or left to air dry depending on the material involved prior to isolation.

Fragile material, such as leaves which are not obviously covered with surface contaminants, can be wiped with IMS prior to isolation.

Isolation of Plant Bacteria

Small pieces of diseased tissue should be selected and aseptically removed from an area where bacterial activity is suspected. This is usually the advancing edge of a necrotic lesion or canker, or a piece of discoloured vascular tissue, although heavily diseased and discoloured areas should be avoided. The tissue should be placed in a small volume of sterile water in a sterile Petri dish or sterilized watch-glass, and teased apart with flamed needles. Tough or woody material can be ground with a sterile pestle and mortar to release the bacteria into solution. The bacteria can be left for several minutes to allow dispersal into the water, although they should not be left for an excessive period of time as the chance of isolating contaminants will increase. A loopful of the resultant suspension should then be removed and streaked over the surface of a suitable agar medium, e.g. Nutrient Agar (see Fig. 2.23). The aim is to spread the suspension thinly so that single bacterial cells are produced, enabling growth into single colonies. To achieve this, the agar surface must be dry, i.e. no free water must be present in which the bacteria can swim and multiply. Dry plates are usually obtained by cooling the molten agar to about 45°C before pouring, and keeping the poured plates at an even room temperature for 24-48h. In very humid climates or for quick drying the plates, set dishes may be dried by placing them open and inverted in a thoroughly cleaned incubator a little above room temperature for 2-3 hours. Care must be taken to avoid air contamination of the agar.

If bacteria are suspected to have colonized internal tissues, and the material is not heavily contaminated with fungi or secondary organisms, it may be possible to subculture directly from the plant tissue onto a suitable agar plate (see Fig. 2.24). Growth should be checked daily and colonies may need to be transferred to fresh agar to maintain purity.

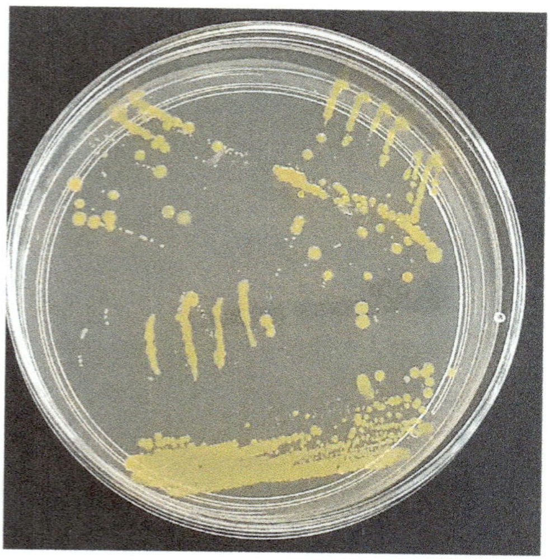

Fig. 2.23. Bacterial streak plate with more than one colony type

The selected isolation medium used must be suitable for growth of the suspect pathogen. Nutrient agar is usually satisfactory for isolation work, but King's Medium B may provide a better option in some cases as it not only supports the growth of most plant pathogens, but also generates useful information about fluorescence.

A few bacteria need more specialized media to grow;

Xanthomonas albilineans and *X. populi* grow on media containing sucrose or glucose and peptone; some *Erwinia* species belonging to the amylovora group grow much better on nutrient agar supplemented with glucose (1%, w/v). *Clavibacter xyli* requires a very complex medium containing hemin and serum albumen.

After streaking, the dishes should be incubated at about 25°C and examined daily for 4-5 days, or up to 10-14 days if a slower-growing pathogen is suspected. If the diagnosis based on the symptoms is borne in mind, the expected pathogen can often be recognized and sub-cultured, even if various other bacteria are present on the plates. With a good specimen of diseased plant and well-chosen pieces of tissue, the pathogen is often the only organism recovered. If more than one type of colony grows, the relative abundance of each should be noted. The most frequently isolated organism, especially if recovered from different sites, is worthy of examination unless readily recognized as a saprophyte. Sub-culturing must only be performed with well separated colonies, preferably on to plates or slopes of the same medium.

It is possible to complete an examination and isolation from material at the same time if care is taken. To achieve this a small strip of tissue should be cut from the edge of a lesion, where bacterial activity is expected to be greatest. If the lesion is very young, a slice can be taken through the centre of the lesion and one half removed. The tissue is then placed on a flamed microscope slide in a drop of sterile water, covered with a gently flamed cover slip and examined at once under the 40X microscope objective. It may be necessary to lower the substage condenser or to close the diaphragm in order to see unstained bacterial cells. If streaming of bacteria from the cut edges of the tissue is seen, the coverslip may be removed and a loopful of the suspension streaked onto suitable agar media.

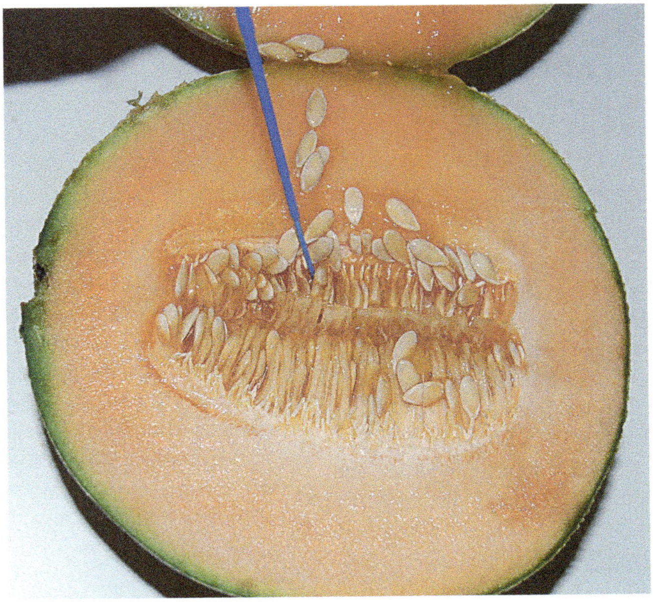

Fig. 2.24. Bacterial isolation from internal tissue of melon

Checking the Purity of Cultures

All cultures should be checked for purity before any identification tests begin. If there is any doubt about the purity of an isolate it is advisable to make a very dilute suspension of some of the cells in sterile water or saline and streaking a loopful on to a dry agar plate (using the same technique as used for an isolation streak). The streak should be examined daily for several days to check that all colonies are identical. If there is any doubt about the purity of a culture at this stage, single colonies should again be sub-cultured onto new agar plates until purity is certain. There is no point in working with mixed cultures, and the results from any tests will be meaningless.

Determination of Colonial Morphology

The streak tests for purity can also be used to observe colonial morphology. The shape, size, texture, colony surface markings, elevation, margin type, consistency, colour, translucency or opaqueness, and rate of growth can be compared to the properties of the target organism. The presence of pigment precipitates or crystals in the medium should also be noted.

Bacterial colony morphology

Once a pure culture has been obtained it may be useful to perform a Gram Stain prior to viewing the cells with a compound microscope. Gram positive cells will stain purple whilst Gram negative will appear red. The age and concentration of cells used for testing must be carefully controlled to guarantee reproducible results. The information gathered from the Gram stain will determine whether the bacteria are Gram positive or negative, but additional information about the size and shape of cells will also be apparent when the slide is examined. The majority of plant pathogenic bacteria are Gram negative, although there are exceptions such as *Curtobacterium* and *Clavibacter*.

Gram Stain and Cell Morphology

Reagents:

(i) Crystal violet (0.5% aq., w/v)
Dissolve in distilled H_2O
Filter through Whatman No.1 filter paper
(ii) Lugol's iodine

Dissolve 2g of potassium iodide in 25ml distilled H_2O, add 1g iodine;
when dissolved make up to 100ml with water.
(iii) Safranin O (0.5% aq., w/v)

Method:
Young, actively growing cultures (usually 24-48 h old) should be used. Place a small drop of, preferably sterile, water on a clean microscope slide. Using a cooled sterile loop remove a part of a single, well-separated colony from the agar surface and smear the bacteria onto the slide. If liquid cultures are to be observed then a loopful of the bacterial suspension will suffice. The smear should be approximately 1 cm^2 and just visible. Air dry the slide then "heat-fix" by passing the slide, organism up, several times through the flame of a Bunsen burner. DO NOT OVERHEAT THE SLIDE. Fixing ensures the bacteria will adhere to the slide surface during the subsequent staining procedure. Flood the slide with crystal violet and leave for 30 s. Pour off excess fluid and wash with water. Flood the slide with Lugol's iodine and leave for 30 s, wash with ethanol until the colour ceases to come away from the smear (a few seconds is enough), wash with water, blot dry and counterstain with Safranin O for about 1-2 min, wash again with water and dry. Examine the slide under 100X with oil immersion.

Note: Gram-positive cells appear as dark purplish, whilst Gram-negative are red.

KOH Solubility Test (Gram stain confirmation)

A simple alternative to the Gram stain is to mix a loopful of bacteria into a solution of KOH on a glass slide. If the cell wall is broken the DNA will be liberated and a slimy string will be visible when an inoculation loop is lifted from the suspension, which is characteristic of Gram-negative bacteria. If no string is produced and the KOH remains watery, it is assumed that the cell walls have not been destroyed, no DNA has been released, and the bacteria are Gram positive. No additional information can be obtained about the bacterial isolate from this test and a slide must be made if cell morphology is required.

Reagent: Potassium hydroxide (3% aq., w/v)

Using a clean Pasteur pipette place a drop of KOH onto a clean microscope slide. Then with a cooled sterile loop remove a part of a single, well-separated colony from a young actively growing agar culture. Mix the bacterial culture into the KOH until an even suspension is obtained. Lift the loop from the slide. If a string of slime is lifted with the loop (*ca.* 5-20 mm in length), the bacterium is Gram negative. The destruction of the cell wall of Gram-negative organisms and subsequent liberation of DNA, which is very viscid in water, produces the string of slime. If a watery suspension is produced and no string of slime observed, the culture is Gram positive. The Gram-positive wall is more resistant to KOH and remains intact, thus no DNA is released. This test is useful in cases of doubtful stain results.

Note: After the test remember that the microscope slide is contaminated with live bacteria and should be disposed of accordingly.

Biochemical and Physiological Testing of Bacteria

The morphological characters of bacteria are of very limited value in their identification. Colony size, rate of growth, colours, texture, opacity etc. are simply not enough to identify a bacterium. In many cases not all of these characters can be used, and in no case are they sufficient for a definitive identification. It is therefore necessary to identify bacteria by other means. Historically a series of well-chosen tests were conducted, based on the presence or absence of certain enzymes. Such tests were selected, set up, used and read with some degree of care and understanding, and the meaning of the results required careful analysis. Positive and negative controls are usually necessary and, in some cases, controls required some modification. It is possible that failure to grow on a given medium may result through vitamin deficiency and not inability to utilize the substrate.

Nowadays the use of individual media-based tests is rare in our labs, and only encountered at CABI if we are investigating specific properties for a project or investigation. Nutrient assimilation and sensitivity tests can be bought in from a number of commercial producers who have miniaturized the testing process; producing multi-well strips and plates which replace the long list of media-based tests previously carried out in diagnostic labs. The desiccated test strips are inoculated with a standardized bacterial suspension of the test organism, and results are recorded and matched to the results of species held in the database. There are different options available for automating the process, or keeping costs down by manually completing the input of results. Examples of the tests that we have used at CABI include the Biolog Phenotype MicroArray and Biomerieux API strips.

Factors Affecting Growth of Bacteria

Some of the factors affecting bacterial growth are controlled primarily by the constituents of the culture medium. These include hydrogen ion activity (pH), water activity, osmotic pressure and viscosity. Other factors are controlled by the external environment and include temperature, oxygen, light and hydrostatic pressure.

The reduction-oxidation potential is controlled by both the medium and the environment. All of these factors can influence the growth rate, cell yield, metabolic pattern, and chemical composition of bacteria. The control of hydrogen ion activity, temperature, and oxygen supply is important with every bacterial culture and in some cases critical. The control of reduction-oxidation potential is of major importance in culturing obligately anaerobic bacteria. Further information can be found in Gerhardt *et al.* (1994) *Methods for General and Molecular Bacteriology*.

Media

Due to their extreme diversity, no one culture medium is perfect for all bacterial species. However, many laboratories use Nutrient Agar (NA) or similar, as it supports the growth of the majority of bacteria, even if not optimally. Preferences for particular media are often developed over many years and evolve as a result of experience.

Most plant pathogenic bacteria prefer a neutral or slightly alkaline pH and will grow sufficiently well on NA. A small number of organisms do not grow on Nutrient Agar; or make unsatisfactory growth, e.g.,

- sugar cane pathogens, *Xanthomonas albilineans*, *Xanthomonas populi*,
- some *Clavibacter* spp.,
- *Erwinia* spp. some of the *E. amylovora* group.

For these organisms one or more of the following media should be tried: GP, YDC, SNA or KB. Media formulae for growth media for bacteria are given in Appendix B, this volume.

Specific growth conditions for some bacteria

Many bacteria can be cultivated in the laboratory using general growth conditions (aerobically at room temperature). However, some bacteria inhabit extreme environments, and therefore have very specific growth requirements which must be fulfilled to establish growth. A few examples are:

- Autotrophs that may not tolerate organic compounds and require low pH, e.g. *Nitrobacter*, *Thiobacillus*
- Strict thermophiles that require temperatures higher than 55°C, e.g. *Alicyclobacillus acidocaldarius*
- Obligate anaerobes that do not tolerate the presence of any oxygen, e.g. *Clostridium*, *Butyrivibrio* and *Bacteriodes*
- Parasitic organisms that need suitable hosts to grow on, e.g. *Bdellovibrio* bacteriophage
- Halophiles and marine organisms that grow only in media containing high salt concentrations
- Recombinant organisms that grow in media formulated to maintain inserted characteristics
- Nutritionally fastidious organisms that require a number of nutrients such as vitamins, blood components, or specific organic compounds, e.g. *Lactobacillus* spp.

Gerhardt *et al.* (1994) provides extensive insight in the choice of media.

pH

Most known bacteria grow over a relatively narrow pH range (usually near neutrality at pH 7.0). However, an ever-increasing number of extremophiles continue to be recognized and isolated from the environment (Schlegel and Jannasch, 1992). Many grow optimally at a very low (acidophiles) or very high pH (alkaphiles). The isolation source often indicates the preferred pH of the target organism, which should be adjusted accordingly.

Water activity

All organisms require water for metabolism and growth, but the amount required varies widely. However, the mere presence of water in a medium does not ensure its availability, which is determined by the water activity (a_w) of the medium. The majority of bacteria need high levels of available water, but some specialized bacteria have specific requirements. For example, halophiles grow optimally on media containing high concentrations of sodium chloride. Variations in the levels of available water may affect growth rates, cell composition and metabolic activities (Gerhardt *et al.* 1994).

Temperature

The incubation temperature dramatically affects the growth rate of bacteria, and affects the rates of all cellular reactions. Although all bacteria have an optimal temperature at which they function best, their growth range is often quite wide, e.g., marine bacteria isolated at 10°C can often grow in the range 5 - 30°C and clinical organisms isolated at 37°C can often grow at or below 20°C. Most of the environmental organisms grown in our labs are incubated at 20 - 28°C, although there are some exceptions. Generally, bacteria not only grow more slowly but die rapidly at temperatures markedly above the optimum for growth, and cultures should be incubated at a temperature below their optimum to an extent determined by the variability of the incubator. Organisms that prefer to grow at low temperatures (<20°C) are called psychrophiles, those that grow at ambient (20 - 37°C), mesophiles and those that prefer higher growth temperatures (≥45°C) are thermophiles.

Metabolism and growth rates of bacteria are often dependent on the concentration of gas in solution. Among the most prevalent concerns in this regard are aerobic and facultative bacteria, where growth rate and yield may depend critically on the concentration of oxygen in solution. There are a number of ways in which the availability of oxygen can be maximized, e.g. ensuring that culture vessels have wide openings, or by forcing air through liquid culture medium. See Gerhardt *et al.* (1994) *Methods for General and Molecular Bacteriology* for more details.

Although not encountered in CABI's DAS work, some bacteria can also grow in the absence of oxygen, a condition known as anaerobiosis. Anaerobic organisms are defined as bacteria that are unable to grow in the presence of oxygen, and fall into two classes:

- Non-stringent anaerobes – these are able to grow on the surface of agar plates with low but significant levels of oxygen in the atmosphere (Gordon *et al.*, 1953).
- Stringent anaerobes – these die, or are inhibited almost immediately on exposure to an environment containing oxygen.

In general, cultures to be preserved should be grown under optimum conditions into late log or early stationary phase. Bacterial spores survive preservation conditions well, and so wherever possible spore producing organisms should be grown on media that will elicit sporulation (media usually low in nutrients so as to initiate the spore forming survival mechanism of some bacteria, e.g., *Clostridium* and *Bacillus*).

Sub-culturing

One of the oldest and most traditional methods for maintaining bacterial cultures is by continuous sub-culturing. Organisms should be grown on their optimum medium, and whilst some species require transfer after days or weeks, others may be transferred after several months or years. The advantages and disadvantages of this form of preservation are outlined in Table 2.1 below.

Culturomics

The European Marine Biological Research Infrastructure Cluster (EMBRIC - https://www.embrc.eu/collaborative-projects/embric-european-marine-biological-research-infrastructure-cluster) project points out that cultivation of candidate organisms for screening is often problematic, with the yet uncultured and slow growing strains giving low biomass production and high variability even within the same species. Research Infrastructure partners follow harmonized best practices to minimize variation and engage culturomics to design improved culture conditions and coordinate efforts (Brennecke *et al.*, 2018). Culturomics is the exhaustive application of culture media and growth conditions for the maximum recovery of cultivable microorganisms from a biological sample (Lagier *et al.*, 2012). Smith *et al.* (2019) discuss the difficulty in recovering organisms from the marine environment where there is huge potential to find organisms new to science with interesting and useful properties. During the project over 250 difficult to grow organisms were recovered from the Pacific and screened using the culturomic approach. Dedicated isolation protocols are required, e.g. selective media: to date, less than *c.* 1% of all microorganisms can be cultivated (Huang *et al.*, 2012; Overmann *et al.*, 2017).

Growth and Examination of Yeasts

Yeasts are economically important organisms, especially in the context of brewing and baking, and are also associated with the spoilage of food and beverages. Over the years CABI has regularly worked with isolates originating from environmental habitats or food products, but we have rarely received strains via the DAS lab. Yeasts are not likely to cause plant disease, and the majority of strains have been secondary organisms, contaminants or isolated from fruits displaying post-harvest problems.

Media

The majority of the yeast isolates cultivated in our labs are able to grow adequately on our standard fungal culture media, and don't have complex or specialized requirements. Often yeast colonies appear unexpectedly on isolation plates from plant material, and may sometimes be mistaken for bacterial colonies. Confirmation of a yeast isolate would be made by making a slide and viewing it using a compound microscope. If required, identification would involve MALDI-ToF or sequencing analysis. We no longer complete any media-based tests, BIOLOG plates or API strips, but these have all been used previously, and we would consider a combination of these options if molecular methods were not available to us.

Sub-culturing

New cultures can be propagated simply by transferring a small quantity of the culture to fresh medium. At CABI we treat yeasts in a similar

Table 2.1. Advantages and disadvantages of the frequent transfer method to maintain collections of bacterial isolates.

Advantages	Disadvantages
• Ideal for small collections • Organism viability maintained for many years • Inexpensive method in terms of capital investment • No specialized equipment required • Quick and easy retrieval of isolates	• Risk of loss, reduction or intensification of an organism's characteristics • Risk of mislabelling cultures with incorrect details • Risk of contamination with air-borne spores or mites • Requirement for constant specialist supervision • Risk of contamination • Requirement for regular maintenance • Risk of failure to recover cultures • Labour intensive for large collections

manner to bacteria, and perform a quadrant streak onto appropriate media; allowing single colonies and contaminants to be instantly recognized. As with other microorganisms, it is not always convenient to maintain cultures by frequent serial sub-culture, especially when a large number of strains are being handled. However, sub-culturing continues to be useful, particularly in the short term. It is simple, quick and relatively inexpensive. A major drawback of this technique is that substantial variation may occur. One investigation found that 10% of strains showed changes in flocculation behaviour following maintenance by sub-culturing over a period of 10 years; other morphological and physiological properties have been found to show variation to a greater or lesser extent (Kirsop, 1974). If strain stability is desired, sub-culturing should be minimized, and alternative preservation methods; such as freeze drying or liquid nitrogen used wherever possible.

Temperature

The optimum temperature for growing a yeast varies with the strain being investigated, although most of the yeasts that we encounter grow well at 20°C to 30°C. The optimal temperature for growth is usually nearer to the upper limiting temperature, but care should be taken not to exceed it – and fluctuations due to incubators etc. need to be considered. Information about the requirements of different species is readily available in yeast taxonomy guides (Yarrow, 1999; Kurtzman *et al.*, 2011) or online e.g., National Collection of Yeast Cultures (NCYC, https://www.ncyc.co.uk/). It is interesting to note that only a handful of strains are cited as being capable of growth at temperatures above 45°C.

Aeration

We usually grow our yeast cultures on solid culture media, although studies have shown that yeast communities grow more vigorously when aerated. The large-scale industrial production of yeasts meant that the effects of aeration were widely studied. Agitation of the yeast cells can also increase yield, and in a liquid culture, occurs alongside any aeration that is taking place. *Cryptococcus*, *Rhodotorula* and *Sporobolomyces* are often grown in shake culture (Kirsop and Doyle, 1991).

pH

As with temperature, there is evidence that optimal conditions with respect to pH may vary somewhat with the strain of yeast being studied. Slightly acid conditions favour the growth of yeasts, but varying the pH over a relatively wide range does not adversely affect growth in nutrient rich media. Usually the range over which most active growth occurs is between pH 4.0 and 6.0. However, yeasts are surprisingly acid-resistant and many strains are able to tolerate, if not grow in, medium in which the pH is as low as 2.5.

Light

The effect of light upon the formation of spores by yeasts has received little attention, even though light markedly influences the sporulation process in other fungi. Different dark/light regimes can have an effect on sporulation in different strains of yeast. Investigations have shown that ultraviolet light destroys yeasts in very short time even if they seem somewhat more resistant than bacteria. The detrimental effects of UV light, particularly on DNA, has been extensively investigated (Bainbridge, 1981; Rastogi *et al.*, 2010).

Sporulation

Many yeasts produce spores, and sporulation is of fundamental importance, not only for taxonomic purposes, but also for genetic analysis. Consequently, the ability of yeasts to sporulate makes possible the controlled hybridization of new strains and leads to a better understanding of the biochemistry of parental strains. Obviously, it is often desirable to be able to induce sporulation more or less at will, but this is not always feasible, because many yeasts tend to lose their ability to form spores after prolonged cultivation in artificial media. Near-starvation and other unfavourable conditions were thought to induce sporulation; however, such circumstances might lead to death. More probably there exists a balance between two sets of conditions, one set favouring vegetative reproduction and the other sporulation. A change in the environment may favour one phase of the life cycle to the detriment of the other.

Yeasts sporulate more readily when growth in a nutrient-rich medium is followed by a transfer to an environment containing little or no nutrient. There is a wide range of pre-sporulation media that have been devised by many workers, a fairly comprehensive coverage of media for inducing different spores in yeasts e.g., ascospores, ballistospores, basidiospores and chlamydospores is given by Yarrow (1999). Wide differences of opinion exist concerning the optimal physical conditions required to stimulate spore formation. There are many other aspects, including moisture, pH, temperature and light, to consider. Bearing these factors in mind along with the huge diversity of yeast strains, it is obvious that optimal conditions will differ with regards to the induction of sporulation. For more information concerning sporulation-inducing media see Yarrow (1999).

Identification methods

Traditional identification methods for yeasts involved an extensive range of media-based tests which were used to evaluate an organism's ability to utilize specific compounds, including individual carbohydrate or nitrogen sources.

Some of the earliest microbiological studies revealed that alcoholic beverages could be produced from fruit-based products through the action of yeasts upon relatively simple carbohydrates. Investigations established that two processes are involved in these metabolic activities:

- The breakdown of sugars to yield alcohol and carbon dioxide.
- Growth or the incorporation of some of the carbon of the sugars into the cell substance itself.

This discovery resulted in an extensive testing regime, devised to distinguish unknown strains by their ability to utilize or ferment organic compounds as the sole source of carbon. Additional tests included the utilization of sole sources of nitrogen, requirement for different vitamins, urea hydrolysis, and growth at high osmotic pressure. Yeast laboratory manuals, and web-based identification tools still provide detailed information about the expected results for each of the tests for any given organism. However, this information is now supplemented with sequencing analysis results, and includes CBS Type Strain sequence information.

The yeast isolates identified at CABI are processed using either MALDI-ToF, for a quick preliminary identification, or rDNA sequencing analysis of the ITS or Large Subunit (LSU) regions. Sequences are checked against the FASTA algorithm from EBI and BLAST algorithm from NCBI, as for fungi, but can also be compared to the sequences held in the CBS online yeast identification tool (https://wi.knaw.nl/page/Identification)

References

Aitchison, E.M. and Hawksworth, D.L. (1993) IMI: Retrospect and Prospect. CABI, Wallingford, UK. https://www.pemberleybooks.com/product/imi-retrospect-and-prospect/27792/

Bainbridge, B.W. (1981) The effects of ultraviolet light radiation on the nuclei of yeasts and filamentous fungi. In: Gull, K. and Oliver, S.G. (eds) *The Fungal Nucleus*. pp. 258. Cambridge University Press, Cambridge, UK. https://doi.org/10.1016/0307-4412(82)90194-7

Blight, D. and Ibbotson, R. (2011) CABI a Century of Scientific Endeavour. CAB International, Wallingford, UK, ISBN 978-1-84593-873-4, p. 171. https://www.cabi.org/wp-content/uploads/100-Year-Book.pdf

Brennecke, P., Ferrante, M.I., Johnston, I.A. and Smith, D. (2018) A Collaborative European Approach to Accelerating Translational Marine Science. *Journal of Marine Science and Engineering* 6, 81. https://www.mdpi.com/2077-1312/6/3/81

Butler, E.J. (1921) The Imperial Bureau of Mycology. *Transactions of the British Mycological Society* 7, 168-172.

Chu, D. (1970) Forest pathology, storing of agar slants and cultures. *Bi-monthly Research Notes* 26, 48. https://cfs.nrcan.gc.ca/publications?id=28175

Clark, C. and Dick, M.W. (1974) Long term storage and viability of aquatic oomycetes. *Transactions of the British Mycological Society* 63, 611-612. https://doi.org/10.1016/S0007-1536(74)80116-6

Gerhardt, P., Murray, R.G.E., Wood, W.A. and Kreig, N.R. (1994) Methods for general and molecular bacteriology. pp 803. American Society for Microbiology, Washington DC. https://www.worldcat.org/title/methods-for-general-and-molecular-bacteriology/oclc/29031551

Ghemawat, M.S. (1977) Polychromatic staining with toluidine blue O for studying the host-parasite relationships in wheat leaves of *Erysiphe graminis* f. sp. *triciti*. *Physiological Plant Pathology* 11, 251-253. https://www.semanticscholar.org/paper/Polychromatic-staining-with-toluidine-blue-O-for-in-Ghemawat/0f79de05a139e242cdaf9955d677d6c0f1b67247

Goldie-Smith, E.K. (1956) Maintenance of stock cultures of aquatic fungi. *Journal of Elisha Mitchell Scientific Society* 72, 158-166.

Gordon, J., Holman, R.A. and McLeod, J.W. (1953) Further observations on production of hydrogen peroxide by anaerobic bacteria. *Journal of Pathological Bacteriology* 66, 527-537. https://doi.org/10.1002/path.1700660224

Huang, P.M., Li, Y. and Sumner, M.E. (eds) (2012) Handbook of soil sciences: properties and processes, 2nd edn. CRC Press, London. https://doi.org/10.1201/b11267

Kirsop, B. (1974) The stability of biochemical, morphological and brewing properties of yeast cultures maintained by sub-culturing and freeze-drying. *Journal of the Institute of Brewing* 80, 565-570 https://onlinelibrary.wiley.com/doi/abs/10.1002/j.2050-0416.1974.tb03650.x

Kirsop, B.E. and Doyle, A. (eds) (1991) *Maintenance of Micro-organisms and Cultured Cells: A Manual of Laboratory Methods*. Academic Press, London. https://wellcomecollection.org/works/vum2qdua

Kurtzman, C.P. and Fell, J.W. (eds.) (2011) The Yeasts: a taxonomic study, 5th edn. Elsevier, Amsterdam. eBook ISBN: 9780080931272. https://www.elsevier.com/books/the-yeasts/kurtzman/978-0-444-52149-1

Lagier, J.C., Armougom, F., Million, M., Hugon, P., Pagnier, I., *et al.* (2012) Microbial culturomics: Paradigm shift in the human gut microbiome study. *Clinical Microbiology and Infection* 18, 1185–1193. https://doi.org/10.1111/1469-0691.12023

Overmann, J., Abt, B. and Sikorski, J. (2017) Present and Future of Culturing Bacteria. *Annual Review of Microbiology* 71, 711–730. https://www.annualreviews.org/doi/abs/10.1146/annurev-micro-090816-093449

Pitt, J.I. (1973) An appraisal of identification methods for Penicillium species: Novel taxonomic criteria based on temperature and water relations. *Mycologia* 65, 1135-1157. https://doi.org/10.1080/00275514.1973.12019533

Pitt, J.I. (1980) *The Genus* Penicillium *and its Teleomorphic State*: Eupenillium *and* Talaromyces. Academic Press, London and New York. https://www.worldcat.org/title/genus-penicillium-and-its-teleomorphic-states-eupenicillium-and-talaromyces/oclc/6387857

Raper, K.B. and Thom, C. (1949) *A Manual of the Penicillia*. Williams and Wilkins, Baltimore. https://archive.org/details/manualofpenicill00rape/page/n7/mode/2up

Rastogi, R.P., Richa, Kumar, A., Tyagi, M.B. and Sinha, R.P. (2010) Molecular mechanisms of ultraviolet radiation-induced DNA damage and repair. *Journal of Nucleic Acids*, 2010, *Article ID* 592980. https://doi.org/10.4061/2010/592980

Schlegel, H.G. and Jannasch, H.W. (1992) Prokaryotes and their habitats. In: Balows, A., Tripper, H.G., Dworkin, M. *et al. The Prokaryotes*. Vol. I, pp. 75–125. Springer Verlag., New York, USA. https://link.springer.com/book/10.1007/978-1-4757-2191-1

Smith, D. (2002) Culturing, preservation and maintenance of fungi. In: Waller, J.M., Lenné, J.M. and Waller, S.J. *The Plant Pathologists Pocket Book*, 3rd edn. CAB International, Wallingford, UK. https://www.cabi.org/bookshop/book/9780851994581/

Smith, D. and Onions A.H.S. (1994) *The Preservation and Maintenance of Living Fungi*. 2nd edn. *IMI Technical Handbooks* No. 2, pp 122. CAB International, Wallingford, UK. https://www.cabi.org/bookshop/book/9780851989020/

Smith, D., Buddie, A.G., Goss, R.J.M., Overmann, J., Lepleux, C., Brönstrup, M., Kloareg, B., Meiners, T., Brennecke, P., Ianora, A., Bouget, F.-Y., Gribbon, P. and Pina, M. (2019) Discovery pipelines for marine resources: an ocean of opportunity for biotechnology?. *World Journal of Microbiology and Biotechnology* 35, 107. https://doi.org/10.1007/s11274-019-2685-y

Snyder, W.C. and Hansen, H.N. (1946) Control of culture mites by cigarette paper barriers. *Mycologia* 63, 455-462. https://doi.org/10.1080/00275514.1946.12024069

Taylor, P. (2019) Plantwise Diagnostic Field Guide: A tool to diagnose crop problems and make recommendations for their management. CABI, Wallingford. https://www.plantwise.org/wp-content/uploads/sites/9/2019/11/Diagnostic-Field-Guide.pdf

Waller, J.M., Lenné, J.M. and Waller, S.J. (2002) *The Plant Pathologists Pocket Book*. 3rd edn. CAB International, Wallingford, UK.

Waller, J.M., Ritchie, B.J. and Holderness, M. (1998) *Plant Clinic Handbook*. IMI Technical Handbooks, No. 3. CAB International, Wallingford, UK. https://biblio.co.uk/9780851989181

Webster, J. and Davey, R.A. (1976) Simple method for maintaining cultures of *Blastocladiella emersonii*. *Transactions of the British Mycological Society* 67, 543-544.

Yarrow, D. (1999) Methods for the isolation, maintenance and identification of yeasts. In: Kurtzman, C.P. and Fell, J.W. (eds) *The Yeasts: a Taxonomic Study*, 4th edn. pp 77-100. Elsevier, Amsterdam. eBook ISBN: 9780080542690. https://www.elsevier.com/books/the-yeasts-a-taxonomic-study/kurtzman/978-0-444-81312-1

3

Identification, Characterization and Use of Microorganisms

Introduction

It has been estimated that whilst microorganisms make up around 15% of the earth's biomass (Bar-On *et al.*, 2018), because of their tiny size (particularly the prokaryotes) relative to plant and animal cells means that the microbes are in the majority (Nature Editorial, 2011; Young, 2016). Microbes can be utilized by humankind for a vast range of applications, hence it is important to properly identify and characterize them. The various discriminatory methods and characterizing tests described in this chapter will be useful in this context. The approaches described range from the simplest morphological observations, through biochemical and physiological tests, to recent massive parallel nucleic acid sequencing advances. Such methods make use of key attributes of related organisms, whilst exploiting differences between them. We also need to differentiate between 'identification' and 'characterization': identification is the process of putting a name to an organism whilst characterization is obtaining some characteristics of the organism. These processes may overlap somewhat but characterization tends to be used more broadly to gain additional information such as close relationships between organisms or to demonstrate particular traits that may be exploited in some way.

Characterizing microbes facilitates their selection for potential uses, and is also useful from a preservation perspective where stability during storage can be monitored. Increased demands for exploitable information and strain data have created a need for easily accessible and up-to-date profiles of microorganisms, including their most basic characteristics, in order to facilitate access to the most appropriate resource for experimentation and use (Smith *et al.*, 2019). This chapter provides the reader with a series of methods that will help them to: 1) identify their organisms; and 2) characterize their strains to provide data to enable their selection and use for particular studies and purposes. Further useful information may be found in Appendix A, this volume, which provides lists of example strains that produce specific enzymes, metabolites or antibiotics; lists of food and beverage strains, mycoparasites, biological control agents, nitrogen fixers; and strains isolated from extreme environments with specific tolerances, resistance or sensitivity are provided. Also included are examples of strains with particular suitability for, amongst others, biodegradation, bioremediation, food spoilage and vaccine production.

Screening organisms – or extracts derived from them – may help in the detection of useful products (metabolites, vitamins, enzymes) and open the way to many other potential uses.

There are many methods available for screening anatomical, physiological and biochemical properties. Techniques have been derived from a wide range of different biological disciplines

DOI: 10.1079/9781800622135.0003

including, but not limited to, biochemistry, bacteriology, mycology and ecology. The methods presented herein provide mainly diagnostic characters, but they may be extended to provide broader application in purification, identification and screening of properties of industrial importance. Many of the methods have been tested extensively at CABI and other national and international culture collections, and are generally applicable to a wide range of microorganisms. The procedures given can be roughly divided into two groups, those that determine a single activity (e.g. urease activity or citric acid production) and those that can be used to investigate multiple activities (e.g. chromatography to detect a range of metabolites). Many tests are commercially available in easy to use formats (e.g. APIZYM strips) that allow standardization between laboratories. Advances in molecular biology have resulted in a further increase in the variety of methods available for investigating the uses and properties of microbes. These techniques can also be utilized in taxonomic and evolutionary studies, industrial screening and strain improvement. Bridge *et al.* (2021) provides a broader coverage of molecular technologies and includes the use of MALDI-ToF-MS (Matrix-assisted laser desorption/ionization time-of-flight mass spectrometry) in *Trends in the Systematics of Bacteria and Fungi.*

Every laboratory requires a basic level of equipment, e.g. autoclave, balance, incubators, etc. Specialized procedures (e.g. for advanced biochemical and molecular analyses) may require specific and often expensive apparatus, depending on the precise methodology employed (and the research question to be answered).

It is essential that microorganisms are named correctly, in order that their taxonomic identity and resultant relationships with other organisms can be established.

Before embarking on a discussion of the ways to identify and characterize microorganisms, we need to define a few basic terms: classification; nomenclature; taxonomy.

- Classification is: 'The assigning of objects to a defined category' (Kirk *et al.*, 2008).
- Taxonomy is: 'The science of classification' (Kirk *et al.*, 2008).
- Nomenclature is: 'The allocation of scientific names to units which a systematist considers merit formal recognition' (Kirk *et al.*, 2008).

Type strains (bacteria) or *ex-type* strains (fungi) are representative examples of their species. They can be used for morphological, anatomical, physiological and genetic comparisons with newly isolated strains. The status given to a living culture of a fungus cultured from the holotype that is preserved in a collection is that of '*ex-type* species' (https://www.iapt-taxon.org/nomen/main.php). These are living cultures of fully described, scientifically validated and named cultures where, in the case of fungal strains, the holotype is deposited as a dried specimen or physiologically inactive strain in a reference collection (herbaria or, more correctly, fungaria) (Hawksworth *et al.*, 1995; Kirk *et al.*, 2008). Such strains are widely used by taxonomists and are becoming increasingly important as reference strains particularly for patent deposits and isolates that are released into the environment.

For culture collections and genetic resource collections, correct identification and characterization:

- Ensures the correct taxonomy to allow selection of the correct preservation procedure (e.g. anatomy, morphology);
- Allows the assessment of the suitability of the preservation regime applied and, if necessary, the need to modify and improve preservation procedure;
- Facilitates the monitoring of long-term stability of the organism;
- Provides information to help identify the scope of use.

The terms 'microbiology', 'microbe' and 'microorganism' are useful descriptors to cover the study and properties of organisms that are too small to be seen with the naked eye. However, before proceeding further in this chapter, we need to consider the key attributes of these organisms and how their similarities and/or differences have enabled generations of food producers, medical practitioners, industrialists and, increasingly, biotechnologists, to make use of this disparate 'group' of organisms.

Non-cellular organisms:

Non-cellular organisms/particles are beyond the remit of the current volume but it is worth noting that they do fall within the remit of the microbiologist. Such entities exist at the very limits of what we consider to be 'life'. Of these, the most common are viruses and prions (proteinaceous infectious particles; Prusiner, 1982), which are small particles of protein that cannot exist independently of living cells but, rather require cells for their continued proliferation.

Viruses are extremely small (20-1000 nm in length) and may only be visualized directly by electron microscopy (although their presence may be detected by a broad range of serological, immunological and molecular tests. The group of viruses that infect bacteria may be visualized indirectly by the cleared areas (plaques) that they leave on an agar plate covered by a (susceptible) bacterial lawn. Viruses generally consist of a nucleic acid core (which may be comprised of either DNA or RNA, depending on the actual type of virus) and a protein coat ('capsid') and, in some cases, this capsid is enclosed by a protein/lipid/carbohydrate envelope – although some 'naked' viruses may not require such a capsule or envelope. Viruses are characterized by their morphology, their nucleic acid type, their action and their host(s), existing purely to commandeer the host cell's mechanisms for making more and more viruses and, ultimately, enabling the virus to find a new host. Viruses will only be mentioned, in passing, in this chapter. For more details on viruses, see Louten (2016).

Prokaryotes:

Prokaryotic organisms are the organisms studied by bacteriologists. Whilst they do share some features that differentiate them both from the viruses, prokaryotes may be distinguished from the eukaryotes on a number of features (see Table 3.1). The Prokaryotes may be split into 2 major groups which are very distinct: 1) Archaea; and 2) Bacteria – which comprises most of the prokaryotic organisms held in culture collections.

Archaea:

The Archaea went largely unnoticed until the late 1970s (Woese and Fox, 1977). Prior to that, the few taxa known at that time (mainly methanogens) were considered to be very primitive

Table 3.1. Some examples of microscopical and general microbial stains (information reproduced from Singleton and Sainsbury, 1993)

Target	Stain (or procedure)
General purpose stains	Crystal violet
	Eosin
	Fuchsin
	Methylene blue
	Cotton blue in lactoglycerol
Protein	Amildo black 10B
	Coomassie brilliant blue
Nucleic acids	Acridines
	Chromomycin
	DAPI
	Methyl green-pyronin stain
Lipids	Nile blue A
	Sudan black B
Starch	Iodine
Actinomycetes and fungi in tissue	Methenamine-silver stain
Bacterial capsules	Capsule stain, e.g. aqueous nigrosin
Bacterial endospores	Ziehl-Neelsen strain (modified)
	Malachite green
	Toluidine blue
Bacterial Gram type	Gram stain (procedure with stain crystal violet, mordant Lugol's iodine, a decolouriser such as alcohol and a counter stain such as safranin)
Blood parasites (e.g. protozoan)	Romanowsky stain
	Leishman stain
Cellulose and fibrillar polysaccharides	Congo red
Detection of acid-fast strains	Ziehl-Neelsen stain
	Auramine rhodamine stain
Detection of flagella	Leifson flagella stain
Metachromatic granules	Albert's stain
	Laybourn's stain

bacteria – hence their initial suggested name: 'Archaebacteria'. However, with the advent of broader sequencing technology, more taxa were discovered for this group and phylogenetic analyses showed that they were sufficiently distinct from the rest of the 'true bacteria' – then renamed 'Eubacteria' but now known simply as 'Bacteria' (qv) – to warrant 'domain' status (Woese *et al.*, 1990), on a par with the Bacteria and Eukaryota. Whilst Archaea share some characteristics with the Bacteria, they display key differences. For the purposes of this chapter (and book), we will not be considering the Archaea further.

Bacteria:
Bacteria are (generally) simple unicellular organisms that conform to one of five main morphotypes: rod (bacilli); round (cocci); comma (vibrios); corkscrew (spirochaetes) and spiral (spirilla) (see: https://microbiologysociety.org/why-microbiology-matters/what-is-microbiology/bacteria.html) but with other less common morphologies (such as the star-shaped genus, *Stella*) also existing. Depending on the taxon, they can exist as one, or more, of the following: individuals, pairs, tetrads, chains, clusters; some taxa (i.e. the Actinomycetes) can become filamentous, producing networks of hyphae almost like Fungi. For many years the most fundamental division of the Bacteria was determined by the composition of their cell wall, as could be demonstrated using the Gram stain (qv). Subsequent broader phylogenetic analyses resulted in a reappraisal of its continued utility except for medical purposes. Bacteria are ultra-adaptable, being able to exist in every natural environment (including being able to colonize all living eukaryotes) on earth and even being able to exploit non-natural substrates. They are able to resist desiccation with some groups able to produce spores and have even been shown to survive exposure to the near-vacuum of space for 3 years (Kawaguchi *et al.*, 2020).

Eukaryota:
For the purposes of this book, we limit ourselves to the eukaryotic microorganisms: predominantly the Fungi, but also make reference to algae, Protista and other taxa.

Fungi:
Kingdom Fungi contains the 'true' Fungi which currently comprises nine phyla of diverse heterotrophic eukaryotes (Naranjo-Ortiz and Gabaldón, 2019) but which are considered to share several features including the following (see Adl *et al.*, 2012): i) chitin and β-glucan as cell wall components; ii) for the biosynthesis of lysine they have the amino adipidic pathway; iii) possessing flattened cristae in their mitochondria; and iv) existing as unicellular and/or multicellular mycelial forms (with many *Fungi* able to switch between these morphotypes). A distinction should be noted here between the Fungi and fungi where the former refers solely to members of Kingdom *Fungi* and the latter refers to filamentous eukaryotes or organisms historically studied by mycologists: including Kingdom *Fungi* but also members of the Protists (such as the *Oomycota*). See Adl *et al.* (2012) for more details.

Protists:
The Protists comprise a vast and particularly disparate (and polyphyletic) group of eukaryotic (micro)organisms that share only one major feature: they cannot be placed into the Kingdoms Animalia, Fungi or Plantae. Many are free-living unicellular organisms but some exist in filamentous form. Most of the Protists are poorly characterized but some are of considerable significance in water treatment (e.g. protozoans such as *Giardia lamblia*, *Cryptosporidium* spp., etc.), human health (e.g. *Entamoeba histolytica*, *Trypanosoma cruzi*, *Plasmodium* spp., etc.) and in plant pathology (e.g. the *Oomycota*, e.g. *Pythium* and *Phytophthora*, which were treated as Fungi for many years before being found to be closer to Kingdom Plantae than to the Fungi and reassigned accordingly). Aside from a very brief mention of those Protists that are considered to be 'algae', in the next paragraph, the present chapter (and book) will only consider the *Oomycota* from this 'super-Kingdom'.

Algae:
Algae are mostly photoautotrophs which get their energy primarily by photosynthesis. The term 'alga' has been applied to an assemblage of organisms that have been largely 'lumped together' due to their shared characteristics of being pigmented and existing in or near aquatic environments. At the smallest level, organisms previously termed 'blue-green algae' have long been recognized to be prokaryotic and, as such, were renamed 'cyanobacteria'. They share many of the environmental niches with the eukaryotic algae and can exist in forms from unicellular to filamentous. Cyanobacteria will not be considered further in this chapter or book but, for more details, please see Tomitani *et al.* (2006) and Moreira *et al.* (2013).

Eukaryotic algae mostly belong to the Protista, although green algae, belonging to the phylum Chlorophyta, belong to Kingdom Plantae.

Although they broadly share the aforementioned photoautotrophic life cycle and are mostly aquatic, algae differ in their colouring, photosynthetic pigments, cell arrangement, storage materials and cell wall composition. They will not be discussed further in this chapter or book but, for further details, please see: Bringloe *et al.* (2020) (brown algae); Yoon *et al.* (2006) (red algae); Medlin (2016) (diatoms); Gómez (2012) (dinoflagellates); Bicudo and Menezes (2016) (euglenazoa).

Identification of Bacteria and Fungi (including the *Oomycota* [Heterokonta])

A first step when working with bacteria and fungi is to ensure the material you are working with is correctly identified as giving a specimen a name will open up a wealth of information about it. When receiving materials from other researchers it is wise to confirm its identity anyway, and if isolating from nature, yourself, there are some key points to consider from the outset. There are many characteristics that may be used to identify the organism, including: morphology, metabolites and DNA sequence. Currently, there is a major shift from more traditional, phenotypic methods (e.g. morphology and basic biochemistry) to the use of molecular tools to decipher the most fundamental aspects of the organisms. However, at the time of writing, although there have been great advances in genome sequencing of the bacteria over the past three decades, there exists a considerable sequencing deficit for the eukaryotes with only a small fraction of the fungi described and fewer still sequenced; databases for comparison are far from complete so a polyphasic approach (i.e. the integration of data from a range of biological traits) is required currently for optimal identification. There are many culture collections and microbial service companies that can assist in this process. Similarly, to many other culture collections, CABI offers an identification service for fungi and bacteria.

Methods Typically Used for Bacterial and Fungal Identification

Morphology

For bacteria, this aspect is rather limited in utility for definitive identification – there are few variations on the aforementioned main morphotypes. There are notable differences in the way that the cells of some aggregate or align: e.g. *Streptococcus* spp. form (twisted) chains of cocci whilst *Staphylococcus* spp. form (grape-like) clusters of cocci. However, even to see this requires high power magnification light microscopy (qv). Some bacteria have distinctive colony structures, dimensions, shapes, texture, elevation or pigmentation that may be discerned with the naked eye or by using a low-magnification binocular microscope although this often may be affected by the components of the medium being used and the nutritional requirements of the particular bacterium (e.g. colonies of the human pathogen, *Streptococcus pyogenes* exhibit zones of clearing on Blood Agar due to haemolysis). Depending on the bacterium, colonies may vary in size from pin-point to confluent growth across the entire plate; they may be regular-edged round colonies, irregular edged or some degree of filamentous; elevation refers to the side-on view of the colony above the medium – shapes include flat, raised and umbonate; textures can vary from smooth, shiny or mucoid to wrinkled, crusty or dull; pigmentation can range from none (semi-translucent) through white, buff, yellow, pink and other colours.

Fungi have far greater variety in their anatomical form and structure: this may be seen in the many types of mycelium (hyphae), spores, origin of the spores and their inherent structure and morphology. Morphological analysis formed the basis of mycology right up until the advent of widespread biochemical and molecular testing in the latter half of the twentieth century. The same classes of traits (and many of the same terms) mentioned for observing bacteria may be used in observing and describing fungal colony morphology but with many additional aspects. For example, a bacterium single-point inoculated in the centre of a plate of nutrient agar (NA) will stop growing outwards after a relatively short time (a notable exception to this being the 'swarming' bacterium *Proteus*) whilst many fungi, inoculated in a similar manner on standard lab media such as malt extract agar (MEA) or potato dextrose agar (PDA) will continue to grow radially outwards until the entire plate surface is covered. It is the leading edge of the fungal colony that is growing most rapidly as it seeks out nutrients, whilst the centre of the colony is almost senescent, accumulating storage products and secondary metabolites. Eventually hyphae may produce a mycelial mat and/or start to produce spores.

An example of how to use published keys, is given Appendix I: Taxonomic Exercise in this volume. The book on which this is based, *The Coelomycetes* (Sutton, 1990), is now somewhat out of date but they are still useful to get to species names if a match cannot be found for its sequence in a public sequence database. The valid name for the species can then always be traced via MycoBank (https://www.mycobank.org/) and/or IndexFungorum (http://www.indexfungorum.org/names/names.asp) and its sister site SpeciesFungorum (http://www.speciesfungorum.org/Names/Names.asp). However, it should be noted that molecular tools have enabled the delimitation of previously cryptic taxa that could not be separated by morphological characters. As such, morphology-based keys are mostly obsolete and therefore should be approached with caution. Conversely, although mycology, and microbiology in general, is moving towards a computer-based, molecular science, there remains considerable utility in examining microorganisms at low- and high-magnification under the light microscope as this will instil a connection between the organism and its identity which a DNA sequence, on its own, cannot do. In addition, this approach may help in the discerning of incongruities, as errors in public sequence databases can still be found.

There are several methods of automated systems for identification described in this Chapter, see below, they include fatty acid methyl esterase gas chromatography (FAME-GC) and matrix-assisted laser desorption/ionization time-of-flight mass spectrometry (MALDI-ToF-MS) where profiles are compared and matched to give an identification and molecular techniques such as the sequencing of PCR-amplified part of ribosomal RNA (rRNA) genes and their spacers, using universal primers or whole genome sequencing.

Microscopy and Morphometrics

Many anatomical and morphological properties may be detected using microscopy (Lillie *et al.*, 1977). These include:

- Detection of specific biochemicals/metabolites and macrostructures using staining techniques;
- Visualization of cells and cell organelles;
- Identification of specific taxa, using labelled oligonucleotides of organism specific stains;
- Assessments of viability.

Features of Various Microscopy Techniques

A good quality compound light microscope remains an essential item in every microbiology laboratory. With the availability of relevant filters and/or condensers and appropriate stains, it is also a versatile tool. Very basic light microscopes can be sufficient for simple checks. However, more expensive microscopes are necessary for critical applications, where additional accessories/adaptations (such as those that aim to improve viewing and allow the researcher to adapt their microscope to perform specific research activities) are required. There are several adaptations to the standard microscope that aim to improve viewing and some of these are mentioned below.

Bright field microscopy

Bright field microscopy remains the most commonly used approach, as samples are easy to prepare and a good microscope, limited to this purpose, is not expensive. However, the technique is restricted through poor resolution, requiring oil immersion lens for higher level magnification (i.e. x100 objective magnification). Some specimens are translucent, but can be viewed after staining (see Table 3.1). The major disadvantage of using stains for visualizing cells is that many stains are cytotoxic or cytostatic but sometimes this is the only way to check an organism's morphology – and it may be characteristic. The most well-known use of stain in microbiology remains the Gram stain test, which for many years was a key tool in the identification of bacteria. Molecular tools (and a reappraisal of many taxa, as a result) have reduced the breadth of its application but it remains a useful first step in many clinical and environmental microbiology laboratories. The stain, first described by the Danish bacteriologist, H.C. Gram, in the late 19th century (see Bartholemew and Mittwer, 1952), makes use of a fundamental difference in cell wall structure in bacteria. The modified technique in use today involves creating a heat-fixed smear of bacteria on a (grease-free) microscope slide, then using crystal violet as a primary stain which is then washed off with iodine solution as a mordant or fixative before being itself washed off with decolourising alcohol or acetone and then finally counterstained with safranin or dilute fuchsin solutions. Bacteria which retain the purple crystal violet stain are termed Gram positive, whilst those that have been decolourised and then take up the red counterstain, are termed Gram negative.

In fungi, the most widespread use of bright field microscopy is with stains such as cotton blue in lactoglycerol, which enables both vegetative and reproductive structures to be easily discerned, in many taxa.

Dark field microscopy

Dark field microscopy is used to examine specimens that are too small to view under standard bright field microscopy or are unstained or still alive. In this technique, directly transmitted light is absent, as it requires the standard condenser to be replaced by a dark field one that contains an opaque disc, resulting in the specimen being viewed on a dark background.

Phase contrast microscopy

Phase contrast microscopy is used to view specimens that do not absorb light very well. It is often useful for distinguishing between viable and non-viable cells. Specimens do not have to be stained prior to examination. It is also particularly good for observing internal structures of the cell.

Differential Interference microscopy

Differential Interference microscopy (Nomarski) produces clearer images (with greater resolution) than the phase contrast technique and the thickness of specimens can be elucidated through colour differences within the specimen.

Confocal microscopy

Confocal microscopy is an optical technique that offers excellent resolution and is used to allow the visualization of living cells. Confocal laser scanning microscopy has allowed visualization of the structural and biochemical components of cells in 'real time' and allows scientists to build three-dimensional (3D) pictures of their specimens and their components. The technique has proved particularly useful in the analysis of microbial biofilms where enabling 3D sectioning can help discern the component layers of cells in these important microbial aggregations.

Cryomicroscopy

Cryomicroscopy allows direct visualization, by light microscopy, of the effects of freezing on cells and has been used for optimizing preservation protocols for fungi (Morris *et al.*, 1988; Smith and Thomas, 1998). In addition, it has also been employed with other eukaryotic microbes and cells.

Fluorescence microscopy

Fluorescence microscopy is a powerful technique that selectively stains for cellular components or whole microorganisms. Specimens are treated with a fluorochrome and then viewed on a microscope equipped with a UV light facility.

Some of the techniques used in fluorescence microscopy include:

Immunofluorescence microscopy

Immunofluorescence microscopy is a very powerful technique. Antibodies specific to a cellular component or biochemical are ligated with a fluorescent stain such as fluorescein. Antigens homologous to the fluorescent antibodies are readily identified by regions of fluorescence within the specimen (Singleton and Sainsbury, 1993). A specific application of this is the Fluorescent Antibody Technique (FAT), which is an immunofluorescent protocol used for detecting specific microorganisms in liquid and environmental samples.

Fluorescence in situ hybridization

Fluorescence *in situ* hybridization (FISH) is another technique which utilizes fluorescence microscopy—in this case, to observe *in situ* hybridization to gene probes (Trask, 1991). Viability assessment – Use of fluorescein diacetate can help differentiate viable and non-living cells in a population. Living cells cleave the fluorescein diacetate molecule releasing fluorescent 'fluorescein'.

Examples

Examples of fluorochromes include acridine orange (which distinguishes between single and double stranded nucleic acids), 4', 6' -diamidino-2-phenylindoledihydrochloride (DAPI; for DNA), rhodamine 123 (viable mitochondria), auramine rhodamine (acid fast strains) or fluorescein isothiocyanate (FITC; for proteins). Application of green fluorescent protein (GFP) revolutionized biology, specifically when it started to be used to look at the organizational structure of cells through confocal microscopy. More recently, the development of the LIVE/DEAD™ assay system (ThermoFisher Scientific) has enabled researchers to determine the proportion of viable vs. non-viable cells in a sample of bacteria, using fluorescence microscopy. The method exploits differences in the porosity of cell membranes to two dyes: the first is taken up by all cells, the second only by cells with compromised membranes (i.e. dead, or non-viable, cells). As the two dyes have different absorbance maxima, use of appropriate filters enables results to be noted and the relative proportions to be calculated (e.g. see Manteca *et al*., 2005).

Electron microscopy (EM)

Even the best light microscope will only enable magnification to a maximum of x2000. This is not sufficient for elucidating surface architecture, internal cell ultrastructure or to visualize most viruses. Electron microscopy (EM) has been particularly useful in taxonomic investigations and morphological analysis and characterization. This technology enables much greater levels of magnification (ca. x10,000-x100,000). However, it requires a huge financial outlay to purchase microscopes and, subsequently, is much costlier in maintenance and time-consuming specimen preparation. A further disadvantage of electron microscopy is the possibility of damage to specimens that occurs during preparation. Dehydration and fixing often result in artefacts and disruption that can influence interpretation.

There are two main types of electron microscopy:

Transmission electron microscopy (TEM)

This method is suited to investigate the ultrastructural features of thinly cut specimens that are typically fixed using osmium tetroxide.

Scanning electron microscopy (SEM)

This method is dedicated to looking at surface architecture of organisms which helps to understand morphological characters of specimens which have been coated in gold or platinum.

There are variations of SEM and TEM such as STEM (Scanning Transmission Electron Microscopy) and ESEM (Environmental Scanning Electron Microscopy). ESEM negates the need to pre-treat specimens and allows hydrated specimens to be examined. Techniques that utilize cryo-stages and freeze fracture reduce the chances of damage. The Immuno Electron microscopy technique can allow the detection of specific cellular components to which antibodies have been raised. Electron dense material is attached to the antibodies. When the antibody attaches to its cellular antigen, it can be visualized under the electron microscope. Improvements in technology and software enabled the initial development of cryoelectron microscopy into a burgeoning tool for *in situ* investigation of bacterial proteins, plasmids and other cell components (see Milne *et al*., 2013).

Identification from (Bio)chemical Properties and from Nucleotide Sequences

When moving beyond direct observation as enabled by microscopy and morphometrics, there is a need for complex equipment. Many approaches have been adopted for the purposes of identification but they all rely on one thing: a good quality (and curated) database. In some cases, the database will come with the particular machine and in others, it will be generated 'in-house' from authenticated material. Commercially available curated databases tend to be relatively expensive and entail use of particular protocols and/or reagents. In effect, the purchaser is paying a premium for all the effort the vendor has put in to generating reproducible profiles from authenticated material. Even in the case of sequence data, whilst the most widely used repositories are those from NIH-NCBI, EMBL-EBI and DDBJ, these mirror sites contain many errors and although there is a strong curatorial effort now, the fact that each database considers the sequence submitter to be the owner of the data, it is not simple to get errors rectified, no matter how glaring they are. Hence, some users may opt for the commercial alternatives, depending on their particular circumstances (and as long as the organisms they are studying are likely to be included in the database).

There are too many alternatives for this to be a definitive list but the most widely used platforms are described in a little more detail, below.

Fatty Acid Methyl Esterase (FAME) Analysis by Gas Chromatography (GC)

All cellular life forms possess lipid-rich cytoplasmic membranes. The membrane lipids may be used as markers for the identification and classification of microbes owing to the diversity of their precise chemical structure. Amongst the most important microbial lipid components are amphipathic lipids (e.g. phosphoglycerides and glycolipids) that possess both polar and non-polar regions. The non-polar tails generally consist of long chain fatty acid esters (Hamilton and Hamilton, 1992). Such characteristics are widely used for the identification and characterization of bacteria and yeasts.

Overview of the method

Fatty acid analysis may be considered in the following steps:

1. Cells are grown under standard culture conditions
2. Cells are then saponified to liberate the fatty acids from the cell surface
3. Fatty acids are methylated for greater volatility
4. Analysis of the fatty acids is undertaken by gas chromatography
5. Comparisons are made between the test fatty acid profile obtained in the above steps (1-4) and a database of profiles of known micro-organisms.

In short, the process involves vaporizing the solvent-extracted sample in a carrier gas (e.g. helium). The substance/gas is carried along a heated tube at a characteristic rate (partition coefficient) and then detected as it exits the tube. The output is then compared against a library of profiles and a best match selected.

Uses

When grown and harvested under such defined conditions, many species produce their own characteristic profile. This can be exploited as a taxonomic tool (Tornabene, 1985) and commercial databases such as the MIDI Sherlock™ Microbial ID System are available to identify unknown isolates. Whilst the technique has been applied predominantly to bacteria, refinements enabled its use with yeasts; it has been used with filamentous fungi but these still present a considerable challenge (Marumo and Aoki, 1990; Muller *et al.* 1994). This technique has also been employed for class (or other higher-level grouping) discrimination in ecosystem analyses.

Matrix-assisted Laser Desorption/ionization Time-of-flight Mass Spectrometry (MALDI-ToF-MS)

This technique employs what has been termed soft ionization of samples. With the advent of lower cost, entry-level machines (and associated software) from several manufacturers, including (but not limited to) the Bruker Biotyper® and the Biomerieux Vitek® MS and, possibly more importantly, their respective validated databases, this technology has become the 'go to' approach for rapid pathogen identification in many clinical laboratories (See Fig. 3.1). In the most commonly used applications of this technology, a 'direct transfer' approach is used. The simplest (and most widely used) protocol entails removing a small amount of biomass from a young (ideally 24 h) bacterial colony and transferring it to a spot on a target plate. This is then overlaid with a small volume of 'matrix solution' – a crystalline material in organic solvent which will lyse most bacterial cells and then, when excited by a laser, will vaporize with large proteins attached to negatively charged matrix particles. These ions are then propelled (by electrodes) along a vacuum tube to a very accurate detector. The time of flight and mass/charge ratios of individual proteins enable a unique profile to be obtained for each species. The profiles are checked against the database and identifications are obtained in seconds (assuming the organism's profile is present in the database already). For more difficult to lyse bacteria, an additional lysis step with the addition of e.g. formic acid is recommended.

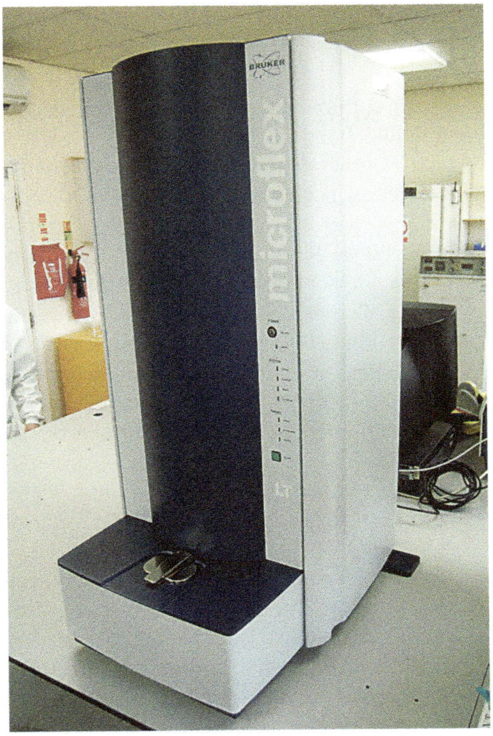

Fig. 3.1. Matrix-assisted laser desorption/ionization time-of-flight mass spectrometry (MALDI-ToF-MS) at CABI

For some bacteria (especially those which may have a H&S risk, an additional, ethanol treatment step is required (this is also required for processing most fungi).

Overview of a simple MALDI method:

1. Cells are grown under standard culture conditions (normally 24 h on standard agar medium for bacteria);
2. Cells are transferred (e.g. by sterile toothpick) to target spot;
3. Matrix solution is added and left to dry/crystallize;
4. Target plate is inserted in MALDI unit;
5. Laser vaporizes samples and the detector captures the spectral profile for that sample;
6. Comparisons are made between the test spectral profile obtained in the above steps (1-5) and a database of profiles of known microorganisms.

Uses

This method has revolutionized clinical bacteriology (e.g. see Lévesque *et al.*, 2015) but has also been extended to other applications such as food microbiology (Nacef *et al.*, 2017; El-Nemr *et al.*, 2019), cyanobacterial profiling (Imanishi *et al.*, 2016), identifying root nodule bacteria (Ziegler *et al.*, 2015), and mycology (Chalupová *et al.*, 2014).

Molecular Methods: DNA Barcoding and Multilocus Sequencing

The move away from morphometric identifications really took hold once the polymerase chain reaction (PCR) had enabled DNA sequencing to become more readily available and to move away from use of radionucleotides. Despite more recent advances with Next Generation Sequencing (NGS), DNA barcoding is still the most common and definitive way of identifying microorganisms. This approach entails the generation of sequence data (reference sequences) from a representative single gene (the barcode) to enable the identification of an 'unknown' organism to species level. This is achieved by comparison of the 'unknown' organism barcode sequence with sequence data for known species (species identifiers) held on a validated/curated sequence database, or library. The gene used for barcode identification scheme differs for different organism groups. Universal primer sets have been constructed for particular groups of organisms. For example, the 16S ribosomal DNA (rRNA) gene is commonly used for bacteria with universal primers 27f, 533r, 907r, 1492r (Lane, 1991; Lane *et al.*, 1985). Weisburg *et al.* (1991) describe a plethora of primers http://lutzonilab.org/nuclear-ribosomal-dna/ the rRNA gene cluster. Some of these fungal rRNA primers are more correctly eukaryotic primers, also amplifying plant and even some animal DNA (e.g. primers ITS1/ITS4 (White *et al.*, 1990) and TW81/AB28 (Curran *et al.*, 1994). As such, many modifications/optimizations have been described (e.g. primer ITS1F is specific for Fungi (Gardes and Bruns, 1993) and ITS4B is specific for Basidiomycota (Gardes and Bruns, 1993)). Although the aforementioned universal barcodes can distinguish some species, in most cases there is insufficient discrimination to get beyond genus level or clade/section/group level. For greater resolution a multilocus sequencing approach is required. In bacteria, considerable use has been made of multilocus sequence typing (MLST) analysis since it was first described for *Neisseria* strains (Maiden *et al.*, 1998). Since then, many 'housekeeping gene' sets have been employed in the related method, mutlilocus sequence analysis (MLSA) for inferring species phylogenies of bacterial genera (e.g. Delétoile *et al.*, 2009 [*Pantoea* spp.]; Cleenwerck *et al.*, 2010 [*Gluconacetobacter* spp.]; Labeda, 2011 [*Streptomyces* spp.]; Joseph *et al.*, 2012 [*Cronobacter* spp.]; Young *et al.*, 2020 [*Xanthomonas* spp.] etc.). A similar approach has been employed for Fungi (see Taylor and Fisher, 2003; also e.g. Hunter *et al.*, 2006 [*Mycosphaerella*]; Summerbell *et al.*, 2011 [*Acremonium*]; Cannon *et al.*, 2012 [*Colletotrichum*]; Hernández-Restrepo *et al.*, 2016 [*Microdochium*]; Kuephadungphan *et al.*, 2020 [*Gibellula*], etc.).

A number of different reference sequence libraries have been constructed by the scientific community and made available for public use. These include the aforementioned repositories hosted by NIH-NCBI, EMBL-EBI and DDBJ; also curated databases such as UNITE (mainly fungi); Ribosomal DNA database project (bacteria), BOLD (Barcoding of Life Datasystems – all organisms), QBOL (Quarantine Organisms Barcode of Life) and many smaller group specific datasets.

Once you know what you have (identification), you may wish to know what it can do (characterization)...

Characterization

Plate Assays

There are many culture-based test methods for the identification of specific properties and characterization of microorganisms. Some of the more useful tests (for example, the simple agar-based plate tests which can enhance the production of pigments) are included here. In addition, media can be supplemented with inhibitory compounds to assess growth.

Assessing Enzyme Activity on Solid Media

Growing cultures on specific substrates in defined and semi-defined media provides a simple way in which a variety of microbial enzymatic activities can be tested. For example, the ability of a species of fungus to grow on media containing sucrose and nitrate salts, as sole sources of carbon and nitrogen, implies invertase and nitrate reductase activities. Complex substrates in culture media, often in conjunction with indicators, can be used to screen for a wide range of enzymatic activities in a similar fashion. Table 3.2. below shows several examples of media and the properties for which they may be used to screen, although many others are available (e.g. Skerman, 1969; Cowan, 1974; Klement *et al.*, 1990; Paterson and Bridge, 1994).

A primary screen is useful when the rapid screening of many fungal strains is required. Volumes and quantities of biomass needed for this method vary with the number of positive results required. If more positives are required then larger quantities of biomass can be used, and/or the final extract can be made more concentrated. After the primary screen it is often desirable to isolate and characterize the active components in the most active extracts discovered. Thin layer chromatography offers an excellent means of doing this and can be carried out at a useful laboratory scale. This can be done using thin and preparative layer chromatography. A further modification is the use of agar overlays (subsequently inoculated with bacteria or fungi) in order to determine zones of inhibition/lysis.

Diffusion tests and susceptibility testing

The diffusion method is a means of measuring the effect of an antimicrobial agent against bacteria or fungi grown in culture.

In the case of antimicrobial susceptibility testing, the goal is to predict the *in vivo* success or failure of antibiotic therapy. Tests are performed *in vitro*, and measure the growth response of an isolated organism to a particular drug or drugs. The tests are performed under standardized conditions so that the results are reproducible. The test results may be used to guide antibiotic choice. In clinical situations, the results of antimicrobial susceptibility testing should be

Table 3.2. Examples of media used to detect enzymic activity (based on Paterson and Bridge, 1994)

Medium	Presumptive properties detected	Reference
Tween 80 medium	Fatty acid esterase activity	Sierra (1957)
Tributyrin lipase medium	Lipase activity	Lima *et al.* (1991)
Casein hydrolysis medium	Protease activity	Skerman (1969)
Gelatin hydrolysis medium	Protease activity	Skerman (1969)
Cellulose medium	Cellulase production	Eggins and Pugh (1962)
Polymeric dye ligninolytic medium	Ligninolytic activity (ligninase)	Glenn and Gold (1983)
Citrus pectin medium	Pectinase activity (esterase, lyase and polygalacturonase)	Cruickshank and Wade (1980)
Starch agar	Amylase activity	Bridge (1985)
Aesculin agar	β-Glucosidase activity	Skerman (1969)

See Appendix B for ingredients, formulation and methodology.

combined with clinical information and experience when selecting the most appropriate antibiotic for the patient.

Assessing antibiotic sensitivity

The standard test for assessing antibiotic sensitivity in bacteria is straightforward. Filter discs, impregnated with extract (antibiotic-impregnated discs are widely available and may be obtained from commercial suppliers), are placed onto a lawn of the test bacterium and incubated. The diameters of the zones of inhibition or antagonistic/synergistic interaction (if different bioassays are conducted on the same plate) are recorded.

This method may be used to assess general microbial resistance to relevant classes of antibiotics and antiseptics. In addition, diffusion tests have been applied to the study of translocation in fungi (e.g. Olsson and Jennings, 1991).

Tolerance, sensitivity and resistance tests

Compounds can be added to nutrient agar to test the tolerance and sensitivity of particular strains. An example of this is varying the pH of the media to help identify and characterize acidophilic and/or alkalophilic organisms. Similarly, growing organisms on nutrient agar that contain high salt or sugar concentrations can help screen for halotolerant/halophilic and osmotolerant/osmophilic isolates respectively.

Variation of temperature, light, aeration and humidity allows assessment of the individual environmental tolerances of strains. Strain resistance to a specific chemical (e.g. heavy metal tolerance) can be tested in the same way.

Biodegradative properties and bioremediation

Growing cultures in association with a chemical pollutant that needs to be broken down or removed can assist in studies targeting the identification of cultures. This also has the potential to be used in biodegradation or bioremediation. For example, growing cultures on a basal medium containing polycaprolactone diol (plastic substrate) (Benedict *et al.*, 1983; Kelley and Yaghmaie, 1988) can be used to assess the ability of a fungus to degrade plastic materials. Most notably, a similar approach has uncovered a strain of *Aspergillus tubingensis* that was shown to degrade polyurethane (Khan *et al.*, 2017). This is important in biodeterioration studies and may be tailored in the development of commercial biodegraders, such as those used in composting. Similarly, other synthetic and organic substrates can be incorporated in growth media to allow an assessment of the biodegradative ability of an isolate.

Modifications to this method can allow screening of microorganisms for their ability to remove heavy metals and/or radionuclides from solution – a process known as biosorption (Gadd, 1994, 2017). The scope for use of biosorption in microbial remediation of contaminated effluents and soil has been demonstrated using fungi and bacteria (McEldowney, 1990; Gadd, 1993; White *et al.*, 1996; Gadd, 2017).

Pathogenicity Tests and Fermentation Studies

It is often necessary or useful to test properties of organisms, for example to assess the suitability of a strain for a particular use or to test its stability after preservation. We consider pathogenicity and fermentation tests here.

Pathogenicity tests

In vitro pathogenicity tests may be used in preliminary studies for the development of biological control agents.

The target organisms are inoculated onto test media with the potential control strain (or metabolite/protein extract from the control strain) and the extent of pathogenicity and mortality of the target organism are recorded.

Data obtained from such experiments may aid the assessment of the fitness of potential biological control strains before their release into the target environment. Additionally, the data are useful in the assessment of the impact of potential biological control strains against indigenous organisms from the target environment. Pathogenicity tests can also be used to monitor the cultural and physiological stability of an organism and to assess whether an isolate is likely to synthesize enzymes or metabolites that may be of wider importance, during pathogenesis. A modified version of this approach may also be employed in the study of plant pathogens, in order to determine their activity.

Interaction/compatibility testing

Growing a strain on a Petri dish in the presence of another organism has also been widely used in microbiology to study the nature of the interaction between them. This can allow assessment of the antagonistic or synergistic activity of particular strains. Antagonistic properties may be due to chemical or vegetative incompatibility. Strains exhibiting antagonistic activity or parasitism may have potential for development as biological control agents.

In mycology, vegetative compatibility group testing of isolates grown on adapted media (i.e. with nitrate) has been used to characterize many fungi including *Fusarium* spp. and *Colletotrichum* spp. (Puhalla, 1985; Leslie, 1987; Brooker *et al.*, 1991). Likewise, interaction tests are useful for genetic and molecular investigations such as mating type experiments (Casselton and Olesnicky, 1998) and parasexuality studies (Hastie, 1981).

One other significant form of microbial interaction is the 'killer yeast' phenomenon (mycocins). Mycocins are toxins produced by many types of yeast that are lethal to other yeasts. All yeast mycocins appear to be proteinaceous and are cytoplasmically-determined. 'Killer' activity is expressed predominantly at pH 3-6. Expression of, and sensitivity to, mycocins have been used in yeast systematics (Golubev, 1998).

Fermentation studies

One of the most commonly-used classes of test for the identification of yeasts is the fermentation of carbohydrates as measured by the production of CO_2 in liquid culture. The basic method employs 'Durham tubes' that contain 2% (w/v) solutions of the appropriate sugar (van der Walt, 1970). The yeast is then inoculated into the broth and fermentation is visualized by the presence of a bubble (of CO_2) in the small inverted test tube. Genera such as *Rhodosporidium* and *Sterigmatomyces* cannot ferment glucose, whilst others such as *Kluyveromyces, Saccharomyces* and *Zygosaccharomyces* are able to do so (Yarrow, 1998).

Carbon Utilization Studies

A very important tool in bacterial identification is the ability of organisms to utilize a wide range of organic molecules as the sole source of carbon and energy, such as carbohydrates, amino acids, carboxylic acids, aromatic compounds and others. Indeed, this has been exploited commercially by the makers of several widely-available kits, including the API® system (Biomerieux, S.A. France) and Biolog plates (Biolog Inc, USA). In some genera of bacteria (e.g. *Methylobacterium*) carbon utilization spectra are the only way of distinguishing between species phenotypically.

There are a variety of commercially available kits for testing carbon source utilization. Biomerieux produce a range of specific bacterial and yeast kits with specific carbon sources, e.g. for the *Enterobacteriaceae*, and the genera *Bacillus, Campylobacter, Candida, Listeria, Staphylococcus* and *Streptococcus*. Tests involve either a simple colour reaction that indicates carbon source utilization (caused by the reaction of a dye with free naphthyl compounds) or, simply, opacity of the growth medium caused by growth of the organism.

Biolog tests (Biolog INC USA) consist of a series of 96-well microtitre plates containing

95 available substrates (e.g. Biolog GEN III Microplate™ for aerobic bacteria and Biolog AN MicroPlate™ for anaerobic bacteria). Incubation of the plates with a live suspension allows the characterization and assessment of bacteria depending on the substrates that are and that are not utilized. Assessment of a positive reaction is facilitated by the presence of a tetrazolium dye that gives a purple coloration. Wildman (1995) extended the use of this technique to fungi, where each well of a Biolog GN plate was inoculated with fungus and, after an incubation period, utilization was assessed by observation of growth in each well. Subsequently, Biolog now market a test plate for filamentous fungi (without the tetrazolium dye incorporated in the bacterial plates: the Biolog FF MicroPlate™). Biolog have expanded their range considerably to enable phenotypic profiling with additional sets for nitrogen sources, phosphorus sources, sulphur sources, osmotic challenges, biosynthetic pathways, pH effects and chemical sensitivities. Biolog also produces a Biolog EcoPlate™ for microbial community analysis.

Nitrogen-source screening

Yeasts are known to be able to utilize a wide range of nitrogen sources. Growth on particular nitrogen sources has been used to help identify yeasts at the specific and sub-specific levels. For example, the genus *Pichia* is distinguished from its former genus, *Hansenula*, by its inability to grow utilizing nitrate as its sole nitrogen source (Yarrow, 1998). The use of nitrogen sources is also important in the discovery of fungal mutants for vegetative compatibility screening.

Enzymes, Proteins and Organic Acids

Microorganisms produce a novel and diverse array of proteins and extra and intra-cellular enzymes from urease (synthesized by strains of the bacterium *Bacillus fastidiosus*) to cellulase and lipase (synthesized by strains of the fungus *Trichoderma viride*). The enzyme and protein profiles of an organism can demonstrate much about its taxon and physiology. The following section provides an overview on the production of enzymes, proteins and organic acids by microorganisms.

Assessing enzymatic activity

Enzymes are biological catalysts necessary for microbial metabolism. They can also be used to help characterize organisms as some enzymes and combinations of enzymes are distinct for specific microorganisms. Many saprophytes excrete extracellular enzymes into the substratum and these can be easily extracted and purified. Other enzymes can only be found intracellularly. Some enzymes are produced on an industrial scale (e.g. amylase from strains of *Aspergillus niger*, *A. oryzae*, *Bacillus polymyxa*, *B. subtilis*, and *Trichoderma viride*). Enzymes can be representative of an organism's mode of life (e.g. insect cuticle-degrading enzymes synthesized by the entomopathogenic fungus *Metarhizium anisopliae*).

Intracellular enzymes are more difficult to extract. Specific enzymes can be purified and their activities and structures determined using a variety of techniques which include: chromatography, spectrophotometry, mass spectrometry and crystallography. Some enzyme systems appear to be generally conserved and are therefore useful for studying relationships between species or species complexes. Other general cytoplasmic enzymes may be extremely variable and may have use in discriminating species and populations.

Enzyme systems studied widely in microbiology have included: esterases, phosphatases and catalases. These were found to be useful for characterization studies on a range of filamentous fungi (e.g. May and Royse, 1982; Micales *et al.*, 1986; Bonde *et al.*, 1991), yeasts (e.g. Lambert and Garcia, 1990) and bacteria (e.g. Gardes *et al.*, 1987; Kubicek *et al.*, 1989) before being superseded by PCR-based molecular tests.

Enzymatic activities of microbial cultures

The enzymatic activities of microbial cultures may be screened directly by testing for a range of intra- and extracellular enzymes. Such methods were used initially in studies of bacteria and yeasts (e.g. Cowan, 1974; Barnett *et al.*, 2000) but were later adapted for analysis of fungi where extracellular enzymes were screened in conidial suspensions and spent culture fluid (Bridge and Hawksworth, 1985; Mugnai *et al.*, 1989). The methods can also be used with diluted intracellular extracts. Intracellular enzymes are retained inside the cells and are detected following the breaking open of the cells and testing the crude or purified extracts.

There are several test kits available commercially for testing enzymatic activity. One of the best-known is the API® ZYM system (Biomeriuex, S.A. France), which has been used in a number of studies with bacteria (e.g. Ling *et al.*, 1994), cell lines (e.g. Nardon *et al.*, 1976), and filamentous fungi (Bridge and Hawksworth, 1984; St Leger *et al.*, 1986), and has been used to characterize and test for bacteria from crude extracts (e.g. Kilian, 1978). A simple colour reaction indicates enzyme activity due to the reaction of a dye with free naphthyl compounds.

Enzyme assays using chromogenic and fluorogenic substrates

There are many substrates available from commercial suppliers which, when acted on by an enzyme, will release a detectable colour or fluorescence. One of the main advantages of using chromophores and/or fluorophores in analyses is that they can be observed easily and may be used to provide (semi-) quantitative data.

Enzymatic activity of extracts and broths

Chromogenic and fluorogenic substrates

Specific enzyme activity can be demonstrated using a wide variety of chromophore-substituted compounds that are available. Perhaps the most widely used, are the large range of *o*- and *p*-nitrophenyl substituted compounds, where cleavage of the substrate liberates free nitrophenol which is a yellow/brown colour. For example, *o*-nitrophenyl β-D-galactopyranoside has been used as a substrate for β-galactosidase (Cowan, 1974). The enzyme activity cleaves the galactopyranoside leaving free nitrophenol, which is yellow. Fluorogenic substrates such as the 4-methylumbelliferyl (4-MU) substituted compounds also have been used on a wide range of microorganisms (e.g. Bradbury, 1977; Grange and Clark, 1977; Slifkin and Gil, 1983; O'Brien and Colwell, 1987; Barth and Bridge, 1989; Claeyssens, 1989). The basic principle is the same as for the chromogenic substrates, except that the free methylumbelliferone is fluorescent, appearing blue/white under long-wave ultra violet light. There are a large number of commercially-available 4-MU substrates that allow screening of a diverse range of enzymes (e.g. β-glucosidase, trypsin and arabinofuranosidase; Merck, Poole, UK). Methylumbelliferyl-substituted compounds have been used, also, in environmental investigations, for example to measure microbial activity in soil (Miller *et al.*, 1998) and as an aid to determining ligninolytic activity (Otsuka *et al.*, 2003).

Analysis of Microbial Proteins

Proteins are compounds made up of amino acid sub-units that are essential to cellular processes such as metabolism and cell replication. Microorganisms produce a fascinating array of proteins (e.g. thermostable DNA polymerase, which is able to withstand high temperatures; this enzyme was sourced from the thermophile *Thermus aquaticus*, leading to the commercial development of PCR technology.) The study of the structure and functions of proteins is widely known as 'Proteomics'.

Proteins can be detected and separated by gel electrophoresis, ultracentrifugation and gas and liquid chromatography. They can be detected using antibody techniques (such as ELISA). Protein structure can be elucidated using x-ray crystallography, mass spectrometry, Nuclear Magnetic Resonance (NMR) and peptide sequencing. Cell breakage (in an appropriate buffer) is required to extract proteins for analysis, for example, physical, chemical or sonification.

Basic Methods for the Characterization of Proteins

Electrophoresis

Electrophoresis can be used to characterize intracellular proteins from a wide range of organisms. The technique involves the separation of proteins through a supporting gel (starch or polyacrylamide). After separation, the proteins can be identified with specific stains. Some enzymes are present in more than one form (isoenzymes or isozymes) and profiles of these have been used to differentiate genera, species and populations of microbes.

Electrophoresis on polyacrylamide gels, with a discontinuous buffer system is normally used with varying conditions. This is dependent upon the type of enzyme or protein system being examined. There are two main classes of protein electrophoresis:

1. Denaturing Gels: These most commonly incorporate the detergent, sodium dodecyl sulphate (SDS). In general, polypeptides bind SDS in a constant weight ratio allowing separation on the polyacrylamide gel on the basis of polypeptide size. Employing a set of standard polypeptide markers, of a known size, allows the calculation of the sample polypeptides to be made (Hames, 1981).
2. Native Gel Electrophoresis: This is the second major class of protein electrophoresis. This form of electrophoresis does not involve the denaturation of the proteins. The separation, therefore, is due to both the size and charge of the polypeptide at a given pH.

Specific proteins may be identified using the Western Blot technique. Following electrophoresis, proteins can be transferred to a nitrocellulose or nylon membrane and are subsequently probed using specific antibodies. Microbial cells must be disrupted to allow extraction of the cytosol. The degree of breakage required—and the subsequent methodology—is dependent upon the nature of the organism being studied and the properties of the protein involved.

Advances in genomics and proteomics have resulted in a dramatic fall in the use of isoenzyme analysis but it is still employed for strain characterization and/or the elucidation of novel enzymes (e.g. Caierão *et al.*, 2016; Abdulaal, 2018; Bernardo *et al.*, 2018).

Cell disruption

The physical methods used to break cells include sonication (of bacteria, yeasts, unicellular algae), manual disruption (of lyophilized or frozen fungal mycelia and algae in the presence of liquid nitrogen or with carborundum powder, using a pestle and mortar) and band/bead beating (e.g. with a cell disintegrator). Chemical and enzymatic means of lysis are also employed (Doonan, 1996).

Crude protein extracts are obtained by centrifugation of the biomass following its homogenization with sterile buffer. Total protein concentrations may be obtained (e.g. Lowry *et al.*, 1951) and may be used for standardization of sample loading. Further purification is not necessary prior to electrophoresis but protease inhibitors may need to be added to the samples to prevent degeneration. Most extracts can be stored, successfully, at -20°C for at least 6 months although there are cases where the enzymes must be analysed immediately after extraction. Repeated freezing and thawing will result in loss of activity and is not recommended.

Production of Organic Acids

Many microorganisms produce acids which are a product of metabolism. Organic acids are generally weak acids that do not dissociate fully in water. Microorganisms are often used commercially to produce such acids. Detection of these acids often involves very simple assay methods.

Methods

Extracellular citric acid production by fungi such as *Aspergillus* spp. can be screened in 10-14 day-old liquid cultures with Altman reagent (Gaffney *et al.*, 1954; Smith, 1969). Presumptive positive culture fluids should be confirmed by thin layer chromatography in butanol:water [120:30:50 (v/v/v)]. Altman reagent will give a pink colour with citric acid, but may also react with ketoglutarate, oxaloacetate and glutaconic acid.

Additional investigations have elicited methodologies for screening for malic acid production by bacteria and fungi (Chibata *et al.*, 1983; Campbell *et al.*, 1987); oxalic acid production by fungi (Takao, 1965) and acetic acid production by *Acetobacter* spp.

Chromatography

Chromatography is a method for separating metabolites. It can also be used for the separation of other biomolecules and chemicals and in the characterization of individual molecules. The principle of the method involves passing a mobile phase that contains a mixture of dissolved chemical constituents (typically in a solvent or gas carrier) over a stationary phase, such as silica gel or paper.

Techniques commonly used include: thin layer (both normal and reversed phase), preparative layer (normal phase), liquid (normal phase), gas and high-performance liquid chromatography.

When associated with Mass Spectrometry, the chemical composition and structure of molecules can be determined.

Thin Layer Chromatography (TLC)

This approach has lost popularity over recent years, being superseded by automated chromatographic procedures that give more quantitative or qualitative data. Standard TLC requires stationary phase cells to be layered over a support plate made of glass, aluminium or plastic. Samples are suspended in a suitable solvent and loaded onto the plate, which, when dry, is placed into a sealed glass TLC tank containing a solvent to saturated atmosphere. This solvent then ascends the TLC plate by capillary action. The choice of solvent depends upon the nature of the chemical being targeted (Stahl, 1969). Thin layer chromatography is a very flexible method, and this flexibility can be increased by the technique of two-dimensional TLC. Applications of 2D TLC include confirming the identity of a metabolite (i.e. if the metabolite runs to the same position as a standard in both solvents then greater confidence in its identity is possible) and determining the purity of individual spots. The utility of this method can be demonstrated through its use in the assessment of conspecificity of fungus strains that produce pigments. *Hobsonia christiansenii* and *Illosporium corallinum* are lichenicolous species that share certain morphological similarities, one of these being the production of a pink/red pigment (Lowen *et al.*, 1986). This method can be used to see if the pigments they are producing are the same, and can be used for other fungi: a 5 x 5cm TLC plate is used, routinely. If spots on a TLC plate are not well-separated after the initial development, it is possible to redevelop the TLC plate using the same solvent, this is referred to as multi-development TLC. Cell wall sugars and amino acids are an important diagnostic tool for coryneform bacteria, TLC can be used to differentiate such molecules and resultant patterns can be used to identify the bacteria. For all methods of TLC, a variety of post-run sprays can be used to enhance the detection of specific metabolites (Stahl, 1969). For example, *p*-anisaldehyde can be used to detect sugars.

Advantages and Disadvantages

TLC is a cost-effective method that does not require expensive equipment and, as such, remains a useful method for preliminary metabolite screening. However, reproducibility and quantification may not be as accurate as automated methods and, as mentioned previously, it is also considered dated.

Techniques:

1. Agar Plug Technique

The Agar Plug Technique (standard method) is most frequently used to detect secondary metabolites from filamentous fungi and was initially developed for terverticillate penicillia (Filtenborg *et al.*, 1983; Paterson, 1986). This method has been used for *Chaetomium*, *Fusarium*, *Metarhizium* and *Verticillium* (Paterson and Bridge, 1994). Growth media, incubation temperatures and growth periods may be varied to optimize metabolite production, but these must be kept constant, thereafter, for comparison purposes. Extracellular metabolites can be detected by direct application of the agar side of a mycelial plug onto the TLC plate. Intracellular metabolites are detected by direct application of the colony to the TLC plate with appropriate solvent (e.g. chloroform/methanol), for extractive purposes. The method can also be used for other organisms that can grow on agar such as bacteria and algae.

2. Extraction from Complete Cultures or Solid Material

Solid food, feed commodities (e.g. grains, peanuts) and agar cultures of fungi, bacteria and yeasts can be screened for metabolites using this method. Metabolites are extracted using a suitable solvent (e.g. methanol); however, different solvents can be used to optimize the extraction of specific metabolites. Chemical extraction may require complex procedures but such extracts may be inoculated directly onto the TLC plates (Paterson and Bridge, 1994) or analysed by HPLC.

3. Analysis of Liquid Samples and Broths

Depending on the liquid to be screened a defatting procedure may, or may not, be required. For substances such as milk, which contain lipid, an *n*-hexane extraction stage will be required. When carrying out this analysis it is important to analyse a presumptive uncontaminated control and have an indication of which microbe is suspected to have contaminated the product to support decisions about the identity of the spots on the TLC plates (Paterson and Bridge, 1994).

4. Multi-mycotoxin Screen
Biological materials suspected of being contaminated by mycotoxins or other secondary metabolites can be analysed using this procedure. A control sample from the non-contaminated material should also be extracted in case the non-contaminated material contains metabolites that have similar TLC characteristics to the fungal products (Paterson and Bridge, 1994).

5. Extraction from Dried Herbarium Cultures
Data on secondary metabolites can be obtained from dried herbarium cultures (Paterson and Hawksworth, 1985). It is possible to obtain metabolites from cultures dried almost half a century previously. Such metabolites may be used as taxonomic characters, or for screening for particular compounds.

6. Metabolite Identification
Colours and Rf values (of metabolite spots on TLC plates) can be used in an attempt to determine the identity of metabolites by comparing these to available information (e.g. from databases, literature or other sources - see Paterson and Bridge, 1994). If Rf values and colours are not available, pure samples of particular fungal metabolites can be obtained from chemical supply companies or from individual laboratories where metabolite research is undertaken. If specific chemical characteristics are known, the chemical of interest can be extracted from the microorganism which produces it. This can, however, be time consuming and experimentally demanding (as well as hazardous: many fungal secondary metabolites may be toxic to man, especially when present in concentrated form). Once characterized, metabolites can be extracted and purified.

7. Ubiquinones
Ubiquinones are a class of terpenoid lipids, consisting of a benzoquinone ring with a non-polar isoprenoid chain. In the electron transport chain, they function as highly mobile electron carriers between the flavoproteins and the cytochromes. Ubiquinones have been used in microbial chemotaxonomy due to the inherent structural variation observed between some taxa (i.e. the length of the isoprenoid chain and its degree of saturation). Partially saturated and unsaturated structures are found in fungi, whereas only unsaturated quinines have been observed in bacteria. Their use as taxonomic characters in yeasts, fungi and bacteria has attracted considerable interest (e.g. Yamada and Kondo 1973; Collins, 1985; Kuraishi *et al.*, 1985; Sugiyama *et al.*, 1988; Lisdiyanti *et al.*, 2003).

Analysis of ubiquinones has suffered in the past, however, due to the time-consuming processes involved, the specification (and expense) of equipment such as HPLC, and the use of rigorous extraction procedures, particularly saponification. A 'direct' method was developed to avoid these problems as it does not require expensive equipment, it is experimentally undemanding and it gives highly reproducible results using a mild extraction procedure. This method was originally developed for characterization of *Legionella* spp. (Mitchell and Fallon, 1990) but was modified, subsequently, for fungal analyses (Paterson and Buddie, 1991). In the modified method all spots that appear on TLC plates are considered to be characters, whether they are ubiquinones or non-ubiquinones (i.e. predominantly phospholipids).

High Performance Liquid Chromatography (HPLC)

HPLC is an automated system for analysing complex mixtures that produces rapid, reproducible results. Small columns with high-resolution power are used, which increase the speed of elution. This is possible because the particle size of the stationary phase is very small. The small columns are eluted with a mobile phase that is pumped through in a highly controlled manner. Microlitre volumes are injected onto the column through an injection port. After the solvent has passed through the column detector systems, the solutes are monitored by the chemical or physical properties of the solutes (e.g. absorbance, fluorescence, etc.). Extractions can be made from whole plates, culture broth or undefined sources, as previously described for the TLC method. HPLC systems primarily utilize one of two separation systems, isocratic HPLC where the effluent composition remains constant and gradient HPLC where controlled alteration of the effluent is used.

Alkylphenone standards
HPLC systems can be standardized using alkylphenone standards (Frisvad and Thrane, 1987; Paterson and Kemmelmeier, 1989). Alkylphenone standards are a homologous series of compounds. They are stable, have a wide range of

polarities and are readily detectable. Retention indices from alkylphenone standards are independent of column efficiency, variations of column efficiency, column flow rate, and temperature. Retention indices may be used between laboratories as a reliable guide to the tentative identification of mycotoxins as the use of alkylphenones reduces variability from chromatographic conditions. There are a number of ways to identify unknown peaks, either by comparison with metabolite standards (e.g. from a collection of fungal metabolites, or by using UV or visible spectroscopy).

UV/Visible Spectroscopy

Chromatographic methods are used to separate secondary metabolites, whereas UV spectroscopy is used to identify them. Spectra can be obtained from metabolites removed from thin or preparative layer chromatography plates. Fractions from preparative HPLC compounds and proteins purified from column chromatography can also be used. The spectra obtained from spectroscopy are associated with transitions between electronic energy levels.

Because the promotion of electrons from the ground state to the excited state of such systems gives rise to absorption, they are used to detect conjugated systems. Diode Array Detection (DAD) combined with the ability to take UV spectra spots from TLC plates for the identification of compounds, facilitates the association between UV spectroscopy and chromatography methods (Frisvad and Thrane, 1987). Comparing spectra of metabolites in neutral solvents to those in strongly alkaline solvents and subtracting the neutral from the alkaline spectrum generates a difference spectrum. This serves to increase the utility of UV spectroscopy for certain classes of compound. A similar principle can be applied using neutral solvents (Paterson and Bridge, 1994).

Uses

Difference curves can be used in three ways:

1. To give quantitative and qualitative information on the compounds in mixtures, but are characteristic only of the ionizable elements of the compounds.
2. To aid the differentiation of metabolites with similar chromophores by using the data and method of UV/Vis spectroscopy.
3. To permit the study of separate chromophores in complex mixtures without the need for physical separation.

Difference curves have advantages over absorption curves for the analyses of ionizable chromophores when the absorbing units are present either as parts of the molecule or in mixtures of various molecules. This may be useful for the rapid determination of distinctive compounds in crude extracts.

Immunological methods

Currently, the most commonly used immunological technique is the enzyme-linked immunosorbent assay (ELISA). Many modifications have been made to the method but the most straightforward approach is the addition of a specific antigen to a microtitre plate. The antigen (usually protein) adheres to the wall of the microtitre plate. Unbound antigen is washed off before the bound antigen is dried. Independently, an antibody-enzyme conjugate is prepared and is diluted in buffer containing inert proteins (e.g. 5% skimmed milk powder in phosphate-buffered saline). The diluted conjugate is added to the antigen in the wells of the microtitre plate and the wells are washed out to remove unbound components. On addition of a particular substrate that changes colour (or fluoresces, for enzyme-linked fluorescent assay) following enzyme catalysis, the plate is incubated. A quenching reagent is added to stop the reaction and the plates are analysed by spectrophotometer. Positive reactions are indicated by the presence of the specific colour (or fluorescence, for ELFA). This system is termed direct ELISA but modifications such as indirect, sandwich and competition ELISA have proved successful, also (Hampton *et al.*,1990). ELISA has been applied to many situations such as confirmation of clinical disease, phytopathogenic disease, analysis of immune responses, antigenic comparison and production of monoclonal antibodies (Crowther *et al.*, 1998).

Additional immunological techniques that have been useful in screening microorganisms include the use of single and double diffusion plates, microagglutination, latex agglutination and Western Blotting (Hampton *et al.*, 1990).

Molecular methods; whole genome sequencing; etc. [but also mentioning some of the older methods])

There are many methods available for the analysis and characterization of nucleic acids. Mention has already been made, in this chapter, of DNA barcoding and multilocus sequence analysis, as used for identification. The techniques described have diverse applications including gene cloning, screening of gene expression, species definition, study of interactions, phylogeny, study of degradation processes, detection of metabolite producers, studies of biodiversity and identification of species in crude samples. General molecular procedures such as restriction endonuclease digestion, cloning, hybridization and sequencing are applicable to microbial DNA. The rate of progress in this field soon renders methodologies obsolete but mention is made of some of the techniques that have been integral to microbial identification and characterization over the past five or six decades. Common techniques that have utilized the Polymerase Chain Reaction (PCR) include: RFLPs (Restriction Fragment Length Polymorphism), RAPDs (Random Amplified Polymorphic DNA), AFLP (Arbitrary Fragment Length Polymorphism), SSCP (Simple Strand Conformation Polymorphism). Fingerprinting techniques have allowed scientists to separate taxa that are difficult to distinguish using anatomical or morphological criteria. Non-PCR techniques such as electrophoretic karyotyping of chromosome size and number may also be applied. Other techniques that amplify nucleic acids that will be mentioned briefly include loop-mediated isothermal amplification (LAMP), nucleic acid sequence-based amplification (NASBA) and recombinase polymerase amplification (RPA).

DNA Extraction Methods

Most techniques for the analysis and characterization of nucleic acids require purified DNA, which must be extracted from cells with minimal mechanical disruption to the DNA. Most methodologies involve either a chemical or enzymatic approach. The most common methods use a cell lysis solution followed by purification and storage in an appropriate buffer. Simple-to-use proprietary kits are widely available to extract bacterial and fungal DNA from pure culture – but also from various environmental substrata and commodities.

Nucleic acids can be more difficult to obtain from fungi than from bacteria, primarily because of the nature of their cell walls. In this case, cetyltrimethyl ammonium bromide (CTAB) is still used widely for extracting total genomic DNA from filamentous fungi (Zolan and Pukkila 1986) especially those that produce excess extracellular polysaccharides. Recently developed DNA release technologies have given encouraging results with fungi. There are many commercially available kits and solutions that make DNA isolation/purification simpler and safer: these options tend to be more expensive and consume more single-use plastics than is the case for 'classical' methods such as CTAB but, in general, the former use more noxious chemicals and are less reproducible from sample to sample. Variability exists across kits with new approaches and improvements still being released. Extraction efficiency depends on a number of parameters including the nature of the organism itself, its age, morphology, substrate and many other aspects (e.g. see Fredricks *et al.*, 2005).

RFLP Analysis

Despite being superseded to a great extent by the arrival of affordable DNA sequencing, restriction fragment length polymorphisms (RFLP) still have a place in the modern microbiology laboratory as they provide a quick and inexpensive method for screening – and comparing – strains. They were used widely in microbial systematics at different taxonomic levels (Bruns *et al.*, 1991; Cooley 1992; Hibbett, 1992). Probes were derived for use with chloroplast, mitochondrial and chromosomal DNA. There are many reported maps and sequences from fungal mitochondrial DNA. Mitochondrial genomes have two advantages over chromosomes which make them attractive to study: 1) they are generally inherited uniparentally (Fort *et al.*, 1984); and 2) they are present in fungi, at least, in multiple copies (Bruns *et al.*, 1991). Single-parent inheritance is important for genetic analysis as it means that mitochondrial-derived recombination products do not occur. Fungal mitochondria possess one additional advantage for study – they are particularly rich in the bases adenine (A) and thymine (T) (Hudspeth *et al.*,

1980; Bruns *et al.*, 1991) with guanine and cytosine levels (G+C) of 24-39 Mol % (Marriott *et al.*, 1984). The result of this base imbalance is that, when total DNA extracts are digested using 4-base GC-containing restriction endonucleases (e.g. *Hae*III which cuts at GGCC), the chromosome is digested into many small fragments (usually below 2kb in size) whereas the mitochondrial DNA is digested less frequently. When these digests are separated on agarose gel electrophoresis, clear mitochondrial bands can be seen, which are distinctly separate from the smear of smaller chromosomal fragments. This technique allows some mitochondrial RFLPs to be visualized quickly, without the need for labelling, hybridization, etc., and can also be used to produce well-separated mitochondrial fragments for the subsequent production of probes.

Currently, the most common use of the RFLP technique is in digestion of PCR generated amplification products such as the (chromosome-located) ribosomal DNA gene cluster but, more recently, also of other functional gene loci (e.g. Rousseaux and Guilloux-Bénatier, 2017; Blanton *et al.*, 2019; Pharkjaksu *et al.*, 2020; Li *et al.*, 2021).

Polymerase Chain Reaction

One of the most significant advancements in molecular biology was the development of the polymerase chain reaction (PCR) (Saiki *et al.*, 1985; Mullis and Faloona, 1987). Although the concept of a chain reaction amplification of DNA had been described in the 1970's (Kleppe *et al.*, 1971), it was not made practicable until the isolation of a thermostable DNA polymerase *Taq* (derived from the thermophilic bacterium *Thermus aquaticus*). This allowed repeated extension of the DNA molecule to progress without the addition of fresh enzyme at the end of each cycle, i.e. continuous cycling without the DNA polymerase being heat-inactivated by the denaturation step.

The PCR Process

PCR reactions can be undertaken in small (e.g. 0.5 ml) microcentrifuge tubes or other specialized vials. The tubes should have thin walls for optimal heat transfer to the contents when placed within a thermal cycler. The thermal cycler is an extremely sensitive block which can be heated and cooled to raise and lower temperatures in seconds. There must also be a carefully defined reaction mixture in the tube as the amplification is a closed-system process (Edel, 1998). The reaction mixture should contain the target DNA to be amplified [ca. 10-100 ng, (Edel, 1998)]; an excess of free deoxyribonucleotides (i.e. dATP, dCTP, dGTP and dTTP); short, specific sections of DNA (the oligonucleotide primers) in order to anneal to specific sections of the target DNA; and the thermostable DNA polymerase that will facilitate the sequential addition of bases onto the growing strand, as it is amplified/replicated. The DNA polymerase will not allow *de novo* DNA synthesis. Hence, the primer DNA is required to give the DNA polymerase an available site on which to hang fresh nucleotides. Some polymerases have proof-reading/editing capability that will repair any errors that arise during the cycled reactions.

General PCR Method

The PCR technique allows the amplification of tiny amounts of DNA (in theory, from one molecule of DNA) by the following steps:

1. Denaturing of the DNA by heating to c. 95°C;
2. Annealing of the primers to complementary target DNA sequences (according to T_m of primers);
3. Extension of the new copies of the DNA (usually at 72°C);
4. Repeated cycles of steps 1-3.

Many modern polymerases require a 'hot start' activation. This is an initial heating step to 95°C for up to 10 minutes, which reduces the incidence of non-specific amplification.

Targeted vs Non-targeted PCR

There have been countless variations on the general PCR method. Essentially there are two basic approaches–Targeted PCR and Non-targeted PCR, as described below.

Targeted PCR

Targeted PCR is used widely in microbial identifications (using mainly rDNA gene cluster and some functional loci such as structural genes or those coding for enzymes), taxonomy and phylogenetic analysis. If the sequence of the gene of interest is known, primer pairs can be created. There are many documented examples in the literature and software is available to aid in the construction of new ones and to optimize reaction conditions. This method is more widely used in medical, veterinary and phytosanitary/quarantine fields.

Non-targeted PCR

Non-targeted PCR is used for DNA/genomic/genetic/PCR fingerprinting or genotyping. Total DNA is extracted and individual PCR primers are used to amplify at multiple regions scattered across the genome; this result is in complex banding patterns that can give an overall picture of the total genome and allow visualization of variation between samples. The primers tend to be short (decamers in the case of RAPDs, 12-18 nucleotides in the case of more stringent methods). Such methods can differentiate below species level and can even allow tracking of strains at the individual level.

PCR Fingerprinting

There are a number of fingerprinting techniques that can be used to characterize genetic variations within microorganisms. A few of them are described below:

RAPD (random amplified polymorphic DNA) is a technique that was used widely in the early days of PCR analyses. The number and nature of bands produced were often difficult to interpret (Bridge *et al.*, 1997) and results were not very reproducible because of low stringency of the technique, due to the relatively low PCR annealing temperature and the short primers. Due to the problems associated with this method, alternatives were developed that were more robust. The next generation of fingerprinting techniques included variable number tandem repeat (VNTR) PCR which was also known as simple sequence repeat (SSR) PCR. Flanking sequences that had previously been the basis of early DNA fingerprinting probes were adapted as primers for this RAPD-like PCR. Simple sequence repeats [e.g. $(CAG)_5$, which is shorthand for the primer sequence: 5'-CAGCAGCAGCAGCAG-3'] were used widely in bacterial and fungal studies. This was refined further for a modification of the SSR approach, entitled inter-SSR (or ISSR), which involved the use of SSR primers with degenerate anchored ends. Primers follow the same general format: e.g. ISSR-CGA is 5'DHB$(CGA)_5$ or, in full, 5'-DHBCGACGACGACGACGA-3'; where B = C or G or T; D = A or G or T; H = A or C or T; with an SSR type sequence anchored to a degenerate end to increase the number of potential bands but allied to a relatively high annealing temperature that avoids the mis-priming that is common in less stringent methods such as RAPDs. Indeed, ISSR-PCR has been shown to recover a similar level of resolution to AFLPs. RAPDs are now considered obsolete whilst use of SSRs and ISSRs is limited to studies involving plants, which have genome sizes that do not lend themselves to NGS-based screening.

AFLP PCR

Amplified fragment length polymorphism-PCR (Vos *et al.*, 1995) is a highly reproducible PCR fingerprinting method that has stood the test of time, although similarly to the early DNA fingerprinting methods described above, it is employed mostly in plant population analyses (one of the few recent exceptions being Arastehfar *et al.*, 2019). The method involves the restriction endonuclease-meditated digestion (generally a 'rare-cutting' 6-base cutter and a 'common-cutting' 4-base cutter) of genomic DNA to produce fragments which are then attached (ligated) to oligonucleotide 'adapters'. A PCR reaction is undertaken with primers that include the adapter sequence and between one and five randomly chosen additional nucleotides. This allows selective amplification of a subset of the restriction fragments in which the nucleotides flanking the restriction site match the additional nucleotides of the primer (Edel, 1998). The AFLP method is considered to be much more reproducible than most of the early fingerprinting methods such as RAPDs. It harnesses the reproducibility advantages of the RFLP method to the utility of PCR-based methods, allowing screening of populations from even small amounts of starting DNA, without the need for labour-intensive large-scale DNA extractions. The original AFLP method resulted in such complex banding patterns that automated DNA sequencers were required for the analyses. The dual-enzyme method was simplified to a single-enzyme method to allow electrophoreses to be undertaken in standard agarose gels instead of the use of expensive DNA sequencers or other polyacrylamide gel apparatus (Mueller *et al.*, 1996).

Other PCR-based approaches:

There are many other ways in which PCR has revolutionized molecular studies. Two of the most significant have added to the complexities of the discipline by sharing the same initials: 'RT-PCR', thereby heaping confusion onto the unwary researcher taking a cursory glance at the literature.

RT-PCR (reverse transcriptase PCR) is the method of choice for studies based on isolation of total RNA or mRNA. The RNA sequence is converted to the appropriate DNA complement (i.e. cDNA) using the enzyme reverse transcriptase (this enzyme plays an essential role in the existence of many RNA viruses such as HIV). It is the method of choice for PCR studies in gene expression.

qPCR (quantitative PCR) allows reactions to be observed in real time, in order to determine the relative amount of a particular organism or group of organisms from a given DNA sample. It can also be used to determine levels of gene expression or copies of particular functional genes. When used with RNA starting product it has the potential to reveal absolute quantification of the target organism or gene. This method allows very sensitive, accurate and robust detection of given microorganisms (e.g. quarantine organisms) in complex samples, particularly when allied with the use of taxon-specific probes (and following the 'gold standard' MIQE guidelines).

Unfortunately, the level of confusion is increased as reverse transcriptase enzymes are also often used in real-time PCR. As its name suggests, real-time PCR allows a PCR reaction to be followed in real time (without the need for post-amplification gel electrophoresis separation of the PCR products). It is also used for gene expression studies. Real-time PCR is used widely for quantification studies (hence, its more correct short name: qPCR) (See Fig. 3.2). Several different methodologies are available from commercial suppliers with their own appropriate chemistries (and associated equipment) and including tailored controls. Necessary equipment includes an optics system for excitation of fluorescent dyes and capture of emissions, a PC to interpret the data and a specialist real-time PCR machine.

Limitations and Advantages of PCR

The major benefit of the PCR technique is the ability to amplify minute amounts of target DNA. This is also its biggest drawback as any contaminating DNA may also be replicated. Care must be taken to avoid unwanted contamination from the environment or even the operator's own skin. In some cases, the contaminant DNA may be replicated instead of the target DNA. Thus, scrupulous attention is needed to good laboratory practice and the use of negative controls (i.e. reactions containing no DNA), to ensure the integrity of the system. This is of particular concern when working with material extracted from sources other than pure

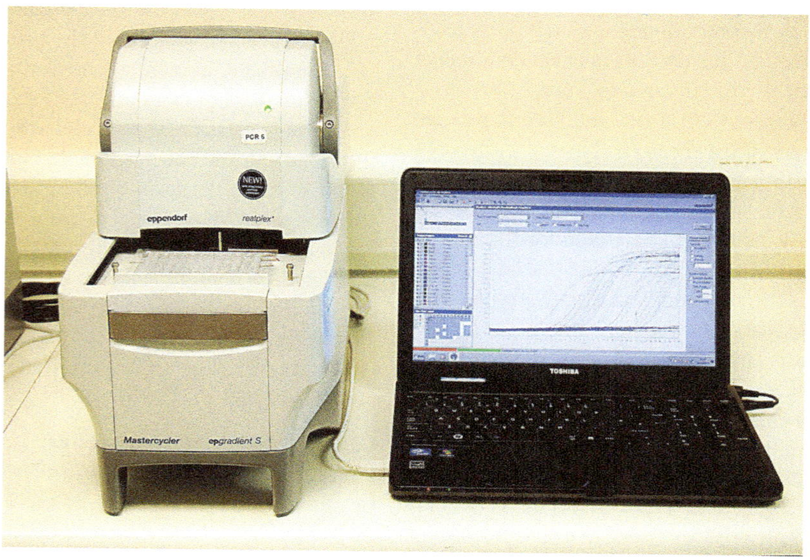

Fig. 3.2. The real-time PCR machine at CABI

culture (e.g. DNA from obligate plant pathogenic fungi where the host plant DNA may be amplified, if methodologies for careful handling of the material and appropriate selection of primers are not followed). Other disadvantages include the high cost of equipment and reagents and the need for highly trained staff and appropriate clean facilities.

Other nucleic acid amplification tests

Whilst PCR revolutionized molecular biology and became, by far, the most used approach, many other methods have been developed for amplification of nucleic acids. Of these, the most notable ones are LAMP, NASBA and RPA, each of which will be outlined briefly, below.

Loop-mediated isothermal amplification (LAMP) is the most widely used alternative to PCR. It was first described at the turn of the 21st century (Notomi *et al.*, 2000) and involves nucleic acid amplification with a set of four (or six) different primers, which bind to six (or eight) different regions on the target gene, thereby making it highly specific. The primer set consists of two outer primers, two inner primers and loop primers (from which it gains the key part of its name). LAMP methodology has been adopted widely as it may be undertaken at isothermal conditions, thereby avoiding the need for expensive thermal cyclers. Conventional LAMP assays have been used in the detection of various human pathogens in a clinical setting, including *M. tuberculosis, S. pneumoniae, B. pertussis* and others (for more details see Wong *et al.*, 2018). Many modifications to the method have been described that broaden its utility – it seems almost as adaptable as PCR itself. Further details may be found in Wong *et al.* (2018) and Becherer *et al.* (2020). Key advantages of LAMP are: its speed (ca. 30 mins for a typical assay versus ca. 90 mins for the most rapid of PCR reactions); no need for expensive thermal cyclers (it can be undertaken in a water-bath or dry-block); easy visualization of results – which tend to be shown by simple colour changes (see Wong *et al.*, 2018; Becherer *et al.*, 2020). Limitations tend to be: around the lack of downstream uses for reaction products (as compared with PCR); complexity of primer design for target specificity; incidence of cross-contamination (see Karthik *et al.*, 2014).

Nucleic acid sequence-based amplification (NASBA) was first described as an alternative to PCR for amplifying nucleic acids (Compton, 1991) and it has since been used most for amplification of RNA targets. Its key strength is that, compared to other methods, it has been shown to generate fewer false positives in the presence of genomic DNA. For RNA studies, it is faster than PCR methods that entail reverse transcriptase steps and has performed at least as well as RT-qPCR (see Compton, 1991; Brink *et al.*, 1998; Nai *et al.*, 2022).

Recombinase polymerase amplification (RPA) was developed by a commercial biotechnology company but then described by Piepenburg *et al.* (2006). It has become a popular alternative to PCR, owing to it enabling accurate, fast and cost-effective identification of organisms (Daher *et al.*, 2016). It does not require extensive sample-preparation or costly thermal cyclers, with typical reaction temperatures of 25-42 °C (Daher *et al.*, 2016). The method does not entail the use of multiple primers. It does not require initially a high temperature denaturation step to start the reaction and is probably the most rapid of all the amplification methods (Lobato and O'Sullivan, 2018). Further details on the precise method and a list of assay applications already developed, including for the human pathogens HIV-1, *Listeria monocytogenes, Plasmodium falciparum, Rickettsia typhi,* and many others, plus phytopathogenic assays may be found in Lobato and O'Sullivan (2018).

DNA Sequencing

DNA sequencing is a time-consuming, but powerful technique that allows scientists to read the nucleic acid make-up of individual regions of DNA. In addition to its uses for identification (as described previously), sequence information is essential for gene cloning and genetic manipulations. Sequence analysis is used for strain characterization and can allow discrimination of taxa. See Fig. 3.3.

Different Techniques

There may not be sufficient variability at the individual strain level, within a normal sequence length read of ca. 400-700bp (the typical length of sequence obtained using Sanger sequencing technology), to characterize individual strains. One approach is to screen for single nucleotide polymorphisms (SNPs) within a particular locus

Fig. 3.3. Sequence analysis at CABI.

but this can be difficult to assess fully. The sequencing process itself is not 100% reliable as even a high-fidelity polymerase can make occasional errors and the machine itself may misread. Screening for SNPs has been employed widely in plant pathology and in clinical studies, in order to screen for resistance to specific antifungal agents.

Alternative fingerprinting methods have been developed that employ such a large number of DNA fragments that they require a polyacrylamide matrix, such as those found within modern automated DNA sequencers, for optimal resolution of fragments (e.g. dual-enzyme sequencer based AFLPS). Another alternative is to utilize strain-unique fragments as markers (e.g. SCAR markers).

SCAR Technique

Sequencing can be carried out from known or random amplified samples. This technique is known as SCAR (sequence characterised amplified region). The technique involves the creation of a complex banding profile on a polyacrylamide gel from which a unique band may be excised, the DNA eluted and sequenced in order to create a 'tag' for a particular strain in order to distinguish isolates at the infraspecific level.

SCAR analysis allows the detection of sequence variation among genes from different organisms. Once sequence data have been obtained, sequences can be compared for similarity with the vast libraries of known sequences that are available over the internet (e.g. at the EMBL or GENBANK databases). It is used primarily to allow tracking of known and/or proprietary strains.

Sequencing has been used in systematics to define groupings at the most basic level through the sequences of specific gene regions such as the subunits of the ribosomal gene clusters, allowing proper phylogenetic analyses of taxa to be undertaken. SCAR markers are still employed as the source of strain-specific PCR primers and/or probes in determining the spread of individual strains.

DGGE/SSCP analysis

Denaturing gradient gel electrophoresis (DGGE; Fischer and Lerman, 1983) and single strand conformation polymorphism (SSCP; Orita *et al.*, 1989) were developed as alternatives to sequencing, although reduction in sequencing costs has rendered them each fairly obsolete. The techniques were useful particularly for ecological studies of mixed populations e.g. from soil, activated sludge, or other environmental samples, and were used with bacteria predominantly. Both are powerful techniques that could detect sequence variations in amplified products. DGGE is especially useful for detecting single base substitutions, it differs from SSCP because samples are not denatured prior to electrophoresis, but are run in a gradient gel under denaturing conditions. SSCP can be used to separate taxa and identify microorganisms in crude samples but was particularly useful for mutation analysis in DNA fragments of 100-300 bases (Sambrook and Russell, 2001). Amplified PCR products are denatured and samples run in a non-denaturing polyacrylamide gel. More complex banding patterns can be obtained if restriction enzymes are used during the procedure.

Electrophoretic karyotyping

Large pieces of DNA, such as whole chromosomes, can be separated by methods such as pulsed field gel electrophoresis (PFGE) (Brody and Carbon, 1989). This technique allows the separation of complete chromosomes or large chromosomal elements. In PFGE, electrodes are arranged around the gel, the current is switched between different opposing pairs, which causes large pieces (Mb size) to move through the gel, the procedure being analogous to the shaking of a sieve. Variations in the placing of the electrodes (and, subsequently, the nature of the electric field) have led to refinements such as contour-clamped homogeneous electric field (CHEF), orthogonal field-alternating gel electrophoresis (OFAGE) and rotating field gel electrophoresis (ROFE) (Burmeister and Ulanovsky, 1992). The 'pulse' time (or switching time) and the running time both increase with increasing fragment size. Most separations of chromosomes from filamentous fungi have involved pulse times of 100-3000 seconds and electrophoresis runs of between 2 and 7 days (bacterial runs are normally about 1 day). It is of great importance to the reproducibility of results that the DNA sample is not physically disrupted and the chromosomes remain intact. To achieve this, protoplasts are prepared from fungi and lysed *in situ* in agarose blocks. The blocks are then set into the electrophoresis gel (Sambrook *et al.*, 1989). Interestingly, whilst its use for karyotyping with fungi has gone out of fashion due to advances in whole genome sequencing, PFGE has maintained its popularity as a robust strain fingerprinting tool for bacteria. This requires use of a very rare-cutting restriction endonuclease to digest the bacterial genome into a 'ladder' of fragments and the runs (and pulse times) are much shorter than for the fungal karyotyping. This approach is still considered useful in clinical approaches such as strain typing of *Mycobacterium* spp., for which it is still considered the 'gold standard' method (Shin *et al.*, 2020).

Next generation sequencing

Next generation sequencing (NGS) is a catch-all term for methods that enable massive parallel sequencing. This has only been made possible by advances in technology and microfluidics. The most well-known platforms currently are: i) Illumina's Fluorescently-labelled sequencing by synthesis; ii) PacBio's Single-molecule real-time sequencing; and iii) Oxford Nanopore Technologies' Nanopore sequencing, see Fig. 3.4.

i) Illumina platform: The Illumina technology makes use of four fluorescently-labelled chemical dyes that resemble the deoxynucleotide triphosphates dATP, dCTP, dGTP and dTTP but which also contain chemically-removable groups that prevent further chain extension by a DNA polymerase. First, the DNA is carefully broken (by enzyme or ultrasound) into fragments of ca. 200-500bp. Then specialized oligonucleotide adaptors are attached to each end of the fragments. A carefully controlled number of molecules is deposited onto a glass slide, giving a single DNA molecule roughly every 1 μm or so on the surface. Then these single molecules are replicated (by enzyme action) on the surface until each site contains thousands of identical DNA molecules (generating different DNA sequence clusters at each site on the array). Then all sites (ca. 25×10^6) are sequenced, in parallel. This approach gives ca. 25 million sequences of up to 600bp. This is ideal for metabarcoding (sequencing a short part of the rRNA gene cluster for fungi [ITS] or bacteria [part of the 16S] to determine all the component taxa in a complex sample, such as a soil sample), and the accuracy and output of the MiSeq/Hiseq units make them the 'go to' approach for such screening but less suitable for whole genome sequencing of a microorganism where a huge amount of computing

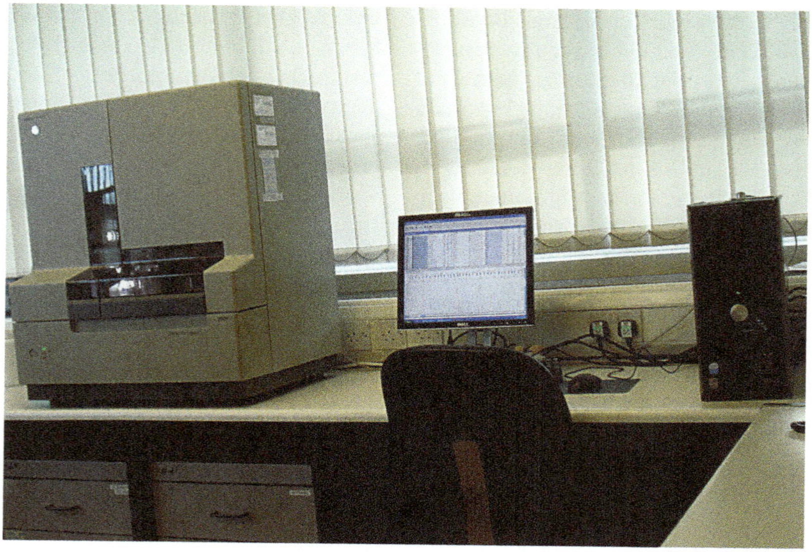

Fig. 3.4. ABI3130 sequencer at CABI

power and bioinformatics is required to assemble the genome from such small fragments, unless a good reference genome is available to act as a 'scaffold' (see Slatko *et al.*, 2018 for more details on this platform). It is the workhorse of the current boom in microbiome analyses.

ii) PacBio platform. This technology is currently the 'gold standard' of NGS approaches for whole genome sequencing, as it enables long read sequencing of a single DNA molecule to a high degree of accuracy. Individual DNA molecules are sequenced in real time in a highly-parallel manner. The methodology was developed by Pacific Biosciences, and employs microfabricated devices known as zero-mode waveguides (ZMWs). The method again makes use of four fluorescently-labelled chemicals that resemble the deoxynucleotide triphosphates dATP, dCTP, dGTP and dTTP but which react relatively slowly. As each nucleotide is added to the DNA, the emitted fluorescence indicates briefly which base was incorporated before a fluorescently-labelled 'pentaphosphate' fragment diffuses from the field of view, ready for the next cycle. Many thousands of ZMWs are observed in parallel. Outputs are contiguous reads of >10kbp across almost half a million sequences (as above, see Slatko *et al.*, 2018 for more details).

iii) ONT Nanopore sequencing. Nanopore sequencers use protein nanopores in an electrically resistant polymer membrane. When a single nucleotide passes through the membrane, a characteristic current change occurs and this is detected. Long, single nucleotide sequences can ratchet through the nanopore and recording is in real time. Current average reads are >10Kb, with a single read of 2Mb recorded. This platform is very portable, being the size of a small smartphone and plugging in to a laptop USB slot. They have been used in the field for DNA sequencing of the Ebola virus in Guinea (Quick *et al.*, 2016) and have been used in Antarctica (e.g. Johnson *et al.*, 2017). However, despite considerable advances in the chemistry and protocols, there remain concerns over the error rate with this platform.

Characterizing mixed samples/ communities

There has been recent interest in gaining an understanding of complex ecosystems and environments such as the microbiomes present in the human (or animal) gut, biofilms, fermented foods, activated sludge and, increasingly soil and rhizosphere samples. Whilst modifications of some of the phenotypic methods mentioned previously (e.g. FAME-GC, BIOLOG plates and even serial dilution plating) may enable the elucidation of some parts of the respective samples, it has only been with the revolution in sequencing capability described above that the true scale of the microbial component has been revealed. Metabarcoding and metagenomic approaches using the massively parallel sequencing platforms (and related bioinformatic/computing advances) have shown not just the organisms present but are also starting to discern their respective function(s) within the system. These new approaches mean that non-culturable microorganisms can be detected and sequenced; if starting from RNA, they may also discriminate the viable organisms from the non-viable 'historic' DNA vestiges of dead cells and may show the actual metabolic function of the relevant taxa within the ecosystem. For further details on this vast and rapidly growing field of research, (see e.g. Mendes *et al.*, 2013 (rhizosphere); Walsh *et al.*, 2016 (kefir); Barko *et al.*, 2018 (mammalian gut); Cafà *et al.*, 2019 (agricultural soil); Callewaert *et al.*, 2020 (human skin); Mukherjee and Reddy, 2020 (polluted environments); etc.)

Food spoilage strains

Microbial spoilage of foodstuffs costs the food industry £millions world-wide every year. Added to this is the significant risk to public health that arises after the consumption of spoiled food, either due to the direct pathogenesis by food-borne bacteria, or the toxins released by the microorganisms during and after substrate breakdown. Many food biodeteriogens are tolerant to added preservatives or have survived sterilization procedures and are therefore useful to scientists whose research is aimed at improving food preservation. Examples of food products from which strains have been isolated include: gelatine (*Geomyces pannorum*), lime flavour carton drink (*Trichoderma viride*), table jelly (*Xeromyces bisporus*), jam (*Wallemia sebi*), canned blueberries (*Penicillium lapidosum*), cheese (*Lactobacillus brevis*), fish (*Shewanella putrefaciens*), milk (*Enterobacter aerogenes*) and meat (*Pseudomonas* sp.).

Harnessing Microbial Resources

The characterization methods described above enable the identification of microbial properties that can lead to valuable products and uses. Microorganisms play a very important role in today's world. It is often their negative side that is drawn attention to, for example they cause diseases in plants, animals and humans, cause damage to artificial environments and spoil and contaminate our food. However, they have a positive side too, and this topic provides an overview of the different ways in which microbes have been harnessed for the benefit of mankind; from food and drinks to antibiotic and immunosuppressant drugs.

The public service culture collections hold a large and diverse number of strains and cell lines. Of these, a significant number have important properties that can be utilized by the wider scientific community. In this chapter we have discussed some of these properties and content should be used in conjunction with Appendix A, this volume, where an extensive list of microbial properties is given citing examples of the organisms and the strains that exhibit them. The information on strain properties in a collection often originates from the depositor and on occasion has been confirmed by the collection. In the process of collection activities and research or that of its parental organization the data may have been generated internally. Occasionally strains may cease to exhibit properties during storage but more often organisms fail to exhibit the property because they are not grown under the prescribed conditions. It is essential that appropriate growth conditions are provided, information that is included in Chapter 2, this volume, and generally available in the published literature or directly from the collection itself. Where a collection is not able to confirm a particular property, it cannot guarantee that it will be present when supplied, for example if an organism is cited as a tree pathogen, rarely has this been confirmed by tests following preservation.

Many cultures are representative of their species and are deposited as 'ex type strains' while others may exhibit a particular morphological, physiological or anatomical property that can be used for research or teaching purposes. The commercial exploitation of microbes, cell lines and their properties has been associated with substantial research on a large number of economically important organisms. Information presented here is designed to help scientists locate cultures with specific chemical or functional properties and provides an insight to the methodology used to determine some of these properties.

Test Strains

Many microorganisms are used to assess the effectiveness of antimicrobials, e.g. antiseptics (*Mycobacterium hirae*), antibiotic susceptibility (*Pseudomonas aeruginosa*) and disinfectants (*Bacillus cereus*, *Mycobacterium fortuitum*, *Staphylococcus aureus*).

Key reference strains are used to test materials against microbial attack and many of these are cited in international quality control standards such as UKAS, MIL, ASTM etc. For example: emulsion paint (*Enterobacter aerogenes*), cutting fluids (*Shewanella putrefaciens*), fabrics (*Klebsiella pneumoniae* subsp. *pneumoniae*), vinyl (*Streptoverticillium reticulum*) and wall covering adhesive (*Aspergillus niger*).

Many reference strains named in technical standards conform to national and international specifications. These include – mould proofing (by strains of *Penicillium pinophilum* and *Rhizopus stolonifer*) and wood preservatives in BS6009:1982 (*Poria placenta*).

Microbial Products Used in Industry

Enzyme producing strains

Microorganisms produce a novel and diverse array of extra- and intra-cellular enzymes from urease synthesized by strains of the bacterium *Bacillus fastidiosus* to cellulase and lipase synthesized by strains of the fungus *Trichoderma viride*. Over 50 enzymes and the strains that produce them in the CABI collection (c. 120) are listed in Appendix A, this volume. Culture collections attempt to characterize their holdings but inevitably they hold many untested representatives: the potential of the uncharacterized strains is enormous. Saprophytic organisms (e.g. many fungi) excrete extracellular enzymes into the substratum and these can be easily extracted and purified. Other organisms produce intracellular enzymes that are more difficult to extract. Many enzymes are produced on an industrial scale (e.g. amylase from strains of *Aspergillus niger*, *A. oryzae*, *Bacillus polymyxa*, *B. subtilis*, and *Trichoderma viride*). Enzymes can also be used to help taxonomically characterize organisms, while some enzymes and combinations of enzymes are distinct for specific microorganisms (e.g. inulinase synthesis by *Sphingobacterium multivorum*). Other enzymes are representative of an organism's mode of life (e.g. insect cuticle degrading enzymes synthesized by the entomopathogenic fungus *Metarhizium anisopliae*). Enzymes and their genes are studied widely within the fields of biochemistry and molecular biology, and the collections hold accessions for genetically modified enzyme producers (e.g. arginase from a modified strain of *Zygosaccharomyces pombe*).

Here are some examples of enzymes produced by microorganisms:

- Amylase from *Aspergillus oryzae* – Used in detergents, textile and paper industries and medicine (Saxena *et al*., 2004)
- Proteases from *Aspergillus* and *Mucor* spp. – Used in detergents, leather industries and medicine (Saxena *et al*., 2004)
- Lipases from *Candida*, *Rhizopus* and *Aspergillus* spp. – Used in the cosmetics, leather and detergent industries (Saxena *et al*., 2004)

See Appendix A, this volume, for further examples.

Metabolite producing strains

The CABI Culture Collection holds a unique and diverse array of metabolite producing strains, almost 400 compounds are listed in Appendix A, this volume, produced by over 1000 organisms, the numbers of which are increasing all the time. Both primary and secondary metabolite producers are included. However, many metabolites are only produced under pre-specified experimental conditions. Microbial metabolites are of enormous economic importance in today's society and many are extracted for use in the food and chemical industries and are produced on an industrial scale (e.g. alcohols by *Rhizopus oryzae* and organic acids by *Aspergillus niger*). Examples of primary metabolites produced by strains include carbohydrates (e.g. cellulose by *Acetobacter aceti*, exopolysaccharide by *Alteromonas atlantica* and mannocarlose by *Penicillium fellutanum*), fats (e.g. produced by *Penicillium indonesiae* [now *P. javanicum*]), amino acids (e.g. tryptophan and lysine produced by strains of *Escherichia coli*) and organic acids which are widely produced by both bacteria and fungi (e.g. glutamic acid by *Corneybacterium* sp. and mycophenolic acid by *Penicillium brevicompactum*). Some microorganisms produce vitamins (e.g. vitamin B12 synthesis by the bacterium *Propionibacterium freudenreichii*), secondary carotenoids (e.g. astaxanthin synthesized by *Haematococcus pluvialis*) and hormones (e.g. abscisic acid synthesized by strains of *Cercospora rosicola*; antheridiol synthesized by strains of *Achlya bisexualis*; gibberellin synthesized by strains of *Fusarium moniliforme*).

Secondary metabolites are of huge economic importance, many are of fungal origin (Turner and Aldridge, 1983). Of these, the antibiotics are of particular economic significance. Some secondary metabolites are highly toxic (e.g. the carcinogen patulin produced by *Penicillium expansum*, aflatoxin produced by *Aspergillus flavus*; anatoxin-a produced by the cyanobacterium *Anabaena flosaquae* and exotoxin produced by the bacterium *Pseudomonas aeruginosa*) while others are beneficial and utilized by food and industry (e.g. a blue pigment produced by *Pseudomonas fluorescens* and a fluorescent red diffusible pigment produced by *Arthrobacter* sp.). Some secondary metabolites are specific to an organism's mode of life, for example the destruxins produced by *Metarhizium anisopliae* which are essential for the induction and maintenance of pathogenesis in the insect host. Some workers have used metabolite profiles to classify fungi. For example, Svendsen and Frisvad (1994) demonstrated that chemosystematics of secondary metabolite profiles could be applied to separate 279 terverticillate *Penicillium* spp. using cluster analysis.

See Appendix A, this volume, for further examples.

Pharmaceutical Industry

Microorganisms have been utilized in the pharmaceutical industry for many years. Since the discovery of Penicillin from *Penicillium notatum* by Alexander Fleming, in the 1930s, the range and effectiveness of antibiotics has been steadily improved and increased.

An antibiotic is a microbial product (typically a secondary metabolite although can be synthetically produced) that in low concentration can suspend or extinguish microbial growth. Although the majority of antibiotic producing strains are antibacterial, there are also antifungal and antiprotozoal producing strains. Other products include biopolymers such as cellulose which are used in drug formulations and also to aid in the healing of wounds. Table 3.3 below displays some examples of pharmaceutical products derived from microorganisms.

Horticulture and Agriculture

Microorganisms play a huge role in horticulture and agriculture through mutualistic symbiosis, from nitrogen fixing bacteria to mycorrhiza. This aids in:

- the establishment, growth and development of plants by improving mineral nutrition,
- increasing drought tolerance,
- supplying growth-regulating substances and vitamins, and
- protecting against pollutants and soil-borne pathogens (Mitchell, 1993).

Mycorrhizal technology also has applications in forestry, desertified ecosystems, bioremediation, biodegradation, recovery of sterilized soils and environmental recovery programmes.

There are many applications of microbes in horticulture. *Agrobacterium* spp. have been used in plant breeding applications (mediated gene transfer) to improve herbicide tolerance, insect resistance, virus resistance and male sterility. *Agrobacterium tumefaciens* and *Agrobacterium rhizogenes* are used as vectors for the production of transgenic plants because the organism is capable of transferring its plasmid DNA directly into plant chromosomes (Evans, 1993).

The bacteria *Rhizobium*, *Bradyrhizobium* and *Azorhizobium* are leguminosae plant symbionts, strains that may be host specific, e.g. *Rhizobium leguminosarum* biovar. *phaseoli* to bean plants. Rhizobia make nitrogen available to the plant host and harvest yields may be increased if *Rhizobium* spp. are inoculated directly into plant rhizosphere (Jones and Lewis, 1993). Further symbiotic associations that make atmospheric nitrogen available to plants include those between: 1) *Gluconacetobacter diazotrophicus* and sugar cane (see James *et al*., 2001); and 2) the cyanobacterium (*Anabaena* spp.) and *Azolla* (water-fern). Many 'free living' cyanobacteria

Table 3.3. Examples of pharmaceutical products derived from microorganisms – antibiotics (adapted from Smith and Ryan, 2008)

Compound	Organism derived from
Antibacterial	
Actinomycin	*Streptomyces* spp.
Borrelidin	*Streptomyces parvulus*
Ceftriaxone	*Acremonium* spp.
Chloramphenicol	*Streptomyces venezuelae*
Ciclosporin	*Tolypocladium inflatum*
Gentamicin	*Micromonospora purpurea*
Penicillin	*Penicillium notatum*, *Penicillium chrysogenum*
Antifungal	
Griseofulvin	*Penicillium griseofulvum*
Amphotericin B	*Streptomyces nodosus*
Antiparasitic	
Ivermectin	*Streptomyces avermitilis*
Examples of pharmaceutical products derived from microorganisms – cholesterol lowering	
Lovastatin, Pravastatin, Simvastatin	*Aspergillus terreus, Monascus ruber*
Mevinolin	*Aspergillus terreus*
Examples of pharmaceutical products derived from microorganisms – anti-tumour and immuno-suppressants	
Borrelidin	*Streptomyces parvulus*
Cephalosporin	*Acremonium strictum* (now *Sarocladium strictum*)
Examples of pharmaceutical products derived from microorganisms – used during and after childbirth	
Ergotamine	*Claviceps purpurea*

(e.g. *Calothrix* spp. and *Gloeotrichia* spp.) also make nitrogen available to plants.

Similarly, fungal mycorrhizal plant symbionts can be beneficial to the horticulturist. Mycorrhiza aid the establishment, growth and development of plants by improving mineral nutrition, increasing drought tolerance, supplying growth-regulating substances and vitamins and protecting against pollutants and soil-borne pathogens (Mitchell, 1993). Inoculating agricultural soils with mycorrhizae may therefore improve harvest yield. There are a number of types of mycorrhizal association: ectomycorrhizal (e.g. *Cantharellus cibarius*, *Boletus edulis*), arbuscular mycorrhizal (e.g. *Glomus*, *Gigaspora*), ericoid mycorrhizal (e.g. *Hymenoscyphus ericae*) and orchidaceous mycorrhizal (e.g. *Armillaria* spp., *Thanatephorus* spp.). Mycorrhizal technology also has applications in forestry, desertified ecosystems, bioremediation, biodegredation, recovery of sterilized soils and environmental recovery programmes.

Concern about the use of chemical agents used in the prevention of animal and crop diseases and/or to control insect pests and weeds, has led to the development of biological control agents (BCAs). Many of these are microbial. BCAs are used in integrated pest management and crop protection.

Table 3.4 below displays some examples of microorganisms important to agriculture and horticulture and explains their role.

Food Production

Microorganisms have been utilized as food for thousands of years. Today microorganisms are not only being utilized for the production of fermented food, baking and cheese production, but also as ingredients in food manufacturing. Furthermore, fungal fruiting bodies (i.e. 'mushrooms') are eaten world-wide. They are commercially produced e.g. the edible mushroom *Agaricus bisporus*, the oyster mushroom *Pleurotus ostreatus*, the shiitake mushroom *Lentinus edodes* and the straw mushroom *Volvariella* spp. (Chang *et al.*, 1993), while others are strictly the domain of the specialist field mycologist (e.g. the Dotted-stem Bolete, *Boletus luridiformis*). More recently, biotechnological development has allowed the mass production of single celled

Table 3.4. Examples of microorganisms important to agriculture and horticulture

Organism	Use
Bacteria	
Agrobacterium tumefaciens; *Agrobacterium rhizogenes*	Used in plant breeding for mediated gene transfer to improve resistance etc.
Anabaena spp.	With *Azolla* spp. (water fern); makes nitrogen available to plants in paddy fields
Bacillus thuringiensis	Used to control insect pests
Ruminococcus spp.; *Bacteroides succinogenes* (now *Fibrobacter succinogenes*); *Butyrivibrio* spp.	Used in fodder to aid cellulose degradation during rumination
Rhizobium spp.	Makes nitrogen available to plants
Calothrix spp.	
Gloeotrichia spp.	
Chromista	
Achlya bisexualis	Used to produce the plant hormone antheridiol
Fungi	
Cantharellus cibarius	Ectomycorrhizal
Boletus edulis	
Glomus spp.	Arbuscular mycorrhizal
Gigaspora spp.	
Hymenoscyphus ericae	Ericoid mycorrhizal
Armillaria spp.	Orchidaceous mycorrhizal
Thanatephorus spp.	
Cercospora rosicola	Used to produce the plant hormone abscisic acid
Fusarium moniliforme	Used to produce the plant hormone gibberellin
Metarhizium anisopliae	Used for the biological control of locusts
Yeast	
Saccharomyces cerevisiae	Cattle fodder (silage)
Saccharomyces boulardii	Probiotic, maintains natural flora of large and small intestines
Yarrowia lipolytica	Bioremediation of palm oil mill effluent

proteins (e.g. from strains of *Saccharomyces cerevisiae*, *Kluyveromyces fragilis marxianus* and *Candida utilis* (Arora *et al.*, 1991)) and the bacterium *Methylophilus methylovorus* which produces single cell protein from methanol. Probably the most well-known example is processed mycelium (Quorn™ mycoprotein) production by *Fusarium venenatum* (Wiebe, 2002).

Many microbial products are used as ingredients in food production (e.g. alginates produced by *Pseudomonas mendocina*), while others are vital components of food manufacturing (Table 3.5). For example, bacteria and fungi are used in the production of fermented food, baking and cheese production. Strains of the yeast *Saccharomyces cerevisiae* are used in the brewing of beers, fermentation of wines and distilling of spirits and the baking of bread. Many foods are fermented including Indonesian Tempe from soybean using *Rhizopus* sp. (Onions *et al.*, 1981) or *Lactobacillus* spp. Other examples include acid fermented vegetables e.g. sauerkraut and pickles using *Aspergillus oryzae*, *Leuconostoc mesenteroides*, *Lactobacillus brevis*, *Lactobacillus plantarum*, *Pediococcus cerevisiae* and *Lactobacillus brevis*; soy sauce using *Aspergillus oryzae* and Japanese miso soybean using *Saccharomyces rouxii* (Steinkraus, 1983). Mould ripened cheeses require cultures of *Penicillium camembertii*, *Penicillium roquefortii* (Beuchat, 1987) and yoghurts require bacterial starter cultures (e.g. *Lactobacillus* spp.). There is also a keen interest in recent years in the use of organisms, such as *Lactobacillus*, in probiotics as well as a growth in actively fermenting microbial products such as kefir (see Walsh *et al.*, 2016) and kombucha (see Marsh *et al.*, 2014).

Bioremediation and Biodegradation

Degraders and utilizers are a distinct and economically important group of microorganisms.

Table 3.5. Metabolic products of microorganisms used in food production

Organism	Metabolite/Enzyme	Product
Xanthophyllomyces dendrorhous	Astaxanthin	Colouring*
Penicillium oxalicum,	Arpink Red	Colouring*
Ashbya gossypii	Riboflavin	Colouring*
Blakeslea trispora	β-Carotene	Colouring*
Corynebacterium glutamicum	Glutamic acids	Flavourings
Aspergillus niger	Organic acids e.g. citric acid	e.g. Beverages
Propionibacterium freudenreichii	Vitamin B12	Vitamin
Aspergillus spp. and *Bacillus* spp.	Amylase	Bread and noodle production
Aspergillus spp. and other microorganisms	Xylanase	Clarification and flavourings
Aspergillus spp. and other fungi	Lipase	Cheese and milk flavours
Pseudomonas mendocina	Alginates	Gelling/thickening agent

**See Dufossé, 2006*
Further examples can be seen in Appendix A, this volume.

Often these microbes possess physiological and biochemical properties that can be utilized by the wider economic community. Substrates degraded include alcohol (*Rhodococcus* spp.), aldehydes (*Amycolata hydrocarbonoxydans* [now *Nocardia hydrocarbonoxydans*]), benzaldehyde (*Pseudomonas* spp.), benzene (*Acinetobacter* spp.), crude oil (*Pseudomonas* spp., *Rhodococcus rhodochrous*), detergent (*Rhizobiaceae* sp.), industrial waste (*Chaetomium jodhpurense*), lignin (*Phanerochaete chrysosporium*), methane (*Methylococcus capsulatus*), methanol (*Alteromonas* spp.), nitriles (*Rhodococcus* spp.), paper (*Phoma violacea*), phenol (*Pseudomonas putida*), rubber (*Nocardia* spp.), salicylic acid (*Acinetobacter* spp.), xylan (*Bacillus subtilis*), cyanide (*Pseudomonas fluorescens*) and cresols (*Pseudomonas putida*).

Biodeterioration, any undesirable change in the properties of a material caused by the vital activities of organisms, (Hawksworth *et al.*, 1995; Kirk, *et al.*, 2008) of synthetic and natural products by microorganisms costs £millions each year. Many biodeteriogens may have important physiological and chemical properties (e.g. enzymes capable of breaking down complex organic or synthetic substrates) that can be utilized by the wider scientific community or in the development of biodegraders or bioremediators. Many others can be used in research aimed at establishing the mechanisms involved in substrate breakdown, the information from which can be used to improve the resistance of materials to microbial attack.

Culture collections hold accessions of many diverse and interesting biodeteriogens including those isolated from aviation fuel (*Amorphotheca resinae, Aspergillus fumigatus*), a binocular lens (*Aspergillus tonophilus*), cellophane paper (*Aspergillus versicolor*), a concrete wall (*Arthrobotrys javinica*), cotton yarn (*Aspergillus tamarii*), creosoted wood (*Amorphotheca resinae*), diesel fuel (*Trichoderma harzianum*), kerosene fuel filter (*Pestalotiopsis aquatica*), metal working fluid (*Scopulariopsis brevicaulis*), paint (*Cladosporium sphaerospermum*), plastics (*Penicillium pinophilum*), polurethane foam (*Penicillium chrysogenum*), a radio set (*Aspergillus ustus*), vinyl wall covering (*Penicillium pupurogenum*), wallpaper (*Aspergillus tamarii*), PVC (*Pseudomonas* spp.), rubber (*Nocardia* spp.), cider (*Zygomonas mobilis* subsp. *pomacii*), steel and cutting fluids (*Shewanella putrefaciens*), and concrete (*Thiobacillus concretovorans*).

Strains used in the biodegradation of wastes include *Acetobacterium woodii, Methanosarcina barkeri* and *Lactobacillus casei* which are used specifically for the anaerobic digestion of food processing wastes (Greenshields, 1989). Agricultural wastes are degraded using *Zooglea ramigera* (Hawker and Linton, 1971). A controlled form of biodegradation is bioremediation, which can be achieved using bioreactors, solid phase treatments, composting, land farming or *in situ* treatment (Balba, 1993). For example: the recovery of heavy metals from aqueous effluents: copper using *Penicillium spinulosum* and *Trichoderma viride* (Greenshields, 1989), gold using *Chlorella vulgaris* and uranium using *Chlorella regularis* and *Aspergillus niger*, or the degradation of the phased-out herbicide 2,4,5-T [(2,4,5-trichlorophenoxy) acetic acid-triethylammonium] with bacteria.

Tolerant Strains

Tolerant strains are often isolated from extreme environments. Their unique biochemistry makes them useful for research in a variety of fields. Microorganisms can be tolerant to many substances. Some examples include benzimidazole (*Venturia inaequalis*), benzoic acid (*Zygosaccharomyces bailii*), copper (*Scytalidium acidophilum*), creosote (*Vibrio alginolyticus*), glucose (*Candida magnoliae*), sodium chloride (*Zygosaccharomyces rouxii*) and silver (*Stenotrophomonas maltophilia*).

Resistant Strains

Resistant strains are useful in research aimed at determining the mechanisms of resistance (i.e. to antibiotics or radiation) and for researching new mechanisms to counter the resistance. Antibiotic resistance is an increasing problem. Examples of resistant strains include erythromycin (*Micrococcus luteus*), aminoglycosides (*Escherichia coli*), ampicillin (*Escherichia coli* resistant phage) benomyl fungicide (*Penicillium hirsutum*), gentamycin (*Escherichia coli*), methicillin (*Staphylococcus aureus*), penicillin (*Staphylococcus aureus*), streptomycin (*Micrococcus luteus*) and tetracycline (*Staphylococcus aureus*).

Comparing such organisms with those deposited in a collection prior to discovery and first use of antibiotics can generate useful information in the combat of resistant diseases. Chemical and physical resistant strains include those resistant to UV (*Deinococcus radiodurans*), formaldehyde (*Pseudomonas putida*), gamma rays (*Rubrobacter radiotolerans*), germicides (*Pseudomonas aeruginosa*), heavy metal ions (*Stenotrophomonas maltophilia, Thermus flavus, Zooglena ramigira*), chlorine (*Flavobacterium* spp.), radiation (*Deinococcus radiodurans*) and phenol mercury 2.5ppm (*Penicillium crustosum*).

Pathogens

Pathogens that are deposited in culture collections provide valuable reference resources to plant and animal pathologists and the wider scientific community. Many pathogens have been developed for use as biological control agents, while others exhibit physiological and biochemical properties, which allow scientists to utilize them in the fight against virulent environmental strains. Additionally, suspensions of killed or live attenuated bacteria can be used as vaccines. There are thousands of pathogens and parasites deposited in culture collections worldwide. Appendix A, microbial properties, lists only examples of mycoparasites and those pathogens employed as biological control agents for insect pests and plant diseases. It is not practical to include all parasite and opportunistic pathogen which would total over 15,000 strains. Users can access the UK national collection holdings via the UKBRCN link to their catalogues (https://www.ukbrcn.org/who-we-are/) of pathogens using organism or additional host-specific search criteria. As mentioned elsewhere in this manual, licences are often required before pathogens are dispatched and certain pathogens will not be released to unknown or unauthorized parties.

There are many and varied microbial plant pathogens that are pathogenic to nearly every identified plant species. Common fungal plant pathogens include *Fusarium oxysporum* (which has over 100 so-called 'special forms'), other *Fusarium* spp., *Elsinoe* spp., *Alternaria* spp., *Aphanomyces* spp., *Ganoderma* spp., *Pleurotus* spp., *Nectria* spp., *Venturia* spp., *Ascochyta* spp., *Gibberella* spp., *Amylostereum* spp. Fungal pathogens of humans include the causative agents of athlete's foot (*Epidermophyton floccosum* and *Trichophyton* spp.), alimentary toxic aleulia mycotoxicosis (*Fusarium tricinctum*) and farmer's lung aspergillosis (*Aspergillus fumigatus*). There are hundreds of pathogenic human bacteria, the most severe include *Shigella dysenteriae*, causative agent of dysentery; *Vibrio choleriae*, causative agent of cholera; *Clostridium tetani*, causative agent of tetanus.

Microbial pathogens are of particular interest and many have potential for development as biological control agents for fungal parasites of crops (see Appendix A, this volume: Biological control agents. Mycoparasites include *Piptocephalis* spp., *Syncephalis* spp. and *Verticillium* spp. Examples of fish pathogens are *Vibrio anguillarum*, *Aeromonas salmonicida* and *Aerococcus viridans*.

Biological control strains

Concern about the use of chemical agents against animal and crop diseases and to control insect pests and weeds has led to the development of biological control agents (BCA), many of which are microbial. BCA should not damage the natural biota of the target environment and generally have no adverse physical effects, making them preferable over potentially harmful chemical agents, which may not be specific to the target organism and may pollute and damage the ecosystem. Many BCA have been developed from cultures directly isolated from the target organisms or from isolates showing an *in vitro* chemical or antagonistic property against the target organism. Indeed, non-pathogenic forms are being used to control pathogenic forms of the same species. BCA are used in integrated pest management and crop protection. Some of the strains listed as pathogens may have the potential for development as biological control agents. The entomopathogenic fungus *Metarhizium anisopliae* has been used to control a wide range of insect pests including: locusts (*Schistocerca gregaria*) (Prior, 1992), spittle bugs (*Manhanarva posticata*) (Risco, 1980), German cockroaches (*Blatella germanica*) (McCammon and Rath, 1994), the Australian pasture pests (*Adoryphorus couloni*) (Rath *et al.*, 1995) and tsetse flies (*Glosina* spp.) (Kaaya and Munyinyi, 1995). Entomopathogenic nematodes of the genera *Steinernema* and *Heterorhabditus* are host to strains of the bacterial genera *Xenorhabdus* and *Photorhabdus*, respectively, which have been proposed as the source of their insect-killing abilities (see Forst *et al.*, 1997). Plant weeds have been controlled with *Phragmidium* spp. and *Puccinia* spp., (Julien, 1992), soil borne plant pathogens have been controlled with *Trichoderma* spp. and *Fusarium* spp. (Hornby, 1990) and plant parasitic nematodes controlled with *Arthrobotrys* spp. and *Dactylella* spp. (Stirling, 1991). Other fungal biocontrol strains include *Cordyceps* spp., *Ascheronia* spp., *Verticillium* spp. and *Beauveria* spp. (see Appendix A, this volume: Biological control agents). Bacterial biocontrol agents are also found in a large range of taxa including actinomycetes, *Bacillus* spp. and fluorescent pseudomonads.

Environmental strains

Microorganisms isolated from extreme and interesting environments often exhibit unique properties, many of which are biologically important and can be utilized in research projects. For example, strains isolated from hot springs (e.g. *Naegleria clarki* and *Sulfolobus acidocaldarius*) may withstand high temperatures (i.e. are thermophilic). Isolates isolated from the Antarctic (e.g. *Heteromita globosa* and *Xanthonema* spp.) may withstand low temperatures (i.e. are psychrophilic). Likewise, organisms isolated from diverse chemical environments can withstand extreme pH and chemical conditions (e.g. the halophile *Marinococcus halophilus*, the alkalophile *Natronobacterium gregoryi* and the acidophile *Lactobacillus acetotolerans*). Other examples are a strain of *Radiosphaera negevensis* isolated from the desert that may be able to survive dehydration, a strain of *Anabaena variabilis* isolated from the deep-sea (70m), which can survive high pressure and *Deinococcus radiodurans* which can survive high levels of gamma radiation. An isolate of *Hemichloris antarctica* isolated from a rock in the Antarctic may be able to withstand low temperature, osmotic and chemical extremes. Similarly, organisms that are isolated from mutualistic associations (symbionts) often exhibit unique or unusual chemical properties (e.g. the lichen symbionts *Coccomyxa pringsheimii* and *Trebouxia crenulata*). Other microorganisms isolated from interesting environments that may have unique properties include the coprophilous fungi (e.g. *Utharomyces* and *Piptocephalis* spp.), river foam isolates (e.g. *Tripospermum camelopardus* and *Varicosporium helicosporum*), aquatic protistan bloom isolates (e.g. *Stentor* spp. and *Chrysochromulina brevitturrita*) and thermostable protease producing *Bacillus* spp. which are used in biological washing powder.

Special properties: morphological and physiological

Although not deposited as ex-type strains, some isolates may clearly exhibit a feature representative of a taxonomic group or of particular interest. This could be an anatomical or morphological feature, a physiological feature or a molecular feature. Anatomical features range from strand formation in the fungus *Serpula lacrymans* and akinete production in the alga *Anabaena variabilis*, to rhapidosomes in the bacterium *Saprospira grandis*. Physiological features include adjuvant effects in the bacterium *Gordonia rubropertincta* and the study of adhesion in *Pseudomonas* sp. to the white mutant of the fungus *Neurospora crassa*. In addition, CABI-IMI has 144 fungal mating strains which can be used for teaching or research. Other UK national collections also list strains suitable for teaching purposes (https://www.ukbrcn.org/who-we-are/).

Special properties: chemical (bioconversion/biotransformation etc.)

Many strains carry out specific chemical processes such as bioconversion or biotransformation, for example ascorbic acid reduction by a strain of *Escherichia coli*, caproic acid oxidation by a strain of *Bacillus sphaericus*, hydrocarbon oxidation by a strain of the genus *Corynebacterium* and steroid hydroxylation by a strain of *Rhizopus arrhizus* (see Appendix A, this volume: Chemical transformation, bioconversion and bioaccumulation).

Genetic strains

The UK's National collections hold a large number of phages, plasmids and genetically modified organisms that have a wide range of properties and applications including mapping strains, suppressor strains, mis-sense suppressors, producer strains, auxotrophs, sensitive mutants, deficient mutants, nuclear division arrest strains, MINI INC -GAL plasmids, plasmids used for molecular weight determinations, transformable strains, transposon donor vectors, fermentation markers, cell division cycle mutants (see https://www.ukbrcn.org/who-we-are/).

Summary

Of the living biomass on the earth, fifty percent is made up of microorganisms. They are an essential component of biological diversity, and serve an important role in maintenance and sustainability of ecosystems. They provide a major source of genetic information to solve many problems in agriculture, industry, plant, animal and human health, and several other biotechnological applications. This topic has described the potential uses of these organisms to make life better for us, including:

- The importance of strain identity (taxonomy) and characterization;
- The potential benefits of microbial resources;
- The use of the different types of strains;
- The commercial use of strains and how to identify potential new ones for exploitation, based on similar properties.

References

Abdulaal, W.H. (2018) Purification and characterization of α-amylase from *Trichoderma pseudokoningii*. *BMC Biochemistry* 19(1), 4. doi:10.1186/s12858-018-0094-8. PMID: 29902965; PMCID: PMC6001013.

Adl, S.M., Simpson, A.G.B., Lane, C.E, Lukeš, J., Bass, D., Bowser, S.S., Brown, M.W. *et al.* (2012) The revised classification of Eukaryotes. *Eukaryotic Microbiology* 59(5), 429-514. https://doi.org/10.1111/j.1550-7408.2012.00644.x

Arastehfar, A., Khodavaisy, S., Daneshnia, F., Najafzadeh, M.J., Mahmoudi, S., Charsizadeh, A., Salehi, M.R., Zarrinfar, H., Raeisabadi, A., Dolatabadi, S., Zare Shahrabadi, Z., Zomorodian, K., Pan, W., Hagen, F. and Boekhout, T. (2019). Molecular identification, genotypic diversity, antifungal susceptibility, and clinical outcomes of infections caused by clinically underrated yeasts, *Candida orthopsilosis*, and *Candida metapsilosis*: an Iranian multicenter study (2014-2019). *Frontiers in Cellular and Infection Microbioogy.* 30;9:264. doi: 10.3389/fcimb.2019.00264. PMID: 31417877; PMCID: PMC6682699.

Arora, D., Mukerji, K.G. and Martin, E.H. (eds) (1991) *Handbook of Applied Mycology: Foods and Feeds* Volume 3. Marcel Dekker Inc., New York.

Balba, N.T. (1993) Microorganisms and detoxification of industrial waste. In: Jones D.G. (ed.) *Exploitation of Microorganisms* pp 371-410. Chapman and Hall, London.

Barko, P.C., McMichael, M.A., Swanson, K.S. and Williams, D.A. (2018) The gastrointestinal microbiome: a review. *Journal of Veterinary Internal Medicine* 32(1), 9-25. doi: 10.1111/jvim.14875. Epub 2017 Nov 24. PMID: 29171095; PMCID: PMC5787212.

Barnett, J.A., Payne, R.W. and Yarrow, D. (2000) *Yeasts: Characteristics and Identification*. Cambridge University Press, Cambridge, UK. https://doi.org/10.1046/j.1525-1470.2001.1862020a.x

Bar-On, Y.M., Phillips, R. and Milo, R. (2018) The biomass distribution on Earth. *Proceedings of the National Academy of Sciences* 115, 6506-6511. https://doi.org/10.1073/pnas.1711842115

Barth, M.G.M. and Bridge, P.D. (1989) 4-methylumbelliferyl substituted compounds as fluorogenic substrates for fungal extracellular enzymes. *Letters in Applied Microbiology* 9, 177-179. https://doi.org/10.1111/j.1472-765X.1989.tb00318.x

Bartholomew, J.W. and Mittwer, T. (1952) *The Gram stain. Bacteriological Reviews* 16(1), 1-29. https://doi.org/10.1128/br.16.1.1-29.1952

Becherer, L., Borst, N., Bakheit, M., Frischmann, S., Zengerle, R. and von Stetten, F. (2020) Loop-mediated isothermal amplification (LAMP) – review and classification of methods for sequence-specific detection. *Analytical Methods* 12, 717-746.

Benedict, C.V., Cook, W.J., Jarrett, P., Cameron, J.A., Huang, S.J. and Bell, J.P. (1983) Fungal degradation of polycaprolactones. *Journal of Applied Polymer Science* 28, 327-334. https://doi.org/10.1002/app.1983.070280128

Bernardo, W.L.C., Silva, J.J.D., Höfling, J.F., Rosa, E.A.R. and Boriollo, M.F.G. (2018) Dynamics of the seasonal airborne propagation of *Staphylococcus aureus* in academic dental clinics. *Journal of Applied Oral Science* 26:e20170141. doi:10.1590/1678-7757-2017-0141

Beuchat, L.R. (ed.) (1987) *Food and Beverage Mycology*. 2nd edn. Van Nostrand Reinhold Company Inc. New York. pp 661.

Bicudo, C. E. de M. and Menezes, M. (2016) Phylogeny and classification of *Euglenophyceae*: a brief review. *Frontiers in Ecology and Evolution* 4. https://www.frontiersin.org/article/10.3389/fevo.2016.00017

Blanton, L.S., Quade, B.R. and Bouyer, D.H. (2019) Differentiation of *Rickettsia felis* and *Rickettsia felis*-like organisms via restriction fragment length polymorphism analysis. *Vector Borne Zoonotic Diseases* 19(8), 637-639. doi: 10.1089/vbz.2018.2428. Epub 2019 Apr 25. PMID: 31021302; PMCID: PMC7366258.

Bonde, M.R., Petersen, G.L. and Maas, J.L. (1991) Isozyme comparisons for identification of *Colletotrichum* species pathogenic to strawberry. *Phytopathology* 81, 1523-1528. https://www.apsnet.org/publications/phytopathology/backissues/Documents/1991Articles/Phyto81n12_1523.PDF

Bradbury, J.M. (1977) Rapid biochemical tests for characterisation of the *Mycoplasmatales*. *Journal of Clinical Microbiology* 5, 531-534. https://doi.org/10.1128/jcm.5.5.531-534.1977

Bridge, P.D. (1985) An evaluation of some physiological and biochemical methods as an aid to the characterisation of species of *Penicillium* subsection *Fasciculata*. *Journal of General Microbiology* 131, 1887-1895. https://doi.org/10.1099/00221287-131-8-1887

Bridge, P.D., Pearce, D.A., Rivera, A. and Rutherford, M.A. (1997) VNTR derived oligonucleotides as PCR primers for population studies in filamentous fungi. *Letters in Applied Microbiology* 24, 426-430. https://doi.org/10.1046/j.1472-765X.1997.00060.x

Bridge, P.D. and Hawksworth, D.L. (1984) The APIZYM testing system as an aid to the rapid identification of *Penicillium* isolates. *Microbiological Sciences* 1, 232-234. https://pubmed.ncbi.nlm.nih.gov/6444135/

Bridge, P.D. and Hawksworth, D.L. (1985) Biochemical tests as an aid to the identification of *Monascus* species. *Letters in Applied Microbiology* 1, 25-29. https://doi.org/10.1111/j.1472-765X.1985.tb01481.x

Bridge, P., Stackebrandt, E. and Smith, D. (eds) (2021) *Trends in the Systematics of Bacteria and Fungi*. CABI, Wallingford, UK 368 pp. 9781789244984; ePDF 9781789244991 I ePub 9781789245004. https://www.cabi.org/bookshop/book/9781789244984/

Bringloe, T.T., Starko, S., Wade, R.M., Vieira, C., Kawai, H., De Clerck, O., Cock, J.M., Coelho, S.M., Destombe, C., Valero, M., Neiva, J., Pearson, G.A., Faugeron, S., Serrão, E.A. and Verbruggen, H. (2020) Phylogeny and evolution of the brown algae. *Critical Reviews in Plant Sciences* 39(4), 281-321. DOI: 10.1080/07352689.2020.1787679

Brink, A.A., Vervoort, M.B., Middeldorp, J.M., Meijer, C.J. and van den Brule, A.J. (1998). Nucleic acid sequence-based amplification, a new method for analysis of spliced and unspliced Epstein-Barr virus latent transcripts, and its comparison with reverse transcriptase PCR. *Journal of Clinical Microbiology.* 36(11), 3164-3169. doi: 10.1128/JCM.36.11.3164-3169.1998. Erratum in: *Journal of Clinical Microbiology* 1999 Nov;37(11):3788. PMID: 9774558; PMCID: PMC105294.

Brody, H. and Carbon, J. (1989) Electrophoretic karyotype of *Aspergillus nidulans*. *Proceedings of the National Academy of Sciences, USA* 86, 6260-6263. https://doi.org/10.1073/pnas.86.16.6260

Brooker, N.L., Leslie, J.F. and Dickman, M.B. (1991) Nitrate non-utilizing mutants of *Colletotrichum* and their use in studies of vegetative compatibility and genetic relatedness. *Phytopathology* 81, 672-677. https://www.apsnet.org/publications/phytopathology/backissues/Documents/1991Articles/Phyto81n06_672.PDF

Bruns, T.D., White, T.J. and Taylor, J.W. (1991) Fungal molecular systematics. *Annual Reviews in Ecology and Systematics* 22, 525-564. https://www.annualreviews.org/doi/pdf/10.1146/annurev.es.22.110191.002521

Burmeister, M. and Ulanovsky, L. (eds) (1992) *Methods in Molecular Biology Vol. 12: Pulsed-field Gel Electrophoresis*. Humana Press Inc., Totowa, New Jersey. https://dergipark.org.tr/tr/download/article-file/121559

Cafà, G., Caggiano, B., Reeve, M.A., Bhatti, H., Honey, S.F., Bajwa, B. and Buddie, A.G. (2019) A polyphasic approach aids early detection of potentially toxigenic aspergilli in soil. *Microorganisms* 7(9), 300. doi: 10.3390/microorganisms7090300. PMID: 31470555; PMCID: PMC6781248.

Caierão, J., Paiva, J.A.C.D., Sampaio, J.L.M., Silva, M.G.D., Santos, D.R.S., Coelho, F.S., Fonseca, L.S., Duarte, R.S., Armstrong, D.T. and Regua-Mangia, A.H. (2016) Multilocus enzyme electrophoresis analysis of rapidly-growing mycobacteria: an alternative tool for identification and typing. *International Journal of Infectious Diseases* 42, 11-16. doi: 10.1016/j.ijid.2015.11.010. Epub 2015 Nov 19. PMID: 26603643.

Callewaert, C., Ravard Helffer, K. and Lebaron, P. (2020) Skin microbiome and its interplay with the environment. *American Journal of Clinical Dermatology* 21(Suppl 1), 4-11. doi: 10.1007/s40257-020-00551-x. PMID: 32910439; PMCID: PMC7584520.

Campbell, S.M., Todd, J.R. and Anderson, J.G. (1987. Production of *l*-malic acid by *Paecilomyces varioti*. *Biotechnology Letters* 9, 393-398. https://link.springer.com/article/10.1007/BF01089003

Cannon, P.F., Damm, U., Johnston, P.R. and Weir, B.S. (2012) *Colletotrichum* - current status and future directions. *Studies in Mycology* 73(1), 181–213. https://doi.org/10.3114/sim0014

Casselton, L.A. and Olesnicky, N.S. (1998) Molecular genetics of mating recognition in basidiomycete fungi. *Microbiology and Molecular Biology Reviews* 62, 55-70. https://www.ncbi.nlm.nih.gov/pmc/articles/PMC98906/

Chalupová, J., Raus, M., Sedlářová, M. and Sebela, M. (2014) Identification of fungal microorganisms by MALDI-TOF mass spectrometry. *Biotechnology Advances* 32(1), 230-241. https://doi.org/10.1016/j.biotechadv.2013.11.002

Chang, S.T., Miles, P.G. and Buswell, J.A. (eds) (1993) *Genetics and Breeding of Edible Mushrooms*. pp 324. Gordon and Breach Publishers, Philadelphia. https://doi.org/10.1201/9780203753682

Chibata, I., Tosa, T. and Takata, I. (1983) Continuous production of *L*-malic acid by immobilized cells. *Trends in Biotechnology* 1, 9-11. https://doi.org/10.1016/0167-7799(83)90019-7

Claeyssens, M. (1989) Chromophoric and fluorophoric glycosides as tools in biochemical and biotechnological research. *Applied Fluorescence Technology* 1, 11-12.

Cleenwerck, I., De Vos, P., & De Vuyst, L. (2010). Phylogeny and differentiation of species of the genus *Gluconacetobacter* and related taxa based on multilocus sequence analyses of housekeeping genes and reclassification of *Acetobacter xylinus* subsp. *sucrofermentans* as *Gluconacetobacter sucrofermentans* (Toyosaki *et al.* 1996) sp. nov., comb. nov. *International Journal of Systematic and Evolutionary Microbiology*, 60(Pt 10), 2277–2283. https://doi.org/10.1099/ijs.0.018465-0

Collins, M.D. (1985) Isoprenoid quinone analysis in bacterial classification and identification. In: Goodfellow, M. and Minnikin, D.E. (eds) *Chemical Methods in Bacterial Systematics*. Society for Applied Microbiology. Technical Series 20 pp. 267-287. Academic Press, London.

Compton, J. (1991) Nucleic acid sequence-based amplification. *Nature* 350(6313), 91–92. https://doi.org/10.1038/350091a0 PMID: 1706072

Cooley, R.N. (1992) The use of RFLP analysis, electrophoretic karyotyping and PCR in studies of plant pathogenic fungi. In: Stahl, U. and Tudzynski, P. (eds) *Molecular Biology of Filamentous Fungi* pp. 13-26. VCH, Weinheim.

Cowan, S.T. (ed.) (1974) *Cowan and Steel's Manual for the Identification of Medical Bacteria*. 2nd edn. Cambridge University Press, Cambridge, UK.

Crowther, R.A., Jakes, R., Spillantini, M.G. and Goedert, M. (1998) Synthetic filaments assembled from C-terminally truncated α-synuclein. *Federation of European Biochemical Societies Letters* 436, 1873-3468. https://doi.org/10.1016/S0014-5793(98)01146-6

Cruickshank, R.H. and Wade, G.C. (1980) Detection of pectin enzymes in pectin-acrylamide gels. *Analytical Biochemistry* 107, 177-181. https://doi.org/10.1016/0003-2697(80)90508-4

Curran, J., Driver, F., Ballard, J.W.O. and Milner, R.J. (1994) Phylogeny of *Metarhizium*: analysis of ribosomal DNA sequence data. *Mycological Research* 98(5), 547-552. https://doi.org/10.1016/S0953-7562(09)80478-4

Daher, R.K., Stewart, G., Boissinot, M. and Bergeron, M.G. (2016) Recombinase Polymerase Amplification for diagnostic applications. *Clinical Chemistry* 62(7), 947-958. doi: 10.1373/clinchem.2015.245829. Epub 2016 May 9. PMID: 27160000; PMCID: PMC7108464.

Delétoile, A., Decré, D., Courant, S., Passet, V., Audo, J., Grimont, P., Arlet, G. and Brisse, S. (2009) Phylogeny and identification of *Pantoea* species and typing of *Pantoea agglomerans* strains by multilocus gene sequencing. *Journal of Clinical Microbiology* 47(2), 300-310. https://doi.org/10.1128/JCM.01916-08

Doonan, S. (ed.) (1996) *Methods in Molecular Biology Vol. 59: Protein Purification Protocols*. Humana Press Inc., Totowa, New Jersey. https://link.springer.com/book/10.1385/0896033368

Dufossé, L. (2006) Food Grade Pigments. *Food Technology and Biotechnology* 44 (3), 313–321. https://www.academia.edu/74070008/Microbial_Production_of_Food_Grade_Pigments

Edel, V. (1998) Polymerase chain reaction in mycology: an overview. In: Bridge, P.D., Arora, D.K., Reddy, C.A. and Elander, R.P. (eds) *Applications of PCR in Mycology*, pp. 1-20. CAB International, Wallingford, U.K.

Eggins, H.O.W. and Pugh, G.J.F. (1962) Isolation of cellulose decomposing fungi from soil. *Nature* 193, 94-95. https://www.nature.com/articles/193094a0

El-Nemr, I.M., Mushtaha, M., Sundararaju, S. *et al.* (2019) Application of MALDI Biotyper system for rapid identification of bacteria isolated from a fresh produce market. *Current Microbiology* 76, 290–296. https://doi.org/10.1007/s00284-018-01624-1

Evans, G.M. (1993) The use of microorganisms in plant breeding. In: Jones D.G. (ed.) *Exploitation of Microorganisms*. pp 225-248. Chapman and Hall, London. https://link.springer.com/content/pdf/bfm%3A978-94-011-1532-2%2F1.pdf

Filtenborg, O., Frisvad, J.C. and Svendsen, J.A. (1983) Simple screening methods for moulds producing intracellular mycotoxins in pure culture. *Applied and Environmental Microbiology* 45, 481-585. https://www.ncbi.nlm.nih.gov/pmc/articles/PMC242327/

Fischer, S.G. and Lerman, L.S. (1983) DNA fragments differing by single base-pair substitutions are separated in denaturing gradient gels: correspondence with melting theory. *Proceedings of the National Academy of Sciences, USA* 80, 1579-1583. https://doi.org/10.1073/pnas.80.6.1579

Forst, S., Dowds, B., Boemare, N. and Stackebrandt, E. (1997) *Xenorhabdus* and *Photorhabdus* spp.: bugs that kill bugs. *Annual Review of Microbiology* 51, 47-72. https://doi.org/10.1146/annurev.micro.51.1.47

Fort, P., Bonhomme, F., Darlu, P., Piechaczyk, M., Jeanteur, P. and Thaler, L. (1984) Clonal divergence of mitochondrial DNA versus populational evolution of nuclear genome. *Evolutionary Theory* 7, 81-90. https://hal.archives-ouvertes.fr/hal-02267608/document

Fredricks, D.N., Smith, C. and Meier, A. (2005) Comparison of six DNA extraction methods for recovery of fungal DNA as assessed by quantitative PCR. *Journal of Clinical Microbiology* 43(10), 5122-5128. https://doi.org/10.1128/JCM.43.10.5122-5128.2005

Frisvad, J.C. and Thrane, U. (1987) Standardized high-performance liquid chromatography of 182 mycotoxins and other secondary metabolites based on alkylphenone retention indices and UV-VIS spectra (diode array detection). *Journal of Chromatography* 404, 195-214. https://doi.org/10.1016/S0021-9673(01)86850-3

Gadd, G.M. (1993) Interactions of fungi with toxic metals. *New Phytologist* 124, 1-35.

Gadd, G.M. (1994) Interactions of fungi with toxic metals. In: Powell, K.A., Renwick, A. and Peberdy, J.F. (eds) *The Genus Aspergillus* pp. 361-374. Plenum Press, New York. https://link.springer.com/chapter/10.1007/978-1-4899-0981-7_28

Gadd, G.M. (2017) Geomicrobiology of the built environment. *Nature Microbiology* 2, 16275. https://doi.org/10.1038/nmicrobiol.2016.275

Gaffney, G.W., Schneier, K., Diferrante, N. and Altman, K.I. (1954) The quantitative determination of hippuric acid. *Journal of Biological Chemistry* 206, 695-698.

Gardes, M. and Bruns, T.D. (1993) ITS primers with enhanced specificity for basidiomycetes - application to the identification of mycorrhizae and rusts. *Molecular Ecology* 2(2) 113-118. https://doi.org/10.1111/j.1365-294X.1993.tb00005.x

Gardes, M., Bousquet, J. and Lalonde, M. (1987) Isozyme variation among 40 *Frankia* strains. *Applied and Environmental Microbiology* 53, 1596-1603. https://www.ncbi.nlm.nih.gov/pmc/articles/PMC203917/pdf/aem00124-0210.pdf

Glenn, J.K. and Gold, M.H. (1983) Decolorization of several polymeric dyes by the lignin-degrading basidiomycete *Phanerochaete chrysosporium*. *Applied and Environmental Microbiology* 45, 1741-1747. https://www.ncbi.nlm.nih.gov/pmc/articles/PMC242532/

Golubev, W.I. (1998) Mycocins (Killer Toxins). In: Kurtzman, C.P. and Fell, J.W. (eds) *The Yeasts: A Taxonomic Study*, 4th edn; pp. 55-62. Elsevier, Amsterdam, Netherlands.

Gómez, F. (2012) A quantitative review of the lifestyle, habitat and trophic diversity of dinoflagellates (Dinoflagellata, Alveolata), *Systematics and Biodiversity* 10(3), 267-275. DOI:10.1080/14772000.2012.721021

Grange, J.M. and Clark, K. (1977) Use of umbelliferone derivatives in the study of enzymatic activities of mycobacteria. *Journal of Clinical Pathology* 30, 151-153. https://www.ncbi.nlm.nih.gov/pmc/articles/PMC476339/

Greenshields, R. (ed.) (1989) *Resources and Applications of Biotechnology: The new wave.* Stockton Press, New York. https://b-ok.cc/book/2670800/33d575

Hames, B.D. (1981) An introduction to polyacrylamide gel electrophoresis. In: Hames, B.D. and Rickwood, D. (eds) *Gel Electrophoresis of Proteins* pp. 1-91. IRL Press, Oxford, UK.

Hamilton, R.J. and Hamilton, S. (eds) (1992) *Lipid Analysis: A Practical Approach*. IRL Press, Oxford, UK.

Hampton, R., Ball, E. and De Boer, S. (eds) (1990) *Serological Methods for Detection and Identification of Viral and Bacterial Plant Pathogens: A Laboratory Manual*. APS Press, St. Paul, Minnesota.

Hastie, A.C. (1981) The genetics of conidial fungi. In: Cole, G.T. and Kendrick, B. (eds) *Biology of Conidial Fungi* Vol. 2 pp. 511-547. Academic Press, New York.

Hawker, L.E.and Linton, A.H. (eds) (1971) *Microorganisms: Function, form, and environment.* Edward Arnold, London.

Hawksworth, D.L., Kirk, P.M., Sutton, B.C. and Pegler, D.N. (1995) *Ainsworth and Bisby's Dictionary of the Fungi*. 8th edn. pp. 616. CAB International, Wallingford, UK.

Hernández-Restrepo, M., Groenewald, J.Z. and Crous, P.W. (2016) Taxonomic and phylogenetic re-evaluation of *Microdochium*, *Monographella* and *Idriella*. *Persoonia - Molecular Phylogeny and Evolution of Fungi* 36, 57-82(26). https://doi.org/10.3767/003158516X688676

Hibbett, D.S. (1992) Ribosomal RNA and fungal systematics. *Transactions of the Mycological Society of Japan* 33, 533–556.

Hornby, D. (1990) *Biological Control of Soil-borne Plant Pathogens.* CAB International, Wallingford, UK.

Hudspeth, M.E.S., Shumard, D.S., Tatti, K.M. and Grossman, L.I. (1980) Rapid purification of yeast mitochondrial DNA in high yield. *Biochimica et Biophysica Acta* 610, 221-228. https://doi.org/10.1016/0005-2787(80)90003-9

Hunter, G.C., Wingfield, B.D., Crous, P.W. and Wingfield, M.J. (2006) A multi-gene phylogeny for species of *Mycosphaerella* occurring on *Eucalyptus* leaves. *Studies in Mycology* 55, 147–161. https://doi.org/10.3114/sim.55.1.147

Imanishi, S.Y., Nakayama, T., Asukabe, H. and Harada, K. (2016) Application of MALDI Biotyper to cyanobacterial profiling. *Rapid Communications in Mass Spectrometry* 31(4), 325-332. https://doi.org/10.1002/rcm.7793

James, E.K., Olivares, F.L., de Oliveira, A.L.M., dos Reis Jr, F.B., da Silva, L.G. and Reis, V.M. (2001) Further observations on the interaction between sugar cane and *Gluconacetobacter diazotrophicus* under laboratory and greenhouse conditions. *Journal of Experimental Botany* 52, 747–760. https://doi.org/10.1093/jexbot/52.357.747

Johnson, S.S., Zaikova, E., Goerlitz, D.S., Bai, Y. and Tighe, S.W. (2017) Real-time DNA sequencing in the Antarctic Dry Valleys using the Oxford Nanopore sequencer. *Journal of Biomolecular Techniques* 28(1), 2-7. doi:10.7171/jbt.17-2801-009

Jones, D.G. and Lewis, D.M. (1993) Rhizobium inoculation of crop plants. In: Jones D.G. (ed.) *Exploitation of Microorganisms* pp 197-224. Chapman and Hall, London.

Joseph, S., Desai, P., Ji, Y., Cummings, C.A., Shih, R., *et al.* (2012) Comparative analysis of genome sequences covering the seven *Cronobacter* species. *PLoS ONE* 7(11): e49455. https://doi.org/10.1371/journal.pone.0049455

Julien, M.H. (ed.) (1992) *Biological Control of Weeds: A World Catalogue of Agents and their Target Weed*. 3rd edn. CAB International, Wallingford, UK.

Kaaya, G.P. and Munyinyi, D.M. (1995) Biocontrol potential of the entomogenous fungi *Beauveria bassiana* and *Metarhizium anisopliae* for Tsetse flies (*Glossina* spp.) at developmental sites. *Journal of Invertebrate Pathology* 66, 237-241. https://doi.org/10.1006/jipa.1995.1095

Karthik, K., Rathore, R., Thomas, P., Arun, T.R., Viswas, K.N., Dhama, K. and Agarwal, R.K. (2014) New closed tube loop mediated isothermal amplification assay for prevention of product cross-contamination. Methods X 1:137–143. https://doi.org/10.1016/j.mex.2014.08.009

Kawaguchi ,Y., Shibuya, M., Kinoshita, I., Yatabe, J., Narumi, I., Shibata, H., Hayashi, R. *et al.* (2020) Damage and survival time course of Deinococcal cell pellets during 3 years of exposure to outer space. *Frontiers in Microbiology* 11. https://www.frontiersin.org/article/10.3389/fmicb.2020.02050

Kelley, J. and Yaghmaie, P.A. (1988) Screening of fungal strains employed in the testing of plastics materials. *International Biodeterioration* 24, 289-298. https://doi.org/10.1016/0265-3036(88)90013-9

Khan, S., Nadir, S., Shah, Z.U., Shah, A.A., Karunarathna, S.C., Xu, J., Khan, A., Munir, S. and Hasan, F. (2017) Biodegradation of polyester polyurethane by *Aspergillus tubingensis*. *Environmental Pollution* 225, 469-480. https://doi.org/10.1016/j.envpol.2017.03.012.

Kilian, M. (1978) Rapid identification of *Actinomycetaceae* and related bacteria. *Journal of Clinical Microbiology* 8, 127-133. https://www.ncbi.nlm.nih.gov/pmc/articles/PMC275167/

Kirk, P.M., Cannon, P.F., Minter, D.W. and Stalpers, J.A. (eds) (2008) *Dictionary of the Fungi*, 10th edn. CABI, Wallingford, UK. https://www.cabi.org/bookshop/book/9781845939335/

Klement, Z., Rudolph, K. and Sand, D.C. (eds) (1990) *Methods in Phytobacteriology*. Akadémiai Kiadó, Budapest, Hungary. https://www.worldcat.org/title/methods-in-phytobacteriology/oclc/22859508

Kleppe, K., Ohtsuka, E., Kleppe, R., Molineux, I. and Khorana, H.G. (1971) Studies on polynucleotides: XCVI. Repair replication of short synthetic DNAs as catalyzed by DNA polymerases. *Journal of Molecular Biology* 56(2), 341-361. https://doi.org/10.1016/0022-2836(71)90469-4

Kubicek, Q.B., Civelero, E.L., Bonde, M.R., Hartung, J.S. and Peterson, G.L. (1989) Isozyme analysis of *Xanthomonas campestris* pv. *citri*. *Phytopathology* 79, 297-300. https://www.apsnet.org/publications/phytopathology/backissues/Documents/1989Abstracts/Phyto79_297.htm

Kuephadungphan, W., Tasanathai, K., Petcharad, B., Khonsanit, A., Stadler, M. and Luangsa-ard, J.J. (2020) Phylogeny- and morphology-based recognition of new species in the spider-parasitic genus *Gibellula* (Hypocreales, Cordycipitaceae) from Thailand. *MycoKeys* 72. https://mycokeys.pensoft.net/article/55088/

Kuraishi, H., Katayama-Fujimura, Y., Sugiyama, J. and Yokoyama, T. (1985) Ubiquinone systems in fungi. I. Distribution of ubiquinones in the major families of ascomycetes, basidiomycetes and deutoromycetes, and their taxonomic implications. *Transactions of the Mycological Society of Japan* 26, 383-395.

Labeda, D.P. (2011) Multilocus sequence analysis of phytopathogenic species of the genus *Streptomyces*. *International Journal of Systematic and Evolutionary Microbiology* 61, 2525–2531. doi:10.1099/ijs.0.028514-0

Lambert, R.H. and Garcia, J.R. (1990) Evidence of morphology-specific isozymes in *Candida albicans*. *Current Microbiology* 20, 215-221.

Lane D.J. (1991) 16S/23S rRNA sequencing. *Nucleic Acid Techniques in Bacterial Systematics* (Stackebrandt E Goodfellow M eds), pp.115–175. John Wiley and Sons, New York. https://www.worldcat.org/title/nucleic-acid-techniques-in-bacterial-systematics/oclc/22310197

Lane, D.J., Pace, B., Olsen, G.J., Stahl, D.A., Sogin, M.L., and Pace, N.R. (1985) Rapid determination of 16S ribosomal RNA sequences for phylogenetic analyses. *Proceedings of the National Academy of Sciences, USA* 82, 6955–6959. https://doi.org/10.1073/pnas.82.20.6955

Leslie, J.F. (1987) A nitrate non-utlizing mutant of *Gibberella zeae*. *Journal of General Microbiology* 133, 1279-1287. https://doi.org/10.1099/00221287-133-5-1279

Lévesque, S., Dufresne, P.J., Soualhine, H., Domingo, M-C., Bekal, S., Lefebvre, B. and Tremblay, C. (2015) A side by side comparison of Bruker Biotyper and VITEK MS: utility of MALDI-TOF MS technology for microorganism identification in a public health reference laboratory. *PLOS One* 10. https://doi.org/10.1371/journal.pone.0144878

Li, J.S., Wu, L.Y., Zhang, H., Song, X.S., Wang, J.X., Zhou, M.G. and Hou, Y.P. (2021) PCR-RFLP for detection of *Fusarium graminearum* genotypes with resistance to phenamacril. *Plant Disease* 105(4), 889-895. doi: 10.1094/PDIS-06-20-1156-RE. Epub 2021 Mar 10. PMID: 33044138.

Lillie, R.D., Stotz, E.H. and Emmel, V.M. (1977) *H. J. Conn's Biological Stains: A handbook on the nature and uses of the dyes employed in the biological laboratory*. 9th ed. Williams and Wilkins, Baltimore.

Lima, N., Teixeira, J.A. and Mota, M. (1991) Deep agar-diffusion test for preliminary screening of lipolytic activity of fungi. *Journal of Microbiological Methods* 14, 193-200. https://www.semanticscholar.org/paper/Deep-agar-diffusion-test-for-preliminary-screening-Lima-Teixeira/84a1fd6be9865c65a1682edce0df52f7ebbb2250

Ling, W.H., Saxelin, M., Hanninen, O. and Salminen, S. (1994) Enzyme profile of *Lactobacillus* strain GG by a rapid APIZYM system: a comparison of intestinal bacterial strains. *Microbial Ecology in Health and Disease* 7, 99-104. https://doi.org/10.3109/08910609409141578

Lisdiyanti, P., Yamada, Y., Uchimura, T. and Komagata, K. (2003) Identification of *Frateuria aurantia* strains isolated from Indonesian sources. *Microbiology Culture Collections* 19, 81-90.

Lobato, I.M. and O'Sullivan, C.K. (2018) Recombinase polymerase amplification: Basics, applications and recent advances. *Trends Analytical Chemistry* 98, 19-35. doi: 10.1016/j.trac.2017.10.015. Epub 2017 Oct 26. PMID: 32287544; PMCID: PMC7112910.

Louten, J. (2016) Virus structure and classification. In: Louten, J. (ed.) *Essential Human Virology*. Academic Press, London, pp 19-29. ISBN 9780128009475. https://doi.org/10.1016/B978-0-12-800947-5.00002-8

Lowen, R., Brady, B.L., Hawksworth, D.L. and Paterson, R.R.M. (1986) Two new lichenicolous species of *Hobsonia*. *Mycologia* 78, 842-846. https://doi.org/10.1080/00275514.1986.12025332

Lowry, O.H., Rosebrough, N.J., Farr, A.L. and Randall, R.J. (1951) Protein measurement with the folin phenol reagent. *Journal of Biological Chemistry* 193, 265-275. https://doi.org/10.1016/S0021-9258(19)52451-6

McCammon, S.A. and Rath, A.C. (1994) Separation of *Metarhizium anisopliae* strains by temperature dependent germination rates. *Mycological Research* 98, 1253-1257. https://agris.fao.org/agris-search/search.do?recordID=GB9516293

McEldowney, S. (1990) Microbial biosorption of radionuclides in liquid effluent treatment. *Applied Biochemistry and Biotechnology* 26, 159-180. https://hero.epa.gov/hero/index.cfm/reference/details/reference_id/1426801

Maiden, M.C., Bygraves, J.A., Feil, E., Morelli, G., Russell, J.E., Urwin, R., Zhang, Q., Zhou, J., Zurth, K., Caugant, D.A., Feavers, I.M., Achtman, M. and Spratt, B.G. (1998) Multilocus sequence typing: a portable approach to the identification of clones within populations of pathogenic microorganisms. *Proceedings of the National Academy of Sciences, USA* 95(6), 3140-3145. DOI: 10.1073/pnas.95.6.3140

Manteca, Á., Fernández, M. and Sánchez, J. (2005) A death round affecting a young compartmentalized mycelium precedes aerial mycelium dismantling in confluent surface cultures of *Streptomyces antibioticus*. *Microbiology* 151(11), 3689-3697. https://doi.org/10.1099/mic.0.28045-0

Marriott, A.C., Archer, S.A. and Buck, K.W. (1984) Mitochondrial DNA in *Fusarium oxysporum* is a 46.5 kilobase pair circular molecule. *Journal of General Microbiology* 130, 3001-3008. https://doi.org/10.1099/00221287-130-11-3001

Marsh, A.J., O'Sullivan, O., Hill, C., Ross, R.P. and Cotter, P.D. (2014) Sequence-based analysis of the bacterial and fungal compositions of multiple kombucha (tea fungus) samples, *Food Microbiology* 38, 171-178. *ISSN* 0740-0020, https://doi.org/10.1016/j.fm.2013.09.003.

Marumo, K. and Aoki, Y. (1990) Discriminant analysis of cellular fatty acids of *Candida* species, *Torulosis glabrata* and *Cryptococcus neoformans* determined by gas-liquid chromatography. *Journal of Clinical Microbiology* 28, 1509-1513. https://www.ncbi.nlm.nih.gov/pmc/articles/PMC267979/

May, B. and Royse, D.J. (1982) Genetic variation and joint segregation of biochemical loci in the common meadow mushroom, *Agaricus campestris*. *Biochemical Genetics* 20, 1165-1173. https://link.springer.com/article/10.1007/BF00498940

Medlin, L.K. (2016) Evolution of the diatoms: major steps in their evolution and a review of the supporting molecular and morphological evidence. *Phycologia* 55(1), 79-103. DOI: 10.2216/15-105.1

Mendes, R., Garbeva, P. and Raaijmakers, J.M. (2013) The rhizosphere microbiome: significance of plant beneficial, plant pathogenic, and human pathogenic microorganisms. *FEMS Microbiology Review* 37(5), 634-663. doi: 10.1111/1574-6976.12028. Epub 2013 Jul 22. PMID: 23790204.

Micales, J.A., Bonde, M.R. and Peterson, G.L. (1986) The use of isozyme analysis in fungal taxonomy and genetics. *Mycotaxon* 27, 405-449.

Miller, M., Palojarvi, A., Rangger, A., Reeslev, M. and Kjoller, A. (1998) The use of fluorogenic substrates to measure fungal presence and activity in soil. *Applied and Environmental Microbiology* 64, 613-617. https://www.ncbi.nlm.nih.gov/pmc/articles/PMC106091/

Milne, J.L.S., Borgnia, M.J., Bartesaghi, A., Tran, E.E.H., Earl, L.A., Schauder, D.M., Lengyel, J., Pierson, J., Patwardhan, A. and Subramaniam, S. (2013) Cryo-electron microscopy – a primer for the non-microscopist. *The FEBS Journal* 280(1), 28-45. https://doi.org/10.1111/febs.12078

Mitchell, D.T. (1993) Mycorrhizal associations. In: Jones, D.G. (ed) *Exploitation of Microorganisms.* pp 169-196. Chapman and Hall, London.

Mitchell, K. and Fallon, R.J. (1990) The determination of ubiquinone profiles by reversed-phase high-performance thin-layer chromatography as an aid to the speciation of *Legionellaceae*. *Journal of General Microbiology* 136, 2035-2041. https://doi.org/10.1099/00221287-136-10-2035

Moreira, C., Vasconcelos, V. and Antunes, A. (2013) Phylogeny and biogeography of Cyanobacteria and their produced toxins. *Marine Drugs* 11, 4350-4369. https://doi.org/10.3390/md11114350

Morris, G.J., Smith, D. and Coulson, G.E. (1988) A comparative study of the changes in the morphology of hyphae during freezing and viability upon thawing for twenty species of fungi. *Journal of General Microbiology* 134, 2897-2906. https://doi.org/10.1099/00221287-134-11-2897

Mueller, U.G., Lipari, S.E. and Milgroom, M.G. (1996) Amplified fragment length polymorphism (AFLP) fingerprinting of symbiotic fungi cultured by the fungus-growing ant *Cyphomyrmex minutus*. *Molecular Ecology* 5, 119–122.

Mugnai, L., Bridge, P.D. and Evans, H.C. (1989) A chemotaxonomic evaluation of the genus *Beauvaria*. *Mycological Research* 92, 199-209. https://doi.org/10.1016/S0953-7562(89)80012-7

Mukherjee, A. and Reddy, M.S. (2020) Metatranscriptomics: an approach for retrieving novel eukaryotic genes from polluted and related environments. *3 Biotech*. 10(2), 71. doi: 10.1007/s13205-020-2057-1. Epub 2020 Jan 27. PMID: 32030340; PMCID: PMC6985312.

Muller, M.M., Kantola, R. and Kitunen, V. (1994) Combining sterol and fatty acid profiles for the characterization of fungi. *Mycological Research* 98, 593-603. https://doi.org/10.1016/S0953-7562(09)80404-8

Mullis, K.B. and Faloona, F.A. (1987) Specific synthesis of DNA in vitro via a polymerase-catalyzed chain reaction. *Methods in Enzymology* 155, 335-350. https://doi.org/10.1016/0076-6879(87)55023-6

Nacef, M., Chevalier, M., Chollet, S., Drider, D. and Flahaut, C. (2017) MALDI-TOF mass spectrometry for the identification of lactic acid bacteria isolated from a French cheese: The Maroilles. *International Journal of Food Microbiology* 247, 2-8. https://doi.org/10.1016/j.ijfoodmicro.2016.07.005

Nai, Y.H., Doeven, E.H. and Guijt, R.M. (2022) An improved nucleic acid sequence-based amplification method mediated by T4 gene 32 protein. *PLoS ONE* 17(3): e0265391. https://doi.org/10.1371/journal.pone.0265391

Naranjo-Ortiz, M.A. and Gabaldón, T. (2019) Fungal evolution: diversity, taxonomy and phylogeny of the Fungi. *Biological Reviews* 94(6), 2101-2137. https://doi.org/10.1111/brv.12550

Nardon, P., Monget, D., Didier-Fichet, M.L. and de The, G. (1976) Comparison of zymogram of three lymphoblastoid cell lines with a new microtechnique. *Biomedicine* 24, 183-190. https://pubmed.ncbi.nlm.nih.gov/186133/

Nature Editorial (2011) Microbiology by numbers. *Nature Reviews Microbiology* 9, 628. https://doi.org/10.1038/nrmicro2644

Notomi, T., Okayama, H., Masubuchi, H., Yonekawa, T., Watanabe, K., Amino, N. and Hase, T. (2000) Loop-mediated isothermal amplification of DNA. *Nucleic Acids Research* 15;28(12):E63. doi:10.1093/nar/28.12.e63. PMID: 10871386; PMCID: PMC102748.

O'Brien, M. and Colwell, R.R. (1987) A rapid test for chitinase activity that uses 4-methylumbelliferyl-N-acetyl-b-D-glucosaminide. *Applied and Environmental Microbiology* 53, 1718-1720. https://www.ncbi.nlm.nih.gov/pmc/articles/PMC203942/

Olsson, S. and Jennings, D.H. (1991) Evidence for diffusion being the mechanism of translocation in the hyphae of three moulds. *Experimental Mycology* 15, 302-309. https://doi.org/10.1016/0147-5975(91)90033-A

Onions, A.H.S., Allsopp, D. and Eggins, H.O.W. (eds) (1981) *Smith's Introduction to Industrial Mycology*. 7th edn. pp 398. Edward Arnold, London. https://doi.org/10.1002/jobm.19830230417

Orita, M., Iwahana, H., Kanazawa, H., Hayashi, K. and Sekiya, T. (1989) Detection of polymorphisms of human DNA by gel electrophoresis as single-strand conformational polymorphisms. *Proceedings of the National Academy of Sciences, USA* 86, 2766-2770. https://doi.org/10.1073/pnas.86.8.2766

Otsuka, Y., Sonoki, T., Ikeda, S., Kajita, S., Nakamura, M. and Katayama, Y. (2003) Detection and characterization of a novel extracellular fungal enzyme that catalyzes the specific and hydrolytic cleavage of lignin guaiacylglycerol beta-aryl ether linkages. *European Journal of Biochemistry* 270(11), 2353-2362. doi: 10.1046/j.1432-1033.2003.03545.x. PMID: 12755689.

Paterson, R.R.M. (1986) Standardised one- and two-dimensional thin-layer chromatographic methods for the identification of secondary metabolites in *Penicillium* and other fungi. *Journal of Chromatography* 368, 249-264. https://doi.org/10.1016/S0021-9673(00)91067-7

Paterson, R.R.M. and Bridge, P.D. (1994) *Biochemical Techniques for Filamentous Fungi*. CAB International, Wallingford, UK https://www.worldcat.org/title/biochemical-techniques-for-filamentous-fungi/oclc/30477452

Paterson, R.R.M. and Buddie, A. (1991) Rapid determination of ubiquinone profiles in *Penicillium* by reversed phase high performance thin layer chromatography. *Letters in Applied Microbiology* 13, 133-136. https://doi.org/10.1111/j.1472-765X.1991.tb00590.x

Paterson, R.R.M. and Hawksworth, D.L. (1985) Detection of secondary metabolites in dried cultures of *Penicillium* preserved in herbaria. *Transactions of the British Mycological Society* 85, 95-100. https://doi.org/10.1016/S0007-1536(85)80158-3

Paterson, R.R.M. and Kemmelmeier, C. (1989) Gradient high performance liquid chromatography using alkylphenone retention indices of insecticidal extracts of *Penicillium* strains. *Journal of Chromatography* 483, 153-168. https://doi.org/10.1016/S0021-9673(01)93118-8

Pharkjaksu, S., Kwon-Chung, K.J., Bennett, J.E. and Ngamskulrungroj, P. (2020) Population diversity and virulence characteristics of *Cryptococcus neoformans/C. gattii* species complexes isolated during the pre-HIV-pandemic era. *PLoS Neglected Tropical Diseases* 14(10):e0008651. doi:10.1371/journal.pntd.0008651. PMID: 33017391; PMCID: PMC7535028.

Piepenburg, O., Williams, C.H., Stemple, D.L. and Armes, N.A. (2006) DNA detection using recombination proteins. *PLoS Biology* 4(7):e204. doi:10.1371/journal.pbio.0040204. PMID: 16756388; PMCID: PMC1475771.

Prior, C. (1992) Discovery and characteristics of fungal pathogens for locust and grasshopper control. In: Lomer, C.J. and Prior, C. (ed.) *Biocontrol of Locusts and Grasshoppers.* pp. 159-180. CAB International, Wallingford, UK.

Prusiner, S.B. (1982) Novel proteinaceous infectious particles cause scrapie. *Science* 216(4542), 136-144. https://www.science.org/doi/10.1126/science.6801762

Puhalla, J.A. (1985) Classification of strains of *Fusarium oxysporum* on the basis of vegetative compatibility. *Canadian Journal of Botany* 63, 179-183. https://doi.org/10.1139/b85-020

Quick, J., Loman, N., Duraffour, S. *et al.* (2016) Real-time, portable genome sequencing for Ebola surveillance. *Nature* 530, 228–232. https://doi.org/10.1038/nature16996

Rath, A.C., Worledge, D., Koen, T.B. and Rowe, B.A. (1995) Long-term field efficacy of the entomogenous fungus *Metarhizium anisopliae* against the subterranean scarab, *Adoryphorus couloni. Biocontrol Science and Technology* 5, 439-451. https://doi.org/10.1080/09583159550039639

Risco, B.S.H. (1980) Biological control of the leaf froghopper, *Mahanarva posticata* Stal, with the fungus *Metarhizium anisopliae* in the state of Alagoas. *Brazil Entomology Newsletter, International Society of Sugarcane-Technologists* 9.

Rousseaux, S. and Guilloux-Bénatier, M. (2017) PCR ITS-RFLP for *Penicillium* species and other genera. *Methods Molecular Biology* 1542, 321-333. doi:10.1007/978-1-4939-6707-0_21. PMID: 27924548.

Saiki, R.K., Scharf, S., Faloona, F., Mullis, K.B., Horn, G.T., Erlich, H.A. and Arnheim, N. (1985) Enzymatic amplification of b-globin genomic sequences and restriction site analysis for diagnosis of sickle cell anemia. *Science* 230, 1350–1354. https://www.science.org/doi/10.1126/science.2999980

Sambrook, J., Fritsch, E. F. and Maniatis, T. (1989) Molecular cloning: a laboratory manua*l*. Cold Spring Harbor Laboratory Press, Cold Spring Harbor, NY, USA. ISBN : 9780879693091

Sambrook, J. and Russell, D.W. (2001) *Molecular Cloning: A Laboratory Manual* 3rd edn. Vols. 1-3. Cold Spring Harbor Laboratory Press, New York. https://www.sigmaaldrich.com/GB/en/product/sigma/m8265

Saxena, S., Yuan, P., Dhar, S.K., Senga, T., Takeda, D., Robinson, H., Kornbluth, S., Swaminathan, K. and Dutta, A. (2004) A dimerized coiled-coil domain and an adjoining part of geminin interact with two sites on Cdt1 for replication inhibition. *Molecular Cell* 15(2), 245–258. https://doi.org/10.1016/j.molcel.2004.06.045

Shin, J.I., Shin, S.J. and Shin, M.K. (2020) Differential genotyping of *Mycobacterium avium* complex and its implications in clinical and environmental epidemiology. *Microorganisms* 8(1), 98. doi:10.3390/microorganisms8010098. PMID: 31936743; PMCID: PMC7022546.

Sierra, G. (1957) A simple method for the detection of lipolytic activity of microorganisms and some observations on the influence of the contact between cells and fatty substrates. *Antonie van Leeuwenhoek* 23, 15-22. https://link.springer.com/article/10.1007/BF02545855

Singleton, P. and Sainsbury, D. (1993) *Dictionary of Microbiology and Molecular Biology*, 2nd edn. Wiley, London. https://doi.org/10.1002/jobm.3620280712

Skerman, V.B.D. (ed.) (1969) *Abstracts of Microbiological Methods*. John Wiley Interscience, *London.* https://doi.org/10.1002/food.19710150238

Slatko, B.E., Gardner, A.F. and Ausubel, F.M. (2018) Overview of next-generation sequencing technologies. *Current Protocols Molecular Biology* 122(1):e59. doi:10.1002/cpmb.59

Slifkin, M. and Gil, G.M. (1983) Rapid biochemical tests for the identification of groups A, B, C, F and G streptococci from throat cultures. *Journal of Clinical Microbiology* 18, 29-32. https://doi.org/10.1128/jcm.18.1.29-32.1983

Smith, D. and Thomas, V.E. (1998) Cryogenic light microscopy and the development of cooling protocols for the preservation of filamentous fungi. *World Journal of Microbiology and Biotechnology* 14, 49-57. https://link.springer.com/article/10.1023/A:1008820432471

Smith, D. and Ryan, M.J. (2008) Fungal sources for new drug discovery. In: *McGraw Hill 2009 Yearbook of Science and Technology* pp. 131-132. https://doi.org/10.1036/1097-8542.YB090090

Smith, D., Buddie, A.G., Goss, R., Overmann, J., Lepleux, C., Brönstrup, M., Kloareg, B., Meiners, T., Brennecke, P., Ianora, A., Bouget, F-Y., Gribbon, P. and Pina, M. (2019) Discovery pipelines for marine resources: an ocean of opportunity for biotechnology? *World Journal of Microbiology and Biotechnology* 35, 107. https://doi.org/10.1007/s11274-019-2685-y

Smith, I. (1969) *Chromatographic and Electrophoretic Techniques*, Volume 1. Pitman Press, Bath, UK.

Stahl, E. (ed.) (1969) *Thin-Layer Chromatography – A Laboratory Handbook*, 2nd edition. Springer-Verlag, Berlin.

Steinkraus, K.H. (1983) *Handbook of Indigenous Fermented Foods.* Pp. 672. Marcel Dekker Inc., New York. https://www.routledge.com/Handbook-of-Indigenous-Fermented-Foods-Revised-and-Expanded/Steinkraus/p/book/9780824793524

St Leger, R.J., Charnley, A.K. and Cooper, R.M. (1986) Enzymatic characterisation of entomopathogens with the API ZYM system. *Journal of Invertebrate Pathology* 48, 375-376.

Stirling, G.R. (1991) *Biological Control of Plant Parasitic Nematodes.* CAB International, Wallingford, UK. https://www.cabi.org/bookshop/book/9781780644158/

Sugiyama, J., Itoh, M., Katayama, Y., Yamaoka, Y., Ando, K., Kakishima, M. and Kuraishi, H. (1988) Ubiquinone systems in fungi. II. Distribution of ubiquinones in smut and rust fungi. *Mycologia* 80, 115-120. https://doi.org/10.1080/00275514.1988.12025508

Summerbell, R.C., Gueidan, C., Schroers, H.J., de Hoog, G.S., Starink, M., Rosete, Y.A., Guarro, J. and Scott, J.A. (2011) *Acremonium* phylogenetic overview and revision of *Gliomastix*, *Sarocladium*, and *Trichothecium*. *Studies in Mycology* 68, 139–162. https://doi.org/10.3114/sim.2011.68.06

Sutton, B.C. (1990) *The Coelomycetes*. Commonwealth Mycological Institute, Kew. ISBN 9780851984469. https://www.worldcat.org/title/coelomycetes-fungi-imperfecti-with-pycnidia-acervuli-and-stromata/oclc/716116732?referer=di&ht=edition

Svensden, A. and Frisvad, J.C. (1994) A chemotaxonomic study of the *terverticillate penicillia* based on high performance liquid chromatography of secondary metabolites. *Mycological Research* 98, 1317–1328. https://doi.org/10.1016/S0953-7562(09)80305-5

Takao, S. (1965) Organic acid production by basidiomycetes. *Applied Microbiology* 13, 732-737. https://www.ncbi.nlm.nih.gov/pmc/articles/PMC1058333/

Taylor, J.W. and Fisher, M.C. (2003) Fungal multilocus sequence typing — it's not just for bacteria. *Current Opinion in Microbiology* 6(4), 351-356. https://doi.org/10.1016/S1369-5274(03)00088-2

Tomitani, A., Knoll, A.H., Cavanaugh, C.M. and Ohno, T. (2006) The evolutionary diversification of cyanobacteria: Molecular–phylogenetic and paleontological perspectives. *Proceedings of the National Academy of Sciences USA* 103(14), 5442-5447. https://doi.org/10.1073/pnas.0600999103

Tornabene, T.G. (1985) Lipid analysis and the relationship to chemotaxonomy. *Methods in Microbiology* 18, 209-234. https://doi.org/10.1016/S0580-9517(08)70476-8

Trask, B.J. (1991) Fluorescence in situ hybridization: Applications in cytogenetics and gene mapping. *Trends in Genetics* 7, 149-154. https://doi.org/10.1016/0168-9525(91)90378-4

Turner, W.B. and Aldridge, D.C. (1983) *Fungal Metabolites II.* Academic Press, London.

van der Walt, J.P. (1970) Criteria and methods used in classification. In: Lodder, J. (ed.) *The Yeasts, A Taxonomic Study*, 2nd edn. pp. 34-113. North Holland, Amsterdam, Netherlands.

Vos, P., Hogers, R., Bleeker, M., Reijans, M., Van de Lee, T., Hornes, M., Frijters, A., Pot, J., Peleman, J., Kuiper, M. and Zabeau, M. (1995) AFLP: a new technique for DNA fingerprinting. *Nucleic Acids Research* 23, 4407–4414. https://doi.org/10.1093/nar/23.21.4407

Walsh, A.M., Crispie, F., Kilcawley, K., O'Sullivan, O., O'Sullivan, M.G., Claesson, M.J. and Cotter, P.D. (2016) Microbial succession and flavor production in the fermented dairy beverage kefir. *mSystems* 1(5):e00052-16. doi:10.1128/mSystems.00052-16.

Weisburg, W.G., Barns, S.M., Pelletier, D.A. and Lane, D.J. (1991) 16S ribosomal DNA amplification for phylogenetic study. *Journal of Bacteriology* 173(2), 697-703. DOI:10.1128/jb.173.2.697-703.1991. https://doi.org/10.1128/jb.173.2.697-703.1991

White, G.F., Snape, J.R. and Nicklin, S. (1996) Biodegradation of glycerol trinitrate and pentaerythritol tetranitrate by *Agrobacterium radiobacter*. *Applied and Environmental Microbiology* 62, 637-642. https://www.ncbi.nlm.nih.gov/pmc/articles/PMC1388782/

White, T.J., Bruns, T., Lee, S. and Taylor, J. (1990) Amplification and direct sequencing of fungal ribosomal RNA genes for phylogenetics. In: Innis, M.A., Gelfand, D.H., Sninsky, J.J. and White, T.J. (eds) *PCR Protocols. A Guide to Methods and Applications*. Academic Press, San Diego. 315-322. http://dx.doi.org/10.1016/b978-0-12-372180-8.50042-1

Wiebe, M. (2002) Myco-protein from *Fusarium venenatum*: a well-established product for human consumption. *Applied Microbiology and Biotechnology* 58, 421–427. https://doi.org/10.1007/s00253-002-0931-x

Wildman, H.G. (1995) Influence of habitat on the physiological and metabolic diversity of fungi. *Canadian Journal of Botany* 73 (Supplement 1), S907-S916. https://www.academia.edu/25644243/Influence_of_habitat_on_the_physiological_and_metabolic_diversity_of_fungi

Woese, C.R. and Fox, G.E. (1977) Phylogenetic structure of the prokaryotic domain: The primary kingdoms. *Proceedings of the National Academy of Sciences USA* 74(11), 5088-5090. https://doi.org/10.1073/pnas.74.11.5088

Woese, C.R., Kandler, O. and Wheelis, M.L. (1990) Towards a natural system of organisms: proposal for the domains Archaea, Bacteria, and Eucarya. *Proceedings of the National Academy of Sciences USA* 87(12), 4576-4579. https://doi.org/10.1073/pnas.87.12.4576

Wong, Y.P., Othman, S., Lau, Y.L., Radu, S. and Chee, H.Y. (2018) Loop-mediated isothermal amplification (LAMP): a versatile technique for detection of micro-organisms. *Journal of Applied Microbiology* 124(3), 626-643. doi:10.1111/jam.13647. Epub 2018 Feb 12. PMID: 29165905; PMCID: PMC7167136.

Yamada, Y. and Kondo, K. (1973) Coenzyme Q system in the classification of the yeast genera *Rhodotorula* and *Cryptococcus*, and the yeast-like genera *Sporobolomyces* and *Rhodosporidium*. *Journal of General and Applied Microbiology* 19, 59-77. https://doi.org/10.2323/jgam.19.59

Yarrow, D. (1998) Methods for the isolation, maintenance and identification of yeasts. In: Kurtzman, C.P. and Fell, J.W. (eds) *The Yeasts: A Taxonomic Study* 4th edn. pp. 77-100. Elsevier, Amsterdam, Netherlands. https://www.elsevier.com/books/the-yeasts-a-taxonomic-study/kurtzman/978-0-444-81312-1

Yoon, H.S., Müller, K.M., Sheath, R.G., Ott, F.D. and Bhattacharya, D. (2006) Defining the major lineages of red algae (Rhodophyta). *Journal of Phycology* 42(2), 482-492. https://doi.org/10.1111/j.1529-8817.2006.00210.x

Young, E. (2016) I Contain Multitudes: *The Microbes Within Us and a Grander View of Life*. ECCO Press (imprint of HarperCollins), New York.

Young, J.M., Park, D.-C., Shearman, H.M. and Fargier, E. (2020) A multilocus sequence analysis of the genus *Xanthomonas. Systematic and Applied Microbiology* 31(5), 366-377. https://doi.org/10.1016/j.syapm.2008.06.004

Ziegler, D., Pothier, J.F., Ardley, J., Fossou, R.K., Pflüger, V., de Meyer, S., Vogel, G., Tonolla, M., Howieson, J., Reeve, W. and Perret, X. (2015) Ribosomal protein biomarkers provide root nodule bacterial identification by MALDI-TOF MS. *Applied Microbiology and Biotechnology* 99(13), 5547-5562. https://link.springer.com/article/10.1007/s00253-015-6515-3

Zolan, M.E. and Pukkila, P.J. (1986) Inheritance of DNA methylation in *Coprinus cinereus. Molecular and Cellular Biology* 6, 195-200. https://www.ncbi.nlm.nih.gov/pmc/articles/PMC367498/

4

Preserving Microorganisms

Introduction

This chapter provides information to help users select the best preservation method for microorganisms and the protocols to do this. The primary objective of preserving and storing an organism is to maintain it in a viable state without morphological, physiological or genetic change until it is required for future use. Ideally, complete viability and stability should be achieved, especially for important reference, research and industrial isolates. However, working, teaching or research collections may have to consider additional logistical factors such as simplicity, available facilities and cost. The task of preserving the hidden microbial diversity in microbial communities is setting new challenges (Ryan *et al.*, 2021). Given the difficulties in preserving some single 'axenic' cultured strains, the retention of functionality of whole communities, the microbiome from an animal, plant, a human or the environment is a test at a much greater level of complexity.

A preservation method should suspend metabolic activities instantly while retaining viability and genetic and physiological stability of the microorganism. This provides the basis for the safe and long-term maintenance of strains of scientific or industrial interest and is a prerequisite for continuous and efficient research and production (Ryan *et al.*, 2019). Availability of strains maintained in a genetic and physiological unchanged state must be guaranteed over many years. For example, important production strains, reference strains for research and testing and other strains with valuable properties should be available for comparative examinations even after several decades. Each of the different preservation and storage methods has its advantages and disadvantages which need to be considered to ensure the right method is applied to meet the special conditions or needs in each circumstance. Some general aspects should be considered with any choice of methodology:

- Viability rate should be maximized
- Genetic and physiological changes should be avoided as far as possible (this is also most likely correlated with a high survival rate)
- The risk of contaminating the preserved culture should be reduced

There are many methods available for the preservation and storage of microorganisms (Smith *et al.*, 2013; Smith and Bussas, 2021) and these can be divided into three groups:

- Continuous growth techniques involve frequent transfer from depleted to fresh nutrient sources, which initially provide optimum growth conditions. The need for frequent sub-culture can be delayed by storing cultures in a refrigerator, freezer (at -10°C to -20°C), under a layer of paraffin oil or in water.

DOI: 10.1079/9781800622135.0004

- Drying of the resting stage (e.g., spores, cysts or sclerotia) of an organism, can be achieved by air drying, in or above silica gel, in soil or sand.
- Suspension of metabolism normally involves reducing the water content available to cells by dehydration or cryopreservation. Freeze-drying (lyophilization) is the sublimation of ice from frozen material at reduced pressure and requires storage in an inert atmosphere either under vacuum or at atmospheric pressure in an inert gas. Cryopreservation generally implies storage at temperatures that suppress chemical reactions of -70°C and below. This can be achieved in mechanical deep freezers (some are capable of reaching temperatures of -150°C) or in/above a cryogenic gas such as liquid nitrogen. The aim is to achieve an adequate suspension of metabolism to a point where no physical or chemical reaction can occur and this requires storage at temperatures of below -139°C (Morris, 1981). The rate of cooling and the use of cryoprotectants are critical to prevent ice damage and enable the preservation of the more delicate or recalcitrant fungi. Where traditional approaches to cooling have not been successful, various alternative approaches have proven successful at CABI including encapsulation in alginate beads based on methods originally developed for microalgae and plant cell lines (Benson *et al.*, 2018) and controlled cooling for optimal survival (Ryan *et al.*, 2014). Vitrification, the use of highly concentrated cryoprotective solutions has also proven useful in the cryopreservation of the difficult to preserve fungi (Benson, 2008; Fonseca *et al.*, 2016; Benson *et al.*, 2018).

Desiccation has been used successfully for the preservation of many microorganisms. The removal of water suspends metabolic processes within the cell. Fungal spores have a lower water content than vegetative hyphae and are able to withstand desiccation, reviving and recovering when water becomes available. Drying can be carried out using many techniques. Air-drying is achieved by passing dry air over the culture or spores, this speeds the drying process by evaporation. Air drying has been successful for some *Aspergillus* and *Penicillium* species (Smith and Onions, 1983, 1994). Drying can also be achieved using soil, silica gel or other desiccants. Silica gel storage was first employed in CABI in 1970 and organisms stored at this time are periodically tested to determine optimal shelf life. Soil storage was initiated in the late 60's and over 900 *Fusarium* and related genera are stored at CABI by this method. The best way to preserve strains by desiccation is by freeze-drying (lyophilization), first extensively used with fungal cultures by Raper and Alexander (1945). Methods and machinery have developed over the years to produce a reliable and successful preservation technique for sporulating micro-fungi. Stability and long storage periods are the main advantages of freeze-drying though the expense of the modern and quite complex machinery can be a deterrent.

After a suitable preservation technique is selected and the strains successfully stored, a distribution stock for general use and distribution of the strain and a seed stock for re-preservation are produced. The size of the distribution stock depends upon the anticipated distribution. The seed stock is held in reserve for use when distribution stocks are depleted to produce subsequent distribution stock. This ensures that strains have undergone a minimum number of transfers from the original. Enough replicates of seed stock must be maintained to provide the start point for all subsequent preservation batches dependent on the shelf-life of the preserved materials, i.e. if the estimated shelf-life is 50 years and you wish to re-preserve every 10 years, a minimum seed stock of 5 ampoules is needed. Wherever possible, an original should be preserved without sub-culturing or at the very least the number of manipulative steps used in cultivation must be limited. The seed stock should be stored separately from the distribution stock. An inventory control system should be used to ensure that cultures remain in stock for distribution or use. After preservation, the viability, purity and identity should be re-checked and compared with the original results before the culture is made available outside the collection. It is also advisable to keep a duplicate collection in another secure building or site as a reserve and protection against a disaster or loss. The organism may be sent to the depositor for confirmation of properties to ensure authenticity to the originally deposited strain.

The viability, purity and stability of strains must be monitored and assessed before and after preservation and at various time-points during storage. Simple viability assessment is usually carried out by using both a percentage recovery of propagules followed by a growth test or assessment of described characters. The culture should be grown on the most suitable culture medium, or host, to give optimum growth. Cultures should be monitored to ensure that characteristics have not altered from those noted at the time of deposit. Careful microscopic examination must be carried out to ensure that the culture is not mixed. It may be necessary to grow the culture under special conditions to determine if there are contaminants present. Cultures from single cell isolations can be prepared to give a better chance of growing a pure culture although this carries a risk of unintentional selection. To determine whether the cells remain stable during storage pre- and post-preservation comparisons should be made. Morphology, pathogenicity, genetic profiles, assay properties and biochemical properties can be checked where appropriate (see Chapter 3, this volume for tests that can be performed pre- and post-preservation to indicate stability of properties).

Morphology can be checked by comparison with organism descriptions or photomicrographs of the culture. For pathogens, pathogenicity testing is not always possible and licences are often needed for infecting hosts along with appropriate containment and quarantine facilities. Genomic profiles using a variety of molecular tools such as sequencing, assay of properties and biochemical properties can be checked where appropriate, e.g. using specific carbon sources for the presence of enzymes.

It is essential to record as much information and data concerning the organism's properties as possible so that this information can be compared after preservation and storage. Additionally, all observations on recovery of strains from preservation methods and storage are equally important and such data should be careful recorded and stored in an appropriate database. Important data elements are:

- Organism descriptions including anatomical and morphological features
- Photomicrographs and electronic images from cultures and prepared microscope slides
- Metabolic profiles e.g. HPLC, MALDI-TOF profiles
- Sequence data

The hypothesis for preservation allows an organism to be maintained in a viable state without bringing about any major changes to its stability, deviation from this would require the samples to be re-preserved. All of the preservation and storage techniques described in this chapter can be applied to most microbes with minor modifications.

Preservation Methods for Bacteria

Bacteria, are unicellular, sometimes filamentous and have 'a simple structure' but are quite diverse in their properties and function. They are grown and utilized in huge numbers in laboratories around the globe and are the key to many research programmes and industrial processes. As with all microbes, bacteria must be maintained without change to ensure reproducibility and sustainability. However, their sometimes rapid growth rate and increased likelihood of mutation informs the need to select the correct storage and preservation methods and techniques in their preservation. As with several microorganisms, bacteria can be maintained by sub-culture. However, this is not ideal because of the chances of strain drift (the latter can result from growth in artificial culture conditions) or contamination by other bacteria or fungal spores. Freeze-drying is often the method of choice as it gives excellent recovery and storage in ampoules that are easy to distribute and have simple storage conditions. Another method sometimes used in culture collections, is the gelatin disc technique which also allows easy distribution (Winters and Winn, 2010) after a method first described by Stamp (1947). A harvest of bacterial growth is suspended in melted nutrient gelatine, drops of which are allowed to solidify in Petri dishes. The drops are freeze-dried, or dried over a desiccant, and the resultant flat discs are stored over silica gel. When required, a single disc is placed in warmed broth and the resulting suspension plated onto a suitable growth medium.

This method is invaluable for the storage of a limited number of frequently used strains, such as those used for quality control of media or reagents. However, it is not suitable for storage over long periods. Advantages of storage using gelatine discs include their ease of use and storage (30 or 40 discs can be kept in a 14mm screw-capped glass vial). Because the discs are kept dry there is no opportunity for growth of any contaminants introduced during sampling. Characters remain relatively stable, and because bacteria do not grow, there is no opportunity for mutation and selection. Various species of Enterobacteriaceae and Staphylococci, and strains of *Pseudomonas aeruginosa* and *Corynebacterium diphtheriae* have been successfully preserved for at least 4 years, although this method has not been successful with more delicate species such as *Neisseria* or *Haemophilus*. However, Obara *et al.* (1981) using a method described by Yamai *et al.* (1979) have reported successful preservation of *Neisseria*, *Haemophilus* and *Bacteroides*, using a gelatin-disc method based on a different suspending medium. A number of collections provide standard strains of microorganisms in this form at a cheaper rate than freeze-dried cultures, particularly for use in quality control processes or for use in diagnostic laboratories.

Occasionally, cultures are distributed as active cultures resuscitated from cryopreservation onto a suitable nutrient media. Cryopreservation, especially to ultralow temperatures (-135 and -196°C), is used for isolates that are difficult to preserve, but is considered to be the least damaging preservation technique for bacteria for many years (Gilmour *et al.*, 1978). Lower storage temperatures confer better long-term survival and stability. Storage at -20°C is not generally recommended and storage in -70°C or -80°C freezers can be used if liquid nitrogen is not available. Many collections preserve plasmids, bacteriophages and genetically modified organisms in or above liquid nitrogen. Preserving them in this way reduces the chance of potential release as aerosols produced during lyophilization or when opening ampoules.

In a university setting, bacteria are frequently maintained by continual sub-culture, but this can be incredibly labour intensive with some bacteria needing to be re-cultured after just one week. Storing cultures in a refrigerator or freezer can extend intervals between sub-cultures. Although nutrient agar is widely used to maintain bacteria, many isolates require specialist nutrient requirements. As with all sub-culture regimes, long-term culture carries risks of unintentional selection and contamination. Therefore, continuous culture techniques should not be used for bacterial cultures unless absolutely necessary and in the short-term. Transfer from spent to fresh media is typically undertaken by the transfer of cells from a young colony using a sterile inoculating loop. Culture plates or slopes in Universal tubes are streaked and incubated at an optimal growth temperature. Plates should be monitored to ensure that cultures remain pure and free of contaminants. A regime for the frequency of transfer should be established.

Freeze-drying (lyophilization) is the universal method employed for the preservation of bacteria although there are a number of modifications to the basic freeze-drying procedure that can be used. Variations of the method are required for facultative and stringent anaerobes (Tindall, 2007). The majority of bacteria tolerate freeze-drying well, but a few species can sometimes give disappointing results particularly flagellate bacteria. In some cases, this may be due to difficulties in obtaining adequate pre-drying growth. Species that often prove more difficult than others include *Aquaspirillum serpens*, several *Clostridium* species *including C. botulinum*, *Helicobacter pylori* and *Peptococcus heliotrinreducans*. Additionally, problems have also been recorded with *Bacteroides melaninogenicus*, *Haemophilus canis*, *H. suis*, *Leptotrichia buccalis*, *Mycobacterium microti* and *Neisseria gonorrhoeae* (Smith *et al.*, 2001).

Centrifugal or spin freeze-drying (see Fig. 4.1) is a two-stage process that involves the suspension of cells in a buffer or lyoprotectant such as skimmed milk with the addition of a sugar such as inositol or trehalose before being dispensed into ampoules (Wessman *et al.*, 2013). The centrifugation is used to prevent bubbling 'boiling' of suspension as it degasses. A wedge is formed in each ampoule that increases the surface area for sublimation of water. The suspension will 'flash freeze' under reduced pressure. Frozen water is removed during the primary stage. The latter normally takes around three hours depending on the equipment used and the volumes being preserved. The residual moisture content at this stage is significantly reduced to around 5% of the original and the ampoules are plugged

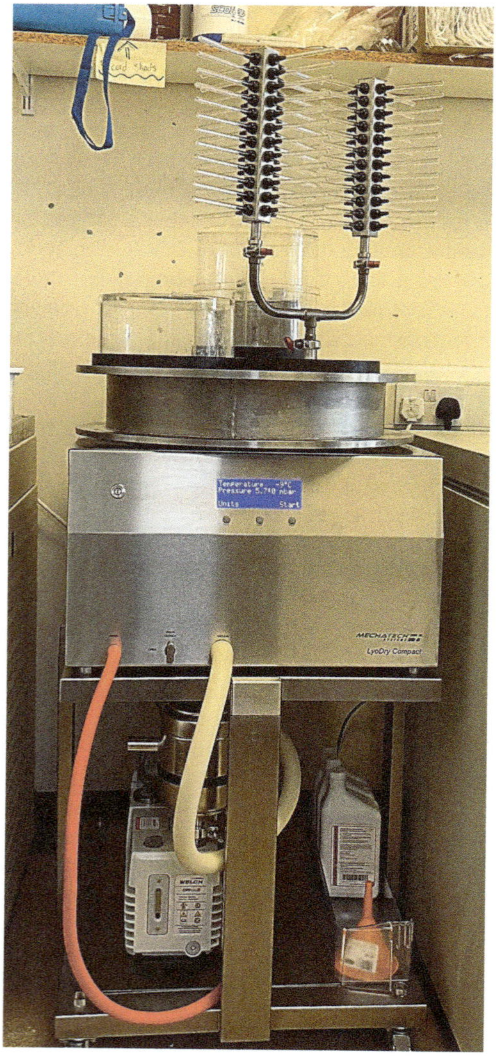

Fig. 4.1. Centrifugal freeze drier

loosely with cotton wool before the secondary drying that reduces water content to between 1 and 2% and the ampoules sealed under vacuum (Hubalek, 1996; Day and Stacey, 2007).

Liquid drying (L-drying as first described by Annear, 1958) is a useful alternative method of vacuum drying used in some collections for the preservation of bacteria such as *Spirilla* and *Azomonas insignis* that are particularly sensitive to damage occurring during the initial freezing stage of the normal lyophilization process. The intrinsic feature of this process is that cultures are prevented from freezing; drying occurs direct from the liquid phase. L-dried cultures are reported to have survived with good recovery levels for up to fifteen years (Smith *et al.*, 2001).

Microdrying is a modification of the freeze-drying method used for preserving bacteria and has been successfully applied to various Gram-negative and Gram-positive bacteria. It has several advantages over centrifugal freeze-drying because fewer manipulations are involved. Resuscitation requires less skill and there is less risk of contamination as the resuspending stage is omitted. The method is very similar to freeze-drying, except the cell suspension is absorbed into a strip of thick filter paper and is then freeze-dried. The filter strip enables the product to be tipped out easily into a resuspending or growth medium. The results obtained indicate that, in general, the long-term survival and recovery of microdried cultures are comparable to those obtained with conventionally freeze-dried cultures. (Smith *et al.*, 2001).

Cryopreservation is often used for bacteria, particularly those which do not survive freeze drying or L-drying (Smith and Ryan, 2012). This method is also used for sensitive mutants, genetically manipulated strains and all bacteriophages, which are preserved to reduce the risk of contaminating equipment and consequently other cultures with phage particles. Most genera survive cryopreservation well. Indeed, survival and recovery rates are comparable with conventional freeze-drying although certain obligate methylotrophic bacteria have been found to lose viability (Malik, 1990) and preservation of anaerobic bacteria can present additional problems. Despite an initial loss of titre of approximately one order of magnitude due to the freezing process, most coliphages will survive for at least 10 years without further significant loss of titre (Puapermpoonsiri *et al.*, 2010). An exception is coliphage MS2, which does not survive for more than 1.5 years. *Pseudomonas aeruginosa* phages have been shown to survive for at least 10 years and *Staphylococcus aureus* phages for 9 years (Merabishvili *et al.*, 2013). Marine phages tend to have shorter shelf lives. For example, *Pseudomonas* phage does not survive for more than 2-3 years (Puapermpoonsiri *et al.*, 2009).The methods for cryopreservation are similar to those described for fungi below.

Preservation Methods for Fungi

The fungi include some of the most important organisms, both in terms of their ecological and economic roles (Ryan *et al.*, 2012; Ryan *et al.*, 2019), from breaking down dead organic material to ensure the cycle of nutrients through ecosystems or in a symbiotic role, to supply essential nutrients to vascular plants. Fungi provide numerous drugs (such as penicillin antibiotics and cholesterol lowering drugs) and foods like mushrooms etc. Because fungi are such a diverse group several methods of preservation are required to ensure the viability and morphological, physiological and genetic integrity of the cultures over time.

Collectively fungi are ubiquitous; including species able to grow in a wide variety of environments utilizing a vast array of natural and man-made substrates. Some are host specific or have fastidious growth requirements and cannot be grown in culture. Generally, fungi grow best on media that are formulated from the natural materials from which they were isolated. Fungi produce structures that enable them to survive adverse conditions (e.g., spores, sclerotia, asci, and thickened hyphae). Such propagules can be readily stored in the laboratory to retain viability. However, methods of storage that allow growth and reproduction may allow the organism to change and adapt to artificial laboratory conditions. Avoidance of selection of variants from within the population, strain deterioration and contamination are important when growing strains for use and essential when maintaining cultures for long periods. There are a number of preservation techniques suitable for fungi, however many criteria should be assessed before preserving an isolate (Ryan *et al.*, 2000). No preservation technique has been successfully applied to all fungi, although storage in liquid nitrogen appears to approach the ideal. However, changes in physiology and genetic stability may occur in some isolates (Ryan, 1999), so optimal preservation protocols may have to be established (Smith and Thomas, 1998; Ryan *et al.*, 2019). Centrifugal freeze-drying allows only the more robust spores to survive. Methods often used are:

- Cryopreservation in liquid nitrogen: Most fungi that grow well in culture survive cryopreservation in liquid nitrogen; including non-sporulating fungi. Recalcitrant fungi that grow poorly tend to do less well. The cryopreservation technique will also allow the preservation of fungal pathogens in infected tissue, those that do not grow in culture, although care must be taken to ensure that a mixed population of organisms is not preserved.
- Lyophilization (freeze-drying): Centrifugal approaches allow only the more robust spores to survive. Some sclerotia and other resting stages, and even in a few cases sterile mycelia, have been known to survive freeze-drying.
- Silica gel storage: Only sporulating fungi survive well in silica gel storage. Spores with thin walls and high water content or those with appendages do less well.
- Storage in mineral oil, soil or water: Storage in mineral oil, soil or water may be of use for a wide range of organisms but the period of storage may be quite short.

Methods of Preservation Applied to Taxonomic Groups of Fungi

The taxonomic position of a fungus should be considered before it is preserved. Knowledge of the identity of a fungus can greatly help when considering how to preserve it. However, one must bear in mind that the survival of a particular technique often depends on the strain, as some strains are more robust or express different physiological characteristics.

OOMYCOTA. The Oomycota are traditionally studied by mycologists, although taxonomically they are members of the Chromista. Unlike most fungi, their mycelium is aseptate and they produce flagellate zoospores. They are best stored in liquid nitrogen using a cooling rate of c. -10°C min^{-1} with a cryoprotectant such as 10% w/v glycerol or DMSO. Some strains do not survive freezing, while others may be best stored as a mycelial plug with a colonized 'Hemp' seed. Storage under mineral oil may be satisfactory for periods up to six months (three strains of *Phytophthora* have survived 41 years at CABI). Cultures can also be kept viable in water storage and transferred every 2 years.

ZYGOMYCOTA. As sporulating organisms, most members of the Zygomycota such as *Rhizopus* and *Mucor* can be easily maintained by a variety of methods, with liquid nitrogen storage being the recommended method. Most isolates can even be freeze-dried. Some species are preservation recalcitrant, and will not survive dehydration, particularly in silica gel, for example *Coemansia*, *Martensiomyces*, *Condiobolus*, *Entomophthora*, *Piptocephalis* and *Syzygites*. Of these genera, only *Piptocephalis* and *Coemansia* strains have been freeze-dried successfully at CABI.

ASCOMYCOTA. Conidial ascomycetous fungi are typically the easiest group of fungi to preserve as they readily produce conidia and resting spores in culture. The majority of *Ascomycetes* which can be grown in culture can be freeze-dried or cryopreserved in liquid nitrogen. The majority of genera and species that do not sporulate well in culture survive dehydration techniques poorly. However, most of these have survived long-term storage in liquid nitrogen. *Aspergillus*, *Penicillium* and *Paecilomyces* species can even tolerate storage for 6 months to 2 years on agar slants in a freezer at -18°C. but are likely to be subject to strain drift and mutation.

BASIDIOMYCOTA. Basidiomycota generally grow only as mycelium in continual culture. They can only be preserved by serial transfer on agar with or without oil, or stored in liquid nitrogen. Those fungi that produce thick-walled hyphae can be freeze-dried but their viabilities are usually low. Basidiospores harvested from fungi growing in their natural environment can usually be freeze-dried, but spores will have been produced by meiotic events so are likely to exhibit variation from the mother culture.

RUSTS. Rust fungi do not normally grow in culture, but living collections can be maintained in good condition in liquid nitrogen in association with the host (Ryan and Ellison, 2003) or as harvested spores. *Ustilaginales* produce very disappointing cultures and survive best in liquid nitrogen though it is possible to keep them by other means. If the spores can be harvested successfully, some survive freeze-drying quite well.

YEASTS. The single-celled vegetative yeasts tolerate cryopreservation and freeze-drying, although the best technique is in or above liquid nitrogen. Unlike other Basidiomycetous fungi, the *Basidiomycete* yeasts generally survive freeze-drying. Most species survive silica gel storage but recovery is usually very low. In general, they are considered to be robust, tolerant of unfavourable conditions, nutritionally undemanding, and readily managed in industry. It is incorrectly assumed that they are easy to maintain as a number of preservation and maintenance methods result in poor viability and instability of properties. Factors affecting survival are becoming better understood at the subcellular level, but many strains from a wide range of species remain difficult to preserve. The relatively poor performance of yeasts post preservation may be partly attributed to the large size of cells compared with bacteria and the absence of the resistant spore types produced by many of the higher fungi. In the light of present knowledge, high survival rates can best be achieved by careful attention to the techniques used for preservation.

Percentage survival of the total population following sub-culture and drying is generally low, although 'cultures may appear viable' (Kirsop, 1974). Viability following freeze-drying is also frequently poor, but may be improved for some strains by careful selection of the suspension medium (Berny and Hennebert, 1989). In contrast, survival following storage in liquid nitrogen is high, often reaching levels of 100% (Hubalek and Kochova-Kratochvilova, 1978). There is no apparent relationship between survival and taxonomic position, and the factors determining survival are strain specific. Therefore, a preservation method that is satisfactory for one strain of a species may be unsuitable for others. If strain stability is of paramount importance, the choice of maintenance method becomes critical. Any method that enables cell division to occur during storage should be rejected. It has been shown that the morphological, physiological and industrial characters of yeasts generally remain unchanged following freeze-drying (Kirsop, 1974), although other authors have found that genetic changes can occur (Souzu, 1973). Some workers have found that yeasts dried on silica gel may show substantial changes, whereas others report satisfactory results. However, there is now growing evidence that most strains stored in liquid nitrogen survive well and remain stable (Wellman and Stewart, 1973; Hubalek and Kochova-Kratochvilova, 1978; Pearson *et al.*, 1990; Bond, 2007).

LICHEN-FORMING FUNGI. Lichen-forming fungi do not survive for long periods by serial transfer leaving only cryopreservation as the method for successful preservation of this group. They respond well to cooling rates of -10 to -20°C min^{-1} in 10% w/v glycerol and storage in or above liquid nitrogen. These fungi tend to produce mycelium only in culture.

Basic Preservation Methods for Fungi

Many different microorganisms are preserved using similar techniques, often with special adaptations for the different types. We will now look at these techniques and assess the similarities:

SERIAL SUB-CULTURE (CONTINUOUS CULTURE). Serial sub-culture is widely used and is perhaps the simplest and most cost-effective method for a small laboratory, especially if cultures are required frequently and quickly. Most laboratories will retain some cultures by this method, commonly maintained on agar slopes rather than on Petri dishes and stored under controlled temperature depending on the genus. Refrigeration below room temperature is often used as it extends the sub-culture interval, thereby reducing the number of transfers required as a result of the suppression of the metabolic rate. Limiting nutrients can also have the effect of reducing growth rate and extending periods between transfers. When maintaining large numbers of organisms or those of greater hazard it is advisable to work in an appropriate microbiological safety cabinet to protect the worker and strict aseptic technique must be observed to protect the organism (Smith and Onions, 1994).

Fungal transfer by sub-culture can be potentially disadvantageous, as frequent sub-culturing could result in contamination from other microorganisms such as bacteria or air-borne spores of other fungal species and the likelihood of strain drift is significant. The choice of medium is an important factor, as some fungi are notoriously difficult to culture (e.g., many mycorrhizal fungi). Most fungi will survive on Malt Agar (MA) or Potato Carrot Agar (PCA) (Smith and Onions, 1994), but others require more specialized media. Dermatophytes, for example, may grow better on a substrate of hair (Al-Doory, 1968). Additives such as growth factors may be added to the growth medium for specific fungi (Smith, 1993). Variation of the nutrient source may prevent the permanent adaptation and modification of the strain to a specific medium. Media should not encourage excessive sporulation or fructification as meiotic or mitotic crossing may promote the formation of recombinants that may differ from the parental genotypes (Smith and Onions, 1994). When inoculating fresh plates, it is recommended to sub-culture from the periphery of the fungal colony, i.e. the region of actively growing mycelium. The precautions mentioned above should ensure that, as far as possible, fungi maintain the characteristics exhibited upon isolation from nature and do not mutate or show selection. It is a feature of the opportunistic nature of fungi to easily adapt to the environment. However, despite the best management, this will inevitably happen if isolates are maintained in culture for long periods. Asexual processes such as conidiogenesis and sexual processes that result in genetic recombination enhance the likelihood of selection and mutation (Burdsall and Dorworth, 1994). Characteristics may unintentionally be 'selected out' from fungal cultures if workers sub-culture from atypical sectors on a plate (Smith and Allsopp, 1992), often where a technician may be unfamiliar with the organism they are working with. The use of cold storage can slow the rate of metabolism and thus increase the intervals between sub-culture. Storage at 4-7°C in a refrigerator, or cold room, can extend the transfer interval to 4-6 months from the average period of 2-4 months. Storage in a deep freeze (-7 to -24°C) will allow many fungi to survive 4-5 years between transfers, though freezing damage may occur in some and the agar may deteriorate making further sub-culture difficult. Furthermore, mites do not invade cultures at these lower temperatures.

During storage at ambient temperatures, cultures should be routinely sealed with air-permeable tape to prevent invasion from mites such as *Tyrophagus* and *Tarsonemus* spp. Mites not only damage cultures by utilizing the fungus as a food source but carry contaminants such as bacteria or fungal spores from plate to plate, thus rendering cultures unusable. A mite infestation can be extremely costly, as all cultures may need to be destroyed. Important cultures can be recovered using a combination of antibiotics, freezing and careful sub-culturing (Smith and Onions, 1994). However, irreversible damage could be caused to the fungus due to the pressures exerted by the restorative methods. Frequent cleaning of laboratory work surfaces with bleach or acaricides can deter mites, but may pose health and safety risks to staff.

DRY AGAR SLOPES. When storing fungi by this method, technicians should be alert for mite infestation. All bottles should be sealed with cigarette papers before incubation and during storage. Cultures will need transferring to fresh media at least every six months, maybe more frequently. The main disadvantages are that frequent transfer can accelerate morphological change, loss of pathogenicity and decrease sporulation.

MINERAL OIL. This method is generally only used for yeasts and filamentous fungi but can be applied successfully to bacteria. It involves covering cultures with mineral oil to prevent dehydration and to slow down the metabolic activity and growth through reduced oxygen tension. The method was first extensively used by Buell and Weston (1947), and subsequent reports have indicated its wide application and success (Dade, 1960; Fennell, 1960; Little and Gordon, 1967; Smith *et al.*, 1970; Onions, 1971, 1977; Smith and Onions, 1994). Mature, healthy, well-sporing cultures grown on a 'short' agar slope (30° to the horizontal) in 30ml Universal bottles are covered by 10mm of sterile (achieved by autoclaving twice at 121°C for 15min) mineral oil (liquid paraffin or medicinal paraffin specific gravity 0.830-0.890). If the oil is deeper than 10mm the fungus may not receive sufficient oxygen and may die, while if the depth is less, exposed mycelium or agar on the sides of the container may allow moisture to evaporate and the culture to dry out. At CABI Bioscience, Universal bottles are stored with their caps loose in aluminium segmented racks. in a temperature controlled (15-18°C) room (See Fig. 4.2). Retrieval is relatively easy and involves the removal of a small section of the colony with a mounted needle. Excess oil is drained away and the inoculum streaked onto a suitable agar medium. More than one sub-culture may be necessary after retrieval as the growth rate may be reduced because of adhering oil. The fungal mycelium can normally recover when it is re-isolated from the edge of the colony on the first agar plate and transferred to fresh media. Inoculating an agar slope centrally sometimes has better results as excess oil can drain down the slope, allowing the fungus to grow more typically towards its top. There is an added risk of personal contamination because of the spattering of oil containing the fungus when inoculation needles are sterilized in a Bunsen flame (Fennell, 1960). The shelf life of yeast cultures preserved in oil is around 2 to 3 years. There is little documentation regarding the yeast species that have been maintained successfully.

The disadvantages of oil include the possibility of contamination by air-borne spores, retarded growth on retrieval and continuous growth under adverse conditions that could lead to selection. However, preservation under oil is recommended for storage of organisms in laboratories with limited resources and facilities. The advantages of oil storage are long viability of some specimens and survival of species that do not survive other preservation regimes; expensive equipment and consumables are not required. Additionally, mites do not cause infestations in oil cultures as they are unable to escape once they have entered the culture bottle. A wide range of fungi and chromists survive this method, *Saprolegniaceae* and other water moulds survive 12-30 months (Reischer, 1949). Species of *Aspergillus* and *Penicillium* have remained viable for 40 years and some strains of *Phytophthora* and *Pythium* species have also survived 40 years at CABI. Many cultures have shown that they deteriorate under mineral oil and must be transferred regularly to reduce this effect. However, organisms that are sensitive to other techniques can be stored successfully in oil, for example *Cercospora*, *Arthrobotrys*, *Colletotrichum*, *Conidiobolus*, *Corticium*, *Nodulisporium* and

Fig. 4.2. Oil storage at CABI

mycelial *Basidiomycetes* should be transferred every two years.

Equipment and reagents

- Medicinal quality liquid paraffin, specific gravity 0.830-0.890 (autoclaved twice at 121°C for 20min on consecutive days)
- Sterile solid growth medium in 30ml Universal bottles: set at 30° slope with an appropriate growth medium
- Metal segmented trays (375 x 175mm divided into 25 x 25mm squares to take 60 x 30ml Universal bottles (Wesbart)
- Inoculating needle or loop
- Class 2 Microbiological safety cabinet
- Bunsen burner
- Alcohol for flaming

Method

1. Inoculate at least two Universal bottles for each strain to be maintained.
2. Label one culture as reserve stock, the other(s) as working stock.
3. Incubate at optimum growth temperature until the organism has reached maturity.
4. Add 8-10ml of sterile liquid paraffin to cover the slope to a maximum depth of 10mm over its highest point.
5. Store the oiled cultures, with the screw caps loose, in metal divided racks at 15-20°C.

Recovery

1. Remove a portion of the working stock culture using a sterile needle or loop.

2. Drain as much oil as possible from the inoculum.
3. Inoculate fresh growth medium (it is often best to either inoculate a slope or an agar plate at an angle so that the adhering oil can drain and the organism can grow up the slope and away from the oil at the point of inoculation). The reserve stock culture is used only when re-preservation becomes necessary – when all the inoculum has been removed, when it is contaminated or when the shelf-life expiry date set for the organism has been reached*.
4. Once the colony has grown away from the oil, sub-culture again on to fresh medium.

*Although several fungi have survived for 44 years at CAB Bioscience UK Centre (Egham) it is advisable to set a re-preservation date of between 2 and 10 years.

SILICA GEL. The silica gel method has been applied to fungi at CABI and proved to be very successful. Sporulating fungi have been stored for 7-18 years in silica gel and appear to remain morphologically stable after resuscitation (Smith and Onions, 1994). The technique is relatively simple and involves the inoculation of a suspension of fungal propagules onto cold silica gel. The culture will then dehydrate to enable storage without growth or metabolism.

Silica gel storage has a number of advantages, it is cheap, simple and does not require expensive apparatus (Perkins, 1962). Cultures are relatively stable, allowing a wide range of sporulating fungi (including representatives of the *Basidiomycota*) to be successfully preserved. Penetration by mites is unlikely, as they cannot survive the dry conditions encountered. Repeated inocula can be removed from a single bottle. However, it is recommended that a stock bottle is prepared, to be used in case of contamination during retrieval. There are some disadvantages of silica gel storage. It is limited to sporulating fungi and is unsuitable for *Pythium*, *Phytophthora* and other *Oomycota*, mycelial fungi or fungi with delicate or complex spores. This limitation doesn't apply to yeast cultures. There is a possibility of introducing contaminants by repeated retrievals. Survival is often up to 25 years, according to species (Sharma and Smith, 1999).

Equipment and reagents

- Coarse non-indicator silica gel
- Sterile 5% (w/v) solution of non-fat skimmed milk cooled to 5°C
- Sterile Pasteur pipettes
- 30ml glass Universal bottles (a minimum of 2 per culture)
- Waterproof tray (100mm deep) filled to 30mm with water
- Refrigerator
- -20°C freezer
- Airtight storage boxes
- Indicator silica gel
- Class 2 Microbiological safety cabinet
- Bunsen burner
- Alcohol for flaming

Method

1. One-third fill glass Universal bottles (30ml bottles are used at CABI Bioscience, but any heat resistant screw cap bottle will suffice) with medium grain plain 6-22 mesh non-indicating silica gel and sterilize by dry heat (180°C for 3h).
2. Place bottles in a tray of water up to the level of the silica gel and leave overnight in a -20°C freezer (nominal).
3. Prepare spore suspensions in cooled 5% (w/v) skimmed milk.
5. Using a Pasteur pipette add 1ml of suspension to at least two bottles of the silica gel (whilst they remain in the frozen water).
6. After 20min, remove the Universal bottles from the ice and agitate them to disperse the suspension.
7. Label one bottle as reserve stock and the other(s) as working stock.
8. Incubate the bottles at 25°C until the silica gel crystals readily separate when agitated, this may take one or two weeks.
9. Screw the bottle caps down and store in air-tight containers over indicator silica gel to absorb moisture (or include an open Universal containing indicator silica gel) at 4°C (storage between 20 and 25°C is satisfactory).

Recovery

1. Sprinkle a few crystals from the working stock on to a suitable growth medium and incubate under appropriate growth conditions.
2. If the organism fails to grow, attempt again, this time streaking a silica gel crystal over the agar to dislodge the cells and discarding the silica gel crystal before incubation.

WATER. Immersion in sterile water can be used to extend the life of an agar culture (Burdsall and Dorworth, 1994). The method is generally applied to fungi and can be achieved in many different ways. One simple way is to grow the fungus on an agar slope in a Universal bottle and then cover the agar surface with water. An alternative method involves the transfer of mycelial plugs or blocks cut from cultures grown on agar in Petri dishes to Universal bottles containing 10ml of sterile deionized water (Boeswinkel, 1976). The protocol used at CABI is outlined below. To reduce the storage space occupied, cryovials may be used in a similar way (Burdsall and Dorworth, 1994). The "shelf-life" of fungi in water is variable but Figueiredo and Pimental (1975) successfully stored examples of phytopathogenic fungi for ten years by this means. Onions and Smith (1984) stored strains of *Pythium* and *Phytophthora* in water for five years but only 58% of these remained viable. Qiangqiang *et al.* (1998) preserved 78 isolates, belonging to seven genera, in water for 12 years, on resuscitation, 89.7% of isolates were viable. Burdsall and Dorworth (1994) reported that water storage did not significantly affect growth rate, viability or genetic stability in 155 isolates of Basidiomycota stored for 7 years. As with all methods, some fungi are better suited to individual protocols, and notably ectomycorrhizal fungi have been successfully stored by this method (Marx and Daniel, 1976). However, the storage of ectomycorrhizal basidiospore slurries in water was not successful (Torres and Honrubia, 1994). The advantages of storage in water are the low cost and easy application. However, the length of storage is often limited and some fungi will not survive even short periods submerged. As with all methods that allow growth or metabolism during storage there are better methods and it is considered only to be useful for short-term preservation (2-5 years) and should be backed up by longer-term storage methods (Smith and Onions, 1994).

Equipment and reagents

- Sterile distilled water (10ml in 30ml Universal bottles, at least two per culture)
- Mature cultures on agar media in Petri dishes
- Metal segmented trays (375 x 175mm divided into 25 x 25mm squares to take 60 x 30ml Universal bottles (Wesbart)
- Inoculating needle or loop
- Class 2 Microbiological safety cabinet
- Bunsen burner
- Alcohol for flaming

Method

1. Cut (generally from the growing edge) agar blocks ($6mm^3$) through the colony*.
2. Transfer 20-30 agar blocks to 10ml of sterile distilled water in two or more 30ml Universal bottles.
3. Label one bottle as reserve stock and the other(s) as working stock.
4. Screw the caps of the universal bottles tightly and store between 20-25°C.

Recovery

Remove an agar block from the working stock and inoculate (organism face down) on a suitable growth medium and incubate under optimum growth conditions. Use the reserve stock when re-preservation is necessary.

*Alternatively, sporulating or non-filamentous organisms can be harvested without the agar and simply suspended in water. Recover the organism by placing a small amount of the suspension on to suitable growth medium.

SOIL. This technique can be applied to a range of microorganisms that can withstand a degree of desiccation, for example the spores and resting stages of filamentous fungi, bacteria such as *Bacillus* spp. and some microalgae such as *Haematococcus pluvialis*. At CABI Bioscience the method involves inoculation of double autoclaved soil (121°C for 15 min) with 1ml of spore suspension in sterile distilled water and then incubation at 20-25°C for 5-10 days depending on the growth rate of the fungus. This initial growth period allows the fungus to utilize the available moisture, before the induction of dormancy. The bottles are stored in a refrigerator (4-7°C). This method was widely used for the storage of *Fusarium* isolates and related genera. Preservation in sterile sandy loam soil may be one of the most practical and cost-efficient ways to preserve filamentous sporulating microorganisms. Other advantages include good viability of cultures for up to 10 years, a reduced chance of mite infection and the option of obtaining repeated inocula from the same source. This method of storage is very successful with *Fusarium* species (Gordon, 1952; Booth, 1971). Atkinson (1953) obtained good recovery of *Rhizopus, Alternaria, Aspergillus, Circinella* and *Penicillium*. No loss in pathogenicity of *Septoria* species isolated from cereals was observed after

20 months in soil (Shearer *et al.*, 1974), or of *Pseudocercosporella* spp. for 1 year in soil (Reinecke and Fokkema, 1979). However, examination of Gordon's collection showed that 76% of *Fusarium equiseti*, 75% of *F. semitectum* and 50% of *F. acuminatum* isolates had been outgrown by mutant strains (Booth, 1971). Despite this, soil storage should be used in preference to oil storage for the preservation of *Fusarium* species and other fungi that show variation under oil. There are few disadvantages but the method is not suitable for many fungi and variation may occur after storage.

Equipment and reagents

- Garden soil or sand
- Sterile distilled water
- Pasteur pipettes
- 30ml glass Universal bottles (a minimum of 2 per culture)
- Refrigerator
- Class 2 Microbiological safety cabinet
- Bunsen burner
- Alcohol for flaming

Method

1. One-third fill 30ml Universal bottles with garden soil or sand and autoclave twice on consecutive days at 121°C for 20min.
2. Prepare a mycelial/spore suspension in sterile distilled water.
3. Using a Pasteur pipette add 1ml (approx.) of suspension to at least two bottles of sterile soil for each strain.
4. Label one bottle as reserve stock and the other(s) as working stock.
5. Incubate at 20-25°C for 5-10 days depending on the growth rate of the fungus being stored (this initial growth period allows the fungus to use the available moisture and gradually to become dormant).
6. Screw the bottle caps down tightly and store in a refrigerator (4-7°C).

Recovery

Sprinkle particles of soil or sand from the working stock on to a suitable growth medium and incubate under appropriate growth conditions.

FREEZE-DRYING. Freeze-drying (lyophilization) is a highly successful method for preserving bacteria, yeasts and the spores of filamentous fungi. During the freeze-drying process water is removed directly from frozen material by sublimation under vacuum. If carried out correctly, freeze-drying will prevent shrinkage, structural change and help retain viability. There is a vast array of freeze-drying equipment available, ranging from laboratory bench models through to pilot scale and huge industrial installations. Freeze-drying should be optimized for different organisms and cell types. If this is done it should be successful for the majority of bacteria, sporulating fungi and yeasts. It is generally unsatisfactory for eukaryotic microalgae as levels of post-preservation viability are unacceptably low (Day and McLellan, 1995). More protocol development is required to achieve successful lyophilization for algae and protozoa, although cyanobacteria are more likely to survive.

Lyoinjury can occur during the cooling and/or drying stages (Tan, 1997). The phase changes encountered during the drying process can cause the liquid crystalline structure of the cell membranes to degenerate to the gel phase, which disrupts the fluid-mosaic structure of the membrane (Tan, 1997). This causes leakage of the membrane, which may culminate in cell damage. Optimal survival can be improved with the use of a suitable suspension medium. It should be readily available, easy to prepare and provide protection during the freeze-drying process (i.e. to protect the spores/cells from ice damage during cooling and storage problems such as oxidation). Skimmed milk is a suitable protectant for fungi and is sometimes used in combination with inositol. Saccharides such as trehalose (Tan *et al.*, 1995; Tan, 1997) protect membranes by attaching to the phospholipids, replacing water and lowering the transition temperature. Other suspending media can be used when preserving bacteria and yeasts with many collections using their preferred preservation base. For example, Tan *et al.* (1995) suggest that a mix of dextran and trehalose improves the viability of cultures.

The recommended final moisture content following drying is between 1 and 2%(w/v). To monitor freeze-drying a means of measuring vacuum both in the chamber and close to the vacuum pump is required. Comparing the measurements will allow the determination of the end point of the drying process. When the values are equal, water has ceased to evaporate from the material being dried and drying is probably complete. This is confirmed by determining the residual water content. This can be done by dry weight determination or by the use of chemical methods such as the Karl Fischer technique (Baker, 1955). The freezing point of the material should be determined and the temperature monitored during freeze-drying. The sample temperature must not rise above the melting point until most of the water has been removed. To ensure that a high-quality product is produced and maintained the equipment used must be reliable and conditions reproducible from batch to batch.

The technique of centrifugal freeze-drying, which relies on evaporative cooling, can be used successfully for the storage of many sporulating fungi (Smith, 1983a), as well as bacteria and yeasts. However, this is not a method that can be adapted and changed easily, as it is dependent upon the scope of the equipment. Optimization of cooling rate to suit the organism being freeze-dried can be applied using a shelf freeze-drier. The sealing of the ampoules or vials is most important and heat-sealed glass is preferred to butyl rubber bungs in glass vials as these may leak over long-term storage and allow deterioration of the freeze-dried organism. There are many advantages of freeze-drying over other methods, including the total sealing of the specimen and protection from infection and infestation. Cultures generally have good viability/stability and can be stored for many years. Ampoules take up little space and can be easily stored. In addition, cultures do not have to be revived before postal distribution. However, there are disadvantages, some isolates fail to survive the process and others have reduced viability and genetic change may occur (Ashwood-Smith and Grant, 1976; Ryan, 1999) though unless high viability is retained it is difficult to differentiate between this and selection of spontaneous mutants by freeze-drying (Heckly, 1978). Ampoules of freeze-dried organisms must be stored out of direct sunlight and chilled storage will reduce the rate of deterioration and should extend shelf-life. However, the process of lyophilization is relatively complex, can be time-consuming and may be expensive.

At CABI a two-stage centrifugal freeze-drying process has been used since 1966 and an optimizable shelf freeze-drying protocol was introduced in 1982 (See Fig. 4.3). Freeze-drying of sporulating fungi such as the Ascomycota and mitosporic fungi is routinely undertaken, but is not so suitable for the Oomycota and other non-sporulating cultures. Although it is only spores and conidia that are routinely freeze-dried, research has been carried out to establish whether lyophilized hyphae can be revitalized successfully after preservation. In most cases this has met with little success, but hyphae from *Claviceps* spp. (Pertot *et al.*, 1977), a limited range of basidiomycetes (Bazzigher, 1962) and some arbuscular mycorrhizal fungi (Tommerup and Kidby, 1979) have been revitalized successfully. Investigations by Tan *et al.* (1991a, b) gave mixed results. Some cultures did not survive at all and others showed only limited viability. Success with freeze-drying varies between isolates of the same species. In general, those fungi that grow and sporulate well in culture survive the process, while weak or deteriorated isolates tend to fail. It may therefore be misleading to state categorically that one particular species will not survive freeze-drying. In general, the young vegetative hyphae of fungi do not survive freeze-drying. At CABI it has been found that sterile ascomata, chlamydospores, sclerotia and in some few cases stroma and resting mycelium have survived. However, in general it is only the spores (e.g., conidia, ascospores and basidiospores) that survive.

Freeze-drying is a generally accepted method for yeast storage, although viability is generally low, typically between 1 and 30%, as compared to >30% for those of yeast preserved frozen in liquid nitrogen. Some strains have been stored for 30 years at the National Collection of Yeast Cultures, (NCYC) in the UK (Dr Ian Roberts, personal communication, 2012) and for longer periods by other laboratories. There are several yeast genera, including *Lipomyces*, *Leucosporidium*, *Brettanomyces*, *Dekkera*, *Bullera*, *Sporobolomyces* and *Rhodosporidium* that have particularly low survival levels and often may not be successfully freeze-dried by standard protocols. However, improvements were evident when trehalose was used as a protectant (Berny and Hennebert, 1991; Roser, 1991). Survival of yeasts following lyophilization is notably strain specific and generalizations regarding survival levels should be viewed with caution.

The apparatus for freeze-drying is very expensive to purchase and maintain. Experienced, trained staff are required to both operate the system and for preparation of the material for preservation.

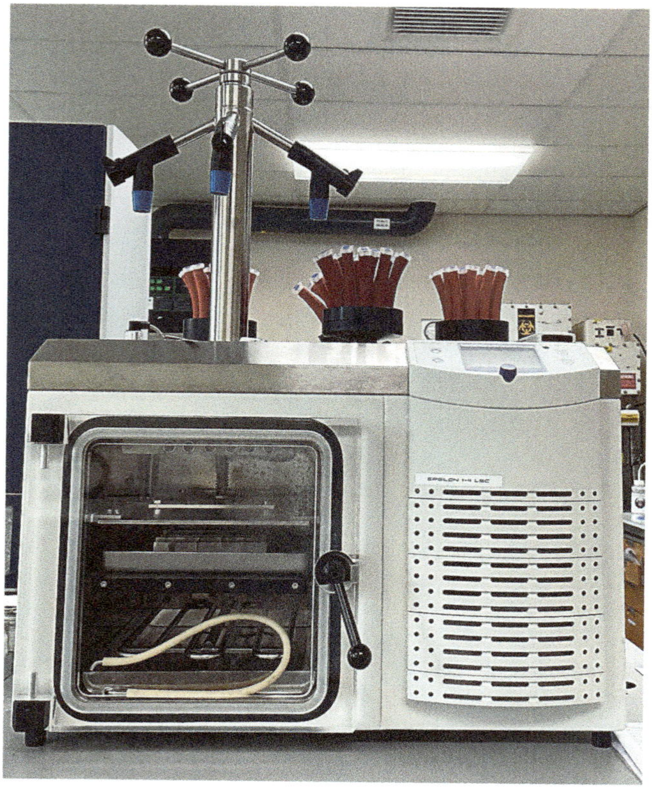

Fig. 4.3. Current shelf freeze drier

TYPICAL METHOD. Cultures are dried from a frozen state by withdrawal of water vapour under vacuum. The dried cultures or spore suspensions are sealed and stored in glass ampoules. Vigorously sporing fungi are best suited to this method. Various recipes for suspending media have been tried for fungi and bacteria. A typical centrifugal freeze-drying procedure would be:

- Prepare cultures in suitable suspending medium
- Plug the ampoules with sterile cotton wool (10-15mm plugs pushed midway down the ampoule)
- Place ampoules in centrifuge plate
- Close drier and switch centrifuge on
- Apply vacuum
- Switch centrifuge off when samples completely frozen (approx. 20 min)
- Primary dry for 3 hours
- Constrict ampoules above the cotton plug and place on manifold
- Apply secondary drying vacuum and leave for approx. 17 hours
- Seal ampoules
- Test to ensure ampoules are correctly sealed with a spark tester

Method for Centrifugal (Spin) Freeze-Drying

Centrifugal freeze-drying is a two-stage process. Several models of freeze-drier are available but the principles of their operation and use are similar.

Equipment and reagents

- Sterile 10% (w/v) skimmed milk + 5% (w/v) inositol (autoclaved at 121°C for 15 min), (different organisms may require different protectants)
- Freeze-drier with spin-freeze and manifold accessories

- 0.5ml (nominal capacity) neutral glass ampoules (Adelphi Healthcare packaging Ltd) heat sterilized (180°C for 2-3h) and labelled with strain number of the organism to be freeze-dried, batch date - date of preservation
- Sterile caps (heat sterilized at 180°C for 2-3h) are fitted to the 0.5ml ampoules either individually or in batches of ampoules of the same strain
- Pasteur pipettes
- Sterile non-absorbent cotton wool
- Air/gas glass constricting torch
- Gas glass sealing torch
- Heat resistant mat
- Phosphorus pentoxide

For revival of strains

- Glass cutter in support handle
- Sterile distilled water
- Class 2 Microbiological safety cabinet
- Bunsen burner
- Alcohol for flaming

Method

1. Prepare a spore suspension in a 10% (w/v) skimmed milk and 5% (w/v) inositol mixture.
2. With a Pasteur pipette add 0.2ml (approx.) of suspension to each sterile ampoule, ensuring that the suspension does not run down the inside of the ampoule.
3. Cover each ampoule with a sterile cap or in batches.
4. Load the ampoules into a spin freeze accessory and place this on the drier.
5. Spin the ampoules for 30min and cool to -40°C.
6. Evacuate the chamber and continue to spin for 30min. (The spore suspension will have frozen into a wedge tapering from the base of the ampoule. This gives a greater surface area for evaporation of the liquid).
7. Leave the ampoules in the chamber and evacuate for a further 3h (at this point the moisture content of the material will be approximately 5%).
8. Admit air into the freeze-drier chamber and remove the ampoules.
9. Remove the caps and plug the ampoules with sterile cotton wool compressed to 10mm in depth, 10mm (approx.) above the top of the freeze-dried material in a microbiological safety cabinet.
10. Constrict the plugged ampoule 10mm above the cotton plug using the air/gas torch. The bore of the constriction should remain greater than 1mm, the outer diameter approximately 2.5mm. [This stage where the freeze-dried material is exposed to atmospheric oxygen and moisture must be kept as short as possible as the exposure of the partially dried material can cause deterioration (Rey, 1977)].
11. Place the constricted ampoules on the secondary-drying accessory of the freeze-drier and evacuate over phosphorus pentoxide desiccant. The ampoules are sealed and removed at the point of constriction after a 17h drying process using a glass sealing torch under a vacuum. At this point the moisture content should be 1-2% by dry weight. This drying period is selected for the convenience of working practices in the laboratory. However, it is possible to reduce the water content of the samples to that required with a 3-6h second stage. The presence of sugars in the suspending media reduces the risk of over-drying.
12. Test the sealed tubes with a high voltage spark tester to ensure the seal is intact. (A purple to blue illumination appearing inside the ampoule indicates low pressure and an intact seal). Any ampoules that fail the test should be discarded.

Recovery

1. Score an ampoule at the centre of the cotton wool plug using a glass-cutter.
2. Heat the tip of a glass rod in a Bunsen burner until red-hot and apply firmly to the score. The heat should crack the tube around the score line.
3. Snap-open the ampoule and remove the cotton plug.
4. With a Pasteur pipette add 2-4 drops of sterile distilled water and replace the cotton plug, leave for 30min to rehydrate the suspension.
5. Inoculate onto a suitable growth medium and incubate under appropriate conditions (record any differences in growth and test viability).

Method (Shelf Freeze-Drying)

Controlled freezing at 1°C min^{-1} has been found to enable the freeze-drying of cultures that do not normally survive the above method (Smith, 1986). Using such a pre-freezing procedure some delicate or non-sporulating fungi can be preserved on the shelf freeze-drier.

Equipment and reagents

- Sterile 10% (w/v) skimmed milk + 5% (w/v) inositol (autoclaved at 114°C for 10min) or other protectant
- 2ml flat-bottomed pre-constricted glass vials labelled with the culture unique identifier and grooved rubber bungs. These are covered with aluminium foil to prevent aerial contamination and sterilized by autoclaving at 121°C for 15min, then dried in a drying oven
- Metal racks for holding 2ml vials
- Shelf freeze drier with programmable shelf temperature control
- Class 2 Microbiological safety cabinet
- Bunsen burner
- Pasteur pipettes
- Sterile non-absorbent cotton plugs to fit the neck of the ampoules
- Air/gas glass blowers torch

Method for Single Phase Shelf Freeze-Drying

All culture work should be carried out in the appropriate microbiological safety cabinet

1. Grow cultures under optimal growth conditions for the species and on suitable media
2. Prepare a spore suspension in sterile 10% (w/v) skimmed milk and 5% (w/v) inositol mixture
3. With a Pasteur pipette add 0.2 ml (approx.) of suspension to each sterile vial, ensuring that the suspension does not run down the inside of the neck of the vial
4. Aseptically insert sterile, grooved rubber bungs into the neck of the vials to the pre-moulded rim so that the groove opening is above the vial lip
5. Place the vials on the pre-cooled shelf (-35°C) of the freeze drier
6. Place a sample temperature probe into a vial containing the skimmed milk and inositol mixture only. When the temperature reaches -20°C, evacuate the chamber, this reduces the temperature of the sample to -145°C as the latent heat of evaporation is removed and this rises again to the shelf temperature
7. Maintain the shelf temperature at -35°C for 3h and then raise to 10°C at 0.08°C/min
8. After 24h drying, from the time the temperature of the sample reaches -45°C, lower the shelf base to push the rubber bungs into the vials and seal them
9. Raise the chamber pressure to atmospheric pressure, remove the vials and heat seal them with the air gas torch, retaining a vacuum of $4x10^{-2}$ mbar
10. When cool, test the sealed vials with a high voltage spark tester to ensure that the seal is intact. (A purple to blue illumination appearing inside the ampoule indicates low pressure and an intact seal.) Any ampoules that fail the test should be discarded
11. Store the vials in appropriate conditions, in the dark. Storing at a low temperature is thought to give greater longevities, 4°C is favoured by many, at CABI freeze-dried fungi have survived almost 50 years at temperatures maintained between 15 and 20°C

Materials for Recovery

- Air/gas glass blowers torch
- Glass cutter in support handle
- Sterile distilled water
- Class 2 Microbiological safety cabinet
- Bunsen burner

Recovery

1. After 4d storage test viability of the preserved strain.
2. Score the glass of the ampoules/vials.
3. Snap the ampoules open with using a heated glass rod.
4. Aseptically add sterile distilled water and leave to rehydrate for 30 minutes.
5. Inoculate appropriate medium and incubate under optimum conditions.

CRYOPRESERVATION. The ability of living organisms to survive freezing and thawing was first realized in 1663 when Henry Power successfully froze and revived nematodes (Morris, 1981). Polge *et al.* (1949) became the first 'modern day' scientists to report the freezing of living organisms when they successfully froze and thawed avian spermatozoa. Liquid nitrogen is the preferred cooling agent for cryopreservation, although other cryogenic substances can be used such as liquid air or carbon dioxide. Lowering the temperature of biological material reduces the rate of metabolism until, when all internal water is frozen, no further biochemical reactions occur and metabolism is suspended (Franks, 1981). Although little metabolic activity takes place below -70°C, recrystallization of ice can occur at temperatures above -139°C (Morris, 1981) and this can cause structural damage during storage. Consequently, the storage of microorganisms at the ultra-low temperature of (-190°C to -196°C) in or above liquid nitrogen is the preferred preservation method of many scientists (Hubalek, 1996; Smith and Thomas, 1998; Ryan and Smith, 2007). Provided adequate care is taken during freezing and thawing, the culture will not undergo change, either phenotypically or genotypically (Smith and Ryan, 2012). See Appendix F, this volume for recommended and tested cryopreservation protocols, and Fig. 4.4.

Choice of cryoprotectant is a matter of experience and varies according to the organism. Cryoprotection is achieved by:

- Non-critical volume loss by the reduction of ice formation.
- An increase in viscosity, which slows down ice crystal growth and formation and solute effects.
- Reduction of the rate of diffusion of water caused by the increase of solutes.

Glycerol 10%(v/v) gives satisfactory results but requires time to penetrate cellular structures; some fungi are damaged by this delay. Dimethyl sulfoxide (DMSO) penetrates rapidly and is often more satisfactory (Hwang and Howells, 1968; Hwang *et al.*, 1976). Sugars and large molecular substances, such as polyvinyl pyrrolidine (PVP) (Ashwood-Smith and Warby, 1971) have been used but in general have been less successful (Smith, 1983b). Trehalose may be better but is comparatively more expensive. Establishing the optimum cooling rate has been the subject of much research (Hwang, 1960; Morris *et al.*, 1988; Smith, 1993). Slow cooling at 1°C min^{-1} over the critical phase has proved most successful (Hwang, 1966, 1968), but some less-sensitive isolates respond well to rapid cooling, preferably without protectant. Slow warming may cause damage owing to the recrystallization of ice, therefore rapid thawing is recommended. Slow freezing and rapid thawing generally give high recoveries for fungi (Heckly, 1978; Smith and Ryan, 2012).

As with other methods of preservation liquid nitrogen cryopreservation has advantages and disadvantages. Advantages include the length of storage, which is considered to be effectively limitless if storage temperature is kept below -139°C. The majority of organisms survive well, giving the method a greater range of successful application. Organisms remain free of contamination when stored in sealed ampoules. Disadvantages of liquid nitrogen storage include the high cost of apparatus such as refrigerators and a continual supply of liquid nitrogen. A regular supply cannot be obtained in some parts of the world and therefore the technique cannot be used. If the supply of nitrogen fails (or the double-jacketed, vacuum-sealed storage vessels corrode and rupture)

Fig. 4.4. Liquid nitrogen storage tank

then the whole collection can be lost. There are also safety considerations to be made, the storage vessels must be kept in a well-ventilated room, as the constant evaporation of the nitrogen gas could displace the air and suffocate workers.

Cryopreservation has been used for the preservation of fungi at CABI since 1968 (Onions, 1971, 1977), although early work involved a very simple procedure. Storage at -196°C in liquid nitrogen or at slightly higher temperatures in the vapour phase is employed. Generally, a cooling rate of -1°C min^{-1} with 10% (v/v) glycerol as a cryoprotectant is applied and to date, over 4000 species belonging to over 700 genera have been successfully frozen. Morphological or physiological change has been rarely observed in the almost 20,000 fungal isolates stored to date. However, some chromists and members of the Basidiomycota survive cryopreservation less well than sporulating fungi and optimized protocols are often needed (Homolka, 2013). These are achieved by employing species-specific cooling rates, alternative cryoprotective agents, encapsulation or vitrification (Smith, 1992; Smith and Ryan, 2012). Protocols for the cryopreservation of fungi can be found in Appendix F, this volume.

Cultures, tissue or spore suspensions are treated with a cryoprotectant such as 10% (w/v) glycerol or inositol, before aseptic transfer into sterile ampoules and frozen to ultra-low temperatures usually in the vapour phase of liquid nitrogen. The cooling rates are critical and the best revivals are achieved when this is done slowly. At ultra-low temperatures metabolism is suppressed. If the organism survives the initial freezing shock, viability should be indefinite. This technique requires large and expensive equipment and a reliable source of liquid nitrogen. Experienced technicians are necessary to ensure optimum standards are maintained for best results. A typical approach may include an array of growth on matrices or pre-treatment before freezing such as:

- Mature cultures washed from agar slopes or mature liquid cultures,
- Agar blocks cut from growing colony,
- Infected leaves, stems or other plant surfaces,
- Spores washed from infected plant tissue,
- Cultures grown on millet grains,
- Harvested and filter concentrated conidia, growth on:
    - Agar discs or
    - Infected lima bean epicotyls or hemp seeds

Most fungi can be preserved using an array of techniques. However, they should also be maintained by very stringent preservation techniques in order to conserve their genomic and physiological integrity. Fungal cultures are more prone to contamination. They also grow best in media containing substances from their natural growth environment. This topic has covered some preservation techniques which are routinely used for microbes with minor modifications to ensure that they are preserved in an optimal manner.

Methods at CABI

There are many protocols suitable for the cryopreservation of microbes, CABI's method provides an example approach but different organism types need adjustments to get the best results:

Safety considerations

It is essential that protective clothing is worn and strict safety procedures are followed in all activities involving the use of liquid nitrogen. The major risks are the intense cold, which causes injury similar to burns, and the issuing gas, which is an asphyxiant. Areas where liquid nitrogen is used or stored must be well ventilated and oxygen content of the air monitored.

Equipment and reagents

- Sterile 10% (v/v) glycerol (other cryoprotectants can be used (Smith and Onions, 1994)
- 2.0 ml sterile graduated cryotubes (Thermo Fisher Scientific, Loughborough, UK) labelled with strain number and batch date
- Controlled rate freezer such as the nitrogen free Via Freeze series (Cytiva Life Sciences) or the nitrogen based Kryo series (Planer plc, Sunbury-on-Thames, UK)
- Liquid nitrogen freezer with metal drawer rack inventory control system (Statebourne Cryogenics, Sunderland, UK)
- Class 2 Microbiological safety cabinet
- Bunsen burner

- Alcohol for flaming
- Safety equipment (to include cryogloves, face shield, forceps and oxygen monitor)

Method

1. Prepare the spore or cell suspensions in sterile 10% (v/v) glycerol or other appropriate cryoprotectant.
2. Pipette 0.5 ml of the cell suspension into 2.0 ml cryotubes. The Pro-Lab Microbank™ system (Pro-Lab Diagnostics, Richmond Hill, Ontario/ Neston, UK/ Round Rock, Texas) can be used, although the stated method does not involve controlled rate cooling; in this the cryotubes are filled with cryoprotectant and porous glass beads.
3. Cool the cryotubes at a suitable rate in a programmable cooler or Stirling cycle freezer (Ryan *et al.*, 2014) (see Fig. 4.5). The cooling rate employed depends upon the organism. A rate of $-1°C\ min^{-1}$ will allow most fungi and bacteria to survive but will not give optimum recovery for all. The rate of cooling is controlled over the critical period from 5°C down to -50°C. The initial cooling rate to 5°C is not critical and this is normally at $-10°C\ min^{-1}$. Alternative cooling rates (e.g. $-0.5°C\ min^{-1}$) can be used for some recalcitrant fungi.
4. Transfer to a metal drawer rack in the liquid nitrogen storage freezer to complete the cooling to the storage temperature. (See Fig. 4.6.)

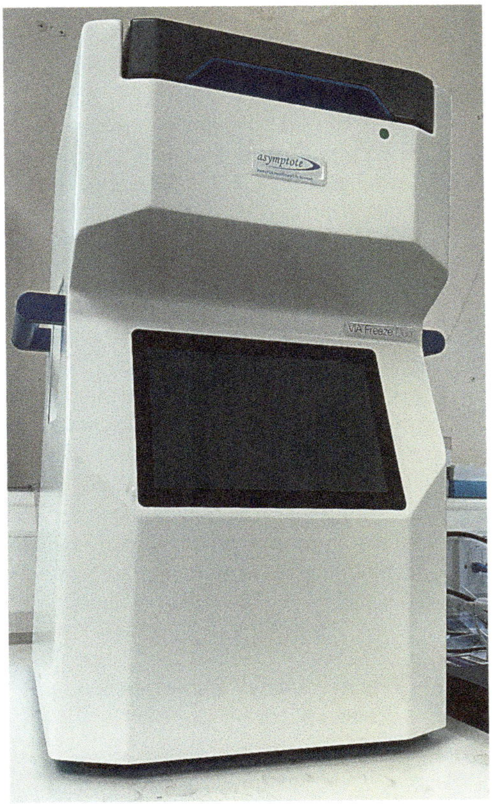

Fig. 4.5. Stirling cycle freezer

Recovery

1. Thaw the cryotube rapidly by placing it in a water bath at 37°C until the contents of the ampoule thaw.
2. Where the Pro-Lab Microbank system is used, recovery of the strain can be achieved without defrosting by chipping off a bead and placing it in an appropriate growth medium.

Similar methods have been in use in other collections for example in the American Type Culture Collection (ATCC) where cryopreservation in liquid nitrogen has been used since 1965 and has given very good results (Hwang, 1966, 1968; Butterfield *et al.*, 1974; Hwang *et al.*, 1976; Alexander *et al.*, 1980).

Preservation of Other Microorganism Groups

Similar techniques are used for the preservation of many different organisms often with special adaptations for the different types (Kirsop and Doyle, 1991; Hunter-Cevera and Belt, 1996). The following sections describe the techniques in basic principles and where there are differences in methods used for different cell types.

As with other microorganisms, algae are maintained under largely artificial conditions of media composition, light and temperature. Such conditions may cause selection or physiological adaptation, especially as algae naturally survive under complex and fluctuating conditions that follow a seasonal life cycle (McLellan *et al.*, 1991). Compared with other groups of microorganisms, relatively little research has been carried out on the development of long-term preservation methods for the microalgae. Freeze-drying does not give good recovery (often <1%) of the original population and prolonged storage may result in a further reduction in viability

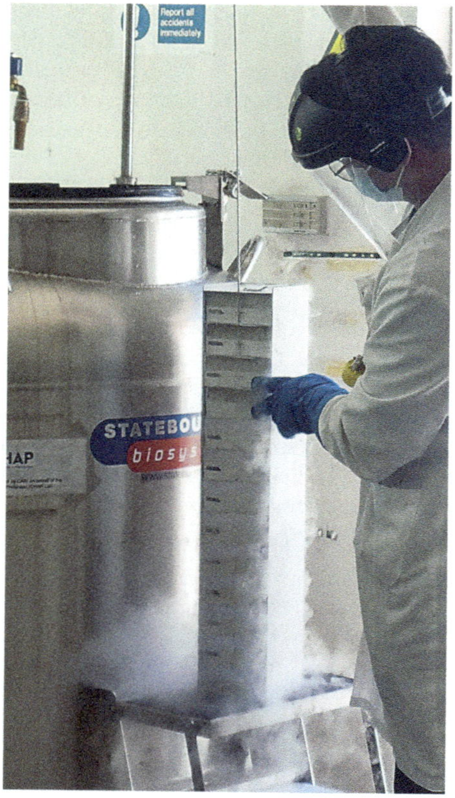

Fig. 4.6. Recovery of ampoule from cryo-storage

(Holm-Hansen, 1967; McGrath *et al.*, 1978; Day *et al.*, 1997; Day, 1998; Day *et al.*, 2000). The successful lyophilization of the cyanobacterium *Nostoc muscorum*, using a method similar to that used for bacteria, has been reported, with no observed reduction in viability after storage for 5 years (Holm-Hansen, 1973). This technique has been adopted by a small number of researchers to preserve selected cyanobacterial strains. Cryopreservation has been successfully employed to maintain algae (Morris, 1976, 1978; McLellan *et al.*, 1991; Day *et al.*, 2000). Cryopreservation is advantageous because once organisms are cryopreserved by a proven method that yields high cell recovery, subsequent viability becomes independent of storage time. For example, some strains have no significant decline in viability >20 years of storage (Day *et al.*, 1997). Using an appropriate protocol, high levels of viability (>90%) may be observed, however, no single protocol has been found to be successful for a wide range of strains.

The algae are frequently maintained on agar slants but most often in liquid media. Transfer by sub-culture can be potentially disadvantageous, as frequent sub-culturing can result in contamination from microorganisms such as bacteria. Environmental parameters such as light and temperature are also critical. The choice of medium is an important factor and is tailored to the requirements of the individual genera. When established, cultures proceed to logarithmic growth before stationary phase is achieved. Eventual exhaustion of the nutrient supply and reduction in the availability of dissolved gases associated with an accumulation of waste products will cause deterioration and subsequent loss of the culture. Transfer to fresh growth medium of viable material in late exponential or early stationary phase is advised (McLellan *et al.*, 1991).

Compared to other groups of microorganisms, relatively little research has been carried out on the development of long-term preservation methods for the microalgae and cyanobacteria. However, the EU Cobra project sought to address this (Day *et al.*, 2005). Cryopreservation has been successfully employed to maintain algae (Morris, 1976, 1978; Day *et al.*, 1997; Day, 1998). Direct immersion in liquid nitrogen, with or without the addition of cryoprotectants, has been successfully employed to preserve a few unicellular Chlorococcales and cyanobacteria (Holm-Hansen, 1963). However, most protocols employ a simple two-step system with the controlled/semi-controlled cooling from room temperature to a sub-zero holding temperature prior to plunging into liquid nitrogen (Box, 1988; Day, 1998). The majority of protocols in current use have been developed empirically with the pre-freezing culture regime, cryoprotectant choice (usually DMSO or methanol) and concentration, cooling rate and thawing regime being manipulated to minimize the amount of damage to the algal cells and hence maximize post-thaw viability levels.

Preservation of Microbiomes and Microbial Consortia

Researchers are currently involved in developing protocols for environmental samples such as soil and microbial consortia. The UK Crop

Microbiome Cryobank project (www.agmicrobiomebase.org) is using both culture dependent and interdependent approaches to culture and proactively preserve material. A review of the needs for biobanking of microbiome samples was conducted in 2021 (Ryan *et al.*, 2021). There are fundamental differences in approaches needed for the cryopreservation of complex samples, the technical challenge of preserving an axenic microbe is difficult enough, translating this to a sample that may contain millions of genetically different organisms is significantly more byzantine. Further, there is a need to not only maintain the genetic stability of organisms in the consortium, but also to conserve the functionality of the microbiome (See Fig. 4.7).

Criteria for the Selection of Preservation Techniques

As we have discussed, there are several methods and techniques available for the storage and preservation of microbiological cultures. However, not all of them are suitable for every kind of microorganism as each microorganism may differ in structure and form. Therefore, it is important to identify the criteria for the selection of the most appropriate preservation technique to keep microbial cultures alive, stable and free from contamination.

Preservation Choices for Collections

Culture collections may be established for many different reasons, for example to store organisms from a specific research project or for a testing laboratory that needs to maintain organisms for its tests. As a result of this, different types of collections have different requirements when it comes to preservation techniques and this is ultimately based on the purpose of each collection. They may be private collections belonging to a university department, or a school, holding small numbers of a wide variety of common organisms maintained for teaching purposes. Such a collection is often useful rather than valuable and isolates can be easily replaced from other collections or re-isolated from the original habitat. Oil storage and the deep freeze should be considered here.

Private collections built up in support of research projects are often very specific. They may consist of many isolates of one species or only a few isolates. This material must be kept

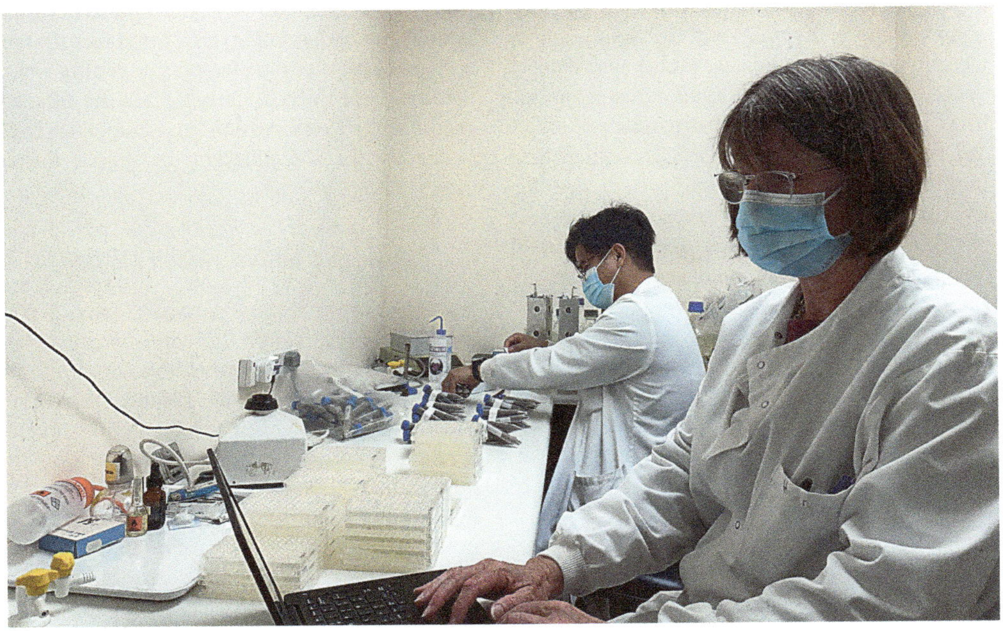

Fig. 4.7. Technicians preparing microbiome and soil samples

unchanged and is of potential long-term interest to the community. Such specialist collections may not require long-term preservation but stability is very important. Time and technical assistance may not be available, funds may be limited and labour may be at a premium. In this instance deep freeze or silica gel storage might prove useful. Freeze-drying and liquid nitrogen are preferable techniques, if finances allow. Regular transfer should still be considered when dealing with a few specialist isolates and careful curation by researchers has kept many valuable private collections in good condition by this means.

Some collections are set up to support a service or an industrial project. They may be required to supply living starter material at regular intervals, in industrial fermentation for example. The culture must always be readily available, active and absolutely stable. The cost in money, equipment and labour is probably small in relation to the importance of stability and availability. Deposit in a national collection is advisable. Regular transfer on suitable media may suffice for the day to day use of the organisms or for the build-up of inoculum. The techniques of freeze-drying and liquid nitrogen are preferential.

Public service collections facilitate the supply of cultures to the scientific community. These are often supported on a national basis and may offer a variety of additional services such as identification of organisms, and they provide a depository for type or ex-type cultures or cultures cited in patents. Many organisms of differing requirements have to be maintained in a highly stable condition for long periods. Specialist supervision is essential and several methods of preservation are desirable, including at least one of high stability and considerable longevity such as liquid nitrogen or freeze-drying, with a back-up of other methods as necessary. There will be some highly sensitive isolates that will require specialist treatment. The relative cost per isolate, despite the sophisticated methods, may not be excessive due to the scale of the operation.

Considering Logistical Criteria

Logistical factors, such as the availability of cash resources and an institution's overall infrastructure, have to be considered before cultures are preserved. Careful consideration should be given as to why an organism needs to be preserved; if it is a very important strain it may be better to have it preserved in a public service culture collection, a list of which is available at the WDCM http://www.wdcm.org/.

Some methodologies may be cheap in terms of consumables but may require frequent maintenance in the longer term adding to labour costs. For example, continuous culture regimes, oil, water, silica gel and soil storage. Other methodologies require apparatus which may be expensive, for example, those required for cryopreservation in liquid nitrogen and freeze-drying.

When considering cryopreservation or refrigeration a reliable guaranteed supply of liquid nitrogen or electricity may be required. Liquid nitrogen can be expensive and complications may arise if the supply is disrupted. These methods are relatively labour intensive, requiring trained personnel to ensure optimal establishment of collections and continuing supervision.

Considering Taxonomic Criteria

Not all microorganisms will tolerate preservation by every method available. Indeed, for some it may be difficult to preserve them at all. These are classed as 'recalcitrant' organisms. Knowledge of an organism's lifecycle and structure is usually important, often, only the stages that result in the production of spores or resting structure can be preserved. Table 4.1 lists the different methods for preservation and indicates their suitability for the different groups of microorganisms.

Choosing a Preservation Method

As discussed previously, when choosing a method of preservation there must be a balance between requirements, cost in material and labour, longevity and genetic stability. In most instances liquid nitrogen storage gives reliable results and infinite longevity. However, for various reasons this method may not be suitable and it is so sensitive to failure of the supply of the liquid that in almost every case there should be a back-up by some other preservation method. Ultra-low temperature freezers are now available that can maintain temperatures below -150°C which can have liquid nitrogen back-up

Table 4.1. The suitability of different methods for preservation of microorganisms

Method	Suitable for	Not or less suitable for
Cryopreservation	Yeasts, bacteria, algae (particularly cyanobacteria, marine pennate diatoms and unicellular chlorophytes), most fungi (different preservation protocols may be required)	Flagellate algae, some chromists
Freeze-drying (lyophilization)	Bacteria, yeasts (single cell vegetative stages), sporing fungi (in particular the Hyphomycetes, Ascomycetes, Zygomycetes)	Algae, non-sporulating fungi and those with high water content, thin walled or complicated multicellular morphology (these include many Basidiomycetes, chromists, Oomycetes, chytrids)
Continual sub-culture	All culturable microorganisms. One of the most commonly used methods for algae	Obligate organisms, organisms subject to strain drift especially where the maintenance of pathogenicity is a concern
Water storage	Fungi, bacteria / yeasts as suspensions	Xerophiles
Storage in sand / soil	Sporing fungi, some bacteria e.g. *Bacillus*, soil-borne algae especially *Haematococcus pluvialis*	Algae, non-sporing fungi, yeasts
Oil storage	Most fungi	Flagellate organisms, bacteria, algae

if power outages occur. Though the stability and longevity of freeze-drying is perhaps not as great as with ultra-low temperature cryopreservation in liquid nitrogen, this method is not vulnerable, but it is initially technically complex and expensive. Not all organisms freeze-dry and some workers in a very few cases cast doubt on its giving reliable genetic stability. However, it must be emphasized that preservation techniques seldom improve an isolate and many reported failures especially for freeze-drying and even for liquid nitrogen result from processing already altered material. A decision-based key has been developed to help facilitate the process of selecting a preservation technique (Ryan *et al.*, 2000) see Box 1 (below).

Box 1. Managing Microbes decision-based key for preservation

Based on the paper:

Ryan, M.J., Smith, D. and Jeffries, P. (2000) A decision-based key to determine the most appropriate protocol for the preservation of fungi. *World Journal of Microbiology & Biotechnology* 16, 183-186. https://link.springer.com/article/10.1023/A%3A1008910006419

The long-term preservation of valuable fungal cultures can be achieved in several ways and the choice of methodology can be problematical. Firstly, there is the decision whether to use a public service culture collection or `in-house' facilities. Secondly, the wide variety of preservation methods available often leads to confusion about which protocol(s) are best suited for specific fungi. No method can be universally applied to all fungi. Some species are notoriously difficult to preserve, whilst other fungi can be preserved by almost any method. A decision-based key has been devised, which uses questions related to fungal characters and user facilities and economics to determine the most appropriate method for long-term preservation of cultures. This key should facilitate the decisions of microbiologists when considering preservation of important fungal cultures.

The key incorporates two sections. Consider section A and then proceed to section B. Section A consists of questions designed to allow the user to consider preservation methods that may be applicable to the particular fungus using species specific information, Section B asks the user a series of questions considering more general aspects of preservation dynamics and resources available.

Continued

Box 1. Continued.

Section A: Species specific criteria

1. Does the fungus readily sporulate asexually in culture?
Yes go to 9
No go to 2
2. Can the fungus be readily maintained by continual sub-culture?
Yes go to 3
No go to 4
3. The following preservation methods may be suitable:
- Cryopreservation in or above liquid nitrogen at –196°C
- Cryopreservation at –70°C or below in a mechanical freezer
- Storage under mineral oil
- Storage in water
- Storage in a freezer at –20°C

4. Is the fungus endomycorrhizal?
Yes go to 5
No go to 6
5. Store as viable inoculum in growth substrate in association with plant material
6. Are sexual structures formed by the fungus?
Yes go to 7
No go to 8
7. Induce sexual reproduction and consider preservation of propagules by cryopreservation or lyophilization (freeze-drying). Note: the inducement of meiotic crossing- over may result in any future cultures being characteristically different from the parent fungus.
8. Consider preservation with substrate from which the fungus was isolated.
9. Does the fungus have motile spores?
Yes (Oomycota) go to 10
No go to 11
10. The following preservation methods may be suitable:
- Cryopreservation in or above liquid nitrogen at –196°C
- Cryopreservation at –70°C or below in a mechanical freezer
- Storage under mineral oil
- Storage in water
- Continual sub-culture

11. The following preservation protocols may be suitable:
- Cryopreservation in or above liquid nitrogen at –196°C
- Cryopreservation at –70°C or below in a mechanical freezer
- Storage under mineral oil
- Storage in water
- Storage in a freezer at –20°C
- Continual sub-culture
- Lyophilization/freeze-drying

Section B: Practical and economic criteria

1. How long is it necessary to store your isolate for?
Short-term (<2 years) go to 2
Long-term (> 2 years) go to 3
2. Consider one of the following:
- Storage in water
- Storage in a freezer at –20°C
- Continual sub-culture
- Storage on silica gel

3. Is your isolate of economical or scientific importance?
Yes go to 5
No go to 4

Continued

Box 1. Continued.

4. Consider:
- Storage under mineral oil
- Storage on silica gel
- Storage in sand or soil

5. Are resources freely available?
No go to 6
Yes go to 7

6. Consider sending isolate to a WFCC affiliated culture collection for preservation and storage if costs permit or send your isolate to a recognized centre for lyophilization (freeze-drying) and arrange for your culture to be returned by post.

7. Consider either or both:
- Lyophilization/freeze-drying
- Cryopreservation

8. Do you have, or have the resources to obtain freeze-drying facilities?
No go to 9
Yes, but don't need to preserve many cultures go to 9
Yes go to 10

9. Send your isolate to a WFCC affiliate collection for freeze-drying and storage

10. Freeze-dry your culture, store at reduced temperature and away from direct light and check for viability every 3-5 years

11. Consider:
- Cryopreservation at –70°C or below in a mechanical freezer
- Cryopreservation in or above liquid nitrogen at –196°C

12. Are electric supplies guaranteed or liquid nitrogen or cold gas back up available to keep temperature stable or low whilst electricity is disrupted?
No go to 13
Yes go to 14

13. Consider alternative or send your isolate for storage at a WFCC affiliate culture collection

14. Cryopreservation at –70°C or below in a mechanical freezer

The early collections were maintained for years by frequent transfer and specialist curation. One culture in the CABI collection, a *Monascus purpureum* (originally in the collection of Krall and Thom) was isolated *c.* 1897. To reduce the labour and time involved, covering cultures with a layer of mineral oil became a well-used method. Mineral oil cultures are generally unpopular as they encourage selection of genetic variants but the method is cheap, it deters mites, and some fungi at the CABI have survived 40 years storage without transfer. Water storage has proven ideal when storing for shorter periods, although some fungi will survive longer in the deep freeze than water. However, it is very convenient and cheap. Storage in silica gel or soil is useful, although stability of fungi has not been widely investigated. Isolates can remain stable for periods of up to 11 years and may be viable for longer retaining certain genetic traits.

All research organizations need to preserve their strains for future work and confirmation of results. However, every organization is limited by funds, space and staff, hence the most cost efficient and practical technique needs to be selected and maintained considering the end use of the organism. CABI will always recommend cryopreservation below -135°C as the preferred method of long-term storage having used it and improved protocols since 1968 (over 50 years successful storage of fungi).

References

Al-Doory, Y. (1968) Survival of dermatophyte cultures maintained on hair. *Mycologia* 60, 720-723. https://www.jstor.org/stable/3757446

Alexander, M., Daggett, P.M., Gherna, R., Jong, S.C., Simione, F. and Hatt, H. (1980) *American Type Culture Collection Methods I. Laboratory Manual on Preservation Freezing and Freeze Drying*. American Type Culture Collection, Rockville, Maryland.

Annear, D.I. (1958) Observations on drying bacteria from the frozen and from the liquid state. *Australian Journal of Experimental Medical Science* 36, 211-221. https://doi.org/10.1038/icb.1958.23

Ashwood-Smith, M.J. and Grant, E. (1976) Mutation induction in bacteria by freeze drying. *Cryobiology* 13, 206-213. DOI: 10.1016/0011-2240(76)90134-6

Ashwood-Smith, M.J. and Warby, C. (1971) Studies on the molecular weight and cryoprotective properties of PVP and dextran with bacteria and erythrocytes. *Cryobiology* 8, 453-464. https://doi.org/10.1016/0011-2240(71)90036-8

Atkinson, R.G. (1953) Survival and pathogenicity of *Alternaria raphani* after 5 years in dried soil cultures. *Canadian Journal of Botany* 31, 542-547. https://cdnsciencepub.com/doi/10.1139/b53-043

Baker, P.R.W. (1955) The micro-determination of residual moisture in freeze-dried biological materials. *Journal of Hygiene* 53, 426-435. https://doi.org/10.1017/S0022172400000930

Bazzigher, G. (1962) A simplified freeze-drying procedure for the preservation of fungal cultures. *Journal of Phytopathology* 45, 53-56.

Benson, E.E. (2008) Cryopreservation of phytodiversity: A critical appraisal of theory and practice. *Critical Reviews in Plant Sciences* 27, 141-219. DOI: 10.1080/07352680802202034

Benson, E.E., Harding, K., Ryan, M., Petrenko, A., Petrenko, Y. and Fuller, B. (2018) Alginate encapsulation to enhance biopreservation scope and success: A multidisciplinary review of current ideas and applications in cryopreservation and non-freezing storage. *Cryoletters* 39, 14-38. https://www.ingentaconnect.com/contentone/cryo/cryo/2018/00000039/00000001/art00003

Berny, J.F. and Hennebert, G.L. (1989) Abstracts, Meeting of Contractors, Biotechnology Action Programme of the Commission of the European Community, Biotechnology Division. http://aei.pitt.edu/94768/1/BAP_PROGRESSS_1987.pdf

Berny, J.F. and Hennebert, G.L. (1991) Viability and stability of yeast cells and filamentous fungus spores during freeze-drying-effects of protectants and cooling rates. *Mycologia* 83, 805-815. https://doi.org/10.1080/00275514.1991.12026086

Boeswinkel, H.J. (1976) Storage of fungal cultures in water. *Transactions of the British Mycological Society* 66, 183-185.

Bond, C. (2007) Cryopreservation of yeast cultures. In: Day, J.G. and Stacey, G.N. (eds) *Cryopreservation and Freeze-Drying Protocols. Methods in Molecular Biology*™, vol 368. Humana Press, New Jersey. https://doi.org/10.1007/978-1-59745-362-2_7.

Booth, C. (1971) *The genus Fusarium*. Commonwealth Mycological Institute, Kew, UK.

Box, J.D. (1988) Cryopreservation of the blue-green alga *Microcystis aeruginosa*. *British Phycology Journal* 23, 385-386. https://doi.org/10.1080/00071618800650411

Buell, C.B. and Weston, W.H. (1947) Application of the mineral oil conservation method to maintaining collections of fungus cultures. *American Journal of Botany* 34, 555-561. https://ur.booksc.eu/book/43687474/56d5f4

Burdsall, H.H. and Dorworth, E.B. (1994) Preserving cultures of wood decaying Basdiomycotina using sterile distilled water in cryovials. *Myxologia* 86, 275-280. https://doi.org/10.1080/00275514.1994.12026408

Butterfield, W., Jong, S.C. and Alexander, M.J. (1974) Preservation of living fungi pathogenic for man and animals. *Canadian Journal of Microbiology* 20, 1665-1673. https://cdnsciencepub.com/doi/10.1139/m74-258

Dade, H.A. (1960) Laboratory methods in use in the culture collection, IMI. In: *Herb IMI Handbook. Commonwealth Mycological Institute*, Kew, UK.

Day, J.G. (1998) Cryo-conservation of microalgae and cyanobacteria. *Cryo-Letters Supplement* 1, 7-14. https://pubmed.ncbi.nlm.nih.gov/19827252/

Day, J.G. and McLellan, M.R. (1995) Cryopreservation and freeze-drying protocols. *Methods in Molecular Biology,* Vol. 38, Humana Press Inc., Totowa, New Jersey. DOI: 10.1385/0-89603-296-5:1

Day, J.G. and Stacey, G. (eds) (2007) *Cryopreservation and Freeze-Drying Protocols*. 2nd edn. Humana Press, Totowa, New Jersey. https://link.springer.com/book/10.1007/978-1-59745-362-2?page=2#toc

Day, J.G., Watanabe, M.M., Morris, G.J., Fleck, R.A. and McLellan, M.R. (1997) Long-term viability of preserved eukaryotic algae. *Journal of Applied Phycology* 9, 121-127. https://zh.booksc.eu/book/11030234/0508b4

Day, J.G., Fleck, R.A. and Benson, E.E (2000) Cryopreservation-recalcitrance in microalgae: novel approaches to identify and avoid cryo-injury. *Journal of Applied Phycology* 12, 369-377. https://www.academia.edu/61210314/Cryopreservation_recalcitrance_in_microalgae_novel_approaches_to_identify_and_avoid_cryo_injury

Day, J.G., Benson, E.E., Harding, K., Knowles, B., Idowu, M., Bremner, D., Santos, L., Santos, F., Friedl, T., Lorenz, M., Lukesova, A., Elster, J., Lukavsky, J., Herdman, M., Rippka, R. and Hall, T. (2005) Cryopreservation and conservation of microalgae: the development of a pan-european scientific and biotechnological resource (the COBRA project). *Cryoletters* 26 (4), 231-238. https://www.ingentaconnect.com/content/cryo/cryo/2005/00000026/00000004/art00003

Fennell, D.I. (1960) Conservation of fungus cultures. *Botanical Review* 26, 79-141. https://www.jstor.org/stable/4353609

Figueiredo, M.B. and Pimentel, C.P.V. (1975) Methods used for the preservation of fungi in the Secao mycology phytopathology library of the Biological Institute. *Summa Phytopathologica* 1, 299-302.

Fonseca, F., Meneghel, J., Cenard, S., Passot, S. and Morris, G.J. (2016) Determination of intracellular vitrification temperatures for unicellular micro organisms under conditions relevant for cryopreservation. *PLoS ONE* 11(4): e0152939. doi:10.1371/journal.pone.0152939 https://journals.plos.org/plosone/article?id=10.1371/journal.pone.0152939

Franks, F. (1981) Biophysics and biochemistry of low temperatures and freezing. In: Morris, G.J. and Clarke, A. (eds) *Effects of Low Temperature of Biological Membranes*. pp 3-19. Academic Press, London.

Gilmour, M.N., Turner, G., Berman, R.G. and Kreuzer, A.K. (1978) Compact liquid nitrogen storage system yielding high recoveries of gram negative anaerobes. *Applied and Environmental Microbiology* 35, 84-88. doi:10.1128/aem.35.1.84-88 https://www.ncbi.nlm.nih.gov/pmc/articles/PMC242782/

Gordon, W.L. (1952) The occurrence of *Fusarium* species in Canada. *Canadian Journal of Botany* 30, 209-251. https://cdnsciencepub.com/doi/10.1139/b52-018

Heckly, R.J. (1978) Preservation of microorganisms. *Advances in Applied Microbiology* 24, 1-53. https://doi.org/10.1016/S0065-2164(08)70635-X

Holm-Hansen, O. (1963) Viability of blue-green algae after freezing. *Physiologia Planta* 16, 530-539.

Holm-Hansen, O. (1967) Factors affecting the viability of lyophilized algae. *Cryobiology* 4, 17-23. https://doi.org/10.1016/S0011-2240(67)80182-2

Holm-Hansen, O. (1973) Preservation by freezing and freeze drying. In: Stein, J.R. (ed.) *Handbook of Phycological Methods*. pp 195-206. Cambridge University Press, Cambridge, UK.

Homolka, L. (2013) Preservation of live cultures of basidiomycetes – Recent methods. *Fungal Biology* 118, 107-125. https://doi.org/10.1016/j.funbio.2013.12.002

Hubalek, Z. (1996) *Cryopreservation of Microorganisms at Ultra-low Temperatures*. Academy of Sciences of the Czech Republic, Prague. pp 286. ISBN80-200-0557-9.

Hubalek, Z. and Kochova-Kratochvilova, A. (1978) Liquid nitrogen storage of yeast cultures I. Survival, and literature review of the preservation of fungi at ultralow temperatures. *Antonie van Leewenhoek* 44, 229-241. https://link.springer.com/article/10.1007/BF00643225

Hunter-Cevera, J.C. and Belt, A. (1996) *Preservation and Maintenance of Cultures used in Biotechnology and Industry*. pp 263. Academic Press Inc., New York. https://www.elsevier.com/books/maintaining-cultures-for-biotechnology-and-industry/hunter-cevera/978-0-12-361946-4

Hwang, S.-W. (1960) Effects of ultralow preservation of fungus cultures with liquid nitrogen refrigeration. *Mycologia* 52, 527-529. https://doi.org/10.2307/3755974

Hwang, S.-W. (1966) Long term preservation of fungus cultures with liquid nitrogen refrigeration. *Applied Microbiology* 14, 784-788. https://www.ncbi.nlm.nih.gov/pmc/articles/PMC1058415/pdf/applmicro00363-0088.pdf

Hwang, S.-W. (1968) Investigation of ultra-low temperature for fungal cultures. I. An evaluation of liquid nitrogen storage for preservation of selected fungal cultures. *Mycologia* 60, 613-621. https://doi.org/10.1080/00275514.1968.12018610

Hwang, S.-W. and Howells, A. (1968) Investigation of ultra-low temperature of fungal cultures II. Cryoprotection afforded by glycerol and dimethyl sulphoxide to eight selected fungal cultures. *Mycologia* 60, 622-626. https://zh.booksc.eu/book/47409400/112864

Hwang, S.-W., Kwolek, W.F. and Haynes, W.C. (1976) Investigation of ultra-low temperature for fungal cultures III. Viability and growth rate of mycelial cultures cryogenic storage. *Mycologia* 68, 377-387. https://doi.org/10.1080/00275514.1976.12019919

Kirsop, B. (1974) The stability of biochemical, morphological and brewing properties of yeast cultures maintained by sub-culturing and freeze-drying. *Journal of the Institute of Brewing* 80, 565-570. https://onlinelibrary.wiley.com/doi/10.1002/j.2050-0416.1974.tb03650.x

Kirsop, B. and Doyle, A. (eds) (1991) *Maintenance of Microorganisms and Cultured Cells*, pp 308. Academic Press, New York. https://www.worldcat.org/title/maintenance-of-microorganisms-and-cultured-cells-a-manual-of-laboratory-methods/oclc/26163794

Little, G.N. and Gordon, M.A. (1967) Survival of fungus cultures maintained under mineral oil for twelve years. *Mycologia* 59, 733-736.

Malik, K.A. (1990) A simplified liquid-drying method for the preservation of microorganisms sensitive to freezing and freeze-drying. *Journal of Microbiological Methods* 12, 125-132. https://doi.org/10.1016/0167-7012(90)90022-X.

Marx, D.H. and Daniel, W.J. (1976) Maintaining cultures of ectomycorrhizal and plant pathogenic fungi in sterile water cold storage. *Canadian Journal of Microbiology* 22, 338-341. https://doi.org/10.1139/m76-051

McGrath, M.S., Daggett, P.M. and Dilworth, S. (1978) Freeze-drying of algae: *Chlorophyta* and *Chrysophyta*. *Journal of Phycology* 14, 521-525. https://doi.org/10.1111/j.1529-8817.1978.tb02480.x

McLellan, M.R., Cowling, A.J., Turner, M.F. and Day, J.G. (1991) Maintenance of algae and protozoa. In: Kirsop, B. and Doyle, A. (eds) *Maintenance of Microorganisms*. Academic Press Ltd., London. pp 183-208.

Merabishvili, M., Vervaet, C., Pirnay, J-P., De Vos, D., Verbeken, G., Mast, J., Chanishvili, N. and Vaneechoutte, M. (2013) Stability of *Staphylococcus aureus* phage ISP after freeze-drying (lyophilization). *PLoS ONE* 8(7): e68797 https://doi.org/10.1371/journal.pone.0068797

Morris, G.J. (1976) The cryopreservation of *Chlorella* 2. Effect of growth temperature on freezing tolerance. *Archives of Microbiology* 107, 309-312. https://link.springer.com/article/10.1007%2FBF00425345

Morris, G.J. (1978) Cryopreservation of 250 strains of Chlorococcales by the method of two step cooling. *British Phycology Journal* 13, 15-24. https://doi.org/10.1080/00071617800650031

Morris, G.J. (1981) *Cryopreservation: An introduction to cryopreservation in culture collections*. Culture Centre of Algae and Protozoa, Cambridge, UK. http://nora.nerc.ac.uk/id/eprint/5049/1/Cryopreservation.pdf

Morris, G.J., Smith, D. and Coulson, G.E. (1988) A comparative study of the changes in the morphology of hyphae during freezing and viability upon thawing for twenty species of fungi. *Journal of General Microbiology* 134, 2897-2906. https://www.microbiologyresearch.org/docserver/fulltext/micro/134/11/mic-134-11-2897.pdf?expires=1642681821&id=id&accname=guest&checksum=03028E169E7F457021CFB70870F23212

Obara, Y., Yamai, S., Nikkawa, T., Shimoda, Y. and Miyamoto, Y. (1981) Preservation and transportation of bacteria by a simple gelatin disk method. *Journal of Clinical Microbiology* 14, 61-66. https://www.ncbi.nlm.nih.gov/pmc/articles/PMC271902/

Onions, A.H.S. (1971) Preservation of fungi. In: Booth, C. (ed.) *Methods in Microbiology 4*, 113-151. Academic Press, London and New York.

Onions, A.H.S. (1977) Storage of fungi by mineral oil and silica gel for use in the collection with limited resources. In: *Proceedings of the Second International Conference on Culture Collections*. World Federation for Culture Collections, Brisbane, Australia.

Onions, A.H.S. and Smith, D. (1984) Current status of culture preservation and technology. In: Batra, L.R. and Iigima, T. (eds) *Critical Problems of Culture Collections*. Institute of Fermentation, Osaka, Japan.

Pearson, B.M., Jackman, P.J.H., Painting, K.A. and Morris, G.J. (1990) The effect of freezing and thawing on plasmid expression in *Saccharomyces cerevisiae*. *Cryoletters* 11, 205-210.

Perkins, D.D. (1962) Preservation of *Neurospora* stock cultures in anhydrous silica gel. *Canadian Journal of Microbiology* 8, 591-594. https://cdnsciencepub.com/doi/10.1139/m62-078

Pertot, E., Puc, A. and Kresmer, M. (1977) Lyophilisation of non-sporulating strains of the fungus *Claviceps*. *European Journal of Applied Microbiology* 4, 289-294.

Polge, C., Smith, A.U. and Parkes, S. (1949) Revival of spermatozoa after dehydration at low temperatures. *Nature* 164, 666. https://www.nature.com/articles/164666a0

Puapermpoonsiri, U., Spencer, J. and van der Walle, C.F. (2009) A freeze-dried formulation of bacteriophage encapsulated in biodegradable microspheres. *European Journal of Pharmaceutics and Biopharmaceutics* 72, 26-33. https://doi.org/10.1016/j.ejpb.2008.12.001

Puapermpoonsiri, U., Ford, S.J. and van der Walle, C.F. (2010) Stabilization of bacteriophage during freeze drying. *International Journal of Pharmaceutics* 389, 168-175. https://doi.org/10.1016/j.ijpharm.2010.01.034.

Qiangqiang, Z., Jiajun, W. and Li, L. (1998) Storage of fungi using distilled water or lyophilisation: comparison after 12 years. *Mycoses* 41, 255-257. https://onlinelibrary.wiley.com/doi/10.1111/j.1439-0507.1998.tb00334.x

Raper, K.B. and Alexander, D.F. (1945) Preservation of molds by lyophil process. *Mycologia* 37, 499-525. https://doi.org/10.1080/00275514.1945.12024010

Reinecke, P. and Fokkema, N.J. (1979) *Pseudocercosporella herpotrichoides:* Storage and mass production of conidia. *Transactions of the British Mycological Society* 72, 329-331.

Reischer, H.S. (1949) Preservation of Saprolegniaceae by the mineral oil method. *Mycologia* 41, 177-179. https://doi.org/10.1080/00275514.1949.12017759

Rey, L.R. (1977) Glimpses into the fundamental aspects of freeze drying. In: Cabasso, V.J. and Regamey, R.H. (eds) *Development in Biological Standardisation.* S. Krager, Basel, Switzerland.

Roser, B. (1991). Trehalose, a new approach to premium dried foods. *Trends in Food Science and Technology* 2, 166-169. https://doi.org/10.1016/0924-2244(91)90671-5

Ryan, M.J. (1999) The Effect of Preservation Regime on the Physiology and Genetic Stability of Economically Important Fungi. PhD Thesis, University of Kent, UK.

Ryan, M. J. and Ellison, C. A. (2003). Development of a cryopreservation protocol for the microcyclic rust-fungus *Puccinia spegazzinii. Cryo letters*, 24(1), 43–48. https://pubmed.ncbi.nlm.nih.gov/12644852/

Ryan, M.J. and Smith, D. (2007) Cryopreservation and freeze-drying of fungi employing centrifugal and shelf freeze-drying. In: Day, J.G. and Stacey, G.N. (eds) *Cryopreservation and Freeze-Drying Protocols. Methods in Molecular Biology*™, vol 368. Humana Press, Totowa, New Jersey. https://doi.org/10.1007/978-1-59745-362-2_9

Ryan, M.J., Smith, D. and Jeffries, P. (2000) A decision-based key to determine the most appropriate protocol for the preservation of fungi. *World Journal of Microbiology and Biotechnology* 16, 183-186. https://link.springer.com/article/10.1023/A:1008910006419

Ryan, M.J., Ritchie, B.J. and Smith, D. (2012) Maintenance and storage of fungal plant pathogens. In: Lane, C., Beales, P. and Hughes, K. (eds) *Fungal Plant Pathogens – Principles and Protocols*. CAB International, Wallingford, UK. ISBN 978 1 84593 668 6 https://www.cabi.org/cabebooks/ebook/20123084569

Ryan, M.J., Kasulyte-Creasey, D., Kermode, A., San, S.P. and Buddie, A.G. (2014) Controlled rate cooling of fungi using a Stirling cycle freezer. *Cryoletters* 35, 63-69. https://www.ingentaconnect.com/content/cryo/cryo/2014/00000035/00000001/art00009#expand/collapse

Ryan, M.J., McCluskey, K., Verkleij, G. and Smith, D. (2019) Fungal biological resources to support international development: Challenges and opportunities. *World Journal of Microbiology and Biotechnology* 35, 139. https://doi.org/10.1007/s11274-019-2709-7

Ryan, M.J., Schloter, M., Berg, G., Kostic, T., Kinkel, L.L., Eversole, K., Macklin, J.A., Schelkle, B., Kazou, M., Sarand, I., Singh, B.K., Fischer, D., Maguin, E., Ferrocino, I., Lima, N., McClure, R.S., Charles, T.C., Souza, R.S.C. de, Kiran, G.S., Krug, H.L., Taffner, J., Roume, H., Selvin, J., Smith, D., Rybakova, D. and Sessitsch, A. (2021) Development of microbiome biobanks – challenges and opportunities. *Trends in Microbiology* 29(2), 89–92. https://doi.org/10.1016/j.tim.2020.06.009

Sharma, B. and Smith, D. (1999) Recovery of fungi after storage for over a quarter of a century. *World Journal of Microbiology and Biotechnology* 15, 517-519. https://link.springer.com/article/10.1023%2FA%3A1008959521962

Shearer, B.L., Zeyen, R.J. and Ooka, J.J. (1974) Storage and behaviour in soil of *Septoria* species isolated from cereals. *Phytopathology* 64, 163-167. https://www.apsnet.org/publications/phytopathology/backissues/Documents/1974Articles/phyto64n02_163.PDF

Smith, D. (1983a) A two stage centrifugal freeze-drying method for the preservation of fungi. *Transactions of the British Mycological Society* 80, 333-337. https://doi.org/10.1016/S0007-1536(83)80018-7

Smith, D. (1983b) Cryoprotectants and the cryopreservation of fungi. *Transactions of the British Mycological Society* 80, 360-363. https://doi.org/10.1016/S0007-1536(83)80025-4

Smith, D. (1986) The Evaluation and Development of Techniques for the Preservation of Living Filamentous Fungi. PhD thesis, London University, UK.

Smith, D. (1992) Optimizing preservation. *Laboratory Practice* 41, 25-28.

Smith, D. (1993) Tolerance to freezing and thawing. In: Jennings, D.H. (ed.) *Stress Tolerance in Fungi.* pp. 145-171. Marcel Dekker Inc., New York.

Smith, D. and Allsopp, D. (1992) Preservation and maintenance of biotechnological important microorganisms. In: Greenshields, R.N. (ed.) *Industrial Biotechnology International 1993*. pp 103-107. Sterling Press, UK.

Smith, D. and Bussas, V. (2021) Preserving reference strains. In: Bridge, P., Stackebrandt, E. and Smith, D. (eds) *Trends in Bacterial and Fungal Systematics*. CABI, Wallingford, UK. pp 55-68. 9781789244984; ePDF 9781789244991; ePub 9781789245004. https://www.cabi.org/bookshop/book/9781789244984/

Smith, D. and Onions, A.H.S. (1983) A comparison of some preservation techniques for fungi. *Transactions of the British Mycological Society* 81, 535-540. https://doi.org/10.1016/S0007-1536(83)80122-3

Smith, D. and Onions, A.H.S. (1994) *The Preservation and Maintenance of Living Fungi*. 2nd edn. *IMI Technical Handbooks* No. 2, pp 122. CAB International, Wallingford, UK. https://www.cabi.org/bookshop/book/9780851989020/

Smith, D. and Ryan, M.J. (2012) Implementing best practices and validation of cryopreservation techniques for microorganisms. *The Scientific World Journal* 2012, Article ID 805659, doi:10.1100/2012/805659. http://downloads.hindawi.com/journals/tswj/2012/805659.pdf

Smith, D. and Thomas, V.E. (1998) Cryogenic light microscopy and the development of cooling protocols for the cryopreservation of filamentous fungi. *World Journal of Microbiology and Biotechnology* 14, 49-57. https://link.springer.com/article/10.1023%2FA%3A1008820432471

Smith, D., Ryan, M.J. and Day, J.G. (eds) (2001) *The UK National Culture Collection Biological Resource: Properties, Maintenance and Management*. pp 382. UK National Culture Collection, Egham. https://www.ukbrcn.org/media/1025/ukncc-biological-resource.pdf

Smith, D., Fritze, D. and Stackebrandt, E. (2013) Public Service Collections and Biological Resource Centres of microorganisms. In: Rosenberg, E., DeLong, E.F., Lory, S., Stackebrandt, E. and Thompson, F. (eds) *The Prokaryotes*. Springer, Berlin, Heidelberg, pp 267-304. https://doi.org/10.1007/978-3-642-30194-0_14

Smith, D.H., Lewis, F.H. and Fergus, C.L. (1970) Long term preservation of *Botryosphaeria ribis* and *Dibotryon morbosum*. *Plant Disease Reporter* 54, 217-218.

Souzu, H. (1973) The phospholipid degradation and cellular death caused by freeze-thawing or freeze-drying of yeast. *Cryobiology* 10, 427-431. https://doi.org/10.1016/0011-2240(73)90070-9

Stamp, L. (1947) The preservation of bacteria by drying. *Journal of General Microbiology* 1, 251-265. https://www.microbiologyresearch.org/content/journal/micro/10.1099/00221287-1-2-251

Tan, C.S. (1997) Preservation of fungi. *Cryptogamie Mycologie* 18, 157-163.

Tan, C.S., Stalpers, J.A. and van Ingen, C.W. (1991a) Freeze-drying of fungal hyphae. *Mycologia* 83, 654-657. https://doi.org/10.1080/00275514.1991.12026065

Tan, C.S., Ingen, C.W. van and Stalpers, J.A. (1991b) Freeze-drying of fungal hyphae and stability of the product. In: Griensven, L.J.D. van (ed.) *Genetics and Breeding of Agaricus. Proceedings of the first International Seminar on Mushroom Science, Mushroom Experimental Station, Horst, Netherlands, 14-17 May 1991*. Pudoc, Wageningen, Netherlands. https://www.semanticscholar.org/paper/FREEZE-DRYING-OF-FUNGAL-HYPHAE-Tan-Stalpers/d78c01327ba822d87a4ac551371393bcc5cf78c3

Tan, C.S., Ingen, C.W. van, Talsma, H., Miltenburg, J.C. van, Steffensen, C.L., Vlug, I.A. and Stalpers, J.A. (1995) Freeze-drying of fungi: influence of composition and glass transition temperature of the protectant. *Cryobiology* 32, 60-67. https://doi.org/10.1006/cryo.1995.1005

Tindall, B.J. (2007) Vacuum-drying and cryopreservation of Prokaryotes. In: Day, J.G.and Stacey, G.N. (eds) *Cryopreservation and Freeze-Drying Protocols. Methods in Molecular Biology™*, vol 368. Humana Press, New Jersey. https://doi.org/10.1007/978-1-59745-362-2_7.

Tommerup, I.C. and Kidby, D.K. (1979) Preservation of spores of vesicular-arbuscular endophytes by L-drying. *Applied and Environmental Microbiology* 37, 831-835. https://www.ncbi.nlm.nih.gov/pmc/articles/PMC243309/

Torres, P. and Honrubia, M. (1994) Basidiospore viability in stored slurries. *Mycological Research* 98, 527-530. https://doi.org/10.1016/S0953-7562(09)80473-5

Wellman, A.M. and Stewart, G.G. (1973) Storage of brewing yeasts by liquid nitrogen refrigeration. *Applied Microbiology* 26, 577-583. https://doi.org/10.1128/am.26.4.577-583.1973

Wessman, P., Håkansson, S., Leifer, K. and Rubino, S. (2013) Formulations for freeze-drying of bacteria and their influence on cell survival. *Journal of Visualized Experiments : JoVE* 78, 4058. https://doi.org/10.3791/4058

Winters, R.D. and Winn, W.C. (2010) A simple, effective method for bacterial culture storage: A brief technical report. *Journal of Bacteriology and Virology* 40, 99-101. https://doi.org/10.4167/jbv.2010.40.2.99

Yamai, S., Obara, Y., Nikkawa, T., Shimoda, Y. and Miyamoto, Y. (1979) Preservation of *Neisseria gonorrhoeae* by the gelatin-disc method. *British Journal of Venerial Disease* 55, 90-93. https://www.ncbi.nlm.nih.gov/pmc/articles/PMC1045596/

5

Regulatory Considerations on the Safe Handling and Distributing of Microorganisms

Introduction

Microorganisms are handled, grown, shared, utilized and stored in many organizations all over the world. Access to such resources facilitates research and development and reduces costs in re-isolation, identification and characterization. Although interesting properties can be harnessed, many organisms are potential pathogens and may even present health hazards, thereby requiring them to be handled with the utmost care. The necessary precautions that must be taken in the storage, handling, distribution and disposal of microbial cultures and best practices have been defined, which offer guidance to collection organizations (Smith *et al.*, 1999). This chapter covers in detail regulatory and in particular, the health and safety requirements that must be taken into consideration while handling microorganisms. The importance of health and safety requirements when working with microorganisms and the impact of regulations and the law is explained and specifically covers:

Health and Safety in Handling Microorganisms *information on the importance of health and safety and the effect of law on safety with regard to microorganisms.*

Classification of Microorganisms on the Basis of Hazard *information on the ACDP hazard group classification of organisms based on their risks and the different containment levels.*

Safety in the Laboratory *information about safety measures regarding equipment, culture media and stains, and chemicals and plant material.*

Regulations Governing Distribution of Cultures *information about the various international regulatory authorities controlling the distribution of organisms.*

The chapter provides guidance on establishing a regulatory compliant and safe environment in which to operate a culture collection and handle microorganisms in the laboratory.

Health and Safety in Handling Microorganisms

Microorganisms are isolated, grown, characterized, preserved and stored, and transported between laboratories all over the world. They are shipped by various means, by mail, courier or by hand, from one laboratory to another within countries and often across borders or continents (Smith and Ryan, 2019). They are sent for identification, reference, research or for production purposes from colleague to colleague, from and to culture collections. All these actions must be carried out safely and in compliance with the legislation and regulations that control these matters (Macreadie, 2019).

The importance of a laboratory's health and safety procedures stretches beyond the

DOI: 10.1079/9781800622135.0005

laboratory to all those who may come into contact with substances and products from that laboratory. A microorganism in transit will put carriers, postal staff, freight operators and recipients at risk if safety and shipping regulations are not followed.

Basic safety requirements for a safe workplace, done in compliance with the law or as duties of a caring employer, are:

- Adequate assessment of risks
- Provision of adequate control measures
- Provision of health and safety information
- Provision of appropriate training
- Establishment of record systems to allow safety audits to be carried out
- Implementation of good working procedures

Good working practice requires assurance that correct procedures are actually being followed and this requires a sound and accountable safety policy. Safety regulations (for example the United Kingdom Control of Substances Hazardous to Health - COSHH) require that every employer makes a suitable and sufficient assessment of the risks to health and safety to which any person, whether employed by them or not, may be exposed through their work (COSHH, 2002). The distribution of microorganisms outside the workplace extends employer duties to protect others but recipients should also make their own risk assessments based on their local laws.

UK legislation typifies the controls needed for handling microorganisms. The COSHH regulations (2002) aim to stimulate and enforce an improvement in Health and Safety in the workplace. Dealing with a known chemical can be much simpler than with a named microorganism. The full metabolic and biochemical potential of a microorganism is rarely known and therefore assessing the risk when the hazard is not clearly defined becomes difficult. This is where the COSHH regulations are realistic leaving room for interpretation. The regulations incorporate terms 'as far as reasonably practicable', 'adequate control', and 'suitable measures' which enable the employer to set relevant safe procedures that can be adjusted to the circumstances whilst providing a safe environment. Microorganisms present different levels and kinds of hazards, leaving an enormous, but necessary, task for microbiologists. A risk assessment for example, must consider the production of potentially hazardous toxins (See Appendix G, this volume, for example risk assessments).

Classification of Microorganisms on the Basis of Hazard

Organisms present various levels and kinds of hazard, which are more difficult to enumerate or measure than chemicals. Virulence and toxicity may vary from strain to strain within a species and additional hazards, such as toxin production and allergenicity, must be considered, e.g. a fungus may produce mycotoxins. Different organisms present different hazards and different size inoculum (infective dose) can be required to cause a problem. Individuals may respond differently to exposure, with some being more sensitive than others. It is therefore critical that the full hazard potential of the organism is considered and that this is related to effects it may have on the particular individual carrying out the work.

There are various classification systems available depending on and under whose jurisdiction you work. These include the definitions for classification by the World Health Organization (WHO); United States Public Health Service (USPHS); Advisory Group on Dangerous Pathogens (ACDP); European Federation of Biotechnology (EFB) and European Community (EC). In Europe, the EC Directive (2000/54/EEC) on Biological Agents sets a common base line, which has been strengthened and expanded in many of the individual member states. The definition and minimum handling procedures of pathogenic organisms are set by appropriate authorities in each country and are often the same or similar for all EC countries. In the UK, the ACDP list four hazard groups 1-4 with corresponding containment levels. Microorganisms are normally classified on their potential to cause disease, their human pathogenicity, into four risk groups (ACDP, 2021):

Risk Group 1
A biological agent that is most unlikely to cause human disease.

Risk Group 2
A biological agent that may cause human disease and which might be a hazard to laboratory workers but is unlikely to spread

in the community. Laboratory exposure rarely produces infection and effective prophylaxis or treatment is available.

Risk Group 3

A biological agent that may cause severe human disease and present a serious hazard to laboratory workers. It may present a risk of spread in the community but there is usually effective prophylaxis or treatment.

Risk Group 4

A biological agent that causes severe human disease and is a serious hazard to laboratory workers. It may present a high risk of spread in the community and there is usually no effective prophylaxis or treatment.

A BRC and indeed any microbiology laboratory must ensure that all strains are assigned to appropriate risk/hazard groups. This includes a positive assignment to risk/hazard group 1 unless otherwise considered hazardous. Hazard information must be recorded and made available to recipients of this material. A safe laboratory is the result of applying good techniques. Containment Level 2 (ACDP, 2018, 2021) is easily achievable and should be standard practice in all microbiological laboratories. Good aseptic techniques applied by well-trained personnel will ensure pure and clean cultures and will minimize contact with the microorganism. However, unexpected incidents or accidents must also be considered when assessing the risk involved.

There are several other pieces of legislation that restrict the distribution of microorganisms; of which a microbiologist must be aware. Employment of good laboratory practice, good housekeeping, workplace and equipment maintenance and ensuring that staff have the relevant information and training, will minimize the risk of accidents. The establishment of emergency procedures to reduce potential harm is a sensible and necessary approach. A Biological Resource Centre (BRC), culture collection or microbiology laboratory should comply with:

- Health and Safety requirements
- Classification of microorganisms on the basis of hazard
- Quarantine regulations
- Ownership of Intellectual Property Rights (IPR)
- Convention on Biological Diversity (CBD) and the Nagoya Protocol
- Regulations governing shipping of cultures
- Control of distribution of dangerous organisms

A microbiology laboratory, BRC or culture collection must put in place procedures to manage the health and safety of all who may be put at risk by its activities. This requires a suitable and sufficient assessment of the risks to health and safety to which any person whether employed by them or not may be exposed to through their work (https://www.hse.gov.uk/coshh/basics/assessment.htm). These assessments must be reviewed regularly, additionally when changes in procedures or regulations demand, and must be recorded. The distribution of microorganisms to others outside the workplace extends these duties to protect others and responsible and best practices require the sharing of safety data to accompany microorganisms supplied to others.

Legislation controls the safe handling and use of organisms and biologists must ensure they keep abreast of existing, new and changing regulations. Health and safety, packaging and shipping and controlled distribution legislation may be extensive and sometimes cumbersome, but it is there to protect both scientists and the wider community. In the interests of the progress of science, biologists must be able to exchange organisms upon which their hypotheses and results are based, but they must do this in a way that presents minimum risk to those who come into contact with the organism.

Hazards Associated with Microorganisms

There are numerous hazards presented by microorganisms. First and foremost is the risk of infection requiring the need to assign them to risk or hazard groups and ensure the appropriate containment measures are in place (ACDP, 2018, 2021). The containment level numbers correlate with the risk group in which the organism falls (i.e. organisms in Risk group 1 require containment Level 1 and so forth).

Infection

Almost 60% of laboratory-acquired infections occur in research laboratories. This is usually because new techniques are being attempted and the full potential of the organism studied is unknown. Risks are minimized by employing good technique.

Occurrence of Laboratory-acquired Infections

Laboratory-acquired infections are not new. Cholera, Yellow Fever and Rocky Mountain Spotted Fever claimed the lives of some of the scientists who first investigated them. Bacteria and viruses account for most reported laboratory-acquired infections and many result in death, whereas fungal infections are fewer and often less serious, although deaths do occur. Of the yeasts, *Candida* infections are quite common. Dermatomycosis is the most common fungal infection (46% of the total) principally caused by species of *Trichophyton*. The top ten laboratory-acquired infections are Brucellosis, Q Fever, Hepatitis, Typhoid fever, Tularaemia, Tuberculosis, Dermatomycosis, Venezuelan equine encephalitis, Psittacosis and Coccidioidomycosis. To infect, the microorganism must be capable of causing disease, enter the body in sufficient numbers to overcome the body's natural defences and use a particular route according to the disease. Staff undergoing certain drug treatments may be particularly at risk and must consult their safety officer or line management for advice.

Microorganisms can enter the body through the mouth, the lungs, broken or unbroken skin and the conjunctivae. However, in the laboratory the route of infection may not be the same as when the disease is acquired naturally. The potential hazards are much greater in the laboratory than in nature. Organisms are grown in vast numbers, transferred from one container to another and manipulated, increasing the risk of infection. It is extremely important to ensure all accidents with microorganisms are recorded in the accident book and reported to the laboratory manager or supervisor. Some infections as a result of laboratory infection will need medical treatment and, in such cases, do not delay, seek treatment immediately. There can be the following routes to infection:

- Accidental inoculation
- Accidental ingestion
- Splashing into the face and eyes
- Spillage and direct contact
- Inhalation of aerosols
- Contaminated clothing

ROUTES OF INFECTION

Accidental Inoculation

This usually happens as the result of pricking, jabbing or cutting the skin with infected instruments or objects such as hypodermic needles, scalpels, and broken, contaminated glassware.

Accidental Ingestion of Microorganisms

This usually happens as the result of mouth pipetting, eating, drinking, smoking or application of cosmetics in the laboratory. **These practices are prohibited in all laboratories.**

Splashing into the Face and Eyes

This can be caused by creation of aerosols, violent separation or discharge, under pressure, of needles and syringes and the rubbing of eyes. Normally laboratory workers know when they are splashed or sprayed and should rinse the affected area well. Procedures that may produce splashes or aerosols must be carried out in a suitable microbiological safety cabinet.

Spillage and Direct Contact

When there is a spillage it may not be noticed or decontamination may not be entirely effective. The contamination can then be transferred by fingers to the mouth or eyes. Microorganisms may enter the bloodstream through cuts or abrasions, some of which may not be visible to the naked eye. Shaking incubators can present an exceptional risk in this respect. Procedures for dealing with spillages are posted near the equipment. If spillage is suspected turn off the power and leave the equipment closed for 30 min. Only open the equipment if you are wearing personal protective equipment (PPE) and have suitable disinfectants and spillage kits to deal with the contamination. Always wash thoroughly after dealing with such spills and ensure that PPE is properly cleaned and disinfected.

Aerosols, Droplets and Infected Airborne Particles

Aerosols can be formed by squirting liquids containing microorganisms through small holes at pressure or by firing a jet of suspension onto a solid surface. Using loops, pipettes, syringes, needles, opening tubes and bottles, use of centrifuges and blenders, freeze-drying and breakage are all processes able to generate aerosols.

Aerosols of 5 μm particles can reach the alveoli in the lungs, larger particles are trapped in the nose but some fungal spores can penetrate. All procedures that generate aerosols containing microorganisms must be carried out in the appropriate microbiological safety cabinet. Droplets vary in size. Smaller ones evaporate into dried droplet nuclei with viable microorganisms in them. Larger ones usually settle out of the air fairly rapidly. Infected airborne particles do not always originate from aerosols. Freeze-dried (lyophilized) cultures, dried bacterial colonies, dried material on stoppers and caps of culture tubes and bottles, dried exudates, fungal and actinomycete spores released when the cultures are opened may provide the source.

Contaminated laboratory coats and other protective equipment could be the source of aerosols when they are removed or changed after use. Coats must be regularly sterilized and cleaned, and manufacturers' recommended procedures for sterilization and disposal of protective equipment must be strictly followed.

Where genetically manipulated or modified microorganisms are concerned, the Scientific Advisory Committee on Genetic Manipulation (SACGM) in the UK prescribes separate but similar regulations for those organisms that have been genetically modified. The assessment of risk in handling Genetically Engineered Micro-organisms (GEMs) or Genetically Modified Organisms (GMOs) is more difficult as the hazards of the donor and recipient have to be considered, as well as those of the resulting GEM. To meet the UK and European legislation, all microbiologists will have to produce a risk assessment on the organisms with which they work or hold in collections. In the case of fungi, it is recognized that many may infect following traumatic inoculation through the skin, or infect a compromised patient, but do not infect healthy individuals. Most fungi from clinical specimens require Containment level 2 (ACDP, 2018, 2021), unless a higher degree of containment is specified.

In the UK, GMOs also require Containment level 2 for handling and all potential work with such organisms must first be referred to the institution's Biological safety officer and/or Biological safety committee.

We have included UK examples as a guide, but similar requirements exist in most countries.

Assessment of risk

The Oxford English Dictionary defines risk as 'the possibility of loss, injury, or other adverse or unwelcome circumstance, a chance or situation involving such a possibility' OED online edition at https://www.oed.com/

Organisms present different levels and kinds of hazard, evaluation of which represents an enormous, but necessary, task for biologists. A risk assessment for example, must consider the production of potentially hazardous toxins or allergenic and systemic impacts. Ultimately, a safe laboratory is the result of applying good techniques, a hallmark of technical excellence. Containment level 2 (ACDP, 2021) is easily achievable and should be standard practice in all laboratories handling organisms that present a risk of infection or of causing other harm. See the guidance documents in the above section *Classification of Microorganisms on the Basis of Hazard*. Good aseptic techniques applied by well-trained personnel will ensure pure and clean cultures and will minimize contact with the organism. However, the possibility of accidents must also be considered when assessing risks. The employment of good laboratory practice and good housekeeping, workplace and equipment maintenance, together with ensuring that staff have relevant information and training, will minimize the risk of accidents (Smith and Onions, 1994). The establishment of emergency procedures to reduce potential harm is an additional and sensible precaution. The containment level numbers correlate with the risk group in which the organism falls (i.e. organisms in Risk group 1 require containment Level 1 and so forth, see Table 5.1 below).

As previously mentioned, SACGM prescribe separate but similar regulations for those organisms that have been genetically modified. Similarly, other European countries have advisory committees, in Germany it is the Zentrale Kommission für die Biologische Sicherheit (ZKBS) within the Federal Office of Consumer Protection and Food Safety (BVL), Berlin. The Trade Corporation Association of the Chemical Industry (BG RCI) advises on how individual Genetically Engineered Micro-organisms (GEMs) should be classified. The assessment of risk in handling GEMs or GMOs is more difficult as the hazards of

Table 5.1. Summary of laboratory containment levels for the UK (ACDP, 2018)

Containment requirement	Containment level 1	2	3	4
Laboratory site: isolation	No	No	Partial	Yes
Laboratory: sealable for fumigation	No	No	Yes	Yes
Ventilation: inward airflow/negative pressure	Optional	Optional	Yes	Yes
Ventilation: through safety cabinet	No	Optional	Optional	No
Mechanical: direct	No	No	Optional	No
Mechanical: independent ducting	No	No	Optional	Yes
Airlock	No	No	Optional	Yes
Airlock: with shower	No	No	No	No
Wash hand basin	Optional	Yes	Yes	Yes
Effluent treatment	No	No	No	Yes
Autoclave site: on site	Yes	–	–	–
in suite	–	Yes	Yes	–
in lab: free standing	–	–	Optional	–
in lab: double ended	–	–	–	Yes
Microbiological safety cabinet/enclosure	No	Optional	Yes	Yes
Class of cabinet/enclosure*	–	Class I	Class I/III	Class I/III

*Guidance on the use of Class II microbiological safety cabinets is given in the Management and operation of microbiological containment laboratories ACDP guidance Appendix 2: Microbiological Safety Cabinets (ACDP, 2018).

the donor and recipient have to be considered, as well as those of the resulting GEM.

The species of bacteria, fungi and other parasites falling into hazard groups 2 and 3 have been defined (ACDP, 2021). Similarly, all bacteria from the Approved List of Bacterial Names (LPSN, DSMZ); https://lpsn.dsmz.de/text/approved-lists) have been assigned to an appropriate hazard group in Germany (Anon, 1997a, b). However, some species of fungi have not been assigned to hazard group 1 (ACDP, 2021). Medically important fungi have been categorized into relevant hazard groups by de Hoog (1996). To meet the UK and European legislation, all microbiologists will have to make a risk assessment on the organisms with which they work or hold in collections. In the case of fungi, it is recognized that many may infect following traumatic inoculation through the skin, or infect a compromised patient, but do not infect healthy individuals. Most fungi from clinical specimens require Containment level 2 (ACDP, 2021), unless a higher degree of containment is specified (see Table 5.1).

The UK procedures for GMOs which also require Containment level 2 for handling are described above but again, legislation can be different in other countries, for example, in Germany some manipulated organisms can be handled at Containment level 1. The COSHH regulations work well and can be easily applied in establishments with designed laboratories but may not work as well in an industrial environment where very large volumes and more hazardous techniques may be used. Total containment is rarely applicable.

To meet the biological agents' legislation and COSHH requirements, a step by step evaluation of a laboratory procedure or an industrial process must be carried out. This is necessary as different organisms present different hazards and different size inoculum can be required to cause a problem. The assessment must cover the procedure from the original inoculum or seed culture to the final product or the point where the organism is killed and disposed of. It is not adequate to say that the microorganism is of ACDP hazard group 2 or less and therefore work can be carried out on the laboratory bench apart from those procedures that may create aerosols. It must be noted that individuals may respond differently to exposure, with some being more sensitive than others. It is therefore critical that the full hazard potential of the organism is considered and that this is related to effects it may have on the particular individual carrying out the work. Example risk assessments from CABI can be found in Appendix G, this volume.

Allergic Reactions

Allergic reactions to fungi, bacteria or the chemical compounds they produce may occur, but only certain individuals will react while the majority of other people remain unaffected. Those who suffer allergic reaction to a particular microorganism must avoid coming into contact with it.

Some allergic reactions may be recognized by a reddening and possible swelling of the skin after contact or, if an aero-allergen, by an increased flow of secretions in the nose and to the eyes and possible respiratory distress such as coughing or difficulty breathing. Sensitization and the development of allergen reaction take a considerable time. Extreme reaction can often lead to severe shock. The person affected should avoid contact with the fungus or chemical concerned.

Sensitivity to Latex Gloves

Sensitivity to latex gloves can develop and even those who do not wear them can be affected. Powdered gloves will release excess powder into the atmosphere when put on or removed. The extractable proteins absorbed by the powder may cause allergic reaction in susceptible persons. Gloves that are powder free and low in allergen content should be used, e.g. nitrile or neoprene gloves. If signs of irritation develop seek advice.

Report ill effects which you think may have been brought on by exposure to microorganisms or chemicals at work to your line manager. Health surveillance is required where staff may come into contact with respiratory sensitizers.

The following has been recommended to prevent latex sensitization:

- Avoid the use of powdered latex gloves where possible.
- Ascertain the allergenicity of the glove.
- Do not use cheap gloves as these are likely to be allergenic.
- Report any irritation that may be related to wearing of gloves immediately.
- Wash hands immediately after wearing gloves.
- Avoid oil-based emollients as this may enhance absorption of the proteins.
- Do not wear latex gloves unless you have to.
- Cotton inner gloves can be worn to prevent direct contact but this will not prevent all the powder from coming into contact with your skin.

Prevention of Allergic Reactions

The spores of fungi (including *Alternaria*, *Aspergillus*, *Cladosporium* and *Penicillium*) and cereal rusts and smuts in the air are able to induce allergy asthma and hay fever in humans. There are a number of special pulmonary conditions caused by fungal spores to which certain industrial workers are subject. These include cheese washer's lung (*Penicillium verrucosum*), maltster's lung (*Aspergillus clavatus*), mushroom worker's lung (*Doratomyces* sp.) and suberosis (*Penicillium glabrum*).

Respiratory sensitizers include airborne spores and other propagules of microorganisms including the fungi mentioned above and the bacterium, *Pseudomonas*. Chemicals are also responsible and these include: mycotoxins, other metabolites (e.g. ampicillin, penicillin, streptomycin sulphate), cobalt salts; dusts, mists, cutting fluids and enzymes are also considered sensitizers. Any fungus has the potential to produce an infection in an immunocompromised individual.

Mycotoxins

One of the better-known hazards associated with fungi is the ability to produce toxic secondary metabolites. There are at least 300 different mycotoxins produced by fungi and there are hundreds of species known to produce them. In general, individual fungi may be able to produce up to five mycotoxins under specific conditions. Potential exposure to mycotoxins in the laboratory is mainly by:

- Ingestion or inhalation of contaminated dust or fungus spores
- Through the skin when the fungi are extracted into solvents

The presence of these in culture media adds to the hazard status of the growing organisms. The toxins produced may be carcinogenic, mutagenic, nephrotoxic, hepatotoxic, haemorrhagic, oestrogenic or cause inflammatory effects. The most commonly known mycotoxin is aflatoxin, which is considered to be carcinogenic, hepatotoxic and potentially mutagenic and is produced by strains of *Aspergillus flavus* and *A. parasiticus*.

Mycotoxicoses are poisonings caused by the ingestion of food contaminated (and sometimes rendered carcinogenic) by toxin producing microfungi (Paterson, 2006). Toxins are also produced by many other fungi (see Table 5.2).

It should always be remembered that many fungi have not been studied chemically and because mycotoxins are not reported for a species does not mean it does not produce them. The handling of materials contaminated by these toxins can lead to their ingestion and subsequent poisoning. Inhalation of mycotoxins can also be dangerous. Toxins from *Aspergillus* and *Fusarium* species have caused problems in patients when inhaled. The first major poisonings of man were reported in 1974 in India when over 1000 cases were diagnosed and 300 deaths occurred (Krishnamachuri *et al.*, 1975).

Bacterial Toxins

Infection patterns caused by bacterial pathogens are very different and depend on the bacterial pathogen and the individual host. Thus, every infection is an extremely individual process.

Factors affecting virulence of bacterial species are invasiveness, production of aggressins and toxigenicity. Most bacterial toxins are capable of damaging or killing normal host cells and are effective upon infection. In contrast, most mycotoxins are effective without invasion or infection. The role played by bacterial invasiveness in damaging the host varies widely: sometimes infection can be extremely localized (e.g., *Corynebacterium diphtheriae*), the toxin diffuses and reaches almost all tissues. Alternatively, pathogens may invade and need to multiply in order to generate enough toxin to cause damage to the host (e.g., *Bacillus anthracis*).

Diseases caused by bacteria may be grouped as follows (Anon, 1998):

- Local infections are manifestation of the pathogen in a localized tissue - examples: *Staphylococcus aureus, Neisseria gonorrhoeae*
- Local infections with production of a potent toxin have low invasiveness, and general diffusion of the toxin through the lymphatic system and bloodstream - examples: *Clostridium tetani, Corynebacterium diphtheriae*
- Acute generalized infections result from highly invasive distribution of the pathogens after infection leading to possible septic-toxic shock. They sometimes have tissue specific (organotroph) manifestation - examples: plague, typhoid, brucellosis, some types of tuberculosis

There are two classes of designated bacterial toxins, which can be distinguished by their chemical nature. Examples of toxin producers are given in Table 5.3.

Classes of Bacterial Toxins

Exotoxins

These are toxins which exhibit protein-like structure and excrete a toxin into their environment such as diphtheria, tetanus, botulinum toxins and enterotoxins.

Endotoxins

These are more complex structures consisting of protein, lipid and polysaccharide elements. The toxin is intrinsically part of the structure of the cell, usually the cell wall, and is released when the cell undergoes lysis. Most endotoxins occur in Gram-negative bacteria and are relatively non-specific, in that they will release the toxin regardless of the location of the bacteria. Some endotoxins may also exhibit pyrogenicity. The most well-known of these are the enteric bacteria of the genera *Escherichia, Salmonella* and *Shigella*.

Some endotoxins also cause capillaries to become more permeable through releasing enzyme-like substances, called aggressins, which support the pathogen's invasion by damaging host tissue.

Non-Infectious Toxin Producers

In addition to those bacteria that produce toxins during infection, there are also those that are non-infectious toxin producers. Of these, the cyanobacteria, of which there are *ca.* 2000 species, are currently considered as hazard group 1. Within these groups of organisms there are three main classes of toxins: liposaccharide toxins, (generally considered to be of low toxicity); peptide hepatotoxins (possessing significant risk on

Table 5.2. Some common mycotoxins and examples of fungi producing them

Mycotoxin	Fungus
Aflatoxin	*Aspergillus flavus, A. parasiticus*
Aflatrem	*Aspergillus flavus*
Altenuic acid	*Alternaria alternata*
Alternariol	*Alternaria alternata*
Austdiol	*Aspergillus ustus*
Austamide	*Aspergillus ustus*
Austocystin	*Aspergillus ustus*
Bentenolide	*Monographella nivalis*
Brevianamide	*Aspergillus ustus*
Citrinin	*Aspergillus carneus, A. terreus, Penicillium citrinum, P. hirsutum, P. verrucosum*
Citreoviridin	*Aspergillus terreus, Penicillium citreoviride*
Cochliodinol	*Chaetomium cochliodes*
Crotocin	*Acremonium crotocinigenum*
Cytochalasin E	*Aspergillus clavatus*
Cyclopiazonic acid	*Aspergillus versicolor*
Destruxin B	*Aspergillus ochraceus*
Fumagilin	*Aspergillus fumigatus*
Fusarin	*Fusarium moniliforme*
Gliotoxin	*Aspergillus fumigatus*
Islanditoxin	*Penicillium islandicum*
Malformin	*Aspergillus niger*
Maltoryzine	*Aspergillus* spp.
Moniliformin	*Fusarium moniliforme, F. oxysporum, F. equiseti*
Ochratoxin	*Aspergillus ochraceus, Penicillium viridictum*
Oxalic acid	*Aspergillus niger*
Patulin	*Aspergillus clavatus, Penicillium expansum, P. roquefortii, P. claviforme, P. griseofulvum*
Penicillic acid	*Aspergillus ochraceus*
Penitrem	*Penicillium crustosum*
Roridin	*Myrothecium roridum, M. verrucaria, Dendrodochium* spp., *Cylindrocarpon* spp., *Stachybotrys* spp.
Rubratoxin	*Penicillium rubrum*
Rubroskyrin	*Penicillium* spp.
Rubrosulphin	*Penicillium viridicatum*
Rugulosin	*Penicillium brunneum, P. kloeckeri, P. rugulosum*
Satratoxin	*Stachybotrys chartarum*
Slaframine	*Rhizoctonia leguminicola*
Sterigmatocystin	*Aspergillus flavus, A. nidulans, A. versicolor, Penicillium rugulosum*
Trichodermin	*Trichoderma viride*
Trichothecin	*Trichothecium roseum*
Trichothecenes T2 toxin, deoxynivalenol (vomitoxin), nivalenol, diacetoxyscirpenol, fusarenone, 3-acetyldeoynivalenol, 15-acetyldexoynivalenol	*Fusarium acuminatum, F. roseum, F. sporotrichioides*
Tryptoquivalene	*Aspergillus clavatus*
Verrucarin	*Myrothecium verrucaria, Dendrodochium* spp.
Verruculogen	*Aspergillus fumigatus*
Viopurpurin	*Trichophyton* spp., *Penicillium viridicatum*
Viomellein	*Aspergillus* spp., *Penicillium aurantiogriseum, P. crustosum, P. viridicatum*
Viriditoxin	*Aspergillus fumigatus*
Xanthocillin	*Eurotium chevalieri*
Zearalenone	*Fusarium culmorum, F. graminearum, F. oxysporum, F. roseum*

Data from the CABI Bioscience Genetic Resource Collection database and Smith and Moss (1985).

Table 5.3. Some bacterial exotoxins and examples of bacterial producers (after Stanier *et al.*, 1987)

Toxin	Bacterium
Neurotoxins	
Botulinum toxins: A, B, C1, C2, D, E, F, G	*Clostridium botulinum*
Tetanospasmin	*Clostridium tetani*
Tetanolysin	
Cytotoxins	
Lecithinase	*Clostridium perfringens*
Necrotic factors	
Hemolysin	
Collagenase	
Diphtheria toxin	*Corynebacterium diphtheriae*
Streptolysin O	*Streptococcus pyogenes*
Streptolysin S	
Enterotoxins	
Enterotoxin	*Clostridium perfringens*
	Staphylococcus aureus
'Enterotoxin'	
Shiga toxin	*Shigella dysenteriae*
Cholera toxin	*Vibrio cholerae*
'Guinea pig toxin'	*Yersinia pestis*
Heat-labile enterotoxin (LT)	*Escherichia coli*
Heat-stable enterotoxin (ST)	*Escherichia coli*

both short-term high level and long-term low-level exposure) and alkaloid neurotoxins, (generally highly toxic). Toxicity of individual strains may vary with environmental conditions but toxicity is invariably associated with bloom formation and generally involves strains of *Microcystis aeruginosa*, *Anabaena flos aquae*, *Oscillalonia (Planktothrise) agardhii* or occasionally other species.

Preventing exposure to microbial toxins

Individuals who are immunocompromised or with other relevant medical conditions may be at greater risk and several of the compounds may cause or promote cancers. Pregnant women and children might also be at particular risk. The effect of low levels of mycotoxins or of mixtures on humans is largely unknown. It is worth noting that even the most poisonous mycotoxins found in food are a factor of about one million times less toxic than the most virulent of the botulinum toxins from the bacterium *Clostridium botulinum*. However, the effects of exposure to low levels over long periods must also be considered.

Individual Risk and Decontamination Procedures
How to Minimize Exposure to Mycotoxins

- Guard against aerosol formation from liquid systems.
- Direct contact with the pure and diluted toxins must be avoided by the use of safety equipment such as mechanical pipetting devices.
- Microbiological safety cabinets should be used whenever possible and decontaminated after use.
- Personal protective equipment such as laboratory coats, gloves, and masks should be used at all times (NB: vinyl gloves can allow solvents to penetrate therefore they should be changed immediately if they become contaminated, or alternative gloves should be used, such as nitrile or neoprene).
- Disposable laboratory coats may be used or laboratory coats should be regularly decontaminated.
- Commodities containing fungi ('mouldy commodities') or commodities suspected of contamination by mycotoxins should never be handled without protective gloves.

- Procedures involving grinding of cultures or commodities must ensure operator safety, particularly from inhalation of particles.
- All work surfaces and equipment used during extraction should be correctly decontaminated after use, and hazardous wastes must be correctly disposed of.
- Analytical equipment such as Thin Layer Chromatography (TLC) plates and High-Performance Liquid Chromatography (HPLC) columns may retain contamination, and have to be disposed of properly. Similarly, equipment used for other forms of analysis (e.g. electrophoresis gels) may become contaminated with mycotoxins and require decontamination before disposal.
- Workers should be aware of scientific literature on the mycotoxins produced by the fungi with which they are concerned irrespective of whether mycotoxins are the reason for studying the particular fungus.

Decontamination Procedures

The following procedures have been developed for use with aflatoxin and may not be effective against all mycotoxins. Glassware which has been used to extract fungi should be soaked in 5% sodium hypochlorite. All exposed surfaces must be swabbed with 5% sodium hypochlorite. Fungal material should be autoclaved prior to the addition of 5% sodium hypochlorite. Spills of pure toxin should be absorbed with diatomaceous earth, carefully collected and placed in a bucket. Clean the spillage area with 5% sodium hypochlorite. In the case of aflatoxin, add 5% sodium hypochlorite followed, after 30 min, by acidification to pH 7.8 - 8.8 with 2M hydrochloric acid and then the addition of acetone to a final concentration of 5% by volume, this should be left for 30 min. The contaminated material should be treated similarly prior to safe disposal.

Alternative methods of decontamination have been published (Castegnaro *et al.*, 1990; WHO, 2020) which involve analysis for decontamination by HPLC. A method using alkaline potassium permanganate is effective against citrinin, ochratoxin A, patulin, sterigmatocystin, and aflatoxin B1, B2, G1, G2. Effectiveness against other mycotoxins has yet to be determined. However, this method is not effective for TLC plates, solutions in DMF and DMSO, and solutions in non-volatile organic solvents miscible with water. The method is suitable for protective clothing, and for the removal and subsequent decontamination of mycotoxins in solvents. This avoids the risk of transporting the waste to an incinerator site. Incineration of contaminated TLC plates may be more convenient than laboratory decontamination.

Safety in the Laboratory

Having considered risk, safety is of equal importance and concerns the prevention of accidents. A microbiology laboratory can be a hazardous environment if basic safety precautions, good practice and common sense are not applied when undertaking any kind of laboratory work. A clean and properly maintained laboratory not only shows a professional attitude towards the work in progress but also, most importantly, contributes to the well-being of yourself and colleagues. This section presents general safety practices for the laboratory. It covers safety regarding equipment, organisms and operations to facilitate working safely in the laboratory, introduction of safety measures regarding equipment, culture media, stains and chemicals and for all to be aware of international safety criteria and compliance.

Basic Requirements to Establish a Safe Workplace

It is the responsibility of all workers to have access to, read and understand the assessments on hazards and associated risks of all materials stored and used in their work place. This knowledge and understanding enables safe handling and containment of all these materials. It is important to maintain a safe working environment in relation to all aspects of your work. Specific aspects are detailed later in this module. **Do not take risks with your own health and safety or that of others around you**. Remember it is more than the microorganisms themselves that can put you and others at risk.

BASIC HEALTH AND SAFETY INFORMATION

Assessing risk. Always:

- Read hazard labels
- Look out for damaged or broken equipment
- Read and understand risk assessments and recommended safety precautions
- Know and practice emergency procedures such as evacuation in case of fire
- Know the whereabouts of emergency equipment such as safety showers, fire extinguishers, first aid boxes and eye wash stations
- Know and understand the consequences of workplace alarms e.g. fire alarms, oxygen level monitors
- Do not attempt to handle or move a heavy or bulky load on your own or without manual handling training

Cleanliness and tidiness:

- Keep laboratory benches and other surfaces clean. It is preferable to clean up after each stage of an operation, overcrowding leads to accidents
- Always wash your hands before leaving the laboratory; avoid the risk of infection, poisoning yourself and others
- Never leave equipment such as Bunsen burners unattended; this presents a risk of fire
- Return unused apparatus to its proper storage place in a clean and working condition
- Dispose of broken or contaminated equipment safely

Thoughtful actions and behaviour:

- Smoking, drinking and eating must not be permitted in laboratories; there is a high risk of inhaling aerosols of fungus spores or ingesting chemicals
- Never smell, touch or taste cultures to determine characteristics
- Never run and take care when opening doors
- Never operate equipment unless fully trained and proven competent
- Never block corridors, emergency exits or access to emergency equipment
- Always report accidents and incidents that could have resulted in harm

Use safety and personal protective equipment

- Wear laboratory coats when engaged in laboratory work and launder them regularly, a filthy laboratory coat is unsightly and a health hazard
- Never wear loose jewellery or loose neck ties that may be caught in equipment; tie long hair back or wear a protective hat
- Wear Personal Protective Equipment (PPE) or protective clothing when and where appropriate, if you are not sure what is required for a task, seek advice and instruction from a capable or qualified colleague. Surgical gloves, safety spectacles, face shields, particle or vapour filtration masks, closed footwear are examples of PPE and protective clothing – wearing the right equipment and clothes for the job
- Use step ladders or safety steps to reach high places
- Use automatic controlled volume pipettes - never pipette by mouth
- Use containment equipment such as the correct microbiological safety cabinets for the risk group organism you are handling
- Use a fume cabinet fitted with the correct filter for the chemicals you are using

Safety Regarding Equipment

All work equipment must be safe and suitable for the task for which it is used. It must only be used when training on its use has been given and is understood. Competence on equipment use must be demonstrated as incorrect use can cause damage to the equipment and danger to yourself and others.

All equipment should be part of institutional procedures for the testing of electrical safety.

Autoclave

Hazards

The main hazards of operating autoclaves include the following:

1. Burns from physical contact with the autoclave or its load
2. Steam burns which can be extensive and serious
3. Explosive breakages of glass vessels during opening and unloading, due to temperature stress.

Precautions for use

Ensure formal instruction before use and follow instructions in the equipment manuals and the maintenance procedures:

- Loosen screw caps, load securely, and do not balance things precariously. Do not overload.
- Avoid contact with wall of autoclave and keep well clear of steam issuing from any vent or opening.
- After autoclaving allow pressure to return to atmospheric level before opening (this is automatic in many autoclaves). Make sure automatic locking device or pressure control is open and has remained open for several minutes before attempting to open autoclave.
- Open and unload carefully, avoiding the residual issuing steam. Insulated gloves will reduce the risk of burning hands. Overalls or a laboratory coat will help protect arms. It is safer to wait a few minutes after opening, to allow the contents to cool slightly, before moving anything. Care should also be taken when removing large medical flats or other glass vessels as the bottom may fracture, spilling their hot contents.
- Leave the autoclave in a clean and usable condition for the next user.
- Report any malfunction as soon as possible, especially if it concerns the safety valves or pressure control operations.

Centrifuge

Hazards

Main hazards in the use of centrifuges include the following:

1. Contact with the rotating head;
2. Failure of the rotor caused by overloading or corrosion;
3. Severe vibration caused by improper balancing;
4. The use of damaged glass/plastic centrifuge tubes.

Precautions

- The manufacturer's instructions must be carefully followed, especially with regard to balancing tolerances, assembly and operating speeds and loads.
- While in use, remember to:
    - Balance the load carefully, and ensure that buckets contain only the rubber cushions (and not spilled liquid or broken glass)
    - Close the lid before switching on, and do not open until the rotor has stopped
    - Ensure that the buckets are properly seated and swing freely, and also properly matched before use
    - Increase speed slowly, and stop the machine if abnormal symptoms are noticed
    - The centres of gravity of paired tubes must be equidistant from the axis of rotation, for example it is not correct to balance 10g water against 10g mercury
- Bench top centrifuges should be watched while in operation, as vibration can cause them to 'wander' on the bench. This can result in them falling over the edge onto the floor with serious damage and contamination
- Always choose the correct type of centrifuge tube for the work required.

Hand Tools

Only experienced personnel should use hand tools, whether powered or not. You should take the following precautions:

- Ensure that you know the correct, accepted and safe method of use.
- Always make sure that tools with handles, for example hammers, have them securely attached.
- Always check tools to make sure that they are in good order prior to use.
- Wearing eye protection is vital when using tools that produce fragments such as lathes, power saws, grinders, etc.
- Power tools must have undamaged cables and plugs and must be earthed where recommended. They should not be used in wet or moist conditions.
- Power tools should be regularly inspected for satisfactory working.
- Electric plugs should only be wired by qualified workers or staff trained in electrical safety.

- The guards on electric saws (bench and hand) must not be removed and must always be in place when the saws are in use.

Electrical equipment

As a general precaution: Familiarize yourself with the position of isolating switches for each laboratory to enable isolation of any faulty equipment or to cut off the power source in case of an emergency.

Precautions

- Do not trail cables along the floor, close to heat sources, beneath or around sinks, or along benches that may become wet or where corrosive chemicals are used.
- Check for damaged cables, that fuses in plugs are the correct rating, and that sheathed pin plugs are used. Any piece of equipment that generates heat is susceptible to premature breakdown of electrical insulation. Ensure earthing is adequate, and that plugs and wiring are all correctly fitted.
- Wherever possible, pull out plugs at night. Some equipment such as incubators, refrigerators, etc. must **not** be turned off; these are normally connected directly into a fused switch. Where such apparatus is plugged in there should be a label to indicate that they are to be left on and not unplugged.
- Do not overload sockets. Never use faulty equipment; always check before use and report any damage or fault. Portable electric equipment must be regularly checked for earth leakage, deterioration and damage. Portable electric equipment should be regularly tested, usually annually and carry a label when declared safe.
- Be vigilant and always carry out a brief visual check yourself, it is better to be safe than sorry.

Safety Regarding Chemicals

Chemicals should be stored in a manner that ensures their purity, the safety of laboratory personnel and the environment.

When handling chemicals, the following safety measures are required:

- Material Safety Data Sheets (MSDS) must be available for all chemicals and the information fully understood and recommended procedures implemented.
- Full risk assessments must be carried out on all laboratory chemicals; on receipt of chemicals that have never been used before in your laboratory the first action is to carry out a risk assessment using the MSDS recommendations and paying particular attention to the way in which it will be handled and used.
- When a particularly hazardous chemical is being used, every effort must be made to:
    - Replace it with a less hazardous material.
    - Reduce the number of staff that may come into contact with the chemical.
    - Plan the staffing, based on who may specifically be at risk i.e. sensitive or allergic, with every effort made to prevent contact with the chemical.
- All chemicals should have the laboratory receipt date written on the label, this enables a check to be maintained on 'shelf-life'.
- Laboratory reagents and chemicals should be placed on the appropriate storage shelves immediately after use with their labels to the front.
- Vessels containing unlabelled and unidentifiable liquids or solids must not be left to others for disposal.
- Appropriate spillage kits must be available to treat spills safely and facilitate safe disposal.

Safety Measures for Other Chemicals used in Microbiology

Flammable Liquids

Flammable liquids must be stored in a metal cabinet that has a containment floor; this allows spills to be contained and, in case of a fire, will prevent flammable liquids from leaking out and spreading the fire. Keep only small amounts, no more than 250ml, of such liquids on the bench for everyday use.

Chemicals with Noxious Fumes

Chemicals with noxious fumes require storage in a cabinet that is vented outside the building. Work with such chemicals should only be carried out in a fume hood; if a fume hood is not part of the laboratory equipment, consideration must be given as to whether the task should be performed at all.

Carcinogens

Some types of cancer may develop in people who have been repeatedly, but not necessarily continuously, exposed to small amounts of certain chemicals. There may be a long interval between the last exposure and the development of cancer.

Tar, pitch, some oils, arsenic and X-rays can all cause cancer of the skin. Certain aromatic amines may cause cancer of the bladder and urinary tract. Asbestos and nickel may cause cancer of the lung. Benzene can cause leukaemia or cancer of the blood cells.

A risk assessment should include details of:

1. The nature of the hazard and the nature and extent of exposure
2. Whether substitution with a less hazardous substance is reasonably practicable
3. The control measures to be applied to prevent or reduce exposure
4. Relevant operating and maintenance instructions and procedures to minimize exposure
5. Precautions under non-routine conditions, including emergencies
6. Use of personal protective equipment
7. Monitoring procedures
8. Health surveillance procedures
9. Arrangements for consultation, reporting defects in equipment, information provision and training

Prevention or control of exposure:

1. Substitution
2. Total enclosure or other effective controls such as Local Exhaust Ventilation (LEV)
3. Reduction of volumes stored and handled
4. Introduction of measures to minimize spread of contamination and use in confined areas
5. Keep the numbers of people liable to be exposed to a carcinogen to a minimum

Additional precautions:

1. All procedures involving carcinogens must be re-assessed regularly and include checking for changes in Occupational Exposure Limits (OEL) and Workplace Exposure Limits (WEL), Time Weighted average (TWA) and Short-Term Exposure Limit (STEL)
2. Accurate records should be made of the quantities of any known carcinogens used, by whom and in what procedure
3. Known carcinogens must be stored in closed containers which are clearly labelled, describing the contents and stating that they can cause cancer and stored in a locked cupboard
4. Carry out work in a fume cupboard but, if a filter is fitted, ensure it will absorb the chemical you wish to use. Every effort must be made to reduce the number of staff that can potentially be exposed
5. Working areas should be clearly defined, well-ventilated, have washing facilities, and the bench should ideally be covered with a suitable, disposable material
6. Gloves and suitable overalls must be worn. Rubber and plastic items should be washed immediately after use. Any contaminated items should be disposed of so as not to put others at risk
7. Take great care when using volatile substances and powders not to inhale fumes or dust. If possible, use powders in suspension or paste form. Avoid creating aerosols. Whenever possible work with them in a fume cupboard
8. Take care not to ingest these substances. Do not lick labels (or anything else around the laboratory). Do not pipette by mouth, or use mouth-operated wash bottles. Do not smoke, eat or drink, bite fingers or nails, or apply cosmetics in a laboratory
9. If the skin is contaminated, soap and cold water are usually the safest cleaning agents. Before you start work ensure you know what to do in case of spillage or contact with the chemical
10. Disposal and clearing of spillages must be carefully considered. The best method is by chemical destruction by a competent person. It must be checked that incineration of a carcinogenic substance really destroys it and does not merely release its vapour into the atmosphere. The MSDS will provide information on disposal, clearing after a spillage and destruction of waste
11. Lactophenol fumes may be carcinogenic, avoid heating slide preparations to the extent that fumes are produced. If fumes are accidentally formed ensure that they are not inhaled, use the safer alternative wherever possible i.e. lactic acid and cotton blue in water

Disinfectants

1. *Clear phenolics* - These compounds are effective against vegetative bacteria (including mycobacteria), fungi, and lipid-containing viruses (including Lassa and Marburg agents). They are inactive against most spores and non-lipid-containing viruses. Most phenolics are active in the presence of considerable amounts of protein but are inactivated to some extent by rubber, wood and plastics. They are not compatible with cationic detergents. Laboratory uses include discard jars and disinfection of surfaces. Clear phenolics should be used at the highest concentration recommended by the manufacturers for 'dirty situations', i.e. where they will encounter relatively large amounts of organic matter. This is usually 2-5%. Dilutions should be prepared daily and diluted clear phenolics may be effective for more than seven days. Skin and eyes should be protected.
2. *Hypochlorite* - The activity is due to chlorine, which is very effective against vegetative bacteria (including mycobacteria), spores of fungi and both lipid-containing and non-lipid containing viruses. Hypochlorites are considerably inactivated by protein and to some extent by natural non-protein material and plastics and they are not compatible with cationic detergents. Their uses include discard jars and surface disinfection but care is necessary as they corrode some metals. They should not be used on the metal parts of centrifuges and other machines which are subjected to stress when in use. For general purposes and for pipette and other discard jars a chlorine concentration of 2500 ppm is recommended. This is a 0.0025% solution. Sodium hypochlorite contains 10-14% available chlorine. Approximately 1ml of sodium hypochlorite or 3ml of domestic bleach in 4 litres of water will give you this concentration. However, for heavily contaminated areas (fungus growing on surfaces) a 4% solution of domestic bleach is recommended (4ml in 100ml of water). Hypochlorites decay rapidly in use, although the products as supplied are stable. Diluted solutions should be replaced after 24 hours. The colouring matter added to some commercial hypochlorites is intended to identify them; it is not an indicator of activity. Hypochlorites may cause irritation of skin, eyes and lungs.
3. *Alcohol and Alcohol Mixtures* - Ethanol and propanol, at concentrations of about 70-80% in water are effective, albeit slowly, against vegetative cells and lipid-containing viruses. They are not especially de-activated by protein and other material of detergents. Effectiveness is enhanced by the addition of formaldehyde, using a mixture of 10% formalin in 70% alcohol. It can also be enhanced by the addition of hypochlorite in 70% alcohol to give 2000 ppm available chlorine (add approximately 0.2ml of hypochlorite to 1 litre). Alcohols and alcohol mixtures are useful for disinfecting clean surfaces and, alcohol-hypochlorite mixtures excepted, for balancing centrifuge buckets. They are relatively harmless to skin but may cause eye irritation. Benches cannot be decontaminated by washing with industrial denatured spirits (IDA) as this is ineffective against fungal spores. It is therefore recommended that freshly prepared dilutions of hypochlorite (1ml in 4 litres of water) or domestic bleach (3ml in 4 litres of water) be used. IDA can be used for washing benches during mite control procedures.
4. *Aldehydes* – formaldehyde and glutaraldehyde have a broad spectrum of activity, including being active against bacterial endospores. The odour of these chemicals limits their wider use
5. *Amphoteric/ Biguanides/Quaternary ammonium* – a variety of these disinfectants are available as general laboratory disinfectants and cleaning agents, with activity against vegetative bacteria, yeast and some fungal spores, but not bacterial endospores. In-use concentrations are usually <5%, with many presenting a low level of chemical hazard
6. *Peroxides* – these have a wide spectrum of activity, including being sporicidal but they can also be corrosive, with special attention paid to PPE.

Cryogenic gases e.g. liquid nitrogen

Cryogenic gases present a risk of burns, or in the event of spillage or failure of equipment – asphyxiation. It is essential you have someone in a position to observe and take action if an accident occurs. The following information gives very good reasons why you should never work alone with liquid nitrogen.

1. Always handle the liquid carefully and avoid contact both with the liquid and any surface with which it has been in contact. At its extremely low temperature, it can produce an effect on the skin similar to a burn. When spilled on a surface it tends to cover it completely and intimately and therefore cools a large area. The gas issuing from this liquid is also extremely cold and can produce burns. Delicate tissues, such as those of the eyes, can be damaged by exposure to these cold gases which is normally too brief to affect the skin of the hands or face.

2. Stand clear of boiling and splashing liquid and its issuing / dissociating gas. Boiling and splashing always occur when charging a warm container or when inserting objects into the liquid. Always perform these operations *slowly* to minimize splashing. Whenever you handle the liquid, be sure there is a large open container of water nearby. Use the water to wash off any area of the body that is accidentally splashed with liquid. If a large area is exposed use an emergency shower.

- *NOTE: The cloudy vapour that appears when liquid nitrogen is exposed to the air is condensed moisture, not the gas itself. The issuing gas is invisible.*

Never allow any unprotected part of your body to touch non-insulated pipes of vessels containing liquid nitrogen; the extremely cold metal may stick fast and tear the flesh when you attempt to withdraw from it. Use tongs to withdraw objects immersed in liquid and handle the tongs and the object carefully. In addition to the hazard of burns or skin sticking to cold materials, objects that are soft and pliable at room temperatures usually become very hard and brittle at the temperature of liquid nitrogen and are very easily broken.

Wear Protective Clothing:

Safety goggles or a face shield should be worn to protect eyes. Insulated or leather gloves should be worn when handling anything that is, or may have recently been, in contact with liquid nitrogen. Gloves should fit loosely, enabling quick removal in case of splashes into them. Trousers without turn-ups should be worn outside footwear and footwear should cover the feet, open footwear such as sandals or canvas shoes should never be worn. Fastened laboratory coats must be worn at all times in a laboratory when liquid nitrogen is being handled.

Ventilation:

Maintain good ventilation in work and storage areas. Ensure the laboratory ventilation system is running when starting work in any room with liquid nitrogen. An oxygen monitor should be installed to warn personnel of increasing nitrogen levels in the room, in addition to personal oxygen monitors which must be made available to persons entering rooms where cryogenic gases are handled and stored.

When enough nitrogen gas evaporates from the liquid in an unventilated space, the percentage oxygen concentration in the air can become very low. Anyone working in such a place can become unconscious without sensing any warning symptoms, such as dizziness. If that person remains long enough in a low oxygen atmosphere, it can result in death. The oxygen content of the air must never be allowed to fall below 16%. Alarms should be set to trigger at 18% O_2.

Transient exposure to cold nitrogen gas may cause an asthma attack in susceptible people.

Liquid nitrogen should not be exposed to the air for long periods of time as oxygen will condense from the air and a potential liquid oxygen layer can result, which may accelerate any fire.

If a person seems to become groggy or loses consciousness while working with liquid nitrogen, get them to a well-ventilated area immediately. If breathing has stopped, start chest compressions (artificial ventilation), whenever a person loses consciousness, summon someone who knows first aid and a doctor immediately.

If any of the liquid comes in contact with the skin or eyes, immediately wash with cold water and then apply cold compresses. If the skin is blistered or if there is any chance that the eyes may be affected, get the patient immediately to a doctor for treatment. Burn areas larger than the palm of a hand will require hospital treatment.

Antibiotics and Fungicides added to Culture Media

Where antibiotics are used to kill unwanted bacteria from culture plates, they can be added to culture plates. In addition to the advice below, users should be aware of the risk of inadvertently producing resistant strains that could present a risk to human health.

- If required use fume hoods to avoid contaminating the laboratory with volatile chemicals

- Use protective equipment to avoid inhaling or ingesting potentially hazardous materials
- If a recipe instructs incorporation of ingredients in a specific order, do so to avoid unstable reactions occurring
- Label containers with dates of preparation for stains prepared in the laboratory; this enables a check to be maintained where limited 'shelf-life' is important
- Label containers with appropriate warning signs to remind users of potential hazards

Safety Equipment

It is important to follow basic safety precautions, best practices and common sense when engaged in any kind of laboratory work. A clean and properly maintained laboratory not only shows a professional attitude towards the work in progress but also consideration to the well-being of colleagues. Laboratory benches and other surfaces should be kept clean. It is preferable to clean up after each stage of an operation as overcrowding leads to accidents. Personnel and equipment are valuable and it is best to ensure that neither is put at risk.

Types of Safety Equipment

Fire Extinguishers

In the UK and European Union (EU), all fire extinguishers are red with a panel of no more than 10% to describe the type. A black panel is used for Carbon dioxide (CO_2), cream for foam and blue for powder. All red signifies a water extinguisher. The colour convention for fire extinguishers may vary in different countries Make sure you know what the colour signifies and what each extinguisher can be used for. Users should be aware that contents will be stored under pressure.

Carbon dioxide

This can be used on fires caused by inflammable liquids or gases and electrical faults. They are safe to use on all fires, but those involving paper, wood, textiles and fabrics may re-ignite. When using this type of extinguisher avoid contact with the dispensing valve and horn as these will become intensely cold and may cause cold burns.

Powder

These produce a cloud of fine powder and are safe to use on any fire.

Water

These produce a jet of water and can be used on fires involving paper, wood, textiles and fabrics. They must not be used on electric or inflammable liquid fires.

Foam

These produce a layer of foam that is used for inflammable liquid, paper, wood and textile fires.

MICROBIOLOGICAL SAFETY CABINETS. These cabinets are graded to the level of containment based upon an organism's risk group and whether there is a need to keep the sample safe. They are designed to prevent microbial contamination being deposited on the worker and have filters to remove microorganisms.

Class I microbiological safety cabinets will keep the operator safe but do not filter the incoming air, and so should not be used where the sample needs to be free of airborne contaminants

Class II or III microbiological safety cabinets can be used for sub-culturing sporulating microorganisms or when engaged in techniques which generate aerosols. No type of cabinet or hood should ever be used as a substitute for good aseptic practice.

Class III cabinets incorporate barrier protection between the operator and the sample.

Laminar flow or 'clean air' benches provide a gentle flow of filtered air into a partly enclosed chamber or hood and must only be used for tasks which will not create aerosols of microorganisms. These are not safety cabinets!

Examples of use: plate-pouring, tissue culture, subculture of non-sporulating microorganisms.

Personal Protective Equipment

Personal protective equipment is an essential part of safety and includes eye protection, gloves, laboratory coats, particle filtration and vapour protection masks etc.

Eye Protection *(Safety Spectacles)*

These are required when tasks may result in aerosol production, spraying (chemicals) or where materials can fracture i.e. handling glass.

Gloves

Disposable Surgical: These must be worn when handling chemicals or microorganisms. It is sometimes necessary to wear stronger rubber gloves for some specific procedures. Remove gloves before leaving the laboratory area, using telephones, touching door handles or any other areas, or equipment. Contaminated gloves should be disposed of safely in an autoclave bag. Be aware that some disposable gloves may cause skin allergies and skin will need to be protected by wearing thin cotton inner gloves under the surgical gloves.

Heat Resistant: These must be worn for unloading hot items from autoclaves and other laboratory areas when handling molten media.

Cold Resistant: These must be worn when handling liquid nitrogen vessels or any items that have been in contact with the liquid.

Solvent and Acid Resistant: These are specially coated to prevent penetration. For use when handling large quantities of solvents, acids or alkalis, also when measuring and weighing hazardous materials.

Laboratory Coats

Must be worn at all times in laboratories to protect against contamination or spillage on clothes or person.

Particle Filtration and Vapour Protection Masks

These must be worn when handling heavily infected materials and equipment either when transporting to a containment cabinet or when no containment facility is available. They must also be worn when aerosols containing microorganisms may be created in laboratory areas. Where the latter is anticipated the work must be carried out in a suitable safety cabinet. Masks must be worn when handling volatile chemicals.

Precautions When Wearing PPE

When carrying out tasks that require the use of protective equipment ensure that others in the vicinity are aware of the hazard and either vacate the area or wear similar PPE to ensure they are protected in the case of an accidental spillage or release of the hazard. In this respect, particular care must be taken when weighing toxic materials, this process may release fine powders into the air.

Precautions when Handling Microorganisms

Microbes may include viable but, as yet, not culturable organisms, cells and tissues. Some important points to note are:

- The dividing line between pathogens and non-pathogens is not completely clear-cut.
- There is a possibility of contamination of growth media by a potential pathogenic, allergenic or toxin producing organism and their multiplication to extremely large concentrations in culture.
- Cross-contamination of cultures might occur. The principles of sterile technique and handling of microorganisms must therefore be thoroughly mastered by all who have to handle microorganisms.
- Work with unknown human or animal pathogens should only be undertaken by fully trained and experienced personnel. Expert advice should be sought in all cases.

Always maintain a high standard of personal hygiene.

- Hands may be contaminated. Wash them before eating.
- Do not put on make-up, eat, drink or smoke in the laboratory.
- Wear a clean laboratory coat.
- Do not pipette by mouth.
- Seal cuts and scratches with a waterproof adhesive dressing before starting work.

Working with Aerosols

Work carefully to prevent or minimize dissemination of the organism. The most common entry route into the body for microorganisms is through the respiratory tract as a result of inhaling microbial aerosols (i.e. infected airborne particles). Aerosols are probably the main hazard in microbiological laboratories. If a procedure is likely to generate aerosols, carry it out in an appropriate microbiological safety cabinet. Aerosols can be generated by operations that cause splashing, frothing and spraying of liquids or by sudden disturbance of powders. Removal of cotton wool plugs, pipetting, centrifuging, shaking, aeration of cultures, etc. are some examples of operations that can give rise to aerosols.

STERILIZE OR DISINFECT. Sterilize or disinfect all used cultures and contaminated apparatus. Used pipettes, Pasteur pipettes, contaminated glassware, etc. should be immersed in disinfectant, or into a container for immediate sterilization. Spilt cultures should be immediately mopped up with suitable disinfectant solution and the materials used to mop up the spills must be disposed of correctly.

REPORT ALL ACCIDENTS. This is particularly important where injury (large or small) is involved and these must be entered in the accident book. If there is reason to believe that an infection or sickness originated from work with microorganisms, inform both your doctor and the work place management without delay.

OPERATIONS THAT CAN GIVE RISE TO AEROSOLS INCLUDE:

- Removal of screw caps when a pressure difference exists between contents and atmosphere
- Snap-on closures
- Removal of cotton wool plugs
- Pipetting, centrifuging, shaking, aeration of cultures
- Homogenization
- Ultrasonic disintegration
- Flaming and cooling loops
- Handling freeze-dried cultures and ampoules, particularly opening them

Safe Disposal of Waste

The person generating the waste has the duty of care to ensure that the waste is disposed of thoughtfully and carefully. A laboratory worker must ensure that waste is minimized and that no other person is put at risk by the waste or during its disposal. All waste materials must be disposed of in suitably labelled containers and safe storage and transportation to the point of disposal must be ensured. Particular care must be taken in the disposal of contaminated, infected and hazardous materials. Before disposal, any biologically infected material, including slides, must be sterilized or disinfected.

SEGREGATION OF WASTES. Safe disposal of waste demands segregation into categories to facilitate the appropriate disposal procedure. This must be done at the point of generation to minimize exposure.

DISPOSAL METHODS. Decontamination, disinfection and safe disposal of infected biological material can be carried out in a number of ways.

You can treat it chemically, by immersing it in a disinfectant for a period of time depending on the disinfectant and material. Chlorine in the form of hypochlorite solution (bleach 4% v/v in water) is generally used. Alternatives available are alcohol acidified to pH2 or halogenated methyl hydantoin. Disinfection reduces the contamination to a 'safe level' but does not always eliminate it totally.

Autoclaving is the use of moist heat and is the preferred method of sterilization of materials. Alternatives that are available are boiling in a 2% solution of sodium carbonate for 10 min, this will kill most resistant spores and the use of dry heat in an oven, for example at 180°C for 2 hours for glassware.

- *Biologically contaminated paper:* must be placed in an autoclave bag solely for this purpose. This is sterilized by autoclaving (121°C for 30min) before disposal off site.
- *Biologically contaminated plastics:* must be placed in an autoclave bag solely for this purpose, to no more than 2/3 full. These are sterilized by autoclaving (121°C for 30min) before disposal off site.
- *Biologically contaminated plastic pipettes and inoculating loops:* must be double bagged in autoclave bags to reduce the possibility of penetration by sharp objects. These are sterilized by autoclaving (121°C for 30min) before disposal off site.
- *Biologically contaminated microscope slides, Pasteur pipettes or sharps (e.g. scalpel and razor blades):* must be placed in a Sharpsafe bin (or similar container). These can be sterilized in the autoclave (121°C for 30min) before disposal off site or go into clinical waste bags for incineration off site.
- *Biologically contaminated glassware:* must be placed upright in rigid labelled containers for transportation to be sterilized by autoclaving (121°C for 30min) before re-use.
- *Biologically contaminated broken glass:* small amounts can be placed in a Sharpsafe bin

and sterilized in the autoclave (121°C for 30min). Larger quantities or large pieces must be disinfected with freshly prepared 10% hypochlorite for 24–48 hours and placed in bins suitable to store broken glass safely for removal to a recycling site.

Operations

Some recommended technical precautions that you can take while performing operations are presented in Table 5.4.

Maintaining a safe environment is an important aspect of working in a laboratory. Operators of microbiology laboratories or culture collections must follow the basic requirements needed to establish a safe workplace. This topic highlighted some general safety measures to be followed in the laboratory. It covered safety regarding equipment, organisms and operations.

Regulations Governing Distribution and Use of Cultures

The distribution of organisms (shipment and receipt) is controlled by numerous regulations. These include postal and shipping regulations, requirements for packaging aimed at protecting handlers and recipients of organisms and quarantine legislation to protect plant health. Some of these are discussed in this section enabling you to:

- Describe quarantine regulations and their function
- Describe the controls that various countries have in place for distribution of dangerous organisms
- Describe regulations that impact on microorganism distribution
- Describe the legislation or procedures BTWC countries have introduced
- List the factors on the basis of which risk assessment is conducted

The collection, isolation, handling, maintenance and distribution of microorganisms and cell lines are controlled by law at the national, regional and international levels. Microbiologists and in particular the providers of living biological resources, must be aware of such legislation and operate within it. This presents additional responsibilities for the collection worker and the consequences for not keeping up to date as legislation continues to develop and change can be catastrophic (Smith, 1996, 2000).

Organisms of hazard groups 2, 3 and 4 (see above Classification of Microorganisms on the Basis of Hazard for definitions of hazard groups) are hazardous substances under the UK Control of Substances Hazardous to Health (COSHH) legislation (COSHH, 2002). They fall under the EU Biological Agents directives and are dangerous goods as defined by the International Air Transport Association (IATA) Dangerous goods regulations (IATA, 2022), where requirements for their packaging are defined. In addition, there are restrictions on distribution imposed by national postal authorities and some prohibit the transport of Infectious, Perishable Biological Substances (IPBS) and, in some cases, Non-infectious Perishable Biological Substances (NPBS), including hazard group 1 organisms. The Universal Postal Union (UPU, 2019) publishes such information. Irrespective of whether organisms are shipped by mail, courier or by hand and whether between or within countries, thought must be given to the regulations that control these matters. The World Federation for Culture Collections (WFCC) attempts to keep abreast of the continuously evolving regulations and inform their membership through newsletters and reports of new and changing rules (Smith, 1996). The EC Directive on biological agents set mandatory control measures for laboratories requiring that risk assessments are carried out on all organisms handled (EC Directive 2000/54/EC). This necessitates the assignment of each strain to a hazard group following a thorough risk assessment including a positive inclusion into hazard group 1 when they are not categorized in hazard groups 2, 3 or 4. Copies of EC Directives are available from the Office for official publications of the European Communities, L-2985 Luxembourg. The risk assessment should include an assessment of all hazards involved, including the production of toxic metabolites and the ability to cause allergic reactions. Organisms that produce volatile toxins or aerosols of spores or cells present a greater risk. It is the responsibility of the microbiologist to provide such assessment data to a recipient of a culture to ensure its safe handling and containment.

Table 5.4. Recommended technical precautions while performing operations

	Recommended technical precautions		
Cause	Hazard	Risk	Precaution
Loops (sub-culturing)	Aerosols Spillage	Inhalation Contact	Use loops of optimum size; internal diameter 2-3mm, length of wire 50-60mm, loop fully closed. Use smooth movements. Cool loop before use.
Mounted needles	Self-inoculation Aerosols	Infection Inhalation	Cool needle after heat sterilization. Use gentle smooth movements.
Pipettes	Aerosols	Inhalation	Do not pipette by mouth, use a controller or bulb. Do not overload. Do not blow out last drop.
	Spillage Sharp or broken pipettes	Contact Injury/infection	Totally immerse in disinfectant after use. Check pipetting devices for damage regularly. Do not use broken or damaged pipettes
Preparing microscope slides	Aerosols - Chemical - Organism	Inhalation infection Contact poisoning	Handle carefully. Do not overheat when removing air bubbles.
Hypodermics	Aerosol Self-inoculation	Inhalation Infection	Do not use damaged needles. Disassemble reusable needles under disinfectant.
Centrifugation	Aerosols	Inhalation	Do not use damaged tubes. Do not overfill tubes. Ensure the tubes are balanced. Check for damage to rotors. Ensure rotor is firmly in place when using solvents, use chemical resistant tubes.
Blending	Aerosols	Inhalation	Ensure potential aerosols are contained if blending materials containing microorganisms is unavoidable.
Opening parcels	Aerosols Spillage	Inhalation Inoculation	Allow trained personnel to open parcels containing microorganisms in a laboratory. They should be opened in a safety cabinet if breakage is suspected or the organisms are unknown.
Opening cultures	Aerosols Condensates Exudates	Inhalation Contact	Open Universal bottles and heavily sporing cultures carefully; do not allow an inrush of air.
Opening ampoules or vials	Aerosol	Inhalation Contact	Crack open ampoules carefully; do not allow an inrush of air. Replace plugs with sterile cotton wool. Remove rubber bungs from vials with forceps, slowly.
Decontamination	Aerosols	Inhalation Contact	Use 4% bleach for all disinfecting. Industrial methylated spirits will only kill vegetative cells.
'Killing' and sterilization of cultures and culture vessels	Aerosols	Inhalation Contact	Place all materials for decontamination disposal into autoclave bags (not more than half full) or appropriate metal autoclave baskets. Do not mix disposables with re-usables. Transport to autoclave in robust sealed containers.

Health and safety precautions are not limited to the laboratory in which the organisms are handled, they extend to all those who may come in contact with substances and products from that laboratory. An organism in transit will potentially put carriers, postal staff, freight operators and recipients at risk, some organisms being relatively hazard free whilst others are quite dangerous. It is essential that safety regulations such as COSHH (2002), and shipping regulations be followed to ensure safe transit. Sound packaging and correct labelling and information must be used to minimize risk.

Regulatory Control of Microbiology

There are regulations in place covering all key actions in microbiology from collecting in the field or importing and exporting organisms, handling, manipulation and growth, genetic manipulation, deposit as part of a patent process, covering storage and their distribution, and some of this is presented in Table 5.5.

Quarantine Regulations

Quarantine regulations control access to and the distribution of particular organisms that present a disease threat to a country. They are designed to reduce the movement of certain pathogens. Work with such organisms requires permits or licences from the relevant authorities.

Lists of such strains are provided by country authorities. You must check these lists, if you wish to work with a particular microorganism in the laboratory. If you are aware it is likely to be a pathogen or a potential controlled organism, you must make enquiries first to get the relevant permissions before you get the strain. If the requestor is unable to produce a copy of the appropriate licence, the request for strains must be refused.

Non-Indigenous Plant Pathogens

Pathogens not present in a country are normally controlled by the department responsible for agriculture, for example DEFRA in the UK and USDA in the USA.

For instance, researchers in the UK who wish to obtain cultures of non-indigenous plant pathogens must first obtain a plant health licence. DEFRA provides a letter of authority to the researcher allowing importation of the named organism(s).

Non-Indigenous Tree Pathogens

Non-indigenous tree pathogens can only be supplied if the researcher holds a current permit issued by the Government department responsible for Forestry.

Animal Pathogens

Animal Pathogens are normally controlled by the country's government department responsible for agriculture.

For instance, the Specified Animal Pathogens Order (2008) makes it an offence to possess or spread a listed animal pathogen (e.g., Brucella) within Great Britain without a licence. It is supplemented by the importation of Animal Pathogens Order (1980), which makes it an offence to import any animal pathogen, or potential or actual carrier, of an animal pathogen from a non-EC country, except under licence.

The Convention on Biological Diversity (CBD)

The Convention on Biological Diversity (CBD), signed in Rio de Janeiro in 1992 and which came into force at the end of 1993 (CBD, 1992) has now 196 Parties (CBD web site - https://www.cbd.int/). The CBD provides sovereign rights over genetic resources to the country of origin and controls access to *in situ* organisms. In the simplest of terms, the CBD requires a biologist who wishes to collect genetic resources to seek prior informed consent from the relevant authorities and to negotiate fair and equitable sharing of benefits that may arise from their use before access can be granted. The Nagoya Protocol enacted in 2014 formalizes the benefit sharing as national governments party to it enact law (Smith *et al.*, 2018). The conditions agreed should include whether strains can be deposited and distributed from public service collections. As a result, the biological resource collection receiving the strain is obliged to inform recipients of cultures of their conditions of its use. The collection is therefore placed in a critical position to ensure there is a link maintained between the genetic resource and the user so that the CBD operates as it was intended. The Conferences of the Parties (COP) to the CBD continue to discuss access to genetic resources and the situation is

Table 5.5. Regulatory control of microbiology

Action	Requirement	Law, regulation, convention	Further information
Collecting in the field	Prior informed consent from a recognized authority	Convention on Biological Diversity (CBD)	https://www.cbd.int/
	Mutually agreed terms on use	Convention on Biological Diversity (CBD)	https://www.cbd.int/
	Consent from the land owner	Property law	
Import	Non-indigenous plant pathogens require licences from country authority	Quarantine regulations	
	Human, animal and plant pathogens can often only be imported to specified laboratories	Health and Safety	
Handling: manipulation; growth	Containment dependent on hazard	Control of Biological Agents - Health and Safety EC Directive 2000/54/EEC on Biological Agents	https://eur-lex.europa.eu/legal-content/EN/TXT/?uri=CELEX%3A02000L0054-20200624
Genetic manipulation	Containment of manipulated organisms	Belgian Service Biosafety and Biotechnology EEC Directives 2009/41/EC /. Contained use of genetically modified microorganisms (GMOs), EEC Directives 2001/18/EC. Release of GMOs, Cartagena Protocol on Biosafety	https://www.biosafety.be/ https://eur-lex.europa.eu/legal-content/EN/TXT/?uri=CELEX%3A02000L0054-20200624 https://eur-lex.europa.eu/legal-content/EN/TXT/?uri=CELEX%3A32009L0041 https://bch.cbd.int/protocol/
Deposit as part of a patent process	Long-term storage and compliance with the Budapest Treaty	Budapest Treaty on the International Recognition of the Deposit of Microorganisms for the Purposes of Patent Procedure	https://www.wipo.int/treaties/en/registration/budapest/
Storage	Appropriate containment	Health and Safety Licence to hold pathogens Security	

Export to another country	Some plant and animal pathogens require export licences	Quarantine regulations	
	Dangerous organisms with potential for dual use or subject to sanctions	Moving prohibited plants, plant pests, pathogens and soil	https://www.gov.uk/guidance/moving-prohibited-plants-plant-pests-pathogens-and-soil/
		Export controls: dual-use items, software and technology, goods for torture and radioactive sources. May include 'know how'	https://www.gov.uk/guidance/export-controls-dual-use-items-software-and-technology-goods-for-torture-and-radioactive-sources
Distribution	Packaging and transport considerations	IATA Dangerous Goods Regulations (DGR),	https://www.iata.org/en/publications/dgr/
		Universal Postal Union Convention (UPU)	https://www.upu.int/UPU/media/upu/files/UPU/aboutUpu/acts/manualsInThreeVolumes/actInThreeVolumesManualOfConventionMaj1En.pdf
		United Nations Sub-Committee of Experts on the Transport of Dangerous Goods (UNSCETDG)	https://unece.org/transport/dangerous-goods
	Sovereign rights over the strains	Convention on Biological Diversity	https://www.cbd.int/
	Access and benefit sharing	Nagoya protocol	https://absch.cbd.int/en/
		Bonn Guidelines	https://www.cbd.int/abs/infokit/revised/web/factsheet-bonn-en.pdf
	Intellectual Property Right ownership	Copyright	
	Customer licensed to receive organism?		
	Dangerous organisms export	EU Council Regulation 428/2009 on the Control of Exports of Dual-Use Goods from the Community	https://eur-lex.europa.eu/legal-content/EN/TXT/?uri=CELEX%3A02009R0428-20211007
			See national Export Offices

ever changing. The Convention and national legislation on access to genetic resources place an enormous duty on the shoulders of the collector, they are not intended to prevent the advancement of science. Current discussions are focussing on digital sequence information (DSI) and whether this should be treated in the same way as the genetic resource itself. This is a broad cross-cutting issue with many implications for research and development. Submissions of views and information from parties and stakeholders have been sent to the CBD Secretariat who have commissioned science-based, peer-reviewed studies to advise country discussions. These covered concept and scope, traceability, databases and domestic measures and were presented at the Ad Hoc Technical Expert Group (AHTEG) held on line in March 2020 - https://www.cbd.int/meetings/DSI-AHTEG-2020-01. Work in the AHTEG continues and outcome of discussions are expected in 2022 and will be posted by the CBD, see https://www.cbd.int/dsi-gr/.

The CBD aims to encourage the conservation and sustainable utilization of the genetic resources of the world and has a number of articles that affect biologists. These cover:

- Development of national strategies for the conservation and sustainable use of biological diversity
- Identification, sampling, maintenance of species and their habitats and the production of inventories of indigenous species
- Encouragement of *in situ* and in-country *ex situ* conservation programmes
- Adoption of economically and socially sound measures to encourage conservation and sustainable use of genetic resources
- Establishment of educational and training programmes and the encouragement of research
- Commitment to allow access to genetic resources for environmentally sound uses on mutually agreed terms and with prior informed consent
- Fair and equitable sharing of benefits and transfer of technology resulting from exploitation of genetic resources
- Exchange of information
- Promotion of technical and scientific co-operation

The CBD requires that Prior Informed Consent (PIC) be obtained in the country where organisms are to be collected before access is granted. Terms, on which any benefits will be shared, must be agreed in advance. The benefit sharing may include monetary elements but may also include information, technology transfer and training. If the organism is passed to a third party it must be under terms agreed by the country of origin. This is enforced by the Nagoya Protocol (https://www.cbd.int/abs/) under which each country party to it is enacting law to ensure benefits are shared. Compliance and traceability will entail the use of material transfer agreements between supplier and recipient to ensure benefit sharing. Many biological resource centres or culture collections have operated benefit sharing agreements since they began, giving organisms in exchange for deposits and re-supplying the depositor with the strain if a replacement is required. However, huge rewards that may accompany the discovery of a new drug are illusory as the hit rate is often reported as less than 1 chance in 250,000. In the meantime, access legislation and the hope for substantial financial returns from isolated strains are restricting the free deposit in public service collections and the legitimate free movement of strains. An EU DG XII project, Micro-organisms Sustainable Use and Access Regulation International Code of Conduct (MOSAICC) developed mechanisms to allow traceability and enable compliance with the spirit of the CBD and with national and international laws governing the distribution of microorganisms, whilst not restricting scientific goals (Davison *et al.*, 1998). The development of such common procedures is an evolutionary process and the co-ordinators of this project have placed the document on their website and will amend it as it develops (http://www.bccm.belspo.be).

There are many concerns that exist and these will take time to resolve. In the meantime, countries are developing legislation to control access to their genetic resources and biologists are struggling to comply. The Access and Benefit Sharing Clearing House (ABSCH - https://absch.cbd.int/en/) provides a platform for exchanging information and is a key tool for facilitating the implementation of the Nagoya Protocol.

The Convention should not affect the functions of public service collections but it increases the importance and extent of their role. The EU Regulation provides the opportunity for a mBRC to become a registered collection (Art. 5 of the EU ABS Regulation). This is a collection which has demonstrated the capacity to apply standardized procedures for exchanging samples of genetic resources (and related information), to supply genetic resources (and related information) to third persons with documentation providing evidence that the genetic resources have been accessed in accordance with applicable ABS legislation, to keep records of samples etc.

Statements are prominently displayed on accession forms and on information accompanying delivery of strains, explaining the implications of, and requesting compliance with, the convention. This draws attention to the requirements, but does not protect the sovereign rights of the country of origin nor any other stakeholder.

- A requirement for depositors to declare in writing that PIC has been obtained and that this includes unrestricted distribution of the materials to third parties or has clearly defined conditions on distribution.
- A requirement for a signed material transfer agreement on supply of material including mutually agreed terms.

These are minimum requirements and should be followed by all. Difficulty lies in defining the beneficiaries and what is a fair and equitable sharing of benefits. It is also difficult at this stage of implementation of the CBD for collections and depositors alike to comply, as in many countries they are still developing law and trying to find approaches that best suit their needs. In such cases, proof can be provided to demonstrate that a depositor has made reasonable efforts to get permission to collect from landowners and a Government Office.

Several organizations have addressed issues on IPR and the CBD and have developed and published their policies. These organizations include large national collections, international organizations and industrial companies, for example, CAB International (CABI), an intergovernmental organization established by treaty, dedicated to improving human welfare through the application of scientific knowledge in support of sustainable development world-wide, with emphasis on agriculture, forestry, human health, conservation of natural resources and with particular attention to the needs of developing countries (http://www.cabi.org). The CABI Genetic Resources Collection (GRC) is based at CABI Bioscience in Egham, UK and is a member collection of the UKBRCN. It is tasked with the collection of organisms to provide a resource for the scientific programmes of CABI and to underpin biotechnology, conservation and science in its member countries. CABI maintains extensive collections that originate from many different countries and has introduced ABS policy (https://www.cabi.org/wp-content/uploads/PDFs/AboutCABI/Cabi-Abs-Policy-Draft-For-Website-May2018.pdf) and introduced best practice (Smith *et al.*, 2018) to ensure compliance with the requirements of the Nagoya Protocol and the CBD.

The CABI policy offers an example of a mechanism to enable compliance with the CBD. CABI complies with national legislation of member country governments concerning rights over natural resources and access to genetic material and operates in a manner consistent with the CBD. It protects the interests of the source country of each element of biodiversity. The work of CABI adds value to the material held, particularly by ensuring authoritative identifications. It makes its reference collections and the information on them available to institutions in the countries of origin and the wider scientific community (CABI, 1996). CABI and the UKBRCN keep the rapidly changing situation under review and will adopt procedures required to maintain operations within the spirit of the convention.

Supply agreements are often put in place but new deposits are equally controlled. For example, before strains can be accepted into the CABI collections, confirmation is required from the depositor that the collector has made reasonable efforts to obtain PIC to collect the organisms and also has permission to deposit them in a public service culture collection. This confirmation forms part of the accession form that must accompany the deposit. Collections also need to know whether there is any restriction on further distribution and if there are conditions that must be included in any material transfer agreement that may accompany the samples when they are passed to a third person. Such information is required from all depositors regardless of the country of origin of the material or the collector.

Biological Resource Collections, like the UKBRCN public service collections, often add value to received and collected biological material. This is done through purification, expert preparation, authoritative identification, description, determination of biochemical and other characteristics, comparison with related material, safe and effective storage/preservation, evaluation of value for specified uses and indication of importance of beneficial and detrimental attributes. They often provide samples of deposited organisms free of charge to the depositor and participate in capacity building projects to help establish facilities and expertise in-country to maintain *ex situ* collections. This plays a role in the utilization of genetic resources and defines a collection as a stakeholder.

There are a number of outstanding issues with the implementation of the Nagoya Protocol that Governments are still negotiating in their Conference of the Parties. Not least is the treatment of Digital Sequence Information that would impact heavily on molecular based taxonomy and work to understand the unculturable majority of microorganisms. Microbiologists must keep abreast and help provide solutions (see the CBD web site https://www.cbd.int/convention/ for current status).

The CBD is not an opportunity for all to benefit financially and prospects of accruing huge profits from exploiting an organism for the country of origin are small. Additionally, the process from sampling to market can take from 8-15 years, therefore nothing will be achieved quickly and is likely to require considerable investment. The CBD was negotiated to protect genetic resources and thus ensure their sustainable use.

The agreement on Trade-related Aspects of Intellectual Property Rights (TRIPs) is thought to conflict with the CBD where there is concern that developing countries are required to allow companies to take out patents on products and processes of biotechnology. There are several forms of intellectual property rights that are relevant to the convention in addition to patents, for example copyright, trade secrets and plant breeder's rights. The CBD requires that terms for technologies subject to IPR protection should recognize and be consistent with adequate and effective protection of IPR (Glowka, 1998). In reality, so long as there is an agreement on mutually agreed terms for benefit sharing with the country of origin, the TRIPs agreement and patenting do not run contrary to the CBD.

The CBD website provides more information, https://www.cbd.int/

THE CBD NAGOYA PROTOCOL. The Nagoya Protocol on Access to Genetic Resources and the Fair and Equitable Sharing of Benefits Arising from their Utilization to the Convention on Biological Diversity is an international agreement which aims at sharing the benefits arising from the utilization of genetic resources in a fair and equitable way. It entered into force on 12 October 2014 and processes for compliance are still being introduced nationally. Although the law of each country is based on the negotiated protocol they differ quite remarkably from country to country and the process for compliance is often different and difficult to locate. The Access and Benefit Sharing Clearing House (ABSCH) is the tool to help do this but be aware it is not always up to date. The focus here is compliance with the EU Regulation No 511/2014 as implemented in the UK through *The Nagoya Protocol (Compliance) Regulations 2015* (UK Statutory Instrument, No. 821) but as of September 2021 there is a total of 68 countries with law, see the ABSCH (https://absch.cbd.int/en/).

The regulation applies to genetic resources accessed in countries that:

- exercise their sovereign rights and have established access measures
- have ratified the Nagoya Protocol

and where the genetic resources are:

- accessed after 12 October 2014
- not already governed by specialized international instruments, e.g. pandemic influenza preparedness (PIP) framework, International Treaty on Plant Genetic Resources for Food and Agriculture (ITPGRFA)

Activities Considered to be In-Scope of EU Legislation

- Research on a genetic resource leading to for example, the isolation of a biochemical

compound used as a new ingredient incorporated into a pharmaceutical product
- Creation or improvement of organisms for example, yeasts, resulting from human action through an R&D process, to be used in manufacturing processes
- Genetic modification – for example the creation of a genetically modified plant containing a gene from another species

Activities Considered to be Out of Scope of the EU Regulation

- Maintenance and management of a collection for conservation purposes
- Handling and storing of biological material and describing its phenotype
- Genetic resources (GR) as testing tools (GR is not the object of the research)

Overview of CABI Access and Benefit Sharing Policy: assessment on use of genetic resources

When considering the use of genetic resources (organisms) in your work you must ensure you comply with provider country requirements under the Nagoya Protocol. This is in addition to checking whether other permissions are needed for example regarding CITES, quarantine, export licences for dangerous organisms etc. and of course land owner/ protected area access.

The three elements of information required for the assessment are: **where** are you collecting the genetic (biological) materials from and if access is regulated (if receiving them from a partner they should be accompanied by a Material Transfer Agreement; (MTA) indicating **where** the material was isolated (the country of origin) and what you can and cannot do with it). **When** were they or will they be collected (isolated from *in situ*); and **what are you going to do with them** - some actions/ observations don't trigger the Nagoya Protocol – e.g. identification/taxonomy but this depends on the country.

Your first port of call should be the ABS Clearing House https://absch.cbd.int/en/countries: to find the source country of the materials in the list and discover if is it party to the Nagoya Protocol.

If not: check if it is party to the CBD and if it has ABS measures, if not there is no further action required other than ensuring other permissions e.g. plant health etc. have been obtained in order for you to access and use it.

If the provider country is party to the Nagoya Protocol you must find out what ABS measures are in place; these are normally listed on the country's profile; also determine when the measures were enacted. If the genetic material (organism) was collected before October 2014 then Nagoya doesn't apply (unless for example in the case of Brazil where it is the date when the material was acquired for the specific use being employed and benefit sharing is triggered by placing a product on the market, see https://portal.fiocruz.br/sites/portal.fiocruz.br/files/documentos/abs_implementation_brazil.pdf). If the collection was after that and the country controls access, you must obtain Prior Informed Consent (PIC) and negotiate Mutual Agreed Terms (MAT) before collection and use. Before you act, check if this has already been done for another project or there is an agreement in place that may be able to help speed up the process and facilitate negotiations.

What will you do with the genetic resource?

The Nagoya Protocol is triggered by utilization - i.e. conducting R&D on the genetic and/or biochemical composition of genetic resources, including through the application of biotechnology.

If the country does not control access (e.g. UK and much of EU) you may collect and use the materials as you wish as long as you are not breaking other laws in doing so (e.g. trespass, IPR, quarantine, shipping etc.). If it is not clear whether access legislation is in place, contact the country National Focal Point or Competent National Authority.

The EU Regulation provides compliance measures for EU users but does not address issues of access in Member States (MS) which is down to them individually. In the EU and the UK there are two checkpoints that may apply to the European user. The Regulator must be informed when:

1. you receive a research grant
2. you are about to place a product on the market.

See EU Guidance document on the scope of application and core obligations of Regulation (EU) No 511/2014 (https://eur-lex.europa.eu/legal-content/EN/TXT/?uri=CELEX%3A52016XC0

827%2801%29). Also see European Commission (2021) 8759 Annex 2 to the Commission Notice which provides further guidance on when genetic resources (falling in temporal, geographical and material scope of the Regulation) are utilized in the meaning of the EU ABS Regulation. Annex covers the value chain, starting from acquisition, via storing, management of collection, identification and characterization, and finishing with product development, product testing and placing of a product on a market (https://www.abs-biotrade.info/fileadmin/Downloads/3.%20TOPICS/ABS%20MECHANISM/4.%20GLOBAL%20PROCESSES%20AND%20NATIONAL%20IMPLEMENTATION/BELOW/5.%20European%20Commission%20-%20Guidance%20on%20ABS%20Compliance%20Measures/Annex_2-EC-Guidance-on-ABS-compliance-measures-2020.pdf).

Advice on ABS in the UK is given at this website: https://www.gov.uk/guidance/abs.

Ownership of Intellectual Property Rights (IPR)

Organisms originating from different habitats all over the world are deposited in collections. Prior to being accepted for deposit, the issue of ownership of intellectual property associated with them must be addressed. The CBD bestows sovereign rights over genetic resources to the country of origin, but intellectual property rights covering their use in processes is another matter. The CBD requires that the country of origin has a share in benefits accruing from such use, but there may be several other stakeholders. These may include the landowner where the organism was isolated, the collector, those involved in purification and growing the organism, the discoverer of the intellectual property, the collection owner where the organism was preserved and the developer of the process. It is clear that all stakeholders do not have an equal stake, this will depend upon the input of each one to the discovery or process. This has implications for the sharing of benefits arising from exploitation of the genetic resource. The collection has a role to play in the protection of IPR even if it is merely informing the recipient of any existing material transfer agreement or the citation of the strain in a patent. The implementation of the CBD is still being discussed by delegates from the countries signatory to it who meet at the Conference of the Parties and their workgroups. Information on the progress of these discussions can be found on the CBD website https://www.cbd.int/.

The general principles of international patent law require that details of an invention must be fully disclosed to the public. Inventions that utilize living organisms present problems of disclosure as a patented process often cannot be tested following the publication of a written description alone. Organisms are not patentable in their natural state or habitat; new species are discoveries not inventions, although genetically manipulated microorganisms are usually considered as a human invention and are therefore patentable. If a process involving an organism has novelty, inventiveness, utility or application and sufficient disclosure it can be subject to patent (Kelley and Smith, 1997). The invention of a product, a process of manufacture or a new use for a known product is an intellectual property owned by the inventor whether it involves an organism or not.

In many cases the organism involved must be part of the disclosure and many countries either recommend or require by law that a written disclosure of an invention involving the use of organisms be supplemented by the deposit of the organism into a recognized culture collection. Most patent lawyers recommend that the organism is deposited, regardless of it being a requirement, to avoid the possibility of the patent being rejected. To remove the need for deposit of organisms in a collection in every country where patent protection is desired, the 'Budapest Treaty on the International Recognition of the Deposit of Micro-organisms for the Purpose of Patent Procedure' was concluded in 1977 and came into force towards the end of 1980. This recognizes named culture collections as 'International Depository Authorities' (IDA) and a single deposit made in any one is accepted by every country party to the treaty. Any collection can become an IDA providing it has been formally nominated by a contracting state and meets certain criteria. There are 48 IDAs (https://www.wipo.int/budapest/en/idadb/) around the world and 7 in the UK which accept patent deposits of human and animal cell lines, algae, bacteria, cyanobacteria, fungi, nematodes, non-pathogenic protozoa, plant seeds and yeasts (Table 5.6).

Table 5.6. Collections designated as IDAs in the UKBRCN

Collection	Material
CABI Bioscience UK Centre (IMI)	Bacteria; fungi; nematodes; yeasts
Culture Collection of Algae and Protozoa	Algae and protozoa
European Collection of Cell Cultures	Animal and human cell cultures, hybridomas; animal viruses; eukaryotic DNA
National Collections of Industrial, Food and Marine Bacteria	Bacteria; phage; plasmids; plant seeds; yeasts
National Collection of Type Cultures	Bacteria
National Institute for Biological Standards and Control (NIBSC)	Animal cell cultures; human cell cultures
National Collection of Yeast Cultures	Yeasts

The Budapest Treaty provides an internationally uniform system of deposit and lays down the procedures which depositor and depository must follow, the duration of deposit and the mechanisms for the release of samples. Eighty-six states are now parties to the Budapest Treaty. The World Intellectual Property Organization (WIPO) publishes data on the numbers of microorganisms deposited in collections under the terms of the Budapest Treaty (1977). Since the treaty's inception, there were 43,711 deposits up to December 2000 and 85,491 deposits from 2001 to December 2020 (see https://www.wipo.int/budapest/en/). Patent protection is covered in Article 16 of the CBD under which parties must co-operate. However, this is subject to national legislation and international law, to ensure patents and other intellectual property rights are supportive of, and do not contravene, the objectives of the Convention (Fritze, 1994). This remains an area of dispute as the article leaves open the possibility that the CBD takes priority over national patent law. Patent law and the CBD are generally compatible but can conflict in cases where exploitation may endanger the resource. In many cases where organisms are grown artificially there is no threat to the existence of the species. Details of the requirements for a collection that relate to the deposit of an organism can be obtained directly from IDA collections and are summarized by Kelley and Smith (1997).

It is quite clear that every intermediary in an improvement or development process is entitled to a share of the IPR, which adds another dimension to ownership. Therefore, it is critical that clear procedures on access, mutually agreed terms on fair and equitable sharing of benefits and sound material transfer agreements are in place to protect the interested parties.

Control of Distribution of Dangerous Organisms

There is considerable concern over the transfer of selected infectious agents capable of causing substantial harm to human health, the environment or the economy. There is potential for such organisms to be passed to parties not equipped to handle them or to persons who may make illegitimate use of them. Pathogens and toxins at Biosafety Levels 3 or 4, but not limited to these, are of special concern. There is specific control legislation in place in different parts of the world, some of which are described below.

- The **Australia Group** (https://www.dfat.gov.au/) have strict controls for movement outside their group of countries but has lower restrictions within.
- The **USA** has rules that include a comprehensive list of infectious agents, registration of facilities that handle them and requirements for transfer, verification and disposal. Contravention of the rules entails criminal and civil penalties.
- In the **UK**, the Security Act controls access and distribution of such strains. All facilities handling organisms in hazard groups 2, 3 or 4 must be registered and there is strict control of hazard groups 3 and 4. The Export Control Joint Unit which is part of the Department for International Trade administers the UK's system of export controls and licensing for military and dual-use

items (https://www.gov.uk/government/organisations/export-control-organisation). It is required that certain infectious agents are exported to members of the Australia Group under an Open General Export Licence (OGEL) which is granted only to registered organizations. Exports of these agents outside the Australia Group require an Individual Export Licence (IEL). Failure to comply with these requirements is a criminal offence. Persons being supplied with these infectious agents should not avoid these regulations by providing subcultures to third parties. See Appendix D, this volume for a compliant process designed for culture collections of microorganisms.

- In **Germany**, permission to import, distribute, store and handle microorganisms allocated to risk group 2 and higher (pathogenic or 'hazardous' biological material able to multiply) are subject to restrictions laid down in the Infectious Diseases Protection Act (Infektionsschutzgesetz, IfSG) (https://perma.cc/B27B-36HD) with its amendments (https://www.loc.gov/item/global-legal-monitor/2020-11-24/germany-amendments-to-infectious-diseases-protection-act-enter-into-force/).
- Always check country controls in the country you are working or sending material to.

Legislation in Germany

In several countries, particularly Germany, a laboratory must be registered with the local health authority. Furthermore, the scientific leader of the responsible institution whether industry, hospital, university etc. or the head of the laboratory must have a personal permit issued by the local health authority. It is not sufficient for an institution to have registered laboratories, additionally there must be at least one authorized qualified person registered. If the person leaves the institution, a new authorized person must be registered. However, the person does not lose authorization (personal authorization is transferable to another institution). Furthermore, in Germany, handling of micro-organisms which are exclusively pathogenic to animals is subject to restrictions according to the Federal Infectious Diseases of Animals Enactment. The position is similar to that with human pathogens; the institution has to have registered laboratories and at least one authorized person. However, it is the district authority that is responsible for granting permission in this case. The district authority is also responsible for permits to laboratories working on genetically manipulated microorganisms.

A similar registration is necessary for handling Genetically Engineered Microorganisms (GEMs) allocated to safety level 1. The laboratory must be registered and there must be a deputy biological safety officer (authorized person as above) and a project leader who is responsible for the genetic engineering project. Additionally, each project involving GEMs must be registered separately with the district authority.

Visit the OECD Biosecurity site for further examples of legislation and practices.

Regulations on Culture Distribution

A laboratory distributing cultures to others (e.g. partners or customers) must follow the packaging and shipping regulations that apply. In particular, the International Air Transport Association (IATA) Dangerous Goods Regulations (DGR) lay down rules on the shipping of organisms by air (IATA, 2022). The IATA DGR (updated annually from 1st January) requires that shippers of hazard groups 2, 3 or 4 microorganisms are trained by IATA certified and approved instructors. Re-certification is required every two years. They also require shipper's declaration forms, which should accompany the package in duplicate and that specified labels are used for organisms in transit by air.

Some cultures represent a health hazard and for post and packaging purposes these are placed into four classes by the Universal Postal Union.

Postal Regulations and Packaging

The distribution of organisms is controlled by numerous regulations. These include postal and shipping regulations, requirements for packaging aimed at protecting handlers and recipients of organisms and quarantine legislation to protect plant health.

Postal Regulations

Most countries have their own regulations governing the packaging and transport of biological material through their domestic mail. International postal regulations regarding the postage of human and animal pathogens are very strict on account of the obvious safety hazard they present. There are several organizations that set regulations controlling the international transfer of such material:

- International Air Transport Association (IATA) https://www.iata.org/
- International Civil Aviation Organization (ICAO) https://www.icao.int/Pages/default.aspx
- United Nations Committee of Experts on the Transport of Dangerous Goods https://www.un.org/ecosoc/en/events/2020/committee-experts-transport-dangerous-goods-and-globally-harmonized-system
- Transportation of dangerous goods in Canada https://tc.canada.ca/en/dangerous-goods/transportation-dangerous-goods-canada?utm_campaign=tc-dangerous-goods-ongoing&utm_medium=vurl&utm_source=tc-gc-ca-tdg
- Universal Postal Union (UPU) https://www.upu.int/en/Home
- World Health Organization (WHO), *Guidance on regulations for the transport of infectious substances 2021-2022* https://www.who.int/publications/i/item/9789240019720

It is common to send microorganisms by post, as this is more convenient and much less expensive than airfreight. However, many countries prohibit the movement of biological substances through their postal services, and it is the responsibility of the sender to check this before a culture is dispatched. The International Bureau of the UPU in Berne publishes all import and export restrictions for biological materials by national postal services. Some countries will not accept human pathogens through the post for carriage overseas and this now includes the UK.

PACKAGING. IATA DGR (IATA, 2022) categorizes infectious substances as either A or B (class 6.2)

Category A infectious substances cover organisms that are commonly assigned to Hazard groups 3 and 4 and require packaging that meets defined standards according to IATA packing instruction 620 (IATA, 2022).

Category B infectious substances cover organisms that are commonly assigned to Hazard group 2 and require packaging that meets defined standards according to IATA packing instruction 650 (IATA, 2022).

Although IATA DGR does not regulate non-category B samples, it would be considered good practice to use similar packaging, which have triple containment (primary, secondary and outer packaging) for hazard group 1 organisms (Anon, 2022).

However, microorganisms that qualify as dangerous goods (class 6.2) and are sent by air must be in UN certified packages. These packages must be sent by airfreight or courier if the postal services of the countries, through which it passes, do not allow the organisms in their postal systems. They can only be sent via airmail if the national postal authorities accept them. There are additional costs above the freight charges and package costs if the carrier does not have its own fleet. This is because the package and documentation will need to be checked at the airport DGR centre.

There are currently very few private carriers that transport dangerous goods internationally. These private carriers do provide assistance in completing the shipper's declaration forms. The shipper is exclusively responsible for the shipment, its correct packaging, documentation, marking and labelling.

THE UNIVERSAL POSTAL UNION. At the invitation of the Swiss government, the first plenipotentiary conference was held in 1874 in Berne, resulting in the signing of the 1874 Treaty of Berne establishing the first collective Convention governing international postal services. The central office (International UPU Bureau) is still located in Berne. There are two kinds of UPU regulations, the Convention itself and the Detailed Regulations. The Convention has a rather broad wording (relevant here: Article 16 on acceptable dangerous goods in international mail) whereas the Detailed Regulations describe for example packaging requirements (relevant here: Articles RL129/RL130). UPU is a specialized agency of the UN and one of the essential features is that it is universal (global), serving the public world-wide. UPU cooperates with several international organizations for example IATA (International Air Transport Association), WCO (World Customs Organization), ISO (International Standard Organization). The UPU Congress Supreme Authority meets regularly, not later than five years after the previous congress. Currently, UPU has 191-member countries. The official postal language is French. The UPU regulations resemble international law enforcement. You can find more information on their website: http://www.upu.int/ (also see UPU, 2018, 2019).

Background

The 'Letter Post Regulations' are based on the UPU 2004 Bucharest Convention, published by the UPU International Bureau, Berne, 2006. UPU is the Universal Postal Union. UPU constantly updates information as submitted by national postal services of the member States. The new Compendium contains detailed lists divided into three categories 'Basic Services', 'Optional Supplementary Services' and 'Other Information'. Admittance of biological materials is published under 'Optional Supplementary Services'; the national postal administrations still have different regulations in place with respect to transport of biological materials. The information given here has been drawn together mainly on the basis of the updated UPU (2018) Convention.

Responsibilities of the Shipper of Biological Materials

Before offering biological substances for transport by any means, shippers must familiarize themselves with export restrictions (dual-use or customs legislation), import permits, quarantine regulations and other possible restrictions according to relevant national and international bio-legislation. Postal transport is usually quick, reliable and cheap. This possibility should not be misused. Breakage of packages has created fear among postal employees and therefore it is necessary to have strict rules in place for safe and efficient transport. For transport of biological materials, it is advisable to choose a mode of transport allowing the shipment to be tracked. The need for tracking the location of a shipment and documentation on route is very important in deciding the service needed when transporting biological materials. The service offered by private carriers is without a doubt an advantage as a registered postal letter only guarantees tracking to a lesser extent. It is important to note that domestic postal mail transport is regulated by the respective national authorities and this is not discussed in this information resource.

Postal Transport of Genetically Engineered Microorganisms (GEMs)

The UPU Compendium (UPU, 2019) has not foreseen or mentioned transport of GEMs; therefore, national postal regulations must be observed. It is most likely that many national postal administrations don't accept GEMs for transport. Trans-border shipments of GEMs should be performed exclusively according to the ICAO/IATA DGR regulations. GEMs might fall under UN2814, UN2900, UN3373 or UN3245, respectively (IATA, 2022) or, they might not be regulated as dangerous goods and can be shipped like Risk Group 1 organisms if the definition applies. There are numerous GEMs of safety level 1 that are not able to alter any other living (micro) organisms and therefore they are not regulated as dangerous goods, this is the majority of e.g. *E. coli* K-12, *Bacillus subtilis* etc. derivatives carrying vectors or similar genetically engineered plasmids. Be aware that all other GEMs fall under regulations for dangerous goods transport (above UN numbers). See WFCC Information Resource on international regulations for transport of GEMs.

Customs Requirements

International shipments containing biological substances are usually subject to customs clearance. Occasionally, samples held by customs may be returned at the expense of the sender or disposed of by national authorities. They usually require the green customs declaration label for postal mail. It should be affixed close to the address label. The type and number of material/receptacles and the value should be entered on the label (e.g., 4 bacterial cultures). Declare 'No commercial value' only in case of free exchange. The wording 'harmless, non-infectious, for research exchange purposes' can speed up customs clearance.

Packaging and Labels

For transporting biological substances by postal mail services, packaging requirements must be carefully observed because minimum safety is given only if the **triple containment principle** is taken into consideration. This principle is internationally agreed and comprises:

1. A leak-proof primary receptacle bearing the biological material
2. A leak-proof secondary packaging (usually screw-capped plastic tubes); if liquids are sent: sufficient absorbent material able to take the whole liquid is inserted between the primary and secondary containment. The secondary packaging must retain its physical integrity during transport.
3. An outer packaging/carton box of sufficient strength and size able to withstand usual mechanical stress during transport and carrying all required labels: sender's and recipient's addresses, green customs label if applicable and biological substances label.

Ready-to-use packaging systems for UN 3373: usually cheap in contrast to the slightly more expensive UN combination packaging for shipping infectious substances, UN2814 or UN2900; all these packaging systems are offered by many suppliers. Such packaging systems are tested by national authorities, fulfilling Packing Instruction PI650 requirements and are permitted for the following range of items:

- Biological substances that are non-infectious (Risk Group 1)
- Infectious substances Category B, UN3373; these are organisms that do not meet the criteria for inclusion in Category A (most Risk Group 2 organisms can be transported under Category B, UN3373)
- Diagnostic specimens if not containing Category A materials. Please note Category A includes organisms capable of causing permanent disability, life-threatening or fatal disease to humans or animals.

Consequently, PI650 is a versatile packaging system and the internationally agreed minimum packaging standard for biological materials bearing no or a low hazard. The preferred information source for detailed packaging and postal transport regulations are the national postal authorities' homepages, they also present illustrations on packaging. Any self-made packaging systems should NOT be used for transport. PI650 packaging is usually quite flat in one dimension (30 mm) and can be sent as 'maxi letters' in several countries (having nothing in common with usual "letters"!).

In contrast, PI620 packaging for Category A shipments (highly pathogenic materials, UN2814, UN2900) must have rigid outer boxes with a minimum external dimension not less than 100 mm. Postal services have to refuse transport of such highly pathogenic materials as IATA prohibits such transport via air mail, see below. Transport as (air) freight is performed by courier services.

Dangerous Goods in Air Mail, IATA DGR (IATA, 2022):

2.4.1 The Universal Postal Union Convention forbids the carriage of dangerous goods in the mail except as permitted in 2.4.2. National postal authorities should ensure that the provisions of the UPU Convention are complied with in relation to the transport of dangerous goods by air.

2.4.2 The dangerous goods listed in this subsection may be accepted in mail for air carriage subject to the provisions of the national postal authorities concerned and the parts of these Regulations which relate to such materials...
(b) Infectious substances, assigned to Category B (UN3373) only, when packed in accordance with the requirements of Packing Instruction 650

Concluding note: **All infectious substances**, including substances allocated to Risk Group 2, are by definition **dangerous goods**, Class 6, Division 6.2, UN classification system. Most Risk Group 2 substances can be classified as Category B, UN3373 and can be transported under deregulated conditions: A Shipper's Declaration form for dangerous goods is not required and Packing Instruction PI650 applies. However, UN3373 substances might be shipped in PI620 combination packaging having much higher quality and strength making transport even safer, especially over long distances (trans-border, trans-continental). Shippers are free to choose a PI602 packaging and a courier for UN3373 transports within their country as an alternative to using PI650 packaging and postal services if possible.

For all transport, the correct applicable labels are vital.

Whenever infectious substances are offered for shipment by air, the ICAO/IATA DGR requirements must be followed. A trained shipper has to sign the Shipper's Declaration form if required (it is not required for UN3373) and a certified UN combination packaging has to be used (see PI650 or PI620, respectively). Furthermore, regulations for road transport (EC Council Directive 2008/68/EC) (**ADR** in Europe), **RID** (railway transport) and **IMO** (International Maritime Organization) are in place.

What does 'perishable biological substance' mean?

(Freeze-) dried biological substances are not considered 'perishable' in the sense of quick postal transportation. With regard to safety aspects, (freeze-) dried non-infectious biological materials should be packed in the same way as 'perishable' non-infectious substances in order to provide adequate mechanical protection during transport. The only difference: material for absorbing liquids is not required between primary and secondary receptacles. However, it is still advisable to send freeze-dried material in approved packaging. The UPU provides information on country requirements provided to them via surveys that include the organization of domestic postal markets and the effect of national or regional policies on the UPU and its Acts and on other agreements reached at UPU level (https://www.upu.int/en/Members-Centre/Policies-Regulation/Extra-Territorial-Offices-of-Exchange-(ETOEs).

If cryovials are dispatched with dry ice then UN 1845 would apply with packaging instruction 954, **in addition** to the relevant UN number of the infectious substance.

Microorganisms and Biosecurity

Many laboratories, particularly Biological Research Centres (BRCs), are entrusted with the maintenance and exchange of hazardous biological resources. With the threat of bio-terrorism the geopolitical landscape has changed. Thus, facilities that handle biological materials and information that could potentially be of 'dual-use' have the added responsibility to make special efforts to secure against loss or theft. Security arrangements should always follow national and local guidelines. Site managers must be aware of the requirements to secure culture collections. It is therefore advisable to keep cultures in a controlled access location and to ensure records are kept as to who is visiting the site. CCTV may also be required.

Biological Weapons Convention

The Convention on the Prohibition of the Development, Production and Stockpiling of Bacteriological (Biological) and Toxin Weapons and on their Destruction (usually referred to as the Biological Weapons Convention, BWC, or Biological and Toxin Weapons Convention, BTWC) was the first multilateral disarmament treaty banning the production of an entire category of weapons. It was the result of prolonged efforts by the international community to establish a new instrument that would supplement the 1925 Geneva Protocol.

The BWC was signed in London, Moscow and Washington on 10 April 1972 and came into force on 26 March 1975. There are currently 162 country signatories of which 18 are still to ratify it. Following the signing of the BWC, countries have introduced new control legislation or procedures to prevent unauthorized access to strains that could be misused in this way.

Details about the Convention on the prohibition of the development, production and stockpiling of bacteriological (biological) and toxin weapons and on their destruction can be found at https://www.un.org/disarmament/biological-weapons/

Export Licensing Measures

Article III of the BWC compels States Parties to not transfer to any recipient nor assist, encourage, or induce anyone to manufacture or otherwise acquire any of the agents, toxins, weapons, equipment or means of delivery specified in article I of the Convention. This is a legally binding obligation. Many countries have employed national export licensing measures as an effective means of implementing these obligations. This assists in the attempt to avoid the possibility of the inadvertent supply of any item which could be used in a BW programme.

Export licences operate to deter proliferation by monitoring trade of relevant materials, and provide authority to stop a sale in the infrequent cases where a prospective export is likely to contribute to a BW programme. It is also in the interest of industry and research institutes to ensure that such firms and institutes are not inadvertently supplying pathogens and dual-use equipment for use in the production of BWs.

Best Practice Guidelines on Biosecurity for BRCs

The Best Practice Guidelines on Biosecurity for BRCs are designed to secure all types of biological materials (e.g. plant, animal, microorganism and human-derived) in proportion to the risk they present. Obstacles that BRCs might face in carrying out their function to provide quality assurance and rapid supply of biological materials for research, public health and economic development are minimized. To address this fundamental point, some countries have adopted a list-based approach to biosecurity, whereby a list of biological materials that are deemed in need of special security measures is established. There is a clear and absolute distinction between those listed organisms that require specified additional security measures. Those not on the list, do not require additional security above and beyond what is normally in place at facilities.

The aim of Best Practice Guidelines on Biosecurity for BRCs (OECD, 2007) is to reduce the probability that dangerous biological material could be obtained by unauthorized persons and deployed to cause harm, without unduly hindering research or being financially burdensome (https://www.oecd.org/health/biotech/38778261.pdf). A code of practice for BRCs has been published to assist them and microbiologists in general with compliance measures (Rohde *et al.*, 2013).

Given the risk of mistakenly excluding a hazardous biological material from such a list, the Best Practice Guidelines on Biosecurity for BRCs adopt an approach consisting of two key components:

(a) Performing a risk assessment of the various biological materials held in collections, and
(b) Recommending risk management practices to reduce the risk of their loss or theft.

The Geneva Protocol was established for the prohibition of the use of biological organisms in war for asphyxiating, poisonous or other gases and of bacteriological methods of warfare.

The European Union has adopted a common position with the Biological and Toxic Weapons Convention. Delivery of microorganisms, which could be used as biological weapons, is controlled by the EU Council Regulation 428/2009 https://eur-lex.europa.eu/legal-content/EN/TXT/?uri=CELEX%3A32009R0428 on the Control of Exports of Dual-use Goods from the Community. More information can be found at https://www.virtualbiosecuritycenter.org/codes-of-ethics/.

France has introduced regulations on highly pathogenic strains. Further information, in French, can be found on the Agence nationale de sécurité du médicament et des produits de santé website: https://solidarites-sante.gouv.fr/ministere/acteurs/agences-et-operateurs/article/ansm-agence-nationale-de-securite-du-medicament-et-des-produits-de-sante.

The US Biological Safety Requirements for Facilities Transferring or Receiving Select Agents was formed in response to a Congressional mandate in 1997, CDC promulgated a regulation, Additional Requirements for Facilities Transferring or Receiving Select Agents. The select agents were drawn from the Australia List and include forty or so microorganisms, a dozen toxins, and certain recombinant DNA molecules.

Institutions were provided with registration packages that included a self-assessing laboratory survey form based on the CDC/NIH publication, Biosafety in Microbiological and Biomedical Laboratories. Each registered facility is to be inspected by personnel from the OHS during a three-year cycle. There are 67 laboratories registered to date. The CDC is currently working to develop a plan to address the role of Public Health to meet the growing national concerns about bioterrorism, etiological agents, hosts and vectors.

Under the Foreign Quarantine guidelines, laid down by the Department of Health and Human Services:

(a) A person may not import into the United States, nor distribute after importation, any etiological agent or any arthropod or other animal host or vector of human disease, or any exotic living arthropod or other animal capable of being a host or vector of human disease unless accompanied by a permit issued by the Director.
(b) Any import coming within the provisions of this section will not be released from custody prior to receipt by the District Director of the U.S. Customs Service of a permit issued by the Director.

Biosecurity Risk Assessment

All laboratories must conduct a risk assessment on the biological materials they handle and hold for the purpose of assigning such materials to biosecurity risk levels. These levels may be assigned as high, moderate, low or negligible. It is important for the level of biosecurity risk to be determined for the biological material. This should be based on the best available information of its potential for intentional misuse (including economic consequences) as well as its virulence.

There is much more information about risk assessment for biological materials available on the WHO website - https://www.who.int/emergencies/risk-assessments#:~:text=WHO%20detects%20public%20health%20events,2018%20to%20100%25%20in%202019. Also see OECD Biosecurity best practice guidelines for BRCs (https://www.oecd.org/health/biotech/38778261.pdf).

Assessment of Virulence and the Potential for Intentional Misuse

Potential for Intentional Misuse

The assessment of the potential for intentional misuse should be based on the following key factors:

Availability: The number of facilities that stock the biological material and their geographical distribution

Amplification: The ease with which the biological material can be replicated, for example, whether it can be grown in culture and its growth rate

Skills and Knowledge: The ease or difficulty of the skills and knowledge necessary to increase and/or genetically modify the biological material

Dispersal: The ease and effectiveness with which the biological material can be dispersed, such as by air, water, food or by other means into the environment. This might include (but may not be limited to) the nature of aerosol production and the inhalation characteristics of the biological material

Environmental Viability: The hardiness of the biological material across a range of temperatures, humidity levels, light exposures

Countermeasures: The existence of and ease of access to prophylaxis, post-exposure treatments and detection and decontamination measures

Economic Consequence: The extent to which the biological material may be used to bring about harmful economic consequences for humans, crops, livestock or infrastructure

Virulence

The assessment of virulence should be based on the following key factors:

Infective Dose: The smallest quantity of the biological material necessary to cause infection

Pathogenicity: The disease-causing ability of the biological material

Lethality: The ability of the biological material to cause death of the host

Transmissibility: The ease with which the biological material can spread either by vector to host, or host to host.

References

ACDP (2018) *Management and Operation of Microbiological Containment Laboratories*. Advisory Committee on Dangerous Pathogens (ACDP), Health and Safety Executive. https://www.hse.gov.uk/biosafety/management-containment-labs.pdf

ACDP (2021) *The Approved List of Biological Agents*. Advisory Committee on Dangerous Pathogens, Health and Safety Executive, Bootle, Merseyside, UK. https://www.hse.gov.uk/pubns/misc208.pdf

Anon (1997a) Safe Biotechnology, Classification of Biological Agents: Bacteria. Trade association of the chemical industry. Leaflet B 006e, 2/97, ZH 1/346. 69021. Jerdermann-Verlag, Heidelberg, Germany.

Anon (1997b) Safe Biotechnology, Classification of Biological Agents: Fungi. Trade association of the chemical industry. Leaflet B 007e, 2/97, ZH 1/346. 69021. Jerdermann-Verlag, Heidelberg, Germany.

Anon (1998) Safe Biotechnology, Classification of Biological Agents: Bacteria. *Leaflet B006*. BG Chemie, Germany.

Anon (2022) Guidance Note Number: 17/2012[REV.7] Transport of infectious substances UN2814, UN2900 AND UN3373. Department for Transport, UK. https://assets.publishing.service.gov.uk/government/uploads/system/uploads/attachment_data/file/917954/dangerous-goods-guidance-note-17-document.pdf

CABI (1996) *CAB International 13th Review Conference Proceedings*. CAB International, Wallingford, UK.

Castegnaro, M., Chernozemsky, I.N., Hietanen, E. and Bartsch, H. (1990) Are mycotoxins risk factors for endemic nephropathy and associated urothelial cancers? *Archive for Tumor Research* 60, 295-303. https://pubmed.ncbi.nlm.nih.gov/2202269/

CBD (1992) The Convention on Biological Diversity 31 I.L.M. 822. https://www.cbd.int/convention/text/

COSHH (2002) *The Control of Substances Hazardous to Health Regulations 2002*. UK Statutory Instruments 2002 No. 2677 Regulation Legislation, UK Government. https://www.legislation.gov.uk/uksi/2002/2677/regulation/7/made

Davison, A., Brabandere, J. de. and Smith, D. (1998) Microbes, collections and the MOSAICC approach. *Microbiology Australia* 19, 36-37.

de Hoog, G.S. (1996) Risk assessment of fungi reported from humans and animals. *Mycoses* 39, 407-417. https://doi.org/10.1111/j.1439-0507.1996.tb00089.x

European Commission (2000) EC Directive 2000/54/EC of the European Parliament and of the Council of 18 September 2000 on the protection of workers from risks related exposure to biological agents at work (seventh individual directive within the meaning of Article 16(1) of Directive 89/391/EEC). https://eur-lex.europa.eu/legal-content/EN/TXT/PDF/?uri=CELEX:32000L0054&from=EN

European Commission (2021) Guidance document on the scope of application and core obligations of Regulation (EU) No 511/2014 of the European Parliament and of the Council on the compliance measures for users from the Nagoya Protocol on Access to Genetic Resources and the Fair and Equitable Sharing of Benefits Arising from their Utilisation in the Union 2021/C 13/01 (OJ C, C/13, 12.01.2021, p. 1, CELEX: https://eur-lex.europa.eu/legal-content/EN/TXT/?uri=CELEX:52021XC0112(02))

Fritze, D. (1994) Patent aspects of the Convention at the microbial level. In: *The Biodiversity of Micro-organisms and the Role of Microbial Resource Centres*. pp 37-43. World Federation for Culture Collections (WFCC), Germany.

Glowka, L. (1998) *A Guide to Designing Legal Frameworks to Determine Access to Genetic Resources*. pp 98. IUCN, Gland, Switzerland and Cambridge, UK. https://portals.iucn.org/library/sites/library/files/documents/EPLP-034.pdf

IATA (2022) *Dangerous Goods Regulations*. 2022 edn. International Air Transport Association, Montreal; Geneva. https://store.iata.org/IEC_ProductDetails?id=9065-63

Kelley, J. and Smith, D. (1997) Depositing Micro-organisms as part of the Patenting Process. In: *European BioPharmaceutical Review* 2, 52-56. Ballantyne Ross Ltd., London.

Krishnamachuri, K.A., Bhat, R.V., Nagarajan, V. and Tilak, T.B.G. (1975) Hepatitis due to aflatoxicosis. *Lancet* 1, 1061-1063. https://doi.org/10.1016/S0140-6736(75)91829-2

Macreadie, I. (ed.) (2019) Sustainable use and preservation of biological resources. *Microbiology Australia* 40(3) https://www.publish.csiro.au/MA/issue/9725

OECD (2007) Best Practice Guidelines for Biological Resource Centres. OECD, Paris. http://www.oecd.org/document/36/0,3343,en_2649_34537_38777060_1_1_1_1,00.html

Paterson, R.R. (2006) Fungi and fungal toxins as weapons. *Mycological Research* 110, 1003–1010. https://doi.org/10.1016/j.mycres.2006.04.004

Rohde, C., Smith, D., Martin, D., Fritze, D. and Stalpers, J. (2013) Code of Conduct on biosecurity for Biological Resource Centres: procedural implementation. *International Journal of Systematic and Evolutionary Microbiology* 63, 2374-2382. https://www.microbiologyresearch.org/content/journal/ijsem/10.1099/ijs.0.051961-0

Smith, D. (ed.) (1996) *Committee on Postal, Quarantine and Safety Regulations Report 1996, Postal, quarantine and safety regulations: status and concerns*. pp39. World Federation for Culture Collections (WFCC), Braunschweig, Germany.

Smith. D. (2000) Legislation affecting the collection, use and safe handling of entomopathogenic microbes and nematodes. In: Navon, A. and Ascher, K.R.S. (eds) *Bioassays of Entomopathogenic Microbes and Nematodes*. CAB International, Wallingford, UK. p295-313. https://www.cabi.org/cabebooks/ebook/20001108781

Smith, D. and Onions, A.H.S. (1994) *The Preservation and Maintenance of Living Fungi*. Second edition. IMI Technical Handbooks No. 2, pp 122. CAB International, Wallingford, UK. https://www.cabi.org/bookshop/book/9780851989020/

Smith, D. and Ryan, M.J. (2019) International postal, quarantine and safety regulations. In: Sustainable Use and Preservation of Microbial Resources. *Microbiology Australia* special issue 40 (3). 10.1071/MA19032 http://microbiology.publish.csiro.au/?paper=MA19032

Smith, D., Rohde, C. and Holmes, B. (1999) Handling and distribution of micro-organisms and the law. *Microbiology Today* 26, 14-16. https://socgenmicrobiol.org.uk/pubs/micro_today/pdf/14_16.pdf

Smith, D., Hinz, H., Mulema, J., Weyl, P. and Ryan, M.J. (2018) Biological control and the Nagoya Protocol on access and benefit sharing – a case of effective due diligence. *Biocontrol Science and Technology* 28, 914-926. DOI: 10.1080/09583157.2018.1460317

Smith, J.E. and Moss, M.O. (1985) *Mycotoxins: Formation, Analysis and Significance*. John Wiley, Chichester, New York. https://doi.org/10.1016/0041-0101(86)90038-3

Stanier, R.Y., Ingraham, J.L., Wheelis, M.L. and Painter, P.R. (eds) (1987) *General Microbiology*, 5th edn pp. 621. Macmillan Education Ltd., Houndmills, Basingstoke, Hampshire, UK. https://www.scribd.com/document/435145671/Stanier-Ingraham-5th-Edition

UPU (2018) *Constitution and General Regulations Manual Rules of Procedure. Legal status of the UPU*. 2018 International Bureau of the Universal Postal Union, Berne. https://www.upu.int/UPU/media/upu/files/aboutUpu/acts/manualsInThreeVolumes/actInThreeVolumesConstitutionAndGeneralRegulationsEn.pdf

UPU (2019) *Compendium of Information*. Berne. International Bureau of the Universal Postal Union. https://www.upu.int/UPU/media/upu/files/UPU/aboutUpu/acts/manualsInThreeVolumes/actInThreeVolumes-ManualOfConventionMaj1En.pdf

WHO (2020) *Laboratory biosafety manual*, 4th edition. World Health Organzation. https://www.who.int/publications/i/item/9789240011311

NB: The EU/EC Directives are available from the Office for Official Publications of the European Communities, 12 rue Mercier, L-2985 Luxembourg.

6

Standards in Microbiology

Introduction

Best practices are derived from many sources, for example, from simple protocols published in peer reviewed journals, and from standards developed by practitioners in the field. Examples of the latter are the International Standards Organization (ISO) standards that are implemented by National Accreditation Bodies. The work of microbiology laboratories can be subject to such standards or best practices. This chapter covers in detail some of the best practices that are available and the differences in their use and application. It will help prepare for the introduction of best practices in the laboratory and comply with applicable standards.

Microbiology is practised in many organizations for many purposes. The activities of Biological Resource Centres or culture collections incorporate most common laboratory activities and best practices that are applied to meet international criteria. These organizations range from small private collections to large service collections, and have widely differing policies and holdings. Public service culture collections are charged with several tasks, which are influenced by access legislation. They are in a unique position as custodians of *ex situ* genetic resources deposited with them by scientists and institutions and therefore have a key role to play in the conservation of genetic resources (Hawksworth and Kirsop, 1988; Kirsop and Hawksworth, 1994; Smith, 2011, 2012; Smith *et al.*, 2013; Ryan *et al.*, 2019).

Organizations have been established to try and co-ordinate collection activities and aid the sharing of knowledge and experience, including the maintenance of organisms, their safe handling and distribution (Stackebrandt *et al.*, 2015; Smith *et al.*, 2020).

Standards and Accreditation

The following are non-exhaustive lists of standards relevant to general microbiology and those specific to culture collections (Table 6.1). The application of these standards co-ordinate collection activities and aid the sharing of knowledge and experience in maintaining collections of organisms, their safe handling and distribution.

Establishing best practice and selection of the most appropriate standard

A survey of best practices and standards was carried out during the preparation phase of the Microbial Resources Research Infrastructure (MIRRI) which expects its member collections to commit to implement a quality management system and/or quality assurance procedures

DOI: 10.1079/9781800622135.0006

Table 6.1. General microbiology and culture collections standards

	Standard	Link
1	Good Laboratory Practice (GLP): An industry standard to ensure good laboratory practices	https://www.oecd.org/chemicalsafety/testing/good-laboratory-practiceglp.htm
2	ISO 9001: 2015: An industry standard to ensure good management practices	https://www.iso.org/standard/62085.html
3	ISO 17025:2017 A standard for testing and calibration laboratories	https://www.iso.org/ISO-IEC-17025-testing-and-calibration-laboratories.html
4	ISO 17034:2016 for reference material producers: An industry standard to ensure good practices in reference material production and supply	https://www.iso.org/standard/29357.html
5	ISO 15190: 2020 – Medical laboratories: An industry standard to ensure good practices in safety in medical laboratories	https://www.iso.org/standard/72191.html
6	ISO 15189: 2012 – Medical Laboratories: An industry standard to ensure good practices for quality and competence	https://www.iso.org/standard/56115.html
7	The World Federation for Culture Collections (WFCC) Guidelines for the Establishment and Operation of Collections of Microorganisms: Formulated in the 1980s to ensure culture collection establishment and operations met acceptable principles and practice	https://wfcc.info/guideline
8	The Microbial Information Network for Europe (MINE) project standards for the member collections (Hawksworth and Schipper, 1989): Formulated for European culture collection partners to this database project to ensure the biological materials on which the data was generated were maintained in good condition	https://link.springer.com/article/10.1007/BF01741757
9	Genebank Standards for Plant Genetic Resources for Food and Agriculture (2014): Formulated for the network of plant germplasm collections (mainly seed banks) and set acceptable and preferred levels of practice to maintain germplasm in good condition	https://www.fao.org/3/i3704e/i3704e.pdf
10	OECD Best Practice for Biological Resource Centres: A compilation of relevant elements from the above standards	https://www.oecd.org/science/publicationsdocuments/bestpracticesguidelines/3/
11	ISO 20387: 2018 Biotechnology — Biobanking: General requirements for biobanking	https://www.iso.org/standard/67888.html
12	ISO 21710: 2020 Biotechnology: Specification on data management and publication in microbial resource centres	https://www.iso.org/standard/71384.html
13	AFNOR NF S96-900 Management System of Biological Resource Centres	https://www.boutique.afnor.org/en-gb/standard/nf-s96900/quality-of-biological-resource-centres-brcs-management-system-of-a-brc-and-/fa169771/1173
14	NIT-Dicla 061 Requirements on the accreditation of testing laboratories and producers of reference materials from Biological Resources Centres	Forti *et al.*, 2016

following appropriate international standards https://mirri.org/wp-content/uploads/2021/03/MIRRI-Partner-Charter-V8-June-2020.pdf. The results of the assessment were published and form a basis for the following approach (Martin *et al.*, 2015). In almost all circumstances there are three key principles to ensuring quality of the microbial resources used in production, application and research and these are **A**uthenticity of the strains used, applying the **B**est preservation and storage conditions to retain stability and **C**onfirmation of associated data and information. The **ABC** of microbial resource quality management. Any system introduced into the laboratory processes must have this at its heart and the practices to ensure it is embedded in the protocols. Quality management systems need to be integrated into processes and not applied to them as the latter will often bring conflict and add another level of complication and additional work. It is also important to remember that the quality management system creates an environment for continuous improvement and any system should be based on what needs to be achieved. Here it is the ABC of microbial resource quality management.

The guidelines and best practice designed for culture collections provide a solid starting point and then the next steps should be influenced by institutional policy and strategy and most importantly user demand. There is not one solution for everyone and although there are standards that purport to be that, you need to have flexibility. There are a number of quality management system compliance determinants, key ones are internal governance, the economic and regulatory environments and the social and political environment (Martin *et al.*, 2015). If the system is built on what the end product needs to deliver, the authentic, stable, reproducible strain with the desired properties, then that is what needs to be put in place. The WFCC guidelines (Anon, 2010) are designed to do this for the typical public service culture collection. If this can be implemented and followed building from the ABC principle above, it provides the ideal starting point. If you have an international standard in mind that is driven by institutional requirements, the industry standard and customer expectancy, make sure that your foundation can lead to this but move a step at a time; you are instigating a continuous improvement environment with the product in mind. Systems need to be traceable so that you can monitor the process to keep consistency and be able to get to the root of any problem that may occur.

Several (>30) culture collections in the WFCC chose the ISO 9000 series as a next step from the WFCC Guidance which can be certified by a qualified third party (Simões *et al.*, 2016). Certification focuses on the organization, service or product conforming to specified requirements whilst accreditation is recognition by an authoritative body of the competence to work to specified standards. The latter is considered better in that it takes competence into consideration as well as the ability to follow a process. The gold standard may be accreditation to the 'industry standard' but it is the user of your services and products that is the final judge.

The key elements of the guidance are provided to enable you to select the most appropriate route to establishing your quality management system.

International Guidance and Coordination: WFCC Guidelines

WFCC guidelines (Anon, 2010) are prepared to provide a framework for the establishment and long-term support of microbiological and cell culture collections as a fundamental part of scientific infrastructure (Smith, 2011). They describe all aspects of the aims, services and support required to operate a culture collection in order to represent the needs of its user community.

The salient features in the guidelines are Organization, Funding, Objectives, Holdings, Staff, Preservation, Deposit Procedures and Culture Authentication, Culture Supply, Publicity and Promotion, Documentation and Computerization, Catalogue, Research, Training, Safety and Quality Standards, National and International Collaboration.

The Guidelines cover:

- The aims of culture collections
- The services culture collections provide to the international scientific community in terms of resources, information and specialist skills

- The long-term support needed to enable culture collections to provide these professional services, including appropriate operational facilities
- Staffing levels to allow operation at a high standard; well trained staff with research expertise related to the aims of the collection
- The contributions made by collections to enhance the research knowledge base in terms of taxonomic studies, preservation, growth and handling procedures and other linked areas
- The capability of collections to meet all relevant national and international regulations concerning the control, transportation and health and safety aspects of resource handling and distribution
- The need to provide support and training in capacity building on a global basis
- The need for international collaboration to enhance the value and quality of biological resources

Salient Features in the WFCC Guidelines

Organization

The parent organization or board under which a service collection is established should be fully aware of and accept the responsibilities inherent in maintaining a public service to appropriate standards.

Funding

It is important to consider the level of funding, both now, and likely to be forthcoming on an on-going basis. This must be adequate to provide the range of services being planned, and at a standard that users would expect.

Objectives

Collections require a clearly summarized general statement of their long-term objectives relating to the scope of their holdings, and to the range of outside services that are envisaged. It is helpful for a collection to have more specific short-term objectives relating to the coming 1-, 3- or 5-year period. These can usefully include the numbers and groups of strains which it is planned to acquire in that time frame, and schedules for installing new facilities and services.

Holdings

In addition to decisions on the groups of microorganisms to be maintained, and the numbers it is envisaged as being retained in the long-term, it is also necessary to have a clearly defined accessions policy on which the acceptance of new strains offered to the collection is to be based. If this is not decided and many unsolicited strains are accepted uncritically without due regard to the collection's objectives, storage capacity and personnel and financial resources can soon become over-stretched. At the same time, the range should not be so strictly defined as to limit the effectiveness of the services provided to the users and a policy should be adopted that balances capability with scientific needs. Stackebrandt (2010) makes a good case for the need to deposit key strains from published scientific works and collections need to work together if sufficient capacity and expertise is to be available to achieve this. If strains are maintained that are potentially pathogenic to man, animals or plants, or that produce toxic or hallucinogenic compounds, those holdings should be clearly labelled and kept secure. Adherence to any safety and control regulations in force, whether institutional, national or international, is mandatory.

Staff

Culture collections are necessarily labour-intensive. When determining the numbers of full- and part-time positions required it is important to consider how time-consuming the routine accessions, preservation, maintenance, and viability checking will become as the collection approaches its target strain numbers. Staff levels need to be sufficient not only for the incorporation and maintenance of cultures, but also to fulfil the anticipated level of culture supply and other services the collection is to offer.

The key staff member(s) recruited would be expected to have a higher degree in an appropriate field and some subsequent direct experience or special training in culture collection curation skills. In order to attract and retain sufficient high calibre staff, arrangements for long term [on-going] employment should be made where possible.

Particular attention should be paid to the qualifications and experience of the person in charge of the collection. The importance of management skills for larger collections should be recognized. Steps need to be taken to provide expertise through collaborative arrangements within and/or outside the collection's parent organization.

Preservation

Different microorganisms often require optimized preservation methods in order to ensure optimal viability, storage, purity and stability for individual strains. For security, and in order to minimize the probability of strains being lost, each strain should whenever practical be maintained by at least two different procedures. At least one of these should be by freeze-drying (lyophilization) or storage in liquid nitrogen (cryopreservation) where applicable, since for most strains these are the best methods for minimizing the risks of genetic change. In some cases, for example cell lines, where only freezing is applicable, duplicates should be stored in separate refrigerators with different electrical supplies.

In order to minimize the risks to important genetic resources from fire, flooding, earthquakes, war or catastrophes, collections should arrange to have duplicates of at least the most important and irreplaceable strains (and also of their associated documentation) securely housed in a different building or ideally at a separate site.

Deposit Procedures and Culture Authentication

Scientists obtain cultures from recognized collections because such material has undergone quality control and authentication testing as part of the routine procedures of the collection. Scientists ordering cultures from collections expect them to be correctly identified. If not, there is a danger of users employing the wrong organism in their investigations which could prove time-wasting, expensive, and lead to invalid published results. Moreover, without proper authentication, noxious organisms could be inadvertently supplied. This places a grave responsibility on collections and demands attention from the time the first cultures are received for preservation.

As part of the deposit procedures, collections must provide accession forms to be completed by all depositors. The forms should include all available information regarding the strain and its origin. This information is important for providing maximum scientific data to future users and also is important for compliance with the requirements of the Convention on Biological Diversity (CBD) and the Nagoya Protocol on Access and Benefit Sharing. All incoming material should be treated as potentially hazardous until its identification is confirmed.

When cultures are recovered from preserved material, either during maintenance, routine re-preservation work, or when they are being dispatched, care should be taken to ensure they conform to the original deposit by carrying out appropriate tests, by comparative study or checking by a specialist.

Culture Supply

If a collection has been established to provide a public service, then it should have the capacity to provide the timely delivery of listed strains. If cultures cannot be delivered in a reasonable time for technical reasons, this should be indicated in catalogues, lists or databases. Monetary arrangements for culture supply will vary according to the financial basis and policies of the legal owners of the collection.

Cultures listed as available in catalogues by service collections should normally be provided without restriction to those requesting them, subject to any import, quarantine or containment regulations that might apply, and to normal credit control procedures where charges are made. Many service collections exchange cultures with other service collections for accession purposes and at no cost. Cultures for teaching and research purposes are also regularly made available at a reduced cost or on other preferential terms compared with ones supplied to industry or commerce.

Cultures which, for any reason, are not available for distribution should not be listed in catalogues or included in publicly accessible databases; cultures with restricted distribution should be clearly marked.

Strains which are pathogenic or toxic to plants, animals or man are often subject to national or international regulations from health and/or agriculture authorities. Scientists requesting strains may need to obtain permits to import material or to handle such cultures.

Collections should maintain detailed records of recipients of cultures showing the material sent (with strain and batch numbers where appropriate), method and date of shipment, and name and address of the person to whom sent. In the case of unsatisfactory results or if it is necessary to supply subsequent information recipients can then be notified. When mailing cultures, attention needs to be given to pertinent postal regulations regarding packaging and labelling.

Publicity and Promotion

Collections need to ensure the widest possible visibility for their holdings and the specialist services they provide for the scientific community.

Visibility can be increased through the preparation of leaflets, participation in workshops and conferences, writing articles and delivering lectures. Collaboration with local or regional organizations can prove beneficial in this regard. Visibility can also be greatly increased by the development of websites and social media.

Other Services

Service culture collections may be well placed to provide a variety of support services to the scientific and industrial community worldwide or in the region they serve. If identification services are to be offered it should be considered whether appropriately trained personnel are available to undertake this demanding task, either in the collection or in an associated institution, major problems can arise as a consequence of misidentifications. If consultancy, advisory or investigation services are to be offered, attention must be given to the provision of appropriate facilities and properly trained personnel and consideration should be given to the need for insurance cover.

Documentation and Computerization

Records need to be kept for each strain held and should include the following categories of information:

- Geographical location [place]
- Substrate or host and date of isolation
- Name of person isolating the strain
- Depositor (and [or] other source of the strain, such as from another collection)
- Name of the person identifying the strain
- Preservation procedures used
- Optimal growth media and temperatures
- Any data on biochemical or other characteristics, and any regulatory conditions applying (relating for example to quarantine, containment levels, and patent status)
- Whenever resources permit, the records should be computerized. For security, duplicate computer files or photocopies of records should be kept separately, perhaps deposited with duplicate strains.

Catalogues

Printed catalogues of the strains available for distribution should be produced at not too infrequent intervals. Computerized catalogues can be readily prepared from machine-readable data and can be distributed via the internet or multimedia format, greatly increasing the visibility of information held by the collection.

Research

As centres of expertise, research programmes should be a part of every collection's activity. Research activities also ensure that staff keep abreast of current developments and are aware of the needs of the user community. Collections are [also] well-placed to develop procedures for screening for the identification of particular organisms or products, quality control tests, preservation protocols for strains difficult to preserve by routine procedures, as well as developing optimal culture media and conditions for growth.

Training

The WFCC organizes training both for individuals and in courses and the education committee can be contacted in relation to training requirements and will seek to find funds to support such needs.

Safety and Quality Standards

Safety aspects of all operations carried out in the collection need to be carefully controlled with respect not only to national health and safety regulations, but also with regard to good laboratory practice. Collections may consider the value of operating within recognized certification standards (e.g. ISO 9001) which indicate a level of quality.

National and International Collaboration

Collections and individual senior staff within collections may join the World Federation for Culture Collections (WFCC). This has work programmes concerned with education, patents, postal/quarantine/safety regulations, endangered collections, and publicity [standards] which all provide information that may be of assistance to new and established collections. The WFCC holds a major international congress every three years which provides a unique forum for the consideration of all aspects of the activity of culture collections. A newsletter is produced, and training schemes and courses are operated. Collection staff should be encouraged to actively participate in the affairs of the WFCC.

Many countries have formal or informal associations or federations of the collections within them. The establishment of formal or informal links with any regional groups active in adjacent countries should be encouraged.

In order to make their holdings widely known, collections are encouraged to register with the WFCC World Data Centre on Microorganisms (WDCM). This international database is available online from the WFCC website. Collaboration provides global visibility for collections. Collaboration with the microbial strain data network (MSDN) is also recommended to facilitate international communication and data exchange.

OECD Biological Resource Centre (BRC) Best Practice

In 2001, in order to meet modern demands for the further advancement of biotechnology and life sciences, the OECD introduced a new concept of repositories and providers (BRCs) of high-quality biological materials and information. BRCs are considered a key element for sustainable international scientific infrastructure and are necessary to underpin successful delivery of biotechnology in all sectors such as health, industry, agriculture and the environment (OECD, 2001).

Experts from OECD and observer countries agreed on a consensus report calling upon national governments in association with the international scientific community to undertake actions to bring the BRC concept into being. These Best Practices are intended as a target for the quality management of collections (OECD, 2007). There was a consensus from the OECD countries that no one collection or single country could do this alone and suggested the establishment of a Global Biological Resource Centre Network (GBRCN - https://www.oecd.org/sti/emerging-tech/towardsaglobalbiologicalresourcecentrenetwork.htm).

Development of Best Practice

Biological materials are the source for scientific investigation. They also enable research and development that leads to discoveries supporting the progress of biotechnology and bioindustries. Proper maintenance and supply of biological resources is essential for future advancements of biotechnology and its contribution to the growth of the bioeconomy.

The Best Practices were developed by a group of national experts brought together as a Task Force on BRCs, under the auspices of the OECD's Working Party on Biotechnology. The Task Force made considerable effort to reach out to the various communities as their thinking progressed. They did this principally in three ways:

- Face-to-face engagement
- Broad, written consultation
- A pilot study on the Best Practice guidelines

Works Undertaken by BRC Task Force

- **Face-to-face engagement**: Task Force experts hosted a number of open special interest meetings at major international conventions for those involved in collection management. These meetings sought to encourage broad awareness of the OECD work on BRCs as well as specific input on the detail of Best Practices.

- **Broad, written consultation:** Comments were sought from a broad range of around 500 stakeholders on practicality, implementation and technical accuracy of BRCs quality guidance.
- **A pilot study on the Best Practice guidelines:** The study tested the validity and operability of the draft Best Practice guidelines. The guidelines were refined on completion of the study. Further details are provided in the next section.

PILOT STUDY ON BEST PRACTICE GUIDELINES. The goal was to test the validity and practicality of the OECD general and domain specific Best Practice guidelines and achieved the following main objectives:

- To ensure broad consultation on Best Practice guidelines with relevant stakeholders and national and international bodies
- To test the validity of the Best Practice guidelines and their applicability to existing culture collections
- To assess the impact of Best Practice guidelines on the functioning of culture collections
- To identify a range of available options for those entities that wish to implement the Best Practice guidelines
- To perform a preliminary cost benefit analysis for collections applying the Best Practice guidelines

The study was undertaken by a stakeholder group including culture collections, research institutes, certification/accreditation agencies and international organizations. All comments and outputs from the study were considered in the Best Practice guidelines. The German Government funded the GBRCN project through the Federal Ministry of Education and Research initially bringing together 23 organizations in16 countries to explore the principles of global exchange through coordinated processes (Fritze *et al.*, 2012). This work resulted in the submission of the preparatory phase for MIRRI (Stackebrandt *et al.*, 2015), influenced the development of the Brazilian BRC Network (Peixoto *et al.*, 2006: da Silva *et al.*, 2020) and collaboration with the United States Culture Collection Network (https://usccn.org/) the Asian Network of BRCs (https://www.acm-mrc.asia/TF/ABRCN.html) and partners in Africa. The seeds of the GBRCN.

General Best Practices for BRCs

Best Practice guidelines assist in the acquisition, maintenance and provision of biological materials and the management of BRCs as defined by the OECD. They can be adopted for use by any laboratory acquiring, using and storing cultures of microorganisms (OECD, 2007).

Purpose

These Best Practice guidelines help to ensure that biological materials are of the highest standard and authenticity. The preservation techniques used should retain the key features of the biological material and ensure its consistency between supply centres. They provide a reliable basis for research and development and contribute towards protection of the health of laboratory personnel, the public and the environment.

Best practices need to cover the following points:

- **Organizational requirements**

These include the need for long-term sustainability, the responsibilities of management, training and qualification of staff, biosafety and data management.

- **All operations in the handling, storage and use of biological materials**

These include the preparation of media and reagents, appropriate documented procedures, quality checks on biological material, preservation and maintenance methods and shipping and packaging.

- **Quality audit and quality review**

These include the purpose, implementation and the method and procedure for quality checks.

BRC Organizational requirements

- Long-term sustainability: Organisms need to continue to be available in the long-term for continuity.

- Responsibilities of management: Laboratories should appoint a Quality Manager whose duties include:
    - Administering and monitoring an efficient up-to-date quality management system
    - Reporting and advising on quality matters
    - Representing the BRC on quality matters when dealing with users, suppliers and outside bodies
- BRC should designate a biosecurity officer, at operational level within the BRC, whose responsibility it is to ensure internal compliance with Best Practice Guidelines on BRC Biosecurity.
- Staff qualifications and training: Training should be in accordance with documented protocols in skills specific for the job and to update skills as new technologies or practices are introduced. Authorization to use specialist equipment should be documented in training records.
- Health and safety (biosafety): All staff should follow the procedures laid down under the appropriate level of containment for the organisms being handled to avoid contaminating samples and risk of infection.
- Premises must be clean and well lit; all usual safety procedures and aseptic techniques must be complied with and appropriate protective clothing worn as required. Appropriate arrangements, in accordance with national and international regulations, for site security should be made to ensure hazardous organisms cannot be released to unauthorized users.
- Equipment management procedures should be laid down in a predefined schedule and should include use, performance control, maintenance and calibration. Service records should be maintained and copies of key documents should be lodged with the Quality Manager.
- Documentation should be correctly updated and ensure compliance with internal documentation. All staff need to be trained in compliance with the prescribed policies and procedures.
- Data management: authentication of data may differ from centre to centre, but a BRC should:
    - Provide traceability of data through a history of modifications (dates and signatures of inputs, validations, modifications and deletions).
    - Give signature for data entry, validation, modification or deletion.

All operations in the handling, storage and use of biological materials

Example areas are:

Preparation of media and reagents should be as per defined standards for all preparations used in the growth and/or maintenance of the living biological materials; these should be documented with appropriate mechanisms to allow updates.

Appropriate documented procedures are needed for receipt and storage of biological materials handled within each BRC. A risk assessment for receipt, handling and storage must be undertaken to determine potential harm to personnel, the public and the environment of any biological material received into the BRC. The risk assessment should be reviewed and updated regularly.

Quality checks on biological material should be performed to determine stability of some key features, growth requirements, and maintenance and preservation methods appropriate to the biological material using relevant technology.

Preservation and maintenance methods should be appropriate, reproducible and key parameters of the process recorded and monitored.

Shipping and packaging must be undertaken in accordance with current postal, IATA and ADR regulations. It must also meet additional requirements imposed by other regulations governing quarantine, biosafety and/or biosecurity.

Quality audit and quality review

Purpose: Periodic audits need to be undertaken by the management to ensure the policies and procedures follow the Best Practice guidelines and any additional specific Best Practice guidelines. External, independent audits should be carried out. This is essential for those organizations with formal accreditation or certification to national or international standards. A process will need to be in place to identify any potential or actual source of non-conformity to the guidelines and to implement any corrective action.

The Quality Manager should be responsible for ensuring that reviews are recorded and that any corrective actions are implemented.

Implementation: At least one audit each year needs to be undertaken according to the schedule in the Rolling Audit Programme. An external independent qualified person should carry out a Third-Party Audit of the procedures, preferably each year. A meeting of relevant staff (audit, BRC, line management) should be held annually to review the audit reports, enquiries and complaints received and discuss potential improvement in procedures and monitoring.

Method and procedure for quality checks: All methods and procedures should be subject to in-use quality checks. For example, the product should be checked for fitness for purpose, i.e. a sample should be selected from a preserved batch and appropriate stability checks carried out. Such checks should be included in the individual documented procedures.

Best Practice for Microorganisms

The OECD Best Practice is provided at two levels:

(a) Those that apply to all BRCs
(b) Those that apply specifically to organism type, for example, microorganisms.

The methods used should ensure that the key features of microorganisms are retained. This provides a reliable basis for research and development in different laboratories. These contribute towards a safe and healthy laboratory, and protect the public and the environment.

Organizational Requirements

STAFF QUALIFICATIONS AND TRAINING. Staff should have relevant qualifications, training and competence to carry out their duties and must follow containment procedures when handling microorganisms. Containment is essential:

- To avoid contaminating samples
- To reduce risks of infection
- To prevent environmental dispersion

Containment levels for microorganisms are defined by the World Health Organization (WHO, 2020) and are interpreted by international law, regulations and policies.

Premises

It is the responsibility of the organization which comprises the BRC, or, within which the BRC is located, to provide an environment that is conducive to handling microorganisms, for example, free from contamination.

Equipment use, calibration, testing and maintenance records

Equipment management procedures should be laid down in a predefined schedule and should include use, performance control, maintenance and calibration. Service records must be maintained and copies of key documents need to be lodged with the Quality Manager.

INFORMATICS. BRCs should follow informatics Best Practice guidelines as set out in General Best Practice Guidelines for all BRCs.

PREPARATION OF MEDIA AND REAGENTS. Accurate preparation and storage conditions of culture media are fundamental steps in the growth and maintenance of biological materials. The BRC should have defined standards for all preparations; media formulae should be documented and procedures put in place to control changes to procedures and for their approval and adoption.

Media batches should be clearly labelled and expiry dates (date after which media and reagents are not to be used) defined and clearly indicated.

Best Practice for Microorganisms: Accession

It is important that all BRCs comply with applicable national and international laws. BRCs need to provide a documented description of the nature of the microorganism and define the level of hazard and containment.

RECEIPT AND HANDLING OF BIOLOGICAL MATERIALS. The BRC should document and implement safe procedures for receipt and storage appropriate to the type of biological materials handled.

PRESERVATION. Long-term preservation of microbial cultures using cryopreservation or freeze-drying methods are the common approach for sustainable preservation. The methods should ensure:

- High viability/recovery of the preserved culture
- The preserved culture is contaminant-free (this does not include any symbiotic or parasitic microorganisms i.e. intended to be present)
- Authenticity of the preserved culture and genomic integrity

AVAILABILITY OF BIOLOGICAL MATERIAL. Freeze-dried or cryopreserved (when supplied frozen) material must be dispatched as soon as possible after necessary licences and/or documentation are provided.

DISTRIBUTION CONTROL OF HAZARDOUS MICROORGANISMS. BRCs must follow the OECD Biosecurity Best Practice Guidelines. There is considerable concern over the transfer of infectious agents capable of causing substantial harm to human, animal or plant health. There is potential for these organisms to pass to parties not equipped to handle them or to people who may make illegitimate/illegal use of them.

To reduce this risk a BRC must have procedures which meet national requirements to check the validity of customers that wish to receive hazardous organisms. European collections have collaborated to produce a code of conduct for compliance with biosecurity regulations and requirements (Rohde *et al.*, 2013).

Best Practice for Microorganisms: Compliance

In the process of isolation, handling, storage and distribution of microorganisms, there are many stages where compliance with the law, regulations or voluntary international conventions is required.

BRC should, for example, comply with:

- Applicable health and safety requirements
- Classification of microorganisms on the basis of risk
- Applicable quarantine regulations
- Intellectual property rights (IPR)
- Requirement for appropriate safety information to be provided to the recipient of microorganisms
- Regulations governing shipping of organisms
- Distribution control of biological material

Whether it is compliance with the law, or duties of a caring employer, essential components for a safe workplace are:

- Adequate assessment of risks
- Provision and implementation of adequate control measures
- Provision of health and safety information and a sound health and safety plan
- Provision of appropriate training and implementation of good working procedures
- Establishment of record systems to allow safety audits to be undertaken

Best Practice for Microorganisms: Classification According to Risk Groups

Various classification systems exist and are implemented nationally. Appropriate authorities in each country define and set minimum handling procedures for pathogenic organisms. The key references are the classification definitions produced by the WHO. The WHO classifies microorganisms into four groups according to the risk they impose to humans.

RISK GROUPS

Risk group 1

No or low individual and community risk - This group will include microorganisms that are unlikely to cause human or animal disease.

Risk group 2

Moderate individual risk, low community risk - This group will include:

- Pathogens that can cause human or animal disease but are unlikely to be a serious hazard to laboratory workers, the community, livestock or the environment
- Laboratory exposures may cause serious infection, but effective treatment and preventive measures are available and the risk of infection is limited

Risk group 3

High individual risk, low community risk - This group will include pathogens that usually cause serious human or animal disease but do not ordinarily spread from one infected individual to another. Effective treatment and preventive measures are available.

Risk group 4

High individual and community risk - This group will include pathogens that usually cause serious human or animal disease and can be readily transmitted from one individual to another, directly or indirectly. Effective treatment and preventive measures are not usually available.

SAFETY INFORMATION PROVIDED TO THE RECIPIENT OF MICROORGANISMS

Safety information should be dispatched with a microorganism indicating which risk group it belongs to and what containment and disposal procedures are necessary. For a microorganism, a safety data sheet should include:

- The risk group of the organism
- A definition of the risks and assessment of the risks involved in handling the organism
- Requirements for the safe handling and disposal of the microorganism
- Containment level and procedures in case of spillage
- Opening procedure for cultures and ampoules
- Appropriate transportation of the microorganism

Approaches to National Accreditation or Certification

To achieve certification, a BRC is required to progress through independent review by a third party to prove successful compliance with OECD's 'Guidance for the Operation of Biological Resource Centres (BRCs)'.

Criteria for Certification

GENERAL BRC CERTIFICATION RULES. Certification bodies need to be acknowledged either through an accredited certification body recognized by national governments or through a transparent certification procedure recognized by or directed by government. Governments would be responsible for ensuring that their respective certification body is independent, qualified and has no conflict of interest with the BRC seeking certification. Certification could be based upon implementation of the applicable operative OECD Best Practices (2007). Future revision of these Best Practice guidelines would be negotiated and approved by the community of certified BRCs.

CERTIFICATION MECHANISM. Any certification process and resultant requirements for certification bodies and third-party auditors should be according to international standards.

GENERAL CRITERIA FOR CERTIFIED BRCS. A national certification procedure would likely require certified BRCs to comply with:

- Their national legislation, regulations and policies concerning acquisition, conservation, utilization, including the fair and equitable sharing of benefits arising from utilization of genetic resources, and distribution of biological resources and data related thereto.
- The regulations of the relevant countries when moving biological materials across national boundaries.
- The relevant national and international agreements, regulations, policies, frameworks and recommendations.

Certified BRCs should have in place a mechanism that updates their knowledge of all of the above information. When certification is granted it should indicate the highest level of hazard, where appropriate, that the BRC is qualified to handle.

Mechanisms to Demonstrate Compliance

The processes of monitoring and recording of actions are not only essential for the proper management of a quality system to enable you to continually improve, they are also necessary to produce an audit trail. This is necessary to demonstrate compliance with the Best Practices or standards in place to auditors or reviewers. You must be able to demonstrate that you have followed the standard and met the required criteria and to show products are within tolerances.

Process for Implementing Best Practice

COMPLIANCE WITH OECD BRC BEST PRACTICE - CHECKLIST. A framework for assessing implementation of the OECD BRCs Best Practice is presented. To achieve quality assurance, common standards of practice have been defined in the *OECD Best Practice for the Operation of Biological Resource Centres (2007)*. These criteria are equally applicable to all microbiology laboratories and they deal with the following major issues:

- Quality management/quality assurance systems that operate across international borders to implement best practices
- Authenticity of biological materials, accuracy of databases and bioinformatics data; accuracy of labelling
- Processing of cultures, cell lines, and genetic constructs, including procedures and standards for ensuring their long-term stability and quality control
- Accuracy of the data collected and supplied
- Expertise and training of human resources, particularly of a new generation of taxonomists able to use molecular techniques and informatics
- Sharing of expertise among centres through coordination and networking
- Traceability of materials, methods and results

The rules and criteria must be rigidly implemented and linked to a mechanism for monitoring and review, based on scientific, technical and administrative expertise.

A BRC ACCREDITATION MECHANISM. A BRC can be certified or accredited by an accreditation body (independent third party), acting on behalf of a government, incorporating the guidance and rules on quality laid down by the BRC documents *OECD Framework and Guidance for BRC Accreditation Implementation* and *Standard for the Operation of Biological Resource Centres (2007)* (Levels 1 and 2). The OECD Best Practice has been converted into a French Standard NF S96-900 which is available from AFNOR (2008) and a laboratory can be certified by AFNOR to this standard. Equally the laboratory may choose to be certified to ISO 9000 series or Good Management Practices (GMP) or accredited to ISO 20387 or ISO 17025, additionally implementing the content of the OECD Best Practice not covered by these standards.

The accreditation process is carried out on a voluntary basis.

It can involve the implementation of the French Standard NF S96-900 'Quality of biological resource centres (BRCs) – Management system of a BRC and quality of biological resources from human or micro-organism origin' available from Association Française de Normalisation 11, rue Francis de Préssensé F – 93 571 La Plaine Saint Denis cedex http://www.afnor.fr . Other mechanisms might include the use of the OECD Best Practice alongside ISO standards such as ISO 17025 or alongside General Laboratory Practice (GLP) or (GMP).

Single-site and institutional membership are not accreditation conditions. It is possible to accredit a collection as multi-site or if managed by several institutions if each one of the BRC components fills the accreditation criteria. However, a BRC must be a legal entity. The applicant commits itself to follow all laws, regulations, principles and general policies in force in its country.

The accreditation procedure must adhere to the following points:

- The accreditation body must ensure that the accreditation procedure is transparent and available to third parties
- A list of accredited BRCs is published
- The procedure must include regular revision and be reproducible
- Those involved in the procedure (experts) must be independent of the applicant body
- The applicant may not call into question either the accreditation body and its competence or the accreditation procedure (BRC documents *OECD Framework and Guidance for BRC Accreditation Implementation* and *Standard for the Operation of Biological Resource Centres* (2007) Levels 1 and 2 are publicly available)
- The applicant must be able to appeal in the event of refusal
- The cost of the accreditation procedure is to be borne by the applicant

AN EXAMPLE OUTLINE OF THE ACCREDITATION PROCEDURE

- Preliminary analysis of the application
- Constitution of an audit team
- Assessment
- Drafting of the audit report
- Analysis and validation of the report
- Decision on accreditation and grant or refusal thereof
- Publication of the BRC description

Operation mode:

- The audit team is made up of a list of experts recognized by their peers
- The accreditation body is responsible for organizing an accreditation committee

Positive results of accreditation procedure:

- Attribution of an accreditation number
- Authorization to present the applicant as an accredited BRC
- Possibility, according to country, to ask for certification
- Inclusion in the international list of accredited BRCs
- Programme of regular review of accreditation status established (normally annual)

Negative results of accreditation procedure:

- Attribution of a 'requested accreditation' number
- Justification of the refusal
- Changes required prior to accreditation being granted

The main features of an accreditation body are its:

- Scientific and technical capacity to judge whether a collection is following the *BRC International Standards: OECD Best Practice for the Operation of Biological Resource Centres* (2007) (levels 1 and 2)
- Capacity to monitor accreditations (audit, etc.)
- Capacity to publish list of accredited BRCs
- Independence from the bodies applying for accreditation

Accreditation bodies should have a level of competence acknowledged by peers at international level. A list of these bodies will be kept up-to-date and published.

AUDIT AND REVIEW. Audit and review are part of accreditation procedures. The regularity of the reviews and the revisions may vary according to countries and fields of activity.

Periodic audits should be carried out by management to ensure that the company policies and procedures are being followed.

Good Management Practices

GMP or current Good Manufacturing Practice (cGMP) is a term recognized worldwide for the control and management of manufacturing and quality control testing of foods, pharmaceutical products and medical devices. GMP takes the holistic approach of regulating the manufacturing and laboratory testing environment itself.

An extremely important part of GMP is documentation of every aspect of the process, activities and operations involved with drug and medical device manufacture. If the documentation showing how the product was made and tested is not correct and in order, then the product does not meet the required specification and is considered contaminated (adulterated in the US). This documentation enables traceability of the whole production chain and in the event of future problems enables product-recall from the market.

GMP requires that all manufacturing and testing equipment has been qualified as suitable for the intended purpose. All operational methodologies and procedures (such as manufacturing, cleaning, and analytical testing) utilized in the manufacturing process have to be validated (according to predetermined specifications), to demonstrate that they can perform their purported function(s).

Collections following GMP ensure a good end product (e.g. its authenticity and purity). This system does not cover every element of BRC Best Practice but the implementation of the additional requirements would not be too onerous.

Self-Audit

There are several levels of audit a laboratory must undergo. The first level is a Self-Audit, this is carried out by institutional staff, an internal process to enable you to self-correct any inconsistencies or non-compliance with the quality management system in place. The criteria for operation of BRCs are used as the basis for this assessment, which are designed to meet *OECD Best Practice for the Operation of Biological Resource Centres* (2007). This is equally applicable to the microbiology laboratory and covers criteria relating to:

- Laboratory-based collections
- The maintenance of biological resources
- The services provided by the BRC or laboratory

The self-audit should prepare the laboratory for an official audit by the certification or accreditation body auditors. All documents need to be available: manuals, instructions, reports, log books etc. as they may be requested by the accreditation body during audit and review. A mechanism for updating changes to these frameworks must be in place. Questions should be raised and answered during the process, e.g. is the laboratory registered to enable it to handle the level of hazard of the materials it works with?

WHAT MUST THE LABORATORY COMPLY WITH?
The laboratory needs to comply with:

- National legislation concerning acquisition, conservation, utilization and distribution of biological resources and data related thereto (including biosafety, biosecurity, ethics)
- The rules and regulations of recipient countries when biological materials are supplied to foreign countries
- The relevant national, regional or international policies, frameworks, agreements and recommendations (e.g. Convention on Biodiversity)
- The detail of any standard that may be implemented institutionally such as ISO standards
- In the case of BRCs - the criteria laid down in the *OECD Best Practice for the Operation of Biological Resource Centres* (2007)

WHAT DOCUMENTATION IS IN PLACE?
The following documents must be available:

- Quality manuals (with quality indicators, preventive actions...)
- Audit procedures (internal, external, senior management reviews)
- Control systems (including corrective actions)
- Day and laboratory books to enable traceability
- Staff training records
- Other documentation required by the quality management system

WHAT PROCEDURES ARE IN PLACE FOR THE RECEIPT OF BIOLOGICAL MATERIALS?
The following procedures should be in place for the receipt of biological materials:

- Biological materials should only be introduced in the laboratory if they are defined and risk assessed and meet the authorized laboratory containment level, e.g. are biological safety level 2 (risk group 2) or below if the facility is authorized to containment level 2
- Appropriate measures must be taken to ensure safety and security
- Biological material should have the minimum information recommended in the standards or Best Practices
- Data should be authentic and stored to provide traceability, it should be backed-up and confidentiality ensured (where applicable)

WHAT LEGAL AND ADMINISTRATIVE PROCEDURES ARE IN PLACE?
The following legal and administrative procedures must be in place:

- An organizational structure must be provided, defining levels of decision making in management processes
- Measures to ensure quality, security and traceability
- The allocation of responsibilities with regard to any activity undertaken must be clearly listed
- The laboratory director must ensure that the centre has the right combination of expertise
- All enquiries/complaints are logged in the enquiry logbook and written records of responses/solutions etc. are stored

IS THE WORKPLACE FIT FOR THE PURPOSE? The following points need to be considered while checking for if the workplace is fit for the required purpose:

- An environment that is conducive to working with biological materials
- Work areas and equipment must be in conformity with all relevant international or national standards and legislation
- A laboratory may not operate beyond the limits of what is safe with the resources at its disposal (e.g. premises, equipment, personnel)
- The premises must be cleaned on a regular basis
- Procedures in place to cope with a major contamination problem or with any threat (e.g. fire, lack of staff)
- All staff must follow the procedures laid down to avoid contaminating samples but also to avoid risk of infection

ARE INSTALLATIONS, EQUIPMENT AND INFORMATION SYSTEMS COMPLIANT? While checking for installations, equipment and information systems compliance, the following points need to be considered:

- Equipment calibration, testing and maintenance records
- All the equipment in the laboratory is used and maintained in accordance with the manufacturer's handbooks/manuals
- Service records are maintained
- The information system description must be available and security safeguards in place

IS A QUALITY MANUAL AVAILABLE AND ARE PROCEDURES MANAGED? The following points should be kept in mind while checking for the management of procedures:

- The quality manual is distributed as required and a system in place to ensure all recipients receive updates as necessary
- Procedures of all the activities of the laboratory have to be written and updated when necessary
- All laboratory staff must adhere to the policies and procedures and are provided additional training when updates are issued
- Departures from documented procedures are not allowed unless senior management has first been consulted. Written permission and justification must be recorded

WHAT PROCEDURES ARE IN PLACE FOR CONTROL OF THE DISTRIBUTION OF BIOLOGICAL RESOURCES? The following procedures must be in place for control of the distribution of biological resources:

- Mechanisms to prove the material despatched is of good quality including the reliability of associated data
- Packaging must meet the relevant postal, IATA and quarantine regulations; any additional obligations met (such as labels or permits)

Procedures for handling complaints and anomalies in place including acknowledgement of communication, investigation and reporting procedures

Internal Vertical Audit

A vertical audit is carried out by staff from outside the specific activity or laboratory and can happen in two ways, top down or bottom up. The vertical audit can start from receipt of a sample or enquiry through to the end of the process or working backwards from a final report to the original receipt of the sample or enquiry.

You could try these two vertical audit activities:

(a) Assess the progress through a microbiology laboratory of a sample that is received for identification. Do the processes from sample arrival to final identity, reporting and sample disposal comply with the laboratory quality system?
(b) Trace an organism that has been purchased for use in a specific test or research from arrival to end use, to ensure organism quality remains undiminished throughout the test procedure.

The criteria described in the self-audit (see Appendix E this volume) should also be used to ensure a full and adequate audit of these elements.

Quality Controls for Microorganisms

The following tables summarize some of the guidance for different organism types. Table 6.2 summarizes example quality controls for microorganisms based on the quality guidelines that were established within the European project CABRI (Romano *et al.*, 2004), available at http://www.cabri.org/guidelines/gl-framed.html. Table 6.3 details preservation methods, distribution forms, and sets of data describing the products as required in the CABRI guidelines. The application of OECD Best Practice has positive impact for the validation of cryopreservation techniques for microorganisms (Smith and Ryan, 2008).

Data Sets suggested for microorganisms

The **Minimum Data Set** (MDS) is the set of data that is asked to describe **each product** from each collection (Tables 6.4, 6.5 and 6.6). Each catalogue field has to be filled in (with unknown or not applicable when no information is given). The MDS is required from new collection centres when they join CABRI.

The **Recommended Data Set**: Collection centres are also asked to supply data for recommended fields; however, if data are not available, the product is still acceptable. Collection centres are free to provide data for other fields, which are part of the **Full Data Set** (See OECD Best Practice for the details for these two data sets).

ISO International Quality Management Standards

ISO 9000 Series

ISO 9000 series are a family of standards for quality management systems, centred around ISO 9001:2015 put in place to ensure procedure controls. These can be assessed by an accredited organization or individual that is validated by a national accreditation body.

ISO 9001 covers all activities from procurement of raw materials through to the process of manufacture and to end product delivery and customer feedback. ISO 9001 is maintained by ISO, the International Organization for Standardization and is administered by accreditation and certification bodies.

The International Organization for Standardization (ISO) introduces a systems approach which enables an organization to effectively identify, monitor and control all its activities. It can easily be adopted to the laboratory environment. An effective management system approach is built on the concept of continual improvement through a cycle of planning, implementing, reviewing and improving the processes and actions that an organization undertakes to meet goals. This is known as the Plan-Do-Check-Act principle.

- Plan: Planning, including identification of hazard and risk and establishing goals,
- Do: Implementing, including training and operational issues,
- Check: Checking, including monitoring and corrective action,
- Act: Reviewing, including process innovation and acting to make needed changes to the management system.

In order to improve quality management, the organization needs to focus on the causes of non-conformities and undesirable events. Systematic identification and correction of system deficiencies leads to improved performance and control of output. It is user driven and processes are inbuilt to ensure that customer needs are met.

ISO Certification/Registration

A company or organization that has been independently audited and certified to conform to ISO 9001 may publicly state that it is 'ISO 9001 certified' or 'ISO 9001 registered'. Certification to an ISO 9001 standard does not guarantee any quality of end products and services; rather, it certifies that formalized business processes are being applied. Indeed, some companies enter the ISO 9001 certification as a marketing tool.

ISO Standards

ISO 9001: SOME REQUIREMENTS Some of the requirements in ISO 9001, which is the main standard in the ISO 9000 family, include:

- A set of procedures that cover all key processes in the business
- Monitoring processes to ensure they are effective
- Keeping adequate records
- Checking output for defects, with appropriate and corrective action where necessary
- Regularly reviewing individual processes and the quality system itself for effectiveness
- Facilitating continual improvement

Some of the other ISO standards are ISO 20387, ISO 17025, ISO 17034, ISO 15190 and ISO 15189.

ISO 17025

- This is the main standard used by testing and calibration laboratories
- Not normally specifically utilized by culture collections it is now often linked with ISO 17034 and employed by reference material producers to ensure the reference materials are produced within defined parameters to produce materials to meet particular specifications
- Only national accreditation bodies can provide accreditation and they review status on an annual basis as with all ISO standards

Table 6.2. Quality controls of microorganisms based on the CABRI quality guidelines

Microorganism	Viability	Purity	Identity	Stability
Plasmids	Confirm presence by growing the host/plasmid combination on appropriate selective medium	Check the texture, the size and the opacity of the colonies grown on selective medium. Check also for homogeneity of the colonies and for fungal and bacterial contaminants	Check plasmid length by determination of the molecular weight of the covalently closed circle (ccc) DNA or by analysis of the restriction site pattern	Confirm presence by growing the host/plasmid combination on appropriate selective medium
Yeasts and Filamentous Fungi	Check growth on appropriate medium	Check for fungal and bacterial contaminants using macro- and microscopic observations on the strain grown on appropriate medium	Identify up to species level using morphological (macroscopic and microscopic) features, and where appropriate physiological and molecular tools dependent on the taxa	Check viability and purity. Confirm identity
Bacteria	Check growth on appropriate medium	Check for contaminants using macro- and microscopic observation on the strain grown on appropriate medium	Identify up to species level using morphological (macroscopic and microscopic), and physiological tools, where appropriate, use molecular tools	Check viability and purity. Confirm identity
Plant Viruses	Test infectivity to indicator plants and propagation hosts	Use electron microscopy observations	Combine host plant reaction, electron microscopic observations and reaction with specific antisera. Where appropriate, use molecular tools	–
Metazoan Cell Lines	Percentage of viable cells	Check for adventitious metazoan cells, mycoplasma, fungal, yeasts and bacterial contaminants. Where appropriate, culture must be screened for viruses	Use at least morphological analysis and isoenzyme analysis	–
Plant Cell Lines	TTC and FDA tests	Check for yeast, fungal and bacterial contaminants	Analysis of culture extracts by UV spectrometric or HPLC methods	–

Table 6.3. Preservation methods, distribution forms, sets of data describing the products as required in the CABRI guidelines

	Preservation	Distribution forms	Useful information
Plasmids	Two of the following methods: • Cryopreservation of the host/plasmid (H/P) below -70°C • Cryopreservation of the H/P in liquid nitrogen • Lyophilization of the H/P • Preservation of the plasmid DNA (preferably precipitated under ethanol) can also be applied as a preservation method	• Actively growing H/P on agar slant or in liquid medium • Cryopreserved H/P in dry ice • Lyophilized H/P • Pure DNA	Plasmids containing genes that may tend to destabilize the physical and/or functional integrity (either by insertion, deletion or point mutation) should preferably be deposited, maintained, tested and delivered as pure DNA
Yeasts and Filamentous Fungi	• Under sterile oil • Under sterile water • Frozen at -70°C • Cryopreservation below –140°C • Lyophilization of the strain	• Actively growing strain on agar slant • Lyophilized material in vials sealed under vacuum or inert gas • Cryopreserved material in dry ice • Suspensions in liquid	Sporulating strains should be maintained by at least two of the four different preservation methods listed (one of which must be cryopreservation or lyophilization). Non-sporulating strains will be maintained under oil or water and cryopreservation
Bacteria	• Cryopreservation in or above liquid nitrogen or in a freezer below –70°C • L-drying	• Actively growing strain on agar slant • Lyophilized material in sealed vials • Cryopreserved material in dry ice	
Plant Viruses	• Virus maintenance *in situ* • Liquid nitrogen • Freeze-drying	• Lyophilized material in sealed vials • Cryopreserved material in dry ice	–
Metazoan Cell Lines	• Liquid nitrogen	• Cryopreserved material in dry ice	–
Plant Cell Lines	• Calluses grown on Petri dishes • Cryopreservation	• Cryopreserved material in dry ice	–

Table 6.4. A general MDS can be applied to each type of collection

Identification	Accession number, name
Origin	Description, depositor, history
Properties	Indication of key properties and where possible references
Culture conditions	Medium, subculture routines
Restrictions	Permits for handling and distribution
Hazard	Biological Risk Group; indication of toxin production

Table 6.5. Minimum data set for bacteria and Archaea

Field Name	Description
Strain number	Should contain: • Collections' own accession number of the strain
Other culture collection numbers	Should contain: • Accession numbers of the strain in other collections *Cannot be given if the strain is unique in the collection
Restrictions	Should contain: • Release conditions/restrictions • Country-specific shipping restrictions • Restrictions imposed by the depositor etc.
Organism type	Should contain: • Value (B) for bacteria or (A) for Archaea
Name	Should contain the full scientific name of the strain: • Genus name and species epithet • Subspecies * • Pathovar * • Authors of the name * Only if applicable
Intrasubspecific names	Should contain: • Variety • Designation • Epithet • Authors and reference * Does not apply to all strains
Status	Should contain: • Nomenclature status of the strain (like (type), (neotype), etc.) * Does not apply to all strains
Other names	Should contain: • Names used for the strain in the past (not synonyms) that may still be in use * Does not apply to all strains
History of deposit	Should contain the history of the element from depositor to isolator • Name of depositor • Institute of depositor • Name of isolator • Institute of isolator • Names and institutes in between
Conditions for growth	Should contain: • Culture medium • Atmospheric and light conditions • Temperature conditions
Form of supply	Should contain: • The form in which the strain will be sent to the customer
Serovar	Should contain: • The serovar name and author * Required for some medically important species * Does not apply to all strains

Table 6.6. Minimum data set for filamentous fungi and yeasts

Field name	Description
Strain number	Should contain: • Collections' own accession number of the strain
Other culture collection numbers	Should contain: • Accession numbers of the strain in other collections * Cannot be given if the strain is unique in the collection
Restrictions	Should contain: • Release conditions/restrictions • Country-specific shipping restrictions • Restrictions imposed by the depositor etc.
Organism type	Should contain: • Value (F) for filamentous fungi or (Y) for yeasts
Name	Should contain the full scientific name of the strain: • Genus name and species epithet • Subspecies * • Variety * • Forma * • Forma specialis * • Authors of the name * Only if applicable
Race	Should contain: • Name of the race of the strain and authors of the race * Applies only to a small subset of the strains
Status	Should contain: • Nomenclature status of the strain (like (type), (neotype), etc.) * Does not apply to all strains
Misapplied names	Should contain: • Names incorrectly applied to the strain (misidentifications) that may still be in use * Does not apply to all strains
History of deposit	Should contain the history of the element from isolator to depositor: • Name of isolator • Institute of isolator • Name of depositor • Institute of depositor • Names and institutes in between
Conditions for growth	Should contain: • Culture medium • Light conditions • Temperature conditions
Form of supply	Should contain: • Form in which the strain will be sent to the customer

ISO/IEC 17025 is the main standard used by testing and calibration laboratories. Originally known as ISO/IEC Guide 25, ISO/IEC 17025 was initially issued by the International Organization for Standardization in 2000. There are many commonalities with the ISO 9000 standard, but ISO/IEC 17025 adds in the concept of competence to the equation. And it applies directly to those organizations that produce testing and calibration results. Since its initial release, a second release was made in 2005 after it was agreed that it needed to have its quality system words more closely aligned with the 2000 version of ISO 9001 and a further release in 2017 that aligned it with ISO 9001:2015.

The standard was first published in 2001 and on 12 May 2005 the alignment work of the ISO committee responsible for it was completed with the issuance of the revized standard. The most significant changes introduced greater emphasis on the responsibilities of senior management, and explicit requirements for continual improvement of the management system itself, and particularly, communication with the customer.

There are two main sections in ISO/IEC 17025 - Management Requirements and Technical Requirements. Management requirements are primarily related to the operation and effectiveness of the quality management system within the laboratory. Technical requirements address the competence of staff, methodology and test/calibration equipment.

Laboratories use ISO/IEC 17025 to implement a quality system aimed at improving their ability to consistently produce valid results. It is also the basis for accreditation from an Accreditation Body. Since the standard is about competence, accreditation is simply formal recognition of a demonstration of that competence.

A prerequisite for a laboratory to become accredited is to have a documented quality management system. The usual contents of the quality manual follow the outline of the ISO/IEC 17025 standard.

PERSONNEL The laboratory management shall ensure the competence of all who operate specific equipment, perform tests and/or calibrations, evaluate results, and sign test reports and calibration certificates.

The management of the laboratory shall formulate the goals with respect to the education, training and skills of the laboratory personnel.

The management shall authorize specific personnel to perform particular types of sampling, test and/or calibration, to issue test reports and calibration certificates, to give opinions and interpretations and to operate particular types of equipment. The laboratory shall maintain records of the relevant authorization(s), competence, educational and professional qualifications, training, skills and experience of all technical personnel, including contracted personnel. This information shall be readily available and shall include the date on which authorization and/or competence is confirmed.

DOCUMENT CONTROL The laboratory shall establish and maintain procedures to control all documents that form part of its quality system (internally generated or from external sources), such as regulations, standards, other normative documents, test and/or calibration methods, as well as drawings, software, specifications, instructions and manuals.

All documents issued to personnel in the laboratory as part of the quality system shall be reviewed and approved for use by authorized personnel prior to issue. Quality system documents generated by the laboratory shall be uniquely identified. Such identification shall include the date of issue and/or revision identification, page numbering, the total number of pages or a mark to signify the end of the document, and the issuing authority or authorities. Changes to documents shall be reviewed and approved by the same function that performed the original.

ISO 15190

- This deals with medical laboratories requirements for safety
- Only addresses this aspect of quality management
- Assessed and awarded by accreditation bodies

ISO 15189

- Medical Laboratories – Particular requirements for quality and competence; applicable to medical laboratories but does not address the requirements of all microbiology laboratories.
- Particularly addresses health and safety issues.
- Assessed and awarded by accreditation bodies.

ISO 17034

- A guidance document for the production of reference material production, originally known as ISO Guide 34.
- Not all accreditation bodies accredit laboratories to this but it is now used in conjunction with elements of ISO 17025 for reference material producers.
- This guide was written for reference materials used for the calibration of measuring equipment and for the evaluation or validation

of measurement procedures such as the pharmacopoeia standards. It is used for the production of some biological reference materials but has yet to be applied to microbiology. The Guide may be suitable for use in the Biological Resource Centre (BRC) context.

- The 2016 revision of ISO 17034 includes all the general requirements that a reference material producer must demonstrate as well as the validation of the reference materials themselves. Some countries have applied the guide on its own but as such is not a full standard. It is often used in conjunction with ISO 17025.

The elements covered in ISO Guide 34 are similar to those of the OECD BRC General Standard

Application of ISO 17034 also requires implementation of ISO Guides 30, 31 and 35. ISO 17034 requires that ISO Guides 30, 31 and 35 should be complied with. These guides were prepared by ISO committee for reference materials (REMCO) to help in the process of preparation, defining, certifying and use of reference materials. Reference materials (RMs) and Certified Reference Materials (CRMs) make possible the transfer of the values of measured or assigned quantities (physical, chemical, biological or technological) between one place and another. These guides can easily be applied to living, reference materials but these were not the prime focus when they were drawn up. In the case of biological RMs they enable the properties to be expressed conveniently in arbitrary units. ISO Guides are intended essentially for internal use in ISO committees or in some cases for the guidance of member bodies when dealing with matters that would not normally be the subject of an international standard.

ISO GUIDE 30:2015 TERMS AND DEFINITIONS IN CONNECTION WITH REFERENCE MATERIALS This Guide recommends terms and the meanings that should be assigned to them when used in connection with reference materials e.g. RM, CRM, primary standard, secondary standard, stability, traceability etc. to achieve uniformity in use. Although they seem primarily addressing physical, chemical or technological reference materials biological references are included with a specific citation of vaccines to which an international unit has been assigned by WHO. Biological materials are examples of RMs and CRMs that have properties which, because they cannot be correlated with an established chemical structure or for other reasons, cannot be determined by exactly defined physical and chemical measurement methods.

There are no intrinsic problems with ISO Guide 30 being applied to living reference materials but this may well be overkill for microbiology laboratories and culture collections.

ISO GUIDE 31:2015 CONTENTS OF CERTIFICATES OF REFERENCE MATERIALS A certificate of reference material is a short document condensing all the information about a provided sample. Certificates are becoming increasingly necessary with the increase in the number of reference material producers. ISO Guide 31 summarizes the content and reason for a certificate. A certificate should communicate information about a reference material from the producer to the user; in essence, this information is a statement of the certified property values, their meaning, and their confidence limits.

The remainder of the information presented is peripheral to this central statement and has two purposes: to describe the general nature and use of the material, and to assure the user of its integrity. In a well-prepared certificate, the information should be presented in a form that enables the user to draw their own conclusions about the integrity of the certified material. The headings of content are: the general particulars of the certifying organization and the reference material (name, sample, number, date of certificate, etc.); a description of the material and its intended use; the certified values; their confidence limits, and an explanation of these values and the techniques for their measurement; the references and the names of participating analysts and certifying officers.

ISO GUIDE 35:2017 CERTIFICATION OF REFERENCE MATERIALS – GENERAL AND STATISTICAL PRINCIPLES ISO Guide 35 states that the process of certification should be carried out using well-characterized measurement methods that have high accuracy as well as precision and provide property values traceable to fundamental units of measurement.

A RM must be suitable for the exacting role it performs, be fit for purpose; stable; homogeneous; have property values sufficient to end use; have clear documentation and preferably certification to ISO Guide 31.

ISO Guide 35 covers the role of RMs in measurement science and refers their role in the International System of Units (SI); it describes connections with analytical chemistry, covers measurement uncertainty giving illustrative examples. The certificate must contain statements of uncertainty that are unambiguous supported by some statistical evidence. Of relevance to microbiology laboratories and collections, it describes certification based on inter-laboratory testing and also on a metrological approach.

A National Culture Collection Quality Management System

Aim of QMS

The Quality Management System (QMS) draws together procedures of all its member collections to provide consistently reliable and reproducible products and services. Some years ago, a UKNCC Quality Policy was defined for the affiliated collections in order to provide a unified quality of service to their customers. The Quality Manual and the Laboratory Procedures Manual were designed to provide working documents to define the ways and means by which a common quality management system could be achieved. The key features of the process and procedures identified during this exercize and built on by work done in collaboration with the formulation of OECD best practice (OECD, 2007), MIRRI's approach (Martin *et al.*, 2015) and observations of the ISO Biobanking Standard ISO 20387:2018 (https://www.iso.org/standard/67888.html) including quality audit and reviews are given below. They formed the basis for each collection to adopt a certification or accreditation system, in general most adopting the ISO 9000 series certification approach. Three levels of documentation are recommended:

The Quality Manual: Deals primarily with policy administration and management aspects

The Laboratory Procedures Manual or Standard Operating procedures (SOPs): Detailed protocols for all procedures in the collection to ensure that new members of staff can be trained and to ensure reproducibility and enable monitoring

Laboratory records: A series of laboratory records are maintained including laboratory books, specified in the individual UKNCC member collection Laboratory Procedures Manual.

ESSENTIALS IN MICROBIAL COLLECTION STANDARDS.

Identified, authenticated and characterized strains: Collecting material without descriptive data will cause unnecessary duplication and waste resources. An organism is not very useful if nothing is known about it. If the organism cannot be fully identified then descriptive data, photomicrographs, metabolic profiles, sequencing data are extremely useful.

Purity: Strains must be pure (or mixtures defined or noted). There would be exceptions where strains cannot be grown without their symbiont, host or food organisms but it would be imperative that this is recorded and defined.

Viability: For fungi the preferred level of viability might be the germination and subsequent development of in excess of 75% of propagules /cells although an acceptable level may be set at 50% or possibly lower for some cell types. Any deviation from the standard would need to be explained and recorded.

Stability of strain properties: A programme of tests to ensure stability of strains must be put in place.

Methodology: To maintain standards, methodology should follow:

- Optimized techniques and standard procedures
- A procedures manual to ensure continuity
- Accepted level of deviation from measurable parameters

Equipment: must be maintained, calibrated and operated to set limits to ensure traceability and reproducibility.

Long-term security: For long-term security there must be:

- Long-term preservation
- Samples stored by a minimum of two techniques

- Working and security stocks
- Samples stored on another site as a disaster avoidance measure

Auditing/ monitoring: It is essential that adherence to set standards is monitored at every level.

Compliance with legislation: Examples of the areas covered by regulations are:

- Packaging, shipping and transport
- Quarantine
- Health and safety
- Patenting
- Access to national genetic resources

Sections in the UKNCC Quality Manual

The quality manual has the following sections:

- Quality policy
- Responsible bodies recognizing the UKNCC Quality Management System
- Aims and form of Quality Management System
- Documentation
- Organization and management of the individual members
- Laboratory environment in which they work
- Accessions to the collection
- Preservation procedures used
- Stock control of the preserved organisms
- Supply
- Confidentiality
- Staff - qualifications and training
- Quality audit and quality review
- Quality control system
- Equipment: calibration, testing and maintenance
- Measurement traceability and calibration
- Methods and procedures
- Laboratory accommodation and environment
- Receipt and handling of organisms
- Record system
- Handling of complaints and anomalies
- Outside support services and supplies
- Site security

Certification or Accreditation standards applied within the UKNCC

- ISO 9001
- ISO 17025

Other Best Practices

ISBER

The desire to preserve biological and environmental specimens for research purposes and to ensure species biodiversity requires the development of methods for long-term storage that will enable their effective future use. Sharing successful strategies for accomplishing this goal was one of the early driving forces for the International Society for Biological and Environmental Repositories (ISBER). In addition, ISBER fosters education and research and promotes quality and safety in all activities relating to specimen collection, storage and dissemination.

ISBER's Best Practices for Repositories (Best Practices) reflect the collective experience of its members to provide repository professionals with a comprehensive foundation for the guidance of repository activities (https://www.isber.org/page/BPR). These Practices reflect input from individuals within and outside ad hoc committees. Best Practices will be reviewed periodically and will be revised to reflect advances in research and technology. All revisions are subject to approval by the ISBER Council. These Practices reflect the most effective approaches to the establishment and running of specimen collection facilities and are not intended as required practices.

ISO 9000 Certification

An industry standard to ensure good management practices.

NCI (National Cancer Institute) for Cancer Cells

The NCI has identified salient guiding principles that define state-of-the-science biospecimen resource practices, promote biospecimen and data quality and support adherence to ethical and legal requirements. The current NCI Best Practices do not comprise detailed lab procedures; they consist of principles by which such procedures should be developed by biospecimen resources.

Genebank Standards

ULTIMATE OBJECTIVE - Long-term safe and sustainable conservation.

Provides targets at two levels: acceptable - minimal levels preferred - higher, safer standards.

The standards acknowledge the difference in the storage of original material not usually for direct distribution and distribution material.

The standards demand set treatments and conditions in many areas:

- Control of environmental conditions
- Standardized drying protocols
- Seed cleaning and health standard criteria for storage containers
- Defined seed storage conditions
- Size of sample preserved
- Viability monitoring
- Minimum information
- Standards for exchange
- Adherence to quarantine and other regulations
- Personnel and training
- Safety and security of the germplasm

ISO 17034

This guide was written for reference materials used for the calibration of measuring equipment and for the evaluation or validation of measurement procedures such as pharmacopoeia standards. It is used for the production of some biological reference materials but has yet to be applied to microbiology. The Guide may be suitable for use in the Biological Resource Centre (BRC) context. The 2000 revision of ISO 17034 includes all the general requirements that a reference material producer must demonstrate as well as the validation of the reference materials themselves.

Impact of QM on Preservation Procedures

The quality management of techniques requires that methods are repeated exactly to specifications, equipment runs in a reproducible manner and data are recorded and checked to allow traceability and audit. In preservation processes such as freeze-drying, simple questions are asked and evidence required giving reliable answers. The principles used would also cover techniques such as cryopreservation.

The following bulleted list identifies elements in this process, it is not exhaustive but it presents the principles involved (evidence provided at every level to prove that what is said will be done is actually performed, and that those carrying out the work are competent and the product is fit for purpose).

Some of the questions in the process of QM in freeze drying are:

- Is the organism identity confirmed or traceable to source?
- Is the organism purity and viability confirmed?
- Are the suspending medium formula and preparation traceable?
- Is equipment use recorded?
- Are the key parameters measured and measurements calibrated to an accepted standard – pressure/temperature?
- Are the post-preservation tests completed satisfactorily?
- When was the work done and by whom?
- Did the persons involved have the correct training?
- Were environmental conditions controlled etc.?

Records should be kept to demonstrate that the correct organism is in the resulting ampoule and it has followed the desired protocol to deliver the required product which has tested satisfactorily.

Impact of QM on Freeze-Drying

Assessing Success – Viability is not Enough

Knowing that the organism is alive does not guarantee that the organism will retain the property for which it is kept. Sometimes it can be screened for a single property after preservation to demonstrate that it still produces it at the same level. However, an organism may be used for other purposes and therefore it is helpful to know whether it remains a true representative of the species or strain. It is not practical to produce a full metabolic profile before and after preservation so a number of tests can be performed to predict stability of the strain.

Confirmation of Organism Identity or Traceable to Source

- Conformity to species descriptions either to morphological or biochemical characteristics or reproducible molecular sequences before and after preservation

- Morphological identifications should be carried out by trained persons with the relevant experience and knowledge – this should be documented in training records
- Biochemical/Molecular test procedures must be documented and followed with the appropriate materials and equipment used and performed by competent personnel

IT IS PURE AND VIABLE

- Utilizing techniques that accurately measure these qualities, i.e. no contaminants found using specific techniques and percentage re-growth of cells after treatment; the equipment used to measure the levels or count the cells etc. must be tested and calibrated to ensure they read accurately and that they do not deteriorate over time

SUSPENDING MEDIUM FORMULA AND PREPARATION IS TRACEABLE

- Formulae are documented, no deviation in use of ingredients occurs, the source and constituents remain unchanged and the medium undergoes quality control testing prior to use

THE EQUIPMENT USED IS RECORDED

- To ensure that batches of materials can be traced to a piece of equipment at a given point in time in case problems are encountered retrospectively
- The procedure used, the cooling rate, temperature and pressure measurements are recorded to document the process accurately so that it is reproducible
- Maintenance of equipment must meet manufacturer's recommendations and be recorded

KEY PARAMETERS MEASURED AND MEASUREMENTS CALIBRATED TO AN ACCEPTED STANDARD – PRESSURE/TEMPERATURE

- All equipment needs to be calibrated to standards, for example temperature recorders and vacuum gauges need to be shown that they read accurately so that it is certain that materials remain frozen during drying and that pressures within the freeze-drying chamber ensure that water sublimes from solid ice to vapour to avoid damage to cells

POST-PRESERVATION TESTS COMPLETED SATISFACTORILY

- It is essential that characteristics of preserved materials match those of the material prior to treatment; this requires that conditions for recording pre-preservation data are reproduced identically for the repeat tests after preservation; in turn requiring detailed recording of parameters and conditions
- Residual moisture contents measured on calibrated equipment and storage conditions of preserved materials recorded

WHEN THE WORK WAS DONE AND BY WHOM?

- Technique reproducibility may well be operator sensitive and knowing when something was done may well be critical because if there was, for example, a contamination event at a particular time, it would be essential to know which product may be affected

DID THE PERSONS INVOLVED HAVE THE CORRECT TRAINING?

- It is essential that operators are competent and able to deliver the results needed

ENVIRONMENTAL CONDITIONS CONTROLLED ETC

- Techniques are often environmentally sensitive and these must be monitored and controlled

Summary

It is crucial that the organisms for use in biotechnology are maintained in a way that will retain their properties. Biological resource collections must ensure a quality product providing standard reference material that will give reproducible results. To achieve this, collections must apply quality control and assurance measures to maintain these standards, consider the needs of users and of the facilities and resources available. As discussed there are several standards and accreditation mechanisms used in microbiology culture collections and it is essential that the most appropriate is chosen for the specific needs of the collection. It is important to start with the international guidance specifically designed for collections by the WFCC and ensure that whatever system is chosen, this forms the basis. A collection may be part of a larger organization and quality management systems should be consistent across it to avoid unnecessary complications and overburdensome processes.

References

AFNOR (2008) French Standard NF S96-900 "Quality of biological resource centres (BRCs) – Management system of a BRC and quality of biological resources from human or micro-organism origin". Association Française de Normalisation 11, rue Francis de Préssensé F – 93 571 La Plaine Saint Denis cedex http://www.afnor.fr

Anon (2010) *The WFCC Guidelines for the Establishment and Operation of Culture Collections (Online)*, http://www.wfcc.info/index.php/wfcc_library/publication/.

Forti, T., da S.S. Souto, A., Roberto C., do Nascimento, S., Nishikawa, M.M., Hubner, M.T.W., Sabagh, F.P., Temporal, R.M., Rodrigues, J.M. and da Silva, M. (2016) Evaluation of a fungal collection as certified reference material producer and as a biological resource center. *Brazilian Journal of Microbiology* 47(2), 403-409. DOI: 10.1016/j.bjm.2016.01.021. https://www.elsevier.es/en-revista-brazilian-journal-microbiology-490-articulo-evaluation-fungal-collection-as-certified-S1517838216000708

Fritze, D., Smith, D. and Martin, D. (2012) *Final report on the GBRCN Demonstration Project*. GBRCN Secretariat, Germany. ISBN 978-3-00-038121-8

Hawksworth, D.L. and Kirsop, B. (1988) *Living Resources for Biotechnology*. Cambridge University Press, UK. https://blackwells.co.uk/bookshop/product/Filamentous-Fungi-by-D-L-Hawksworth-B-E-Kirsop/9780521352260

Hawksworth, D.L. and Schipper, M.A.A. (1989) Criteria for consideration in the accreditation of culture collections participating in MINE, the Microbial Information Network Europe. *World Journal of Microbiology and Biotechnology* 5, 277–281. https://link.springer.com/article/10.1007/BF01741757

Kirsop, B.E. and Hawksworth, D.L. (eds) (1994) *The Biodiversity of Microorganisms and the Role of Microbial Resource Centres*. World Federation for Culture Collections, Braunschweig, Germany.

Martin, D., Stackebrandt, E. and Smith, D. (2015) MIRRI Promoting Quality Management Systems for Microbiology. *EC Microbiology* 2.2 (2015): 278-287. https://www.ecronicon.com/ecmi/microbiology-ECMI-02-000036.php

OECD (2001) *Biological Resource Centres: Underpinning the future of life sciences and biotechnology*. OECD Publications, Paris, France. pp 66. https://www.oecd.org/sti/emerging-tech/2487422.pdf

OECD (2007) *Best Practice Guidelines for Biological Resource Centres* (June 2007), https://www.oecd.org/sti/emerging-tech/38777417.pdf

Peixoto, A., Barbosa, M., Menezes, M. and Maia, L. (2006) *Guidelines and Strategies for the Modernization of Brazilian Biological Collections and the Consolidation of Integrated Information Systems on Biodiversity*. Center for Management and Strategic Studies: Ministry of Science and Technology, Brasília, 324 pp. https://www.cgee.org.br/documents/10195/734063/Livro+Biocomplexidade_4399.pdf/6ad794fb-f37e-4b16-985f-d79e986e89c2?version=1.2

Rohde, C., Smith, D., Martin, D., Fritze, D. and Stalpers, J. (2013) Code of conduct on biosecurity for Biological Resource Centres: procedural implementation. *International Journal of Systematic and Evolutionary Microbiology* 63, 2374-2382. https://doi.org/10.1099/ijs.0.051961-0

Romano, P., Aresu, O., Manniello, M.A. and Parodi, B. (2004) Interoperability of CABRI Services and Biochemical Pathways Databases. *International Journal of Genomics* 5, Article ID 549183, 4 pages. https://doi.org/10.1002/cfg.376

Ryan, M.J., McCluskey, K., Verkleij, G. and Smith, D. (2019) Fungal biological resources to support international development: Challenges and opportunities. *World Journal of Microbiology and Biotechnology* 35, 139. https://doi.org/10.1007/s11274-019-2709-7

da Silva, M., Chame, M. and Moratelli, R. (2020) Fiocruz Biological Collections: strengthening Brazil's biodiversity knowledge and scientific applications opportunities. *Biodiversity Data Journal* 8: e53607. https://doi.org/10.3897/BDJ.8.e53607

Simões, M.F., Dias, N., Santos, C. and Lima, N. (2016) Establishment of a Quality Management System based on ISO 9001 Standard in a public service fungal culture collection. *Microorganisms* 4, 21. https://doi.org/10.3390/microorganisms4020021

Smith, D. (2011) Biological Resource Research Infrastructures to drive innovation in microbiology. *Microbe* 6(11) (November) p482.http://www.microbemagazine.org/index.php/11-2011-home/4035-research-resources-help-to-drive-innovation-in-microbiology

Smith, D. (2012) Culture collections. *Advances in Applied Microbiology* 79, 73–118. https://doi.org/10.1016/B978-0-12-394318-7.00004-8

Smith, D. and Ryan, M.J. (2008) The impact of OECD best practice on the validation of cryopreservation techniques for microorganisms. *Cryoletters* 29, 63-72. PMID: 18392291

Smith, D., Fritze, D. and Stackebrandt, E. (2013) Public Service Collections and Biological Resource Centres of microorganisms. In: Rosenberg E., DeLong, E.F., Lory, S., Stackebrandt, E. and Thompson, F. (eds) *The Prokaryotes*. Springer, Berlin, Heidelberg. https://doi.org/10.1007/978-3-642-30194-0_14

Smith, D., Kermode, A., Cafà, G., Buddie, A., Caine, T. and Ryan, M.J. (2020) Strengthening mycology research through coordinated access to microbial culture collection strains. *CABI Agriculture and Bioscience* 1, 2. https://doi.org/10.1186/s43170-020-00004-9

Stackebrandt, E. (2010) Diversification and focusing: strategies of microbial culture collections. *Trends in Microbiology* 18, 283–287. https://doi.org/10.1016/j.tim.2010.05.001

Stackebrandt, E., Schüngel, M., Martin, D. and Smith, D. (2015) The Microbial Resource Research Infrastructure MIRRI: Strength through coordination. *Microorganisms* 3, 890-902. https://doi.org/10.3390/microorganisms3040890

WHO (2020) *Laboratory Biosafety Manual*. 4th Edition. World Health Organization, Geneva, Switzerland. https://www.who.int/publications/i/item/9789240011311

7

Biological Resource Centre Business Plan

Introduction

The long-term sustainability of a culture collection, a biological resource centre (BRC), will require the generation of a robust business plan. The previous chapters have demonstrated that there are many aspects to running a culture collection, it is not simply the preservation of cultures and putting them into storage until they are needed. It is a different challenge if all you wish to do is retain a number of cultures to the end of a three-year project than if you wish to ensure they continue to be available long into the future for further work or they are critical production strains on which the business depends. You need to first consider the type and number of strains you wish to keep, understand what it is that you wish to achieve by maintaining them, consider the resources you have and any alternatives to keeping the organisms yourself, for example depositing them in a public service collection. The previous chapters have presented what running a collection entails and provided options for the key operations, using this knowledge a realistic plan covering all the required elements specific to your needs can be put together and costed.

An accession policy is needed that provides an understanding of where the organisms will come from and a programme of acquisition designed at a rate that your resources can manage, leading to the size and coverage needed for the work to be serviced by the collection. When acquiring the microorganisms, you must consider the regulatory environment around access to them, handling them, their use and if you are allowed to make them available to others. As you acquire them you will need the finances to store them and you will need to plan for the expansion of the collection, the skills, resources, facilities and space needed. You may then start to consider where the latter will come from. The initial investment would need a business case to explain the benefits that will come from making that investment. The solution might be a cheaper one if you paid someone else to do it for you.

The costs of running a collection are high and most project funding does not cover such activities. Even public service culture collections have struggled to remain sustainable and have had to develop innovative means to finance their activities. Many have disappeared over the years especially those initiated in a research laboratory relying on a single scientist. What happens when the scientist retires? Is it clear what is in that collection and how valuable it is for future work and has it been maintained correctly? These will be needed by anyone that such collection may be offered to or the institutional managers that will decide on its future.

A culture collection, whatever its size, needs a plan for the future, a goal, a strategy to deliver

DOI: 10.1079/9781800622135.0007

that goal and a financial plan to make it happen. If it is to continue to operate, its content and value needs to be apparent to institutional policy makers while if it is to be passed on to others, income lines to provide the needed revenue or other sources of funding need to be identified. Traditionally these have been culture sales, institutional funding, contracted services, contracted research and sponsorship. In addition to this, exploitable Intellectual Property (IP) generated through the characterization and screening of the collection holdings could provide significant income through commercial utilization as long as this is compatible with the CBD, institutional policy and the terms under which the strains were received. Culture sales together with preservation contracts income would normally be insufficient to cover the collection running and maintenance costs and to break even additional support or revenue lines would be needed. Competition for culture provision is severe worldwide not only from other biological resource centres but because more organisms are exchanged between individual scientists free of charge than are obtained from collections. To support a living collection through culture sales would take a major shift in scientist behaviour, which is impossible for one collection to achieve on its own. Culture sales will remain one source of income but other mechanisms will have to be found.

To attain a level of income that will allow expansion and development in the medium term would require for example, a portfolio of new products and services, mechanisms for funding a conservation role and increased competitiveness in securing research funding. Future financial security for the collection may depend upon it becoming a recognized Biological Resource Centre as defined by the OECD Biological Resource Centre initiative providing much more than the biological resource with information, expertise, facilities and its role in knowledge generation. The BRC must find a mechanism to support all its activities to enable it to maintain its collections, expertise and fundamental research. As previously mentioned, screening for potential properties for exploitation could provide opportunities for further revenue but this will require investment and time and until this can be done systematically the collection must utilize a range of income opportunities for financial self-sufficiency.

This chapter outlines the medium-term mechanisms for achieving a sustainable business strategy for a BRC and may I suggest any modest sized collection which purports to achieve the same goals. Remember the ABC of BRCs? The culture collection needs to provide **a**uthentic stable strains with reproducible properties preserved by the **b**est techniques and provide **c**onfirmed and validated information to facilitate their access and use. It is a matter of scale. There is not one solution but there are many commonalities in securing the most appropriate financial plan for the sustainability of your collection (Smith *et al.*, 2014). Just as in Chapter 6 when it comes to implementing a quality management system the collection can develop in a stepwise manner and as the available investment or revenue lines allow, the collection can grow with it. Best practice dictates that you have the vision at the start of the process and have a clear idea where you are heading. A sound plan can be put together to deliver this vision but make allowances as plans can change as new technologies, opportunities and challenges will appear along the way.

Key assumptions

- The collection owners allow it to operate commercially and that it has initial core funding and investment.
- The long-term operation and the stored biological resource collection can be financially supported by the commercial utilization of biodiversity.
- Funding is made available for appropriate research programmes and for countries to meet their Convention on Biological Diversity (CBD) obligations to conserve and sustainably utilize their biodiversity.
- The OECD Biological Resource Centre initiative does form a platform for development of BRCs and countries provide an improved environment for BRCs to operate, enabling them to provide the microbial resources needed to underpin discovery and help find solutions to our food, healthcare, carbon recycling, societal and economic challenges of today.
- Exploitable IP is discovered within the BRC resources.

Outputs and Benefits a Culture Collection Delivers

A sound business case requires a clear understanding of the benefits that having your own organized and resourced collection will deliver. It must have a clear understanding of processes, resources and facilities it needs and the costs involved. The fundamental reason for considering establishing a collection and not using the resources and services of those that currently exist that offer a public service is that there are significant numbers of strains involved, they may need specialist treatment and be readily accessible as working cultures. There are so many more advantages that you gain from holding a collection that can help justify the investment (NSF, 2020).

Benefits from a quality managed culture collection

It is crucial for the benefit of science, future study and the confirmation of published research that the strains on which the published science hinges are maintained in a stable condition for future use (Stackebrandt *et al.*, 2014). The deposit in a public service collection or an in-house culture collection that meets industry standards (Biobank/ BRC guidance or standard see Chapter 6) provides security and other benefits such as:

- Protection of the investment made in isolation and characterization of the strain making them available for future use without having to re-isolate and re-identify (if that is even still possible with for example land use change or pollution)
- Coordinate the microbiological resources to serve the company or institutional needs from a single internal source and enable the introduction of optimal process and functionality
- Preserve strains isolated and characterized by the organization without having to meet an external collection's accession criteria and capacity restrictions
- Reproducible and consistent strain properties for continuity of production and applications
- Viable and stable inocula for future research and development for example to discover new properties and product lines
- Strains available for confirmation of research results and to underpin good science
- Stocks to share with colleagues for internal research projects and partners for external joint ventures
- Immediate and ready access for continuity of work
- Legal clarity for use
- Samples available for training and education
- Reference materials for comparative purposes

Value from CABI's living collection of fungi and bacteria

CABI has built its collections from its very foundation in 1920 and since then has amassed a collection of almost 30,000 strains; 28,000 fungi and 2000 bacteria from the fields of agriculture and the environment. It became a National Collection in 1947 and has offered access to the cultures and expertise to the scientific community as a fee-based service for over 70 years (Report, 1947; Aitchison and Hawksworth, 1993).

The collection has intrinsic value and its potential for microbial interventions in agricultural systems is yet to be fully realized. It is unique in terms of its strain origin and what is currently known about the contents of the CABI collection is but the tip of the iceberg. For example, sequencing the collection could open up many more opportunities through gene hunting, natural products and facilitate an understanding of the microbiome. It stands as an example of the potential a collection can have and therefore provides valuable information for a business case.

The GRC has a designated role as a public service National Collection, as such and through funded projects it has commitments, to:

- Receive and preserve strains from UK researchers to ensure continuity of their research;
- Supply reference strains to support UK capacities in education, taxonomy, proficiency testing and the bio-based-economy (including material resilience in industry);
- Host the 'National Reference Collection of Crop Pathogens' on behalf of the CHAP

(Centre for Crop Health and Protection) Agricentre;
- Established as the BBSRC BBR crop cryobank for microbiomes with database for associated data.

Examples of outputs from the collection

- Retention of 'Green Muscle™' biopesticide production strains
- Retention of well-known strains under patent and safe deposit for key clients
- Holds Conidia test kit strains; a wide range of *Hormoconis resinae* strains and other fuel contaminants collected over decades on which the Conidia FuelStat kits technology is based
- Maintenance of Fleming's penicillin production strain
- Comparative genomics studies e.g., analysis of *Fusarium* wilt pathogens for current PhD study
- Reference strains for in-house mould resistance testing
- Reference strains for identification service, consultancies and validated sequence databases
- Authentic representative strains of UK fungi for Darwin Tree of Life and other international studies
- Safe storage of strains in commercial development
- Cryostorage of key obligate biocontrol agents for weeds and arthropods including rust fungi

Current revenue lines

- Culture supply including individual, organizational, project resource
- Training courses including novel e-learning - Managing Microbes
- Consultancies in microbial resource-based microbiology
- Industrial consultancies that are underpinned by CABI's applied microbiology expertise
- Project portfolio based on microbial products and resources
- Central support to maintain, add value and supply the CABI living collection (GRC) on behalf of the Member Countries

Future opportunities

- Screening programmes: exploit characterized strains and generated data
- Broader project portfolio via CABI themes and existing and new strategic partners
- Cryopreservation/ supply of microbial communities, natural and constructed microbiomes
- Research and Information platform for the microbiome
- Collaborate with strategic partners in joint business ventures utilizing collection products
- Solicit interesting and rare strains of value to CABI from scientists, in compliance with the Nagoya Protocol

A business model can be developed based upon such an example ensuring revenue lines provide sufficient profit margin to maintain, characterize and expand the collection as a resource for your institution's microbiological science.

Mechanisms of Financial Support to Cover Collection Costs

The most common culture collection business model is a fee for service and product base supported by host institution or government support. Generally, funding is limited and it is often the case that where it is provided it is often balanced against the income received for the various services and products offered by the collection and in many cases the additional income above plan is returned to the funder (Smith *et al.*, 2014). Where budgets are so finely balanced, there is very little financial support for collection development to enable the collections to improve their coverage and incorporate new and advancing technologies. There are few collections that are self-financing but they are able to remain sustainable through a variety of routes. The collection or its host may have opportunities for other types of cost recovery activities and these often revolve around expertise and facilities available. Other kinds of funding sources include support from industry, grants from agencies that support research, development of databases and other tools that complement the core role of collections. Even funding from charitable sources, especially those associated with public health or sustainable development are sources of support. Despite there not being one model for their operational and financial sustainability we can learn a lot from the experience of existing culture collections. It is difficult to estimate these costs as they vary so much dependent upon the difficulty in handling and preserving some microorganisms and the extent to which the biological material is characterized (Smith *et al.*, 2014).

Costs associated with collections vary enormously, in 2012 the cost of accessioning a strain into a World Federation for Culture Collections (WFCC)-affiliated collection was shown to vary enormously (Smith, 2012); the most expensive quoted was £3000 and the average was then £350. Characterization was not included in the estimations. Cost for an accession of a fungus to the CABI collection was close to £760 and £775 (€918) was an estimate for the DSMZ collection of microorganisms in 2017 (Overmann and Smith, 2017). The latter costs include the preservation, characterization and confirmation of identity of the strain but could also include isolation and purification.

Costs also vary due to the technologies used, sequencing costs or other methodologies such as metabolic or protein profiling for checking authenticity (Smith, 2012; Smith *et al.*, 2013) or molecular tools such as Ribotyping or MALDI-TOF (Matrix Assisted Laser Desorption/Ionization, Time of Flight) mass spectroscopy (Overmann and Smith, 2017). Maintenance costs for strains vary depending upon the methodology used and the validated life span of the strain. Costs for ampoules stored in liquid nitrogen depend on how many are being stored and the size of the collection as costs for liquid nitrogen can be negotiated for greater amounts used. Using CABI as an example it costs around £12,500 in liquid nitrogen to store a collection of 20,000-30,000 strains per year. As estimated by Smith *et al.* (2014) a collection of 5,000 strains growing at 500 strains per year and supplying 2,000 strains per year should require at least three members of staff to cope with the authentication, preservation and distribution. If an identification service is to be provided then this will have additional staffing needs and associated costs, as would the addition of any other specialist service. Add to this a collection manager/business developer and researchers, the costs rise quickly. Thus, the cost to run a collection with minimal services and research capacity (the latter to improve service to users) may be over €500,000 per year in Europe (Smith *et al.*, 2014) but will vary depending upon the national and local financial conditions (see Table 7.1 adapted from Smith *et al.*, 2014). The CABI collection has running costs of almost £1 million per annum; it holds almost 30,000 strains of fungi and bacteria, authenticated using DNA sequencing (generally the ITS region for fungi, although to differentiate some genera specific genes are sequenced; for bacteria the 16S rRNA). CABI member countries and its host institution and these subsidies cover about 10% of costs.

Table 7.1. Collections' costs (adapted from Smith *et al.*, 2014)

Cost item	Detail	Cost per year (Thousand €)
Staff	3-10 staff (number depends upon the phylogenetic depth, breadth and related activities)	185-300
Acquisition/deposit of strains	500/year at US$500 per strain	185
Maintenance of strains	5000 strains	30
Distribution of strains	2000 strains	30
Data management	Proprietary database	1
Accommodation, utilities etc.	Laboratories etc.	90
Consumables	Ampoules, chemicals	20
Total		**540-730**

Potential sources of funds

Commercial

- Development and ownership of spin-off biotechnology companies
- Sale of products and services
- Bioinformatics – production of a database to provide microbial solutions to world problems
- Provision of consultancy/training services

Research Programme funding

- A series of projects to meet donor requirements
- Development and funding of several large projects that bring together the full range of the BRC competencies and links them to strategic partnerships to achieve national and international Biodiversity Plans
- Engage Research Programme Funders to protect their investments by paying for deposits in the BRC

Government department support

- Provision of services to Governments to help them achieve their conservation and utilization of biodiversity commitments, their environmental policies and their commitment to millennium goals.

Sponsorship

- Seek a single or a consortium of investors to establish a characterization laboratory with facilities and expertise to characterize the BRC holdings and identify exploitable IP to ensure long-term sustainability.

The Market for Microbial Genetic Resources

Biotechnology has harnessed living organisms in bioremediation or detoxification of polluted sites, biodegradation of waste, biocontrol, production of useful chemicals such as enzymes, biocides, drugs and dyes and in processes such as biotransformation and as food. Micro-organisms will provide even more solutions to our problems in the environment, health, agriculture and the economy. At a time when natural resources are being depleted and alternative sources of energy and food are being sought, microorganisms will provide answers. Action must be rapid because we are also in a period of exploration of microbial potential against a background of rapid species depletion.

The economic value of biodiversity

Data from ten Kate and Laird (1999) demonstrates the value of the commercial use of biological diversity. Pharmaceuticals: more than half of all prescriptions filed in the USA in 1993 contained at least one major active compound now or once derived or patterned after compounds derived from biological diversity. 42% of sales of the 25 top-selling drugs worldwide are biologicals, natural products or entities derived from natural products. The annual global sales of medical drugs are currently £200 billion a year. However, of every 5000–10,000 products screened only one becomes an approved drug. Biotechnology may speed bioassays and the production of new drugs, explain more accurately how drugs act in the human system and may reduce the vast costs of pharmaceutical research and development. The annual market for industrial detergent enzymes is £0.5 billion. Bioremediation of soil in the EC in 2000 was considered to be worth £40 billion. There are huge revenues from fungal derived drugs e.g. cyclosporin, £0.8 billion; clarithromycin, £0.9 billion and amoxicillin, £1 billion. As it is not uncommon to find that 15-30% of fungi found in the tropics or in unexplored environments are unknown to science the potential is enormous.

Savills European Life Sciences 2021 report (https://pdf.euro.savills.co.uk/portugal/european-life-sciences-2021.pdf) states that in 2020 patent applications data from the European Patent Office (EPO) showed a 10.2% increase in pharmaceutical and a 6.3% increase in biotechnology applications and overall it demonstrates the continued importance of the biotechnology industry in economic growth and human welfare. A recent report from the McKinsey Global

Institute analysed the profound economic and social impact of biological innovation and found that biomolecules, biosystems, biomachines and biocomputing could collectively produce up to 60% of the physical inputs to the global economy. Despite the Covid pandemic they reported that between 2019 and 2020, biotech saw double-digit annual growth in fundraising from Venture Capitalists and deals such as partnerships, co-developments and joint ventures (https://www.mckinsey.com/industries/life-sciences/our-insights/whats-ahead-for-biotech-another-wave-or-low-tide). Horizon Europe was launched in January 2021 and it extends until 2027, funding from this programme by the European Commission will be worth over €100 billion to boost European research and innovation (https://ec.europa.eu/info/research-and-innovation/funding/funding-opportunities/funding-programmes-and-open-calls/horizon-europe_en). There are many sources of information on funding opportunities. The collection should initiate horizon scanning. An example is the Knowledge Transfer Networks where for example you can find out what the latest calls for proposals of relevance to industrial biotechnology are (https://ktn-uk.org/news/industrial-biotechnology-funding-round-up/).

Opportunity: To harness microbes with potential and find funding for research and development of products
Need: To characterize strains accessed by the collection to realize their potential and secure IPR

This will be a long-term goal as it can take many years to take new products to the market place. This will also require significant investment. The potential should attract funders to enable the discovery of products but the BRC must maintain significant ownership in order to plough back a major part of profits into supporting its biological collections, characterization and its biosystematics activities. The generation of data at the phenotypic and genotypic levels to help us understand and utilize biodiversity better is of great value. The BRC needs to identify a specific niche and acquire some of this funding to finance information generation.

Seeking the technology or property on which to base products of a spin-off company and creation of the bioinformatics database rely on the same basis: facilities and manpower to characterize the holdings of the BRC. Whilst seeking the mechanisms to achieve this on the large scale the collection must identify a mid-term strategy for financial security. This can be partially based on the continuation of funding agreed through the project. Utilizing the mechanisms for financial support for BRCs outlined above and the strengths and opportunities outlined below, an action plan for sustainability can be identified.

Justification for income levels

Various means should be explored to utilize the potential of the collection holdings to generate revenue lines and new products. Projects related to country obligations to implement the Convention on Biological Diversity (CBD) and placing the BRC to take a lead role for their host country in the OECD Global BRC Initiative, or in implementing Biodiversity Laws open the opportunity to secure project funding for the immediate future. The BRC will need to operate to recognized standard procedures putting into practice the OECD BRC Initiative recommendations. In the long term the new products and added collection value generated income will come online. Learning from experiences of established collections will no doubt enable the BRC to spin off new biotech companies to provide added revenues to support the Science at the BRC. Funding sources for implementing biotechnology solutions in local industries and utilizing the new Biotech Funding that is available world-wide can provide the pump priming investment. A significant investment will be required to characterize the entire collection holdings but this can be done in targeted bites.

The Argument for Government Support

There are several national and international initiatives that require Government action to secure the future long-term advantage and enhanced quality of life that centre on the understanding, conservation and utilization of biodiversity. They include:

OECD Biological Resource Centre Initiative

This OECD Initiative was established in 1999 to try and secure the future of microbial resource collections. Since then the definition has broadened to include a wider range of organisms. A report of the first phase was published in March 2001, entitled *Biological Resource Centres Underpinning the Future of Life Sciences and Biotechnology* and made a call for action by OECD countries and beyond. The report recommends that governments, the scientific community and the private sector work together to achieve five goals:

(i) Establish national BRCs
Selectively seek to strengthen existing *ex situ* collections of biological data and materials and, when needed, create new collections, including in non-OECD countries, and raise those collections to the quality required for accreditation as national BRCs.
(ii) Develop an accreditation system for BRCs based on international criteria
Support the development of an accreditation system for BRCs based upon scientifically acceptable objective international criteria for quality, expertise and financial stability.
(iii) Create international linkages among BRCs
Facilitate international co-ordination among national BRCs by creating an agreed system of linkage.
(iv) Co-ordinate standards, rules and regulations taking BRCs into account
Consider the objectives and functioning of BRCs when establishing and harmonizing national or international rules and regulations.
(v) Establish a global BRC network
Support the establishment of a global BRC network that would enhance access to BRCs and foster international co-operation and economic development.

The second phase has been initiated to put together an implementation plan.

The Convention on Biological Diversity (CBD)

The CBD aims to encourage the conservation and sustainable utilization of the genetic resources of the world and has a number of articles that affect biologists. These cover:

- Development of national strategies for the conservation and sustainable use of biological diversity
- Identification, sampling, maintenance of species and their habitats and the production of inventories of indigenous species
- Encouragement of *in situ* and in-country *ex situ* conservation programmes
- Adoption of economically and socially sound measures to encourage conservation and sustainable use of genetic resources
- Establishment of educational and training programmes and the encouragement of research
- Commitment to allow access to genetic resources for environmentally sound uses on mutually agreed terms and with prior informed consent
- Fair and equitable sharing of benefits and transfer of technology resulting from exploitation of genetic resources
- Exchange of information
- Promotion of technical and scientific co-operation

The CBD requires that Prior Informed Consent (PIC) be obtained in the country where organisms are to be collected before access is granted. Terms, on which any benefits will be shared, must be agreed in advance. The benefit sharing may include monetary elements but may also include information, technology transfer and training. If the organism is passed to a third party, it must be under terms agreed by the country of origin. This will entail the use of material transfer agreements between supplier and recipient to ensure benefit sharing with, at least, the country of origin. Many biological resource centres or culture collections have operated benefit sharing agreements since they began, giving organisms in exchange for deposits and re-supplying the depositor with the strain if a replacement is required. However, huge rewards that may accompany the discovery of a new drug are illusory as the hit rate is often reported as less than 1 chance in 250,000. In the meantime, access legislation and the hope for substantial financial returns from isolated strains are restricting the free deposit in public service collections and the legitimate free movement of strains. An EU DG XII project, Micro-organisms Sustainable Use and Access Regulation International Code of Conduct (MOSAICC) is developing mechanisms to allow traceability and enable compliance with the spirit of the CBD and with national and international laws governing the distribution of microorganisms, whilst not restricting scientific goals (Davison *et al.*, 1998; Smith *et al.*, 2018). The development of such common procedures is an evolutionary process and the co-ordinators of this project have placed the document on their website and will amend it as it develops (http://www.bccm.belspo.be).

The CBD places obligations on the signatory countries to conserve and sustainably utilize their biodiversity. This is a driving force for Governmental CBD Focal Points to assist *ex situ* collections meet these national obligations.

Actions to Design and Implement the Collection Business Plan

The collection's future relies upon harnessing its full potential and unique position to achieve financial security. Armed with the information provided here it is possible to design a business plan for a collection. First and foremost a SWOT analysis: Strengths, weaknesses, opportunities, and threats is needed. The collection's capacity, power to influence in terms of expertise, facilities and location must be identified. Identify the goals (objectives) of your collection and the strategies to deliver them:

- Science goals
- Science strategy
- Projects/products/services goals
- Projects/products/services strategies
- Finance goals
- Finance strategy

This will help determine the resources, facilities, technologies and expertise needed to deliver the goals thus enabling costs and the growth necessary to be estimated. These costs then need to be balanced against the funding mechanism which in the first instance would normally be institutional investment with the goal to implement the revenue lines that will take time to establish. It is most likely the key income lines in the short term would be:

1) Culture sales and preservation contracts
2) Development of new products
3) Project income
4) Subscription for some information services, e.g. information to accompany culture sales, etc.
5) Consultancy and training services

1) *Culture sales and preservation contracts*
(i) Establish Promotion and Marketing Strategies for collection products and services
(ii) Distribute electronic and where necessary hard copy catalogues
(iii) Develop business opportunities
2) *Development of new products*
(i) Develop new products and improved delivery and packaging systems for microorganisms
(ii) Initiate isolation programmes to get new and interesting strains into the BRC
3) *Project income*
(i) Submit high quality project proposals to advance the BRC strategy
4) *Preparatory work for future commercial exploitation of BRC holdings*
(i) Develop a plan to characterize the BRC holdings to add value and discover IP for exploitation
(ii) Attract investment to cover the BRC's conservation role
5) *Consultancy and training services*

Table 7.2 identifies additional services and products that might be added to the collection portfolio and provide the revenue for development and expansion.

Summary

The details of the plan can be developed when the focus of the BRC is agreed and the strengths, weaknesses, threats and opportunities are identified. Critical partnerships need to be developed to achieve:

- Trained human resources
- Expert advice
- Filling of gaps in expertise and facilities
- Financial backing
- Business partners

Some suggested initial partners or links:

CAB International (CABI): microbialservices@cabi.org

UK Biological Resource Centre Network (UKBRCN): https://www.ukbrcn.org/

World Federation for Culture Collections (WFCC): http://www.wfcc.info/contact/

National culture collection networks see: http://www.wfcc.info/collections/networks/

Table 7.2. Potential products and services

Service/activity	Benefits
Patent deposit service	This will provide a facility for deposition of strains included in patent applications - a requirement of patent offices in countries that are signatories to the Budapest Treaty Patenting of microorganisms provides protection to the researcher's/ discoverer's intellectual property rights
Safe deposit service	It will provide the microbe-based industries with a facility that would preserve their strains for long-term storage. This would assure the industry that a backup of their strain is kept in a safe and secure facility. Loss of these industrially important strains could mean a tremendous setback in the operations of the company
Microbial diversity R&D	Novel microorganisms with potential applications in agriculture, medicine, industry, etc. discovered and developed through partnerships *Ex situ* conservation of indigenous microorganisms Depository for strains subject to publication by scientists
Preservation by lyophilization or cryopreservation	Prolongs viability of microorganisms and prevents variation changes(mutation) brought about by frequent subculture. This is particularly important when the strains are to be used for vaccine production and other industrial applications Important in the concentration of heat-labile substances needed in the formulation of certain products like antibiotics and other bioactive compounds Prolongs shelf life of samples, which could not be preserved under ordinary temperatures (ambient and refrigeration)
Regional micro-biodiversity Information System - Electronic linkage with other culture collections and institutions	Online access to other culture collections and institution facilitates fast search for microorganisms held elsewhere and information related to the organism being searched. Inventory of India's microbial diversity available for policy makers, stakeholders and researchers. Rapid exchange of technical information among experts Regular information updates to stakeholders (scientists, policy makers, etc.) Increased public awareness through dissemination of re-purposed information Facilitation of literature surveys for background information for studies or information provision to customers Better contact with potential clients/customers
Human resource development	Training courses and other modes of continuing education contributing to the enhancement of quality of research in areas related to patenting, bioprospecting and conservation of microorganisms for various applications in the pharmaceutical, food and other microbiologically based industries Training of trainers, to provide in-country teaching resource for the development of networks of collections
Supply of cultures	Cultures that are being distributed may be used in the production of certain desirable end-products in the food, pharmaceutical and other microbe-based industries and screened for other properties Reference strains are a must in microbial systematics work, thus providing authenticated cultures for instruction and research purposes improves the quality of research and teaching Microbial cultures are also used in strain improvement work to further enhance the organism's desirable traits, e.g. alcohol production, antibiotic production, etc.

Table 7.2. Continued.

Service/activity	Benefits
	Standard test strains with reproducible properties
	Provide controlled and legitimate access with legal clarity for use in research and development
Quality Management System for operations	A quality bench mark for high quality products, enabling international collaboration and a requirement for good business
Complete microbiological analysis/total count	Provides information on the microbial content of a given sample, which is important in the formulation of microbe-based products like biofertilizers, inoculum, etc. Knowledge of the right kinds and number of microorganisms is necessary to produce high-quality products for field application to improve yields of agricultural crops and other products
	Provides information on the quality of raw materials that will be used in production. Knowledge of the types of microflora in the raw material will enable a producer to decide whether to discard a contaminated raw material or modify their production processes
Microbiological analyses of food and water	Provides information on the quality of food and water - whether they are safe for human consumption or are suitable for use in the production of certain products requiring high-quality water
Identification services	Provides value-added information on the morphological, biochemical, and other properties of an unknown isolate. The information will be used as the basis for future application in industry or for further characterization and strain improvement studies
Other services such as:	Provision of added information required by clients for various purposes
	Phytosanitary testing
	Publication of technical information based on the catalogue of strains
	Preparation of teaching aids
	Developing property screens to support Indian bioindustry
	Development of diagnostic kits for the early detection of pathogens

References

Aitchison, E.M. and Hawksworth, D.L. (1993) *IMI: Retrospect and Prospect*. CAB, Wallingford, UK.

Davison, A., Brebandere, J. de and Smith, D. (1998) Microbes, collections and the MOSAICC approach. *Microbiology Australia* 19(1), 36-37.

NSF (2020) National Academies of Sciences, Engineering, and Medicine 2020. *Biological Collections: Ensuring Critical Research and Education for the 21st Century*. The National Academies Press, Washington, DC. https://doi.org/10.17226/25592.

Overmann, J. and Smith, D. (2017) Microbial Resource Centres Contribute to Bioprospecting of Bacteria and Filamentous Microfungi. In: Hawksworth, D., Paterson, R. and Lima, N. (eds) *Bioprospecting – Successes, Potential and Constraints. Topics in Biodiversity and Conservation 16*. Springer. pp 51-79. ISBN: 978-3-319-47933-0 (Print) 978-3-319-47935-4 (Online). http://link.springer.com/book/10.1007/978-3-319-47935-4

Smith, D. (2012) Culture collections. *Advances in Applied Microbiology* 79, 73–118. https://doi.org/10.1016/B978-0-12-394318-7.00004-8

Smith, D., Fritze, F. and Stackebrandt, E. (2013) Public Service Collections and Biological Resource Centres of Microorganisms. In: Rosenberg, E., De Long, E.F., Lory, S., Stackebrandt, E. and Thompson, F. (eds) *The Prokaryotes - Prokaryotic Biology and Symbiotic Associations*. Springer-Verlag Berlin, Heidelberg, pp. 267-304. https://link.springer.com/referenceworkentry/10.1007/978-3-642-30194-0_14

Smith, D., McCluskey, K. and Stackebrandt, E. (2014) Culture Collection funding models and BRC business plans. *SpringerPlus* 3, 81. http://www.springerplus.com/content/3/1/81

Smith, D., Hinz, H., Mulema, J., Weyl, P. and Ryan, M.J. (2018) Biological control and the Nagoya Protocol on access and benefit sharing – a case of effective due diligence. *Biocontrol Science and Technology* 28(10), 914-926. DOI: 10.1080/09583157.2018.1460317

Report (1947) *Specialist Conference on Culture Collections of Micro-organisms*. (S.O. Code No. 47-139.) HMSO, London.

Stackebrandt, E., Smith, D., Casaregola, S., Varese, G.C., Verkleij, G., Lima, N. and Bridge, P.D. (2014) Deposit of microbial strains in public service collections as part of the publication process to underpin good practice in science. *SpringerPlus* 3, 208. http://www.springerplus.com/content/3/1/208

Svensden, A. and Frisvad, J.C. (1994) A chemotaxonomic study of the *terverticillate penicillia* based on high performance liquid chromatography of secondary metabolites. *Mycological Research* 98, 1317–1328. https://doi.org/10.1016/S0953-7562(09)80305-5

ten Kate, K. and Laird, S.A. (1999) *The Commercial Use of Biodiversity; Access to Genetic Resources and Benefit Sharing*. pp. 398. Earthscan Publications Ltd., UK. https://www.routledge.com/The-Commercial-Use-of-Biodiversity-Access-to-Genetic-Resources-and-Benefit-Sharing/Kate-Laird/p/book/9780367357641

Appendix A

Microbial Properties of CABI/IMI Strains

CABI living collection holds strains that were deposited because of their interesting properties and applications. Many have been characterised using the technologies described in chapter 3 of this volume. These strains are listed below under the following headings:

Enzyme producing strains
Metabolite producing strains
Antibiotic producing strains
Food and beverage strains
Biological control agents
Various substrates
Biodeteriogens, Biodegraders/Utilizers
Food spoilage strains
Tolerant strains
Resistant strains
Test strains
Assay strains including assay of antibiotics and vitamins
Special features: anatomical and morphological
Special features: physiological
Mating strains
Special features: chemical transformation, bioconversion and bioaccumulation
Vectors, phages, transposons, genetically modified organisms (auxotrophs, resistant, sensitive, producers), mutants

Because of the large number of strains listed, it has not been possible to cite individual literature references for each strain. Users can easily access bibliographic search databases on the Internet to track down references, if the relevant property and organism are entered as search criteria. The depositors of the strains described many properties of the strains, some of these properties have been confirmed. Other properties have been discovered by the collection staff. CABI accepts no responsibility if a given strain does not exhibit the desired property. Many properties will only be exhibited under specific experimental conditions and microorganisms, by their very nature, often exhibit distinct variability from the original culture. If the desired properties are not obvious immediately, subculturing or maintenance under different cultural conditions may promote production.

Enzyme Producing Strains

Acid protease

Penicillium citreoviride CABI-IMI 169640
Rhizopus oligosporus CABI-IMI 174457

Aconitic hydrolase

Aspergillus niger CABI-IMI 075353

DOI: 10.1079/9781800622135.appa

Amino acid oxidase

Neurospora crassa CABI-IMI 068614ii

Amylase

Aspergillus niger CABI-IMI 096215
Aspergillus oryzae CABI-IMI 052146
Aspergillus oryzae CABI-IMI 052143
Aspergillus oryzae CABI-IMI 052142
Aspergillus oryzae CABI-IMI 044242
Aspergillus foetidus CABI-IMI 041871
Trichoderma viride CABI-IMI 016198

N-aringinase

Aspergillus niger CABI-IMI

Blasticidin S deaminase

Aspergillus flavus CABI-IMI 091856
Aspergillus flavus CABI-IMI 091856iii

Byssochlamyopeptidase

Byssochlamys fulva CABI-IMI 040021

Catalase

Aspergillus niger CABI-IMI

Cellobiohydrolase D

Trichoderma reesei CABI-IMI 192655ii

Cellulase

Trichoderma reesei CABI-IMI 192656ii
Sordaria fimicola CABI-IMI 103759
Trichoderma viride CABI-IMI 016198
Aspergillus awamori CABI-IMI 303389
Aspergillus niger CABI-IMI 096215
Aspergillus niger CABI-IMI 303388
Trichoderma reesei CABI-IMI 192655ii
Trichoderma pseudokoningii CABI-IMI 092027
Neurospora crassa CABI-IMI 053239ii

Amorphous cellulase

Aspergillus fumigatus CABI-IMI 255091

a Amylase

Amorpotheca resinae CABI-IMI 344050
Aspergillus awamori CABI-IMI 303389
Aspergillus niger CABI-IMI 303388
Trichoderma reesei CABI-IMI 192654
Chaetomium globosum CABI-IMI 045550
Chaetomium globosum CABI-IMI 045550iii
Chaetomium globosum CABI-IMI 045550iv
Chaetomium globosum CABI-IMI 045550ii
Myrothecium verrucaria CABI-IMI 045541ii
Myrothecium verrucaria CABI-IMI 045541
Penicillium pinophilum CABI-IMI 371195
Penicillium pinophilum CABI-IMI 371196
Penicillium pinophilum CABI-IMI 371197
Penicillium pinophilum CABI-IMI 371198
Penicillium pinophilum CABI-IMI 371199

Cellulolytic enzymes

Aspergillus niger CABI-IMI 050565ii
Aspergillus niger CABI-IMI 050566

Chitobiosidase

Aspergillus niger CABI-IMI

Detergent

Thermomyces lanuginosus CABI-IMI

Dextranase

Paecilomyces lilacinus CABI-IMI 068202
Penicillium pinophilum CABI-IMI 087160
Paecilomyces lilacinus CABI-IMI 027830
Penicillium funiculosum CABI-IMI 079195
Penicillium pinophilum CABI-IMI 087160ii

Diastase

Aspergillus oryzae CABI-IMI 017299

Endopolygalacturonase

Rhizopus stolonifera CABI-IMI 090609

Esterase

Gliocladium roseum CABI-IMI 304503
Gliocladium roseum CABI-IMI 304501
Gliocladium roseum CABI-IMI 304501

Ferrichrome reductase

Ustilago sphaerogena CABI-IMI 061828

a Fucosidase

Metarhizium anisopliae CABI-IMI

a-o-Glucosidase

Amorpotheca resinae CABI-IMI 344050

b 1,2-Glucanase

Aspergillus quadricinctus CABI-IMI 048583ii
Penicillium papuanum CABI-IMI 040587

b 1-6 Glucanase

Penicillium dodgei CABI-IMI 040592ii
Penicillium dodgei CABI-IMI 040592
Penicillium ochrochloron CABI-IMI 061271

Glucanases

Penicillium melinii CABI-IMI 119893

Glucoamylase

Amorpotheca resinae CABI-IMI 344050
Aspergillus niger CABI-IMI 060286
Aspergillus awamori CABI-IMI

Glucose oxidase

Aspergillus niger CABI-IMI

b Glucosidase

Aspergillus awamori CABI-IMI 303389
Penicillium pinophilum CABI-IMI 371195*
Penicillium pinophilum CABI-IMI 371196*
Penicillium pinophilum CABI-IMI 371197*
Penicillium pinophilum CABI-IMI 371198*

b Glucosidase

Penicillium pinophilum CABI-IMI 371199*

b Glucosidase (intracellular)

Trichoderma reesei CABI-IMI 192655ii

Kynureninase enzymes

Rhizopus stolonifera CABI-IMI 090609
Penicillium roquefortii CABI-IMI 024313

Lipase

Trichoderma viride CABI-IMI 016198
Aspergillus flavus CABI-IMI 057389

Lipolytic enzymes

Aspergillus flavus CABI-IMI 052418ii
Aspergillus flavus CABI-IMI 052148

Maltase

Aspergillus niger CABI-IMI 060286

NAD dependent lactate dehydrogenase

Rhizopus oryzae CABI-IMI 040564

Pectate lyase

Byssochlamys fulva CABI-IMI 040021

Pectinases

Aspergillus foetidus CABI-IMI 050567

Pectinesterase

Byssochlamys fulva CABI-IMI 083277

Penicillin amidase

Fusarium oxysporum CABI-IMI 113138
Mucor hiemalis CABI-IMI 113134
Penicillium spinulosum CABI-IMI 113139
Fusarium culmorum CABI-IMI 113133
Epicoccum nigrum CABI-IMI 127257
Mucor hiemalis CABI-IMI 113136
Alternaria alternata CABI-IMI 127255
Aspergillus versicolor CABI-IMI 127256

Pepsin

Aspergillus parasiticus CABI-IMI 015957

6-Phosphogluconate dehydrogenase

Penicillium griseoroseum CABI-IMI 015378

Polygalacturonase

Byssochlamys fulva CABI-IMI 083277
Byssochlamys fulva CABI-IMI 040021

Protease

Aspergillus awamori CABI-IMI 303389
Aspergillus oryzae CABI-IMI 052145
Aspergillus oryzae CABI-IMI 052144
Aspergillus niger CABI-IMI 096215
Aspergillus niger CABI-IMI 303388
Conidiobolus khandalensis CABI-IMI 102045
Conidiobolus nanodes CABI-IMI 092299
Acremonium chrysogenium CABI-IMI 049137

Protease (extracellular)

Penicillium roquefortii CABI-IMI 024313

Proteolytic enzymes

Aspergillus parasiticus CABI-IMI 087159
Aspergillus parasiticus CABI-IMI 087159ii
Aspergillus parasiticus CABI-IMI 015957
Aspergillus flavus CABI-IMI 052176
Aspergillus flavus CABI-IMI 052148
Aspergillus flavus CABI-IMI 052175
Aspergillus flavus CABI-IMI 052173
Aspergillus flavus CABI-IMI 052418ii

Pullan 4-glucanhydrolase

Aspergillus niger CABI-IMI 091855

Purine beta ribonucleosidase

Penicillium ochrochloron CABI-IMI 061271
Penicillium melinii CABI-IMI 061836
Penicillium glabrum CABI-IMI 040234
Penicillium indonesiae CABI-IMI 039733
Aspergillus ambiguus CABI-IMI 139274

Ribonuclease U1

Ustilago sphaerogena CABI-IMI 061828

Saccharifying enzymes

Aspergillus niger CABI-IMI 060286

Tannase

Aspergillus parasiticus CABI-IMI 087159
Aspergillus parasiticus CABI-IMI 087159ii

Threonine synthase

Neurospora crassa CABI-IMI 075722

Trypsin

Aspergillus parasiticus CABI-IMI 015957

Metabolite Producing Strains

Primary and secondary metabolites including: alcohols, alkaloids, amino acids, peptides and proteins, carbohydrates, hydrocarbons, ketones, lipids and fatty acids, organic acids, pigments, plant hormones and growth regulators, steroids and sterols, toxins, carcinogens, mutagens, teratogens, vitamins, coenzymes and growth factors.

Acetaldehyde

Cochliobolus spicifer CABI-IMI 091972

3-Acetylvomitoxin

Fusarium culmorum CABI-IMI 014764

1'-O-Aetylpaxilline

Emericella striata CABI-IMI 163899

Acidicolin

Verticillium lecanii CABI-IMI 068689

cis-Aconitic acid decarboxylase

Aspergillus terreus CABI-IMI 044243

Acronycine

Cunninghamella blakesleena CABI-IMI 053585
Cunninghamella blakesleena CABI-IMI 063877

Aflatoxin

Aspergillus flavus CABI-IMI 190443
Aspergillus flavus CABI-IMI 015957vii
Aspergillus flavus CABI-IMI 091020
Aspergillus flavus CABI-IMI 242693
Aspergillus flavus CABI-IMI 242684
Aspergillus flavus CABI-IMI 015957vi
Aspergillus parasiticus CABI-IMI 091019bi
Aspergillus parasiticus CABI-IMI 015957

Aspergillus parasiticus CABI-IMI 091019biii
Aspergillus parasiticus CABI-IMI 091019bii
Aspergillus parasiticus CABI-IMI 089717
Aspergillus parasiticus CABI-IMI 091019b
Aspergillus parasiticus CABI-IMI 089717ii
Aspergillus parasiticus var. *globosus* CABI-IMI 120920

Aflatoxin M

Aspergillus flavus CABI-IMI 089717

Aflatoxin S

Aspergillus flavus CABI-IMI 089717

Aflatoxin B1 from sterigmatocystin

Aspergillus parasiticus var. *globosus* CABI-IMI 120920

Alboleersin

Bipolaris leersiae CABI-IMI 089369

Alcohol (industrial production)

Rhizopus oryzae CABI-IMI 309922
Rhizopus oryzae CABI-IMI 309923

Alpha Dicarboxylic acid

Aspergillus fumigatus CABI-IMI 045338

Altenuic acid

Alternaria alternata CABI-IMI 089344

Alternarin

Alternaria alternata CABI-IMI 089343

Alternariol

Alternaria alternata CABI-IMI 089342
Alternaria alternata CABI-IMI 089343
Alternaria alternata CABI-IMI 089344
Alternaria alternata CABI-IMI 089345
Alternaria alternata CABI-IMI 354944
Alternaria alternata CABI-IMI 354943
Alternaria alternata CABI-IMI 354942

Aernarolmethylether

Alternaria alternata CABI-IMI 354943

Di-fumaryl alanine

Penicillium resticulosum CABI-IMI 040227

Alternuene

Alternaria alternata CABI-IMI 354943

Altertoxin

Alternaria alternata CABI-IMI 354943
Alternaria alternata CABI-IMI 354945

Amodin A

Aspergillus amstelodami CABI-IMI 071295

Amodin B

Aspergillus amstelodami CABI-IMI 071295

Anditesin C

Aspergillus variecolor CABI-IMI 060316

Anditomin

Aspergillus variecolor CABI-IMI 060316

Angolide

Pithomyces cynodontis CABI-IMI 101184
Pithomyces sacchari CABI-IMI 102686

Anhydro-fusarubin

Fusarium moniliforme CABI-IMI 316818
Fusarium moniliforme CABI-IMI 316819
Fusarium moniliforme CABI-IMI 316821
Fusarium moniliforme CABI-IMI 316822
Fusarium moniliforme CABI-IMI 316823

Anhydro-fusarubin

Fusarium moniliforme CABI-IMI 316816
Fusarium moniliforme CABI-IMI 316817
Fusarium moniliforme CABI-IMI 316820

Antheridiol (mycohormone)

Achyla bisexualis CABI-IMI 141473

Aphidicolin

Verticillium lecanii CABI-IMI 068689ii

Aphidicolin

Cephalosporium aphidicola CABI-IMI 68689ii

Arugosin

Aspergillus rugulosus CABI-IMI 084338

Aspergillin

Aspergillus niger CABI-IMI 096215

Asperphenamate

Penicillium canadense CABI-IMI 095493

Asperthecin

Aspergillus quadrilineatus CABI-IMI 089348
Aspergillus quadrilineatus CABI-IMI 089349

Asterric acid

Penicillium glabrum CABI-IMI 096659

Asterriquinone

Aspergillus terreus CABI-IMI 044243

Atrovenetin

Penicillium herquei CABI-IMI 112950
Penicillium herquei CABI-IMI 028809
Penicillium melinii CABI-IMI 061835
Penicillium melinii CABI-IMI 061836
Penicillium melinii CABI-IMI 061837

Auranthine

Penicillium aurantiogriseum CABI-IMI 180922

Aurantiamine

Penicillium aurantiogriseum CABI-IMI 357289
Penicillium aurantiogriseum CABI-IMI 357290
Penicillium aurantiogriseum CABI-IMI 357291
Penicillium freii CABI-IMI 357296
Penicillium freii CABI-IMI 357297
Penicillium freii CABI-IMI 357398
Penicillium neoechinulatum CABI-IMI 357302

Aurasperone a

Aspergillus niger CABI-IMI 297707

Aurasperone b

Aspergillus niger CABI-IMI 297707

Aurasperone c

Aspergillus niger CABI-IMI 297707

Aurasperone d

Aspergillus niger CABI-IMI 297707

Aurasperone (iso)

Aspergillus niger CABI-IMI 297707

Aurofusarin

Fusarium culmorum CABI-IMI 089367
Fusarium culmorum CABI-IMI 092033
Fusarium culmorum CABI-IMI 089364
Fusarium culmorum CABI-IMI 089365
Fusarium culmorum CABI-IMI 089366

Auroglaucin

Aspergillus amstelodami CABI-IMI 092026
Aspergillus chevalieri CABI-IMI 091867
Aspergillus niveoglaucus CABI-IMI 091871
Aspergillus pseudoglaucus CABI-IMI 016122ii
Aspergillus pseudoglaucus CABI-IMI 091873
Aspergillus pseudoglaucus CABI-IMI 016122
Aspergillus pseudoglaucus CABI-IMI 091865
Aspergillus repens CABI-IMI 016144ii
Aspergillus repens CABI-IMI 094150
Aspergillus repens CABI-IMI 016131
Aspergillus repens CABI-IMI 016114
Aspergillus repens CABI-IMI 091894
Aspergillus repens CABI-IMI 091869
Aspergillus repens CABI-IMI 091901

Aspergillus repens CABI-IMI 091900

Austdiol

Aspergillus ustus CABI-IMI

Austocystin

Aspergillus ustus CABI-IMI

Averufin

Aspergillus parasiticus var. *globosus* CABI-IMI 120920

Base from lactic acid

Penicillium verrucosum CABI-IMI 297964
Penicillium verrucosum CABI-IMI 297964

Base from creatine

Penicillium verrucosum CABI-IMI 297964

Beauvericin

Beauveria bassiana CABI-IMI

Benzophenone

Aspergillus terreus CABI-IMI 16043

Bikaverin (antiprotozoal)

Fusarium oxysporum CABI-IMI 250086

Bisdechlorogeodin

Penicillium glabrum CABI-IMI 096659

Botrydial

Botrytis cinerea CABI-IMI 124882

Brefeldin

Penicillium aurantiogriseum CABI-IMI 280215

Brefeldin A

Penicillium simplicissium CABI-IMI 378412
Penicillium verrucosum CABI-IMI 297964

Brevianamide A

Penicillium verrucosum CABI-IMI 297964
Penicillium viridicatum CABI-IMI 39758ii
Penicillium viridicatum CABI-IMI 293187
Penicillium viridicatum CABI-IMI 039758
Penicillium brevicompactum CABI-IMI 040225

Brevianamide B

Penicillium brevicompactum CABI-IMI 040225
Penicillium viridicatum CABI-IMI 293187
Penicillium viridicatum CABI-IMI 039758ii
Penicillium verrucosum CABI-IMI 297964

Brevianamide M

Penicillium verrucosum CABI-IMI 297964

Butenolide

Fusarium tricinctum CABI-IMI 309352

Butenolide (on corngrits agar)

Fusarium culmorum CABI-IMI 309344
Fusarium culmorum CABI-IMI 309345
Fusarium sporotrichioides CABI-IMI 309349

Butenolide (on YES agar)

Fusarium trincinctum CABI-IMI 309352
Fusarium trincinctum CABI-IMI 309351

Byssochlamic acid

Byssochlamys fulva CABI-IMI 040021

C16-Ene-diyne

Lentinus degeneri CABI-IMI 110525

C10H10O4

Oidiodendron rhodogenum CABI-IMI 235256

C15H18O4

Oidiodendron rhodogenum CABI-IMI 235256

C21H28O10 triacetate(trichothecene)

Fusarium scirpi CABI-IMI 045490

C19H24O9 diacetate(trichothecene)

Fusarium scirpi CABI-IMI 045490

C27H28N2O4

Aspergillus glaucus CABI-IMI 053242

C29H35NO5

Helminthosporium dematioideum CABI-IMI 074812

C11H10O5

Aspergillus nidulans CABI-IMI 238850

C11H11ClO4

Periconia macrospinosa CABI-IMI 024411

C14H12O6

Cylindrocarpon sp. CABI-IMI 127996

Caffeine

Claviceps sp. CABI-IMI 380290*

Calonectrin

Fusarium culmorum CABI-IMI 014764
Calonectria nivalis CABI-IMI 014764

Candidulin

Aspergillus candidus CABI-IMI 078734

b di Carboxylic acid

Aspergillus fumigatus CABI-IMI 045338

Carlosic acid

Penicillium verrucosum CABI-IMI 297964

2-Carboxy-3,5-dihydroxyphenyl acetyl carbinol

Penicillium verrucosum CABI-IMI 297964

Carlosic acid

Penicillium verrucosum CABI-IMI 297964

Carolic acid

Penicillium verrucosum CABI-IMI 297964

b Carotenes

Blakeslea trispora CABI-IMI 195169

Carotenoids

Neurospora crassa CABI-IMI 075722

Catenarin

Pyrenophora graminea CABI-IMI 089368
Dreschlera catenaria CABI-IMI 91984

Cercosporin

Cercospora beticola CABI-IMI 077043

Cheaotoglobosin C

Penicillium verrucosum CABI-IMI 297964

Chlorine containing

Periconia macrospinosa CABI-IMI 024411

3-Chloro 5,2dihydroxy3,7,8-trimethoxyflavone

Aspergillus candidus CABI-IMI 127260

Cholic acid

Aspergillus nidulans CABI-IMI 227976

Chrysogenin

Penicillium chrysogenum CABI-IMI 039759
Penicillium chrysogenum CABI-IMI 024314
Penicillium griseoroseum CABI-IMI 015378

Chrysophanol

Penicillium islandicum CABI-IMI 092277
Penicillium kloeckeri CABI-IMI 040047

Citreoviridin

Aspergillus terreus var. aureus CABI-IMI 082431

Citric acid

Aspergillus carbonarius CABI-IMI 041873
Aspergillus carbonarius CABI-IMI 041875
Aspergillus foetidus CABI-IMI 041871
Aspergillus foetidus CABI-IMI 015954
Aspergillus niger CABI-IMI 084304
Aspergillus niger CABI-IMI 297707
Chaetomium globosum CABI-IMI 016203
Chaetomium globosum CABI-IMI 045550ii

Citric acid (from beet molasses)

Aspergillus niger CABI-IMI 027809
Aspergillus niger CABI-IMI 075353
Aspergillus niger CABI-IMI 041874

Citric acid (in submerged culture)

Aspergillus niger CABI-IMI 051433

Citrinin

Aspergillus carneus CABI-IMI 135818
Aspergillus terreus CABI-IMI 016043
Penicillium citrinum CABI-IMI 024306
Penicillium citrinum CABI-IMI 092196
Penicillium citrinum CABI-IMI 091961
Penicillium citrinum CABI-IMI 024307
Penicillium expansum CABI-IMI 293192
Penicillium hirsutum CABI-IMI 293183
Penicillium hirsutum CABI-IMI 293206
Penicillium verrucosum CABI-IMI 293203
Penicillium verrucosum CABI-IMI 297964

Citromycetin

Penicillium glabrum CABI-IMI 091944
Penicillium glabrum CABI-IMI 028043
Penicillium glabrum CABI-IMI 091914
Penicillium glabrum CABI-IMI 091945

Clorflavonin (patented strain)

Aspergillus candidus CABI-IMI 127260

Crotocin

Acremonium crotocinigenum CABI-IMI 112775

Culmorin

Fusarium culmorum CABI-IMI 089365
Fusarium culmorum CABI-IMI 014764
Calonectria nivalis CABI-IMI 014764

Culmorone

Fusarium culmorum CABI-IMI 014764

Curvulin

Bipolaris papendorfii CABI-IMI 075861

Curvulinic acid

Bipolaris papendorfii CABI-IMI 075861
Oidiodendron rhodogenum CABI-IMI 235256

Cyclopaldic acid

Penicillium aurantiogriseum CABI-IMI 089375
Penicillium aurantiogriseum var. *album* CABI-IMI 089312

Cyclopenin

Penicillium atramentosum CABI-IMI 089374
Penicillium crustosum CABI-IMI 293182
Penicillium echinulatum CABI-IMI 293180
Penicillium solitum CABI-IMI 293181
Penicillium aurantiovirens CABI-IMI 357292
Penicillium aurantiovirens CABI-IMI 357293
Penicillium cyclopium CABI-IMI 357294
Penicillium cyclopium CABI-IMI 357295
Penicillium freii CABI-IMI 357296
Penicillium freii CABI-IMI 357297
Penicillium freii CABI-IMI 357298

Cyclopenin

Penicillium neoechinulatum CABI-IMI 357302
Penicillium polonicum CABI-IMI 357303
Penicillium polonicum CABI-IMI 357304
Penicillium polonicum CABI-IMI 357305

Cyclopenol

Penicillium aurantiogriseum CABI-IMI 089374
Penicillium echinulatum CABI-IMI 293180
Penicillium verrucosum CABI-IMI 297964
Penicillium aurantiovirens CABI-IMI 357292
Penicillium aurantiovirens CABI-IMI 357293
Penicillium cyclopium CABI-IMI 357294

Penicillium cyclopium CABI-IMI 357295
Penicillium freii CABI-IMI 357296
Penicillium freii CABI-IMI 357297
Penicillium freii CABI-IMI 357298
Penicillium neoechinulatum CABI-IMI 357302
Penicillium polonicum CABI-IMI 357303
Penicillium polonicum CABI-IMI 357304
Penicillium polonicum CABI-IMI 357305

Cyclopeptin

Penicillium aurantiovirens CABI-IMI 357292
Penicillium aurantiovirens CABI-IMI 357293
Penicillium cyclopium CABI-IMI 357294
Penicillium cyclopium CABI-IMI 357295
Penicillium freii CABI-IMI 357296
Penicillium freii CABI-IMI 357297
Penicillium freii CABI-IMI 357298
Penicillium neoechinulatum CABI-IMI 357302
Penicillium polonicum CABI-IMI 357303
Penicillium polonicum CABI-IMI 357304
Penicillium polonicum CABI-IMI 357305

a Cyclopiazonic acid

Penicillium verrucosum CABI-IMI 297964

Cyclopiazonic acid

Aspergillus oryzae CABI-IMI 309454
Penicillium commune CABI-IMI 293201
Penicillium commune CABI-IMI 293179
Penicillium griseofulvum CABI-IMI 293195
Penicillium solitum CABI-IMI 293210

Cyclopolic acid

Penicillium aurantiogriseum var. *album* CABI-IMI 089312
Penicillium aurantiogriseum CABI-IMI 089375

Cynodotin

Cochliobolus lunatus CABI-IMI 061535

Cytochalasin C

Metarhizium anisopliae CABI-IMI 097373

Cytochalasin D

Metarhizium anisopliae CABI-IMI 097373

Cytochalasin E

Rosellinia necatrix CABI-IMI 108006

Cytochalasin G

Pseudeurotium zonatum CABI-IMI 171019

Cytochalasins

Drechslera dematioidea CABI-IMI 074812

Deacetylcalonectrin

Calonectria nivalis CABI-IMI 014764

3Deactylcalonectrin (trichothecin)

Fusarium culmourm CABI-IMI 014764

15Deacetyl8oxocalonectrin

Fusarium culmorum CABI-IMI 014764

Deoxyherqueinone

Penicillium herquei CABI-IMI 028809

Deoxynivalenol (on YES agar)

Fusarium culmorum CABI-IMI 309344
Fusarium gramiearum CABI-IMI 285824
Fusarium gramiearum CABI-IMI 285825
Fusarium gramiearum CABI-IMI 285826

Desacetylpebrolide

Penicillium verrucosum CABI-IMI 297964

Desferritriacetylfusigen

Aspergillus deflectus CABI-IMI 061448

Desmethoxyviridin

Nodulisporium hinnuleum CABI-IMI 214826

Desoxyrosenonolactone

Trichothecium roseum CABI-IMI 129425

Dessferri-ferricrocin

Aspergillus viridinutans CABI-IMI 062875
Aspergillus viridinutans CABI-IMI 062875ii

Diacetoxysciripenol

Fusarium scirpi CABI-IMI 112503

Fusarium scirpi CABI-IMI 045490
Fusarium equiseti CABI-IMI 035100

Diacetoxyscirpenol (on YES agar)

Fusarium sporotrichioides CABI-IMI 309350
Fusarium sporotrichioides CABI-IMI 309349

Dialkylresorcinol

Stemphylium majusculum CABI-IMI 135459

Dihydrobotrydial

Botrytis cinerea CABI-IMI 124882

Dihydrocitrinone

Aspergillus carneus CABI-IMI 135818

7,8-Dihydroxycalonectrin

Fusarium culmorum CABI-IMI 014764

3:5 Dihydroxycarboxyphenylacetylcarbinol acid

Penicillium brevicompactum CABI-IMI 039824

Dihydro paraherquimide

Penicillium sp. CABI-IMI 332995

5,6,Dimethyoxysterigmatocystin

Aspergillus sp. CABI-IMI 069857

3,15DIOH12,13Epoxytrichothec9ene

Fusarium culmorum CABI-IMI 014764

3,15DIOH12,13Epoxytrichothec9en8one

Fusarium culmorum CABI-IMI 014764

7,8-DIOHcalonectrin

Fusarium culmorum CABI-IMI 014764

Dipicolinic acid

Penicillium verrucosum CABI-IMI 297964

Diversonol

Penicillium diversum CABI-IMI 040579
Penicillium diversum CABI-IMI 040579iii

Duclauxin

Penicillium emmonsii CABI-IMI 038805ii
Penicillium emmonsii CABI-IMI 038805iii

Echinulin by isoprenylation of cycl l-alanyl-l-tryptophanyl

Aspergillus amstelodami CABI-IMI 211806

Emericellopsin A

Emericellopsis minima CABI-IMI 069015

Emericellopsin B

Emericellopsis minima CABI-IMI 069015

Emodic acid

Penicillium albicans CABI-IMI 089373

Equisetin

Fusarium equiseti CABI-IMI 309348
Fusarium equiseti CABI-IMI 309347

Erdin

Aspergillus terreus CABI-IMI 044339

Eremofortin

Penicillium roquefortii CABI-IMI 024313

Ergochrome secalonic acid D

Penicillium oxalicum CABI-IMI 039750

Ergokryptine

Claviceps purpurea CABI-IMI 126133

Ergosterol

Gaeumannomyces graminis CABI-IMI 083849
Penicillium aurantiogriseum CABI-IMI 034913ii
Penicillium aurantiogriseum CABI-IMI 034913iii
Penicillium verrucosum CABI-IMI 297964

Ergosterol palmitate

Penicillium italicum CABI-IMI 091959
Penicillium crustosum CABI-IMI 092235

Ergotamine

Claviceps purpurea CABI-IMI 104437

Erythroglaucin

Aspergillus chevalieri CABI-IMI 091867
Aspergillus echinulatus CABI-IMI 091866
Aspergillus echinulatus CABI-IMI 091872
Aspergillus echinulatus CABI-IMI 091860
Aspergillus niveoglaucus CABI-IMI 091871
Aspergillus niveoglaucus CABI-IMI 032050
Aspergillus niveoglaucus CABI-IMI 032050ii
Aspergillus ruber CABI-IMI 091864
Aspergillus ruber CABI-IMI 092030
Aspergillus ruber CABI-IMI 092043
Aspergillus ruber CABI-IMI 094147
Aspergillus ruber CABI-IMI 091863
Aspergillus ruber CABI-IMI 091868
Aspergillus ruber CABI-IMI 091862
Aspergillus umbrosus CABI-IMI 091861

Erythroskyrin

Penicillium islandicum CABI-IMI 092038
Penicillium islandicum CABI-IMI 092270
Penicillium islandicum CABI-IMI 092277

Ethisolide

Penicillium implicatum CABI-IMI 138002

Ethyl acetate

Penicillium digitatum CABI-IMI 091956

Ethyl alcohol

Penicillium digitatum CABI-IMI 091956
Cochliobolus spicifer CABI-IMI 091972

Ethylene oxide

Aspergillus fumigatus CABI-IMI 045338

Ethyleneoxide dicarboxylic acid

Paecilomyces variotii CABI-IMI 058427

Eugenitin

Cylindrocarpon sp. CABI-IMI 127996

Equestin

Fusarium equiseti CABI-IMI 309348

Extracellular polysaccharide

Paecilomyces lilacinus CABI-IMI 280556

F2 Toxin

Fusarium graminearum CABI-IMI 155426
Fusarium graminearum CABI-IMI 152103

Fat

Penicillium indonesiae CABI-IMI 039733

Flavasperone

Aspergillus niger CABI-IMI 297707

Flavicin

Aspergillus flavus CABI-IMI 111023

Flaviolin

Aspergillus foetidus CABI-IMI 015954
Phialophora lagerbergii CABI-IMI 096745

Flavipin

Epicoccum nigrum CABI-IMI 068797
Aspergillus terreus CABI-IMI 089358
Fennellia flavipes CABI-IMI 089347

Flavoglaucin

Aspergillus amstelodami CABI-IMI 092026
Aspergillus chevalieri CABI-IMI 091867
Aspergillus echinulatus CABI-IMI 091860
Aspergillus echinulatus CABI-IMI 091866
Aspergillus echinulatus CABI-IMI 091872
Aspergillus glaucus CABI-IMI 091870
Aspergillus niveoglaucus CABI-IMI 032050
Aspergillus niveoglaucus CABI-IMI 091871
Aspergillus niveoglaucus CABI-IMI 032050ii
Aspergillus pseudoglaucus CABI-IMI 016122
Aspergillus pseudoglaucus CABI-IMI 091865
Aspergillus pseudoglaucus CABI-IMI 016122ii

Aspergillus pseudoglaucus CABI-IMI 091873
Aspergillus repens CABI-IMI 091901
Aspergillus repens CABI-IMI 094150
Aspergillus repens CABI-IMI 016144ii
Aspergillus repens CABI-IMI 016131
Aspergillus repens CABI-IMI 091900
Aspergillus repens CABI-IMI 091894
Aspergillus repens CABI-IMI 091869
Aspergillus repens CABI-IMI 016114
Aspergillus ruber CABI-IMI 091868
Aspergillus ruber CABI-IMI 092030
Aspergillus ruber CABI-IMI 091864
Aspergillus ruber CABI-IMI 094147
Aspergillus ruber CABI-IMI 091863
Aspergillus ruber CABI-IMI 091862
Aspergillus umbrosus CABI-IMI 091861
Aspergillus umbrosus CABI-IMI 091874

Flavoskyrin

Penicillium islandicum CABI-IMI 092277

Fonsecin

Aspergillus carbonarius CABI-IMI 041875
Aspergillus niger CABI-IMI 297707

Fonsecin monomethylether

Aspergillus niger CABI-IMI 297707

Frequentin

Penicillium palitans CABI-IMI 040215
Penicillium verrucosum CABI-IMI 297964

Fucidin

Acremonium fusidioides CABI-IMI 087758
Acremonium fusidioides CABI-IMI 094124
Acremonium fusidioides CABI-IMI 087757
Fusarium coccineum CABI-IMI 087759

Fulvic acid

Penicillium dodgei CABI-IMI 040592ii
Penicillium dodgei CABI-IMI 040592
Penicillium griseofulvum CABI-IMI 092273
Penicillium verrucosum CABI-IMI 297964

Fumagillin

Aspergillus fumigatus CABI-IMI 069714

Fumaric acid

Aspergillus terreus CABI-IMI 089356

Fumigacin

Aspergillus fumigatus CABI-IMI 045338

Fumigatin

Aspergillus fumigatus CABI-IMI 045338
Aspergillus fumigatus CABI-IMI 089353

Fumitremorgin B

Aspergillus caespitosus CABI-IMI 016034ii
Aspergillus caespitosus CABI-IMI 016034

Fungitoxic alcohols and fatty acid

Aspergillus niger CABI-IMI 050565ii

Fumonisin B1

Fusarium moniliforme CABI-IMI 350432
Fusarium moniliforme CABI-IMI 350433
Fusarium moniliforme CABI-IMI 350434
Fusarium moniliforme var. *intermedium* CABI-IMI 350435
Fusarium moniliforme var. *intermedium* CABI-IMI 350436
Fusarium moniliforme CABI-IMI 350437

Fumonisin B2

Fusarium moniliforme CABI-IMI 350432
Fusarium moniliforme CABI-IMI 350433
Fusarium moniliforme CABI-IMI 350434
Fusarium moniliforme var. *intermedium* CABI-IMI 350435
Fusarium moniliforme var. *intermedium* CABI-IMI 350436
Fusarium moniliforme CABI-IMI 350437

Fusarenone

Fusarium crookwellense CABI-IMI 334774

Fusarin C

Fusarium crookwellense CABI-IMI 334774
Fusarium moniliforme CABI-IMI 350432
Fusarium moniliforme CABI-IMI 350433
Fusarium moniliforme CABI-IMI 350434

Fusarium moniliforme var. *intermedium* CABI-IMI 350435
Fusarium moniliforme var. *intermedium* CABI-IMI 350436
Fusarium moniliforme CABI-IMI 350437

Fuscin

Oidiodendron rhodogenum CABI-IMI 235256
Oidiodendron tenuissinum CABI-IMI 089370
Oidiodendron tenuissinum CABI-IMI 089371

Fusidic acid

Verticillium lamellicola CABI-IMI 093439

Galactocarlose

Penicillium fellutanum CABI-IMI 040232

Gentisic acid

Penicillium verrucosum CABI-IMI 297964

Gentisyl alcohol

Penicillium griseofulvum CABI-IMI 034909
Penicillium verrucosum CABI-IMI 297964

Geodin

Aspergillus terreus CABI-IMI 044339

Gibberellic acid

Fusarium avanaceum CABI-IMI 137728
Fusarium lateritium CABI-IMI 250545
Fusarium moniliforme CABI-IMI 058289
Fusarium sambucinum CABI-IMI 273414
Fusarium sambucinum CABI-IMI 267789
Giberella fujikuroi CABI-IMI 300793
Giberella fujikuroi CABI-IMI 300791
Giberella fujikuroi CABI-IMI 300792
Giberella fujikuroi var. *intermedia* CABI-IMI 300796
Giberella fujikuroi var. *intermedia* CABI-IMI 300795
Giberella fujikuroi var. *intermedia* CABI-IMI 300794
Giberella fujikuroi var. *intermedia* CABI-IMI 300789
Giberella fujikuroi var. *moniliformis* CABI-IMI 300787
Giberella fujikuroi var. *subglutinans* CABI-IMI 300788
Giberella fujikuroi var. *subglutinans* CABI-IMI 300785
Giberella fujikuroi var. *subglutinans* CABI-IMI 300784
Giberella fujikuroi var. *subglutinans* CABI-IMI 300783
Giberella fujikuroi var. *subglutinans* CABI-IMI 300782
Giberella fujikuroi var. *subglutinans* CABI-IMI 300781
Giberella sp. CABI-IMI 224130
Giberella tricincta CABI-IMI 314470
Fusarium moniliforme CABI-IMI 112801

Glioresein

Gliocladium roseum CABI-IMI 093065

Gliotoxin

Aspergillus fumigatus CABI-IMI 016030
Gliocladium flavofuscum CABI-IMI 100714

Gliovirin

Gliocladium virens CABI-IMI 283130

Gluco-delta lactone

Aspergillus niger CABI-IMI 041876

Gluconic acid

Aspergillus carbonarius CABI-IMI 041875
Aspergillus carbonarius CABI-IMI 041873
Aspergillus niger CABI-IMI 041876
Aspergillus niger CABI-IMI 041874
Aspergillus wentii CABI-IMI 023010
Leptoxyphium fumago CABI-IMI 089362
Penicillium purpurogenum CABI-IMI 090178

Glycerol

Leptoxyphium fumago CABI-IMI 089362
Aspergillus wentii CABI-IMI 023010
Cochliobolus spicifer CABI-IMI 091972

Helenine

Penicillium funiculosum CABI-IMI 040582

Helminthosporin (iso)

Pyrenophora graminea CABI-IMI 089368

Helvolic acid

Aspergillus fumigatus CABI-IMI 045338

Herqueinone

Penicillium herquei CABI-IMI 089376
Penicillium herquei CABI-IMI 028809

Heveadride

Drechslera hevea CABI-IMI 080137

HT-2 toxin (on YES agar)

Fusarium sporotrichioides CABI-IMI 309350
Fusarium sporotrichioides CABI-IMI 309349

Humicolin

Aspergillus asperescens CABI-IMI 046813

5-Hydroxyaspertin

Penicillium verrucosum CABI-IMI 297964

M Hydroxybenzoic acid

Penicillium verrucosum CABI-IMI 297964

8-Hydroxycalonectrin

Fusarium culmorum CABI-IMI 014764

15-Hydroxyculmorone

Fusarium culmorum CABI-IMI 014764

15-Hydroxyculmorin

Fusarium culmorum CABI-IMI 014764

12-Hydroxyculmorin

Fusarium culmorum CABI-IMI 014764

5-Hydroxyculmorin

Fusarium culmorum CABI-IMI 014764

8-Hydroxyellipticines

Aspergillus alliaceus CABI-IMI 051982

6-hydroxymethyleugenin

Cephalosporium aphidicola CABI-IMI 068689

9-Hydroxy pre helminthosporol

Cochliobolus ativus CABI-IMI 125851

9-Hydroxyellipticines

Aspergillus alliaceus CABI-IMI 051982

Hydroxyprogesterone

Aspergillus niger CABI-IMI 076837

I-Erythritol

Penicillium brevicompactum CABI-IMI 092034

Indigo pigment

Schizophyllum commune CABI-IMI 067963

Iridoskyrin

Penicillium islandicum CABI-IMI 092038
Penicillium islandicum CABI-IMI 092270

Islandicin

Penicillium islandicum CABI-IMI 092038
Penicillium islandicum CABI-IMI 092270

Islanditoxin

Penicillium islandicum CABI-IMI

Itaconic acid

Aspergillus terreus CABI-IMI 126254
Aspergillus terreus CABI-IMI 044243
Aspergillus terreus CABI-IMI 089357

Javanicin

Fusarium solani CABI-IMI 029817

Kojic acid

Aspergillus oryzae var. *effusus* CABI-IMI 016142
Aspergillus oryzae var. *effusus* CABI-IMI 124935
Aspergillus parasiticus CABI-IMI 015957
Aspergillus tamarii CABI-IMI 091888

Aspergillus tamarii CABI-IMI 015956
Penicillium jensenii CABI-IMI 086562
Penicillium lanosum CABI-IMI 090463

L-dopa (by bioconversion)

Aspergillus ochraceus CABI-IMI 061247iv
Gliocladium deliquescens CABI-IMI 040023

Lachnelluloic acid

Lachnellula fuscosanguinea CABI-IMI 250255
Lachnellula fuscosanguinea CABI-IMI 250256
Lachnellulone
Lachnellula fuscosanguinea CABI-IMI 250255
Lachnellula fuscosanguinea CABI-IMI 250256

Lactic acid

Rhizopus oryzae CABI-IMI 040564

Lapidosin

Penicillium lapidosum CABI-IMI 039743

Luteic acid

Penicillium verruculosum CABI-IMI 068239
Penicillium udagawae CABI-IMI 044338

Luteoleersin

Bipolaris leersiae CABI-IMI 089369

Luteose

Penicillium udagawae CABI-IMI 044338

Lysergic acid

Claviceps paspali CABI-IMI 082998
Claviceps paspali CABI-IMI 082999

Macrosporin

Alternaria alternaria CABI-IMI 354944

Malic acid

Cochliobolus spicifer CABI-IMI 091972

Mannitol

Aspergillus candidus CABI-IMI 091897
Aspergillus candidus CABI-IMI 089346
Aspergillus candidus CABI-IMI 091885
Aspergillus nidulans CABI-IMI 091903
Aspergillus ochraceus CABI-IMI 061247iii
Aspergillus wentii CABI-IMI 023010
Byssochlamys fulva CABI-IMI 040021
Cochliobolus spicifer CABI-IMI 091972
Penicillium griseofulvum CABI-IMI 075832

Mannocarlose

Penicillium fellutanum CABI-IMI 040232

Meleagrin

Penicillium melanoconidium CABI-IMI 357299
Penicillium melanoconidium CABI-IMI 357300
Penicillium melanoconidium CABI-IMI 357301

4-methoxytoluquinol

Lentinus degeneri CABI-IMI 110525

3-Methyl-6-methoxy-8-hydroxy-3-4-dihydrosocoumarin

Aspergillus caespitosus CABI-IMI 016034
Aspergillus caespitosus CABI-IMI 016034ii

3-Methylorsellinic acid

Aspergillus terreus CABI-IMI 044243

6-Methylsalicylic acid

Penicillium griseofulvum CABI-IMI 028808
Penicillium griseofulvum CABI-IMI 092273
Penicillium expansum CABI-IMI 297959
Penicillium verrucosum CABI-IM 297964

o-Methylsterigmatocystin

Chaetomium virescens var. *thielavioidium* CABI-IMI 240608

Mevalonate derivatives

Phytophthora cactorum CABI-IMI 021168

Minioluteic acid

Penicillium minioluteum CABI-IMI 089377

Moniliformin

Fusarium moniliforme CABI-IMI 316825
Fusarium moniliforme CABI-IMI 316826
Fusarium oxysporum CABI-IMI 350438
Fusarium moniliforme var. *subglutinans* CABI-IMI 350439

Moniliformin (on YES agar)

Fusarium avenaceum CABI-IMI 309355
Fusarium avenaceum CABI-IMI 309354

Monocerin

Helminthosporium monoceras CABI-IMI 125855

Multicolic acid

Penicillium multicolor CABI-IMI 104602

Multicolosic acid

Penicillium multicolor CABI-IMI 104602

Mycelianamide

Penicillium crustosum CABI-IMI 285510
Penicillium echinulatum CABI-IMI 285508
Penicillium echinulatum CABI-IMI 285503
Penicillium expansum CABI-IMI 285521
Penicillium expansum CABI-IMI 297959
Penicillium griseofulvum CABI-IMI 075832
Penicillium griseofulvum CABI-IMI 028808
Penicillium griseofulvum CABI-IMI 092273
Penicillium janthinellum CABI-IMI 040238
Penicillium solitum CABI-IMI 285509
Penicillium verrucosum CABI-IMI 285522
Penicillium verrucosum CABI-IMI 297964

Mycogone (antagonistic substance)

Acremonium strictum CABI-IMI 210624

Mycophenolic acid

Penicillium brevicompactum CABI-IMI 092274
Penicillium brevicompactum CABI-IMI 094149
Penicillium brevicompactum CABI-IMI 017456
Penicillium brevicompactum CABI-IMI 039824
Penicillium brevicompactum CABI-IMI 092044
Penicillium brevicompactum CABI-IMI 126540
Penicillium brevicompactum CABI-IMI 092262
Penicillium brevicompactum CABI-IMI 040225
Penicillium brevicompactum CABI-IMI 092034
Penicillium roquefortii CABI-IMI 293190
Penicillium roquefortii CABI-IMI 293204
Penicillium verrucosum CABI-IMI 297964

Mycotoxin (unspecified)

Aspergillus repens CABI-IMI 173203
Aspergillus repens CABI-IMI 173204
Aspergillus repens CABI-IMI 173205
Aspergillus repens CABI-IMI 173206
Aspergillus repens CABI-IMI 173207

Nalgiolaxin

Penicillium jensenii CABI-IMI 039804

Nalgiovense

Penicillium jensenii CABI-IMI 039804

Nectriapyronone

Gliocladium vermoesenii CABI-IMI 040231

Neosolaniol (on YES agar)

Fusarium sporotrichioides CABI-IMI 309349
Fusarium sporotrichioides CABI-IMI 309350

Nidulin

Aspergillus nidulans CABI-IMI 085473

Nigrifortine

Penicillium nigricans CABI-IMI 228699

b Nitropropionic acid

Aspergillus oryzae CABI-IMI 052141
Aspergillus oryzae CABI-IMI 052141ii
Penicillium melinii CABI-IMI 061837
Penicillium melinii CABI-IMI 061836

Nivalenol

Fusarium crookwellense CABI-IMI 334774

Nornidulin

Aspergillus nidulans CABI-IMI 085473

Northerqueinone

Penicillium herquei CABI-IMI 089376

Notatin

Penicillium griseoroseum CABI-IMI 015378

Noxiversin

Penicillium griseoroseum CABI-IMI 058138

Ochratoxin

Aspergillus ochraceus CABI-IMI 132429

Ochratoxin A

Penicillium crustosum CABI-IMI 206159
Penicillium expansum CABI-IMI 297959
Penicillium verrucosum CABI-IMI 297964
Penicillium verrucosum CABI-IMI 293203
Penicillium verrucosum CABI-IMI 291193
Aspergillus fumigatus CABI-IMI 376377
Aspergillus fumigatus CABI-IMI 376378
Aspergillus fumigatus CABI-IMI 376380
Aspergillus versicolor CABI-IMI 376379

Organic acids (6)

Penicillium griseofulvum CABI-IMI 075832

Organic acids (8)

Penicillium fellutanum CABI-IMI 040232

Orsellinic acid

Penicillium madriti CABI-IMI 096506
Penicillium madriti CABI-IMI 086563

Oxalic acid

Aspergillus niger CABI-IMI 243400
Aspergillus niger CABI-IMI 243400
Aspergillus niger CABI-IMI 091881ii
Aspergillus terreus CABI-IMI 089356
Aspergillus terreus CABI-IMI 089355

Oxysporone

Fusarium oxuysporum CABI-IMI 211811

Palitantin

Penicillium expansum CABI-IMI 297959
Penicillium palitans CABI-IMI 089379
Penicillium palitans CABI-IMI 089378
Penicillium palitans CABI-IMI 040215
Penicillium palitans CABI-IMI 092276
Penicillium verrucosum CABI-IMI 297964

Paraherquimide

Penicillium sp. CABI-IMI 332995

Paraherquamide

Penicillium paraherquei CABI-IMI 068220

Patulin (clavacin)

Aspergillus clavatus CABI-IMI 015949v
Aspergillus clavatus CABI-IMI 015949iv
Aspergillus clavatus CABI-IMI 015949iii
Aspergillus terreus CABI-IMI 017294
Aspergillus terreus CABI-IMI 017294ii
Aspergillus terreus CABI-IMI 016043
Byssochlamys fulva CABI-IMI 083277
Penicillium claviforme CABI-IMI 293198
Penicillium expansum CABI-IMI 297959
Penicillium expansum CABI-IMI 293192
Penicillium griseofulvum CABI-IMI 039809
Penicillium griseofulvum CABI-IMI 293195
Penicillium griseofulvum CABI-IMI 034908
Penicillium griseofulvum CABI-IMI 028808
Penicillium griseofulvum CABI-IMI 034909
Penicillium roquefortii CABI-IMI 293190
Penicillium roquefortii CABI-IMI 293204
Penicillium verrucosum CABI-IMI 297964

Paxilline

Emericella striata CABI-IMI 163899

Pencolide

Penicillium sclerotiorum CABI-IMI 104602

Penicillic acid

Penicillium atramentosum CABI-IMI 293185
Penicillium aculeatum CABI-IMI 089372
Penicillium aurantiogriseum CABI-IMI 034913iii
Penicillium aurantiogriseum CABI-IMI 040236
Penicillium aurantiogriseum CABI-IMI 034913ii
Penicillium verrucosum CABI-IMI 297964
Penicillium cyclopium CABI-IMI 357294
Penicillium cyclopium CABI-IMI 357295
Penicillium raistrickii CABI-IMI 378416
Penicillium melanoconidium CABI-IMI 357299
Penicillium melanoconidium CABI-IMI 357300
Penicillium melanoconidium CABI-IMI 357301
Penicillium neoechinulatum CABI-IMI 357302

Penicillium pulvivorum CABI-IMI 378411
Penicillium polonicum CABI-IMI 357303
Penicillium polonicum CABI-IMI 357304
Penicillium polonicum CABI-IMI 357305

Penitrem A

Penicillium crustosum CABI-IMI 293182
Penicillium expansum CABI-IMI 297959
Penicillium janczewskii CABI-IMI 228669
Penicillium melanoconidium CABI-IMI 357299
Penicillium melanoconidium CABI-IMI 357300
Penicillium melanoconidium CABI-IMI 357301
Penicillium verrucosum CABI-IMI 297964

Pennigritrem C37H44ClNO6

Penicillium nigricans CABI-IMI 228699

L-Phenylalanine anhydride

Penicillium janczewskii CABI-IMI 104603

Physcion

Aspergillus echinulatus CABI-IMI 091860
Aspergillus niveoglaucus CABI-IMI 032050ii
Aspergillus echinulatus CABI-IMI 091872
Aspergillus echinulatus CABI-IMI 091866
Aspergillus niveoglaucus CABI-IMI 032050
Aspergillus chevalieri CABI-IMI 091867
Aspergillus ruber CABI-IMI 091862
Aspergillus ruber CABI-IMI 091863
Aspergillus ruber CABI-IMI 091864
Aspergillus ruber CABI-IMI 094147
Aspergillus umbrosus CABI-IMI 091861
Aspergillus umbrosus CABI-IMI 091874
Aspergillus niveoglaucus CABI-IMI 091871
Penicillium herquei CABI-IMI 028809

Physcion anthranol

Penicillium herquei CABI-IMI 028809

Pimaydolide

Pithomyces maydicus CABI-IMI 046232
Pithomyces maydicus CABI-IMI 098084

Pink-Mauve pigment (exposed to long wave UV)

Epicoccum nigrum CABI-IMI 336106

Pithoimycolides

Pithomyces chartarum CABI-IMI 074473

Polysaccharide

Penicillium udagawae CABI-IMI 095152

PR toxin

Penicillium roquefortii CABI-IMI 024313
Penicillium expansum CABI-IMI 297959
Penicillium verrucosum CABI-IMI 297964

Protein

Trichoderma reesei CABI-IMI 192655ii

Pseudonigeron(1-3)alpha-D-glucan

Aspergillus niger CABI-IMI 050566ii

Puberculic acid

Penicillium aurantiogriseum CABI-IMI 034913iii
Penicillium aurantiogriseum CABI-IMI 034846
Penicillium aurantiogriseum CABI-IMI 034913ii

Pulvinone derivatives

Aspergillus terreus CABI-IMI 044243

Pyrenophorin

Stemphylium radicinum CABI-IMI 105654

Quadrilineatin

Aspergillus quadrilineatus CABI-IMI 089313
Aspergillus quadrilineatus CABI-IMI 089348
Aspergillus quadrilineatus CABI-IMI 089351
Aspergillus quadrilineatus CABI-IMI 089349
Aspergillus quadrilineatus CABI-IMI 089350

Questin

Penicillium glabrum CABI-IMI 096659

Questinol

Penicillium glabrum CABI-IMI 096659

Qurantiogliocladin

Gliocladium roseum CABI-IMI 093065

Radarins A

Aspergillus sulphureus CABI-IMI 211397

Radarins D

Aspergillus sulphureus CABI-IMI

Radicinin

Alternaria radicina CABI-IMI 105653
Alternaria radicina CABI-IMI 105654

Resorcinol

Penicillium brevicompactum CABI-IMI 092034

Resorcinol derivatives

Penicillium brevicompactum CABI-IMI 092034

Riboflavin

Eremothecium ashbyi CABI-IMI 031269
Eremothecium ashbyi CABI-IMI 014783

Roquefortine A

Penicillium hirsutum CABI-IMI 293183

Roquefortine B

Penicillium expansum CABI-IMI 297959
Penicillium verucosum CABI-IMI 297964

Roquefortine C

Penicillium roquefortii CABI-IMI 293190
Penicillium griseofulvum CABI-IMI 293195
Penicillium chrysogenum CABI-IMI 293188
Penicillium claviforme CABI-IMI 293198
Penicillium hirsutum CABI-IMI 293183
Penicillium expansum CABI-IMI 293192
Penicillium roquefortii CABI-IMI 293189
Penicillium hirsutum CABI-IMI 293184
Penicillium roquefortii CABI-IMI 293204
Penicillium hirsutum CABI-IMI 293206
Penicillium crustosum CABI-IMI 293182
Penicillium italicum CABI-IMI 293196
Penicillium expansum CABI-IMI 297959
Penicillium verrrucosum CABI-IMI 297964

Rosenonolactone

Trichothecium roseum CABI-IMI 129425

Rosololactone

Trichothecium roseum CABI-IMI 129425

Rubratoxin

Penicillium rubrum CABI-IMI 040036

Rubratoxin A

Penicillium rubrum CABI-IMI 136128
Penicillium rubrum CABI-IMI 136127
Penicillium rubrum CABI-IMI 136126

Rubratoxin B

Penicillium rubrum CABI-IMI 112715
Penicillium rubrum CABI-IMI 136128

Rubrofusarin

Fusarium culmorum CABI-IMI 089365
Fusarium culmorum CABI-IMI 089364
Fusarium culmorum CABI-IMI 089366
Aspergillus carbonarius CABI-IMI 041875
Aspergillus niger CABI-IMI 297707
Fusarium culmorum CABI-IMI 089367
Fusarium culmorum CABI-IMI 092033

Rubroglaucin

Aspergillus umbrosus CABI-IMI 091861
Aspergillus ruber CABI-IMI 092043
Aspergillus ruber CABI-IMI 091868
Aspergillus niveoglaucus CABI-IMI 032050
Aspergillus ruber CABI-IMI 091862
Aspergillus echinulatus CABI-IMI 091860
Aspergillus ruber CABI-IMI 094147
Aspergillus echinulatus CABI-IMI 091866
Aspergillus ruber CABI-IMI 092030
Gliocladium roseum CABI-IMI 093065

Rubroskyrin

Penicillium islandicum CABI-IMI 092270
Penicillium islandicum CABI-IMI 092038

Rubrosulphin

Aspergillus sulphureus CABI-IMI 128939

Rubrotiorin

Penicillium hirayamae CABI-IMI 078255

Rugulosin

Penicillium kloeckeri CABI-IMI 040047
Penicillium rugulosum CABI-IMI 040041
Penicillium rugulosum CABI-IMI 089381
Penicillium rugulosum CABI-IMI 092223
Penicillium rugulosum CABI-IMI 089382
Penicillium rugulosum CABI-IMI 089380

Rugulovasine

Penicillium expansum CABI-IMI 297959
Penicillium verrucosum CABI-IMI 297964

Salnopyrone A (in presence of chick pea tissue)

Ascocyhta rabiei CABI-IMI 331919

Salanopyrone C (in presence of chick pea tissue)

Ascocyhta rabiei CABI-IMI 331919

Sambucinol

Fusarium culmorum CABI-IMI 014764

Sambucoin

Fusarium culmorum CABI-IMI 014764

Sclerin

Aspergillus carneus CABI-IMI 135818

Sclerotiorin

Penicillium sclerotiorum CABI-IMI 039742
Penicillium sclerotiorum CABI-IMI 092040
Penicillium hirayamae CABI-IMI 078255
Penicillium sclerotiorum CABI-IMI 040574
Penicillium sclerotiorum CABI-IMI 092039
Penicillium sclerotiorum CABI-IMI 040569

Scytalone

Phialophora lagerbergii CABI-IMI 096745

Secalonic acid (ergochrome)

Penicillium oxalicum CABI-IMI 039750

Shamixanthon

Aspergillus variecolor CABI-IMI 112543

Sirodesmin A

Sirodesmium diversum CABI-IMI 102519

Sirodesmin B

Sirodesmium diversum CABI-IMI 102519

Sirodesmin C

Sirodesmium diversum CABI-IMI 102519

Sirodesmin G

Sirodesmium diversum CABI-IMI 102519

Skyrin

Penicillium rugulosum CABI-IMI 040041
Penicillium rugulosum CABI-IMI 092223
Penicillium rugulosum CABI-IMI 089380
Penicillium islandicum CABI-IMI 092277
Penicillium rugulosum CABI-IMI 089382
Penicillium kloeckeri CABI-IMI 040047
Penicillium rugulosum CABI-IMI 089381

Spiculisporic acid

Penicillium minioluteum CABI-IMI 089377

Spinulosin

Penicillium spinulosum CABI-IMI 091950
Aspergillus fumigatus CABI-IMI 089354
Penicillium spinulosum CABI-IMI 091954
Aspergillus fumigatus CABI-IMI 045338
Penicillium spinulosum CABI-IMI 091955

Sporidesmins

Pithomyces chartarum CABI-IMI 074473

Sporidesmolides I

Pithomyces chartarum CABI-IMI 074473
Pithomyces chartarum CABI-IMI 074473

Sporidesmolides III

Pithomyces chartarum CABI-IMI 074473

Sporidesmolides IV

Pithomyces chartarum CABI-IMI 074473
Pithomyces maydicus CABI-IMI 098084

Sporidesmolides V

Pithomyces chartarum CABI-IMI 074473

Stellatin

Aspergillus variecolor CABI-IMI 075219

Stemlon

Pleospora herbarum CABI-IMI 130782

Sterigmatocystin

Chaetomium virescens var. *thielavioideum* CABI-IMI 240608
Aspergillus rugulosus CABI-IMI 136775
Aspergillus versicolor CABI-IMI 041924
Aspergillus versicolor CABI-IMI 016139
Aspergillus parasiticus var. *globus* CABI-IMI 120920

o-Methyl-sterigmatocystin

Chaetomium virescens var. thielavioideum CABI-IMI 240608

Steroids

Paecilomyces lilacinus CABI-IMI 027830

Sterol binding polysaccharides

Penicillium roquefortii CABI-IMI 024313

Stipitatic acid

Penicillium emmonsii CABI-IMI 038805ii
Penicillium emmonsii CABI-IMI 038805iii

Succinic acid

Aspergillus terreus CABI-IMI 089356
Aspergillus terreus CABI-IMI 017294
Aspergillus terreus CABI-IMI 017294ii
Aspergillus terreus CABI-IMI 091911
Cochliobolus spicifer CABI-IMI 091972
Leptoxyphium fumago CABI-IMI 089362

trans-2,3-epoxysuccinic acid

Lentinus degeneri CABI-IMI 110525

Sulochrin

Penicillium glabrum CABI-IMI 096659
Aspergillus terreus CABI-IMI 016043

Synnematin B

Acremonium chrysogenium CABI-IMI 091579
Acremonium chrysogenium CABI-IMI 049137
Emericellopsis minima CABI-IMI 058330
Emericellopsis minima CABI-IMI 092625

T-2 toxin (on YES agar)

Fusarium sporotrichioides CABI-IMI 309349
Fusarium sporotrichioides CABI-IMI 309350

Tajixanthone

Aspergillus variecolor CABI-IMI 112543

Tardin

Penicillium rugulosum CABI-IMI 028333

Tenuazoic acid

Alternaria alternata CABI-IMI 089344
Alternaria alternata CABI-IMI 089345
Penicillium aurantiogriseum CABI-IMI 293185
Penicillium chrysogenum CABI-IMI 293188
Penicillium crustosum CABI-IMI 285510
Penicillium expansum CABI-IMI 297959
Penicillium griseofulvum CABI-IMI 285525
Penicillium roqueforti CABI-IMI 293190
Penicillium solitum CABI-IMI 285509
Penicillium verrucosum CABI-IMI 297964
Penicillium verrucosum CABI-IMI 285523
Penicillium verrucosum CABI-IMI 285522
Peniciullium expansum CABI-IMI 285521

Terramide A

Aspergillus terreus CABI-IMI 044339

Terramide C

Aspergillus terreus CABI-IMI 044339

Terrein

Aspergillus terreus CABI-IMI 044339

Terrestric acid

Penicillium crustosum CABI-IMI 089386

Penicillium crustosum CABI-IMI 089385
Penicillium crustosum CABI-IMI 293182
Penicillium crustosum CABI-IMI 028040
Penicillium expansum CABI-IMI 297959
Penicillium hirsutum CABI-IMI 293206
Penicillium hirsutum CABI-IMI 293184
Penicillium hirsutum CABI-IMI 293183
Penicillium verrucosum CABI-IMI 297964

Tetronic acid (antitumour / antibacterial activity)

Aspergillus panamensis CABI-IMI 019393iii
Aspergillus panamensis CABI-IMI 019393ii

Toxin

Fusarium graminearum CABI-IMI 183761
Fusarium graminearum CABI-IMI 183762
Cephalotrichum microsporus CABI-IMI 173202
Fusarium culmorum CABI-IMI 202042
Fusarium sulpureum CABI-IMI 206026

Trichodermin

Trichoderma polysporum CABI-IMI 140624
Trichoderma sporulosum CABI-IMI 040624

Trichodermol (roridin C)

Trichoderma polysporum CABI-IMI 140624
Trichoderma sporulosum CABI-IMI 040624

Trichothecin

Trichothecium roseum CABI-IMI 050661
Trichothecium roseum CABI-IMI 050660
Trichothecium roseum CABI-IMI 057973
Trichothecium roseum CABI-IMI 129425

Tricothecenes

Fusarium culmorum CABI-IMI 014764

Ustic acid

Aspergillus ustus CABI-IMI 089359
Aspergillus ustus CABI-IMI 089361
Aspergillus ustus CABI-IMI 089360

Uuroglaucin

Aspergillus glaucus CABI-IMI 091870

Varianose

Penicillium funiculosum CABI-IMI 040586

Veratryl alcohol (in lignin degredation)

Phanerochaete chrysosporium CABI-IMI 232175

Veratryl alcohol

Phanerochaete chrysosporium CABI-IMI 310787

Vermiculine

Talaromyces wortmannii CABI-IMI 275985
Talaromyces wortmannii CABI-IMI 364857

Vermopyrone

Gliocladium vermoesenii CABI-IMI 040231

Verruculogen

Aspergillus caespitosus CABI-IMI 016034
Aspergillus caespitosus CABI-IMI 016034ii

Verrucofortine

Penicillium aurantiovirens CABI-IMI 357292
Penicillium aurantiovirens CABI-IMI 357293
Penicillium cyclopium CABI-IMI 357294
Penicillium cyclopium CABI-IMI 357295
Penicillium polonicum CABI-IMI 357303
Penicillium polonicum CABI-IMI 357304
Penicillium polonicum CABI-IMI 357305
Penicillium tricolor CABI-IMI 357306

Verrucosidin

Penicillium melanoconidium CABI-IMI 357299
Penicillium melanoconidium CABI-IMI 357300
Penicillium polonicum CABI-IMI 357303
Penicillium polonicum CABI-IMI 357304
Penicillium polonicum CABI-IMI 357305
Penicillium aurantiogriseum CABI-IMI 357289
Penicillium aurantiogriseum CABI-IMI 357290
Penicillium aurantiogriseum CABI-IMI 357291
Penicillium melanoconidium CABI-IMI 357301

Versiconal acetate

Aspergillus parasiticus var. *globosus* CABI-IMI 120920

Versimide insecticide activity

Aspergillus versicolor CABI-IMI 129488
Aspergillus versicolor CABI-IMI 129489

Viomellein

Aspergillus sulphureus CABI-IMI 128939
Penicillium cyclopium CABI-IMI 357294
Penicillium cyclopium CABI-IMI 357295
Penicillium expansum CABI-IMI 297959
Penicillium freii CABI-IMI 357296
Penicillium freii CABI-IMI 357297
Penicillium freii CABI-IMI 357298
Penicillium tricolor CABI-IMI 357306
Penicillium tricolo CABI-IMI 357307
Penicillium verrucosum CABI-IMI 297964
Penicillium viridicatum CABI-IMI 357308
Penicillium viridicatum CABI-IMI 293187
Penicillium viridicatum CABI-IMI 293205

Viopurpurin

Aspergillus sulphureus CABI-IMI 128939

Viridamine

Penicillium viridicatum CABI-IMI 357308

Viridic acid

Penicillium viridicatum CABI-IMI 357308

Viridicatic acid

Penicillium crustosum CABI-IMI 206159

Viridicatin

Penicillium aurantiogriseum CABI-IMI 293186
Penicillium solitum CABI-IMI 293181
Penicillium aurantiogriseum CABI-IMI 293185
Penicillium aurantiovirens CABI-IMI 357292
Penicillium aurantiovirens CABI-IMI 357293
Penicillium cyclopium CABI-IMI 357294
Penicillium cyclopium CABI-IMI 357295
Penicillium freii CABI-IMI 357296
Penicillium freii CABI-IMI 357297
Penicillium freii CABI-IMI 357298
Penicillium polonicum CABI-IMI 357303
Penicillium polonicum CABI-IMI 357304
Penicillium polonicum CABI-IMI 357305

Viridicatol

Penicillium aurantiovirens CABI-IMI 357292
Penicillium aurantiovirens CABI-IMI 357293
Penicillium cyclopium CABI-IMI 357294
Penicillium cyclopium CABI-IMI 357295
Penicillium frei CABI-IMI 357296
Penicillium freii CABI-IMI 357297
Penicillium freii CABI-IMI 357298
Penicillium neoechinulatum CABI-IMI 357302
Penicillium polonicum CABI-IMI 357303
Penicillium polonicum CABI-IMI 357304
Penicillium polonicum CABI-IMI 357305

Viridin

Trichoderma viride CABI-IMI 024039

Viridiol

Nodulisporium hinnuleum CABI-IMI 214826
Gliocladium deliquescens CABI-IMI 101523

Viriditoxin

Aspergillus viridinutans CABI-IMI 062875
Aspergillus viridinutans CABI-IMI 062875ii

Vomitoxin

Fusarium moniliforme CABI-IMI 316829

Wortmannin

Penicillium wortmannii CABI-IMI 044277

Wortmannolone

Penicillium wortmannii CABI-IMI 044277

Xanthomegin

Penicillium aurantiogriseum CABI-IMI 293186
Penicillium viridicatum CABI-IMI 293205
Penicillium viridicatum CABI-IMI 293187
Aspergillus sulphureus CABI-IMI 128939
Penicillium cyclopium CABI-IMI 357294
Penicillium cyclopium CABI-IMI 357295
Penicillium freii CABI-IMI 357296
Penicillium freii CABI-IMI 357298
Penicillium tricolor CABI-IMI 357306
Penicillium tricolor CABI-IMI 357307
Penicillium viridicatum CABI-IMI 357308

Zearalenone

Fusarium crookwellense CABI-IMI 334774

Fusarium culmorum CABI-IMI 309345
Fusarium equiseti CABI-IMI 309348
Fusarium equiseti CABI-IMI 309347
Fusarium moniliforme CABI-IMI 316832
Fusarium moniliforme CABI-IMI 316831

Zearalenone (on YES agar)

Fusarium culmorum CABI-IMI 309344

b-Trans-Zearalenol

Fusarium crookwellense CABI-IMI 334774

trans-Zearalenol

Fusarium crookwellense CABI-IMI 334774

Antibiotic Producing Strains

Antimicrobials to fungi, bacteria, protozoa and viruses, also see metabolites

Alamethicin

Trichoderma viride CABI-IMI

Aminoglycoside G-418

Emericellopsis minima CABI-IMI 069015

Antitumour antibiotic

Penicillium emmonsii CABI-IMI 038805ii
Penicillium emmonsii CABI-IMI 038805iii

Anti viral agent

Penicillium brevicompactum CABI-IMI 143520
Penicillium funiculosum CABI-IMI 040582

Caldariomycin

Leptoxyphium fumago CABI-IMI 089363
Leptoxyphium fumago CABI-IMI 089362

Cephalosporin C

Acremonium chrysogenium CABI-IMI 049137
Emericellopsis minima CABI-IMI 092625
Cephalosporium sp. CABI-IMI 049137

Cephalosporin N

Emericellopsis minima CABI-IMI 069015
Cephalosporium sp. CABI-IMI 049137
Acremonium chrysogenium CABI-IMI 091579

Cephalosporin P

Acremonium chrysogenium CABI-IMI 091579
Cephalosporium sp. CABI-IMI 049137

Funicin

Aspergillus funiculosus CABI-IMI 054397ii G-418

Griseofulvin

Khuskia oryzae CABI-IMI 122490
Penicillium italicum CABI-IMI 293196
Penicillium crustosum CABI-IMI 206159
Penicillium janczewskii CABI-IMI 96660
Penicillium griseofulvum CABI-IMI 075832
Penicillium raistrickii CABI-IMI 137808
Penicillium raistrickii CABI-IMI 187604
Penicillium janczewskii CABI-IMI 228669
Penicillium janczewskii CABI-IMI 191499
Penicillium raistrickii CABI-IMI 378416
Penicillium radiolobatus CABI-IMI 378410
Penicillium radiolobatus CABI-IMI 378412
Penicillium radiolobatus CABI-IMI 378413
Penicillium radiolobatus CABI-IMI 378414
Penicillium radiolobatus CABI-IMI 378415
Penicillium radiolobatus CABI-IMI 378416
Penicillium griseofulvum CABI-IMI 293195

Lincomycin sulphates

Aspergillus niger CABI-IMI 041876

Lincomycin sulphoxides

Aspergillus carbonarius CABI-IMI 041875

Penicillic acid *see metabolites*

Penicillin (Fleming isolate)

Penicillium chrysogenum CABI-IMI 024317

Penicillin

Penicillium chrysogenum CABI-IMI 026211
Penicillium chrysogenum CABI-IMI 293188

Penicillin (commercial)

Penicillium chrysogenum CABI-IMI 037767

Penicillin (in submerged culture)

Penicillium griseoroseum CABI-IMI 015378
Penicillium chrysogenum CABI-IMI 026210
Penicillium chrysogenum CABI-IMI 040233
Penicillium griseoroseum CABI-IMI 017968

Penicillin (in surface culture)

Penicillium chrysogenum CABI-IMI 017969

Penicillin N

Acremonium chrysogenium CABI-IMI 091579

Food and Beverage Strains

Including strains used directly for food and strains used in food fermentation, process and production.

Animal feed

Aspergillus foetidus CABI-IMI 015954
Aspergillus foetidus CABI-IMI 041871

Awamori alcoholic beverage

Aspergillus awamori CABI-IMI 015953 ii
Aspergillus awamori CABI-IMI 015953

Baking strains

Saccharomyces cerevisiae CABI-IMI 039916

Beef flavouring strain

Thamnidium elegans CABI-IMI 283871

Beef tenderising strain

Thamnidium elegans CABI-IMI 283871

Beer (Whitbread strain)

Saccharomyces cerevisiae CABI-IMI 061263

Mushrooms (fruiting bodies)

Pleurotus ostreatus NCWRF 40C
Agaricus sp.
Pleurotus ostreatus CABI-IMI 341687
Lyophyllum sp. CABI-IMI 377721

Oyster mushrooom

Pleurotus djamor CABI-IMI 341688

Penicillium powder for bluing cheese

Penicillium crustosum CABI-IMI 143338

Single cell protein

Chaetomium virescens CABI-IMI 185905

Soy sauce

Rhizopus oryzae CABI-IMI 021602
Rhizopus japonicus CABI-IMI 021600
Aspergillus parasiticus CABI-IMI 087159
Aspergillus parasiticus CABI-IMI 087159ii

Mycoparasites (Fungal Pathogens)

Agaric mushroom

Cladobotryum varium CABI-IMI 374366*

Agaricus bisporus

Fusarium sp. CABI-IMI 381120*

Agaricus bisporus

Verticillium fungicola CABI-IMI 268001

Agaricus sp.

Hormiactis alba CABI-IMI 098734
Hormiactis alba CABI-IMI 142892
Hormiactis alba CABI-IMI 142893
Hormiactis alba CABI-IMI 119312
Aspergillus oryzae
Penicillium purpurogenum CABI-IMI 091926
Nectriopsis broomeana CABI-IMI 71367

Mushroom

Melanospora damnosa CABI-IMI 262513
Verticillium fungicola CABI-IMI 290546

Mushroom sporophores

Verticillium fungicola CABI-IMI 188936

Penicillium frequentans

Nematogonium ferrugineum CABI-IMI 246537
Nematogonium ferrugineum CABI-IMI 246538

Penicillium purpurescens

Piptocephalis xenophila CABI-IMI 205708

Penicillium waksmanni

Pitocephalis xenophila CABI-IMI 056692ii

Pilobolus crystallinus

Syncephalis sphaerica CABI-IMI 212171

Rhizoctonia

Trichoderma viride CABI-IMI 170657

Russula sp.

Myrothecium inundatum CABI-IMI 158885
Myrothecium inundatum CABI-IMI 008983iii
Myrothecium inundatum CABI-IMI 008983i

Biological Control Agents

There are many pathogens that have potential for use as biological control agents, only a few representive examples have been included below, more can be found on the UKNCC website (www.ukncc.co.uk) by entering either strain or target organism information into the search criteria fields (i.e. host, order, genus etc.).

Acridids (potential)

Metarhizium flavoviride CABI-IMI 351848

Citrus weevil (potential)

Beauveria brongniartii CABI-IMI 354855

Coffee disease (potential)

Verticillium lecanii CABI-IMI 337420

Grasshopper / locust

Beauveria bassiana CABI-IMI 331274

Gryllidae (potential)

Metarhizium anisopliae CABI-IMI 351846

Locust

Metarhizium anisopliae CABI-IMI 074660

Onion white rot

Trichoderma longibrachiatum CABI-IMI 297702

Orthoptera (potential)

Metarhizium anisopliae CABI-IMI 351847

Pythium ultimum

Gliocladium virens CABI-IMI 283130

Root borer (potential)

Beauveria bassiana CABI-IMI 348041

Root grub (potential)

Beauveria densa CABI-IMI 168426

Scapanes australis

Metarhizium anisopliae CABI-IMI 298095

Sclerotinia sclerotiorum

Trichoderma harzianum CABI-IMI 361087
Trichoderma koningii CABI-IMI 361084
Trichoderma koningii CABI-IMI 361086

Sclerotinia sp.

Trichoderma harzianum CABI-IMI 361092

Sitona discoideus (potential)

Beauveria bassiana CABI-IMI 351847

Teleogryllus (potential)

Metarhizium anisopliae CABI-IMI 351831

Unspecified (potential)

Trichoderma harzianum CABI-IMI 361076
Trichoderma harzianum CABI-IMI 361077

Trichoderma harzianum CABI-IMI 361080
Cochliobolus heterostrophus CABI-IMI 311811
Sphaerulina mimosae-pigrae CABI-IMI 352347
Verticillium lecanii CABI-IMI 337422
Verticillium lecanii CABI-IMI 339529

Various Substrates

Alkali (from pulp manufacture)

Trichoderma viride CABI-IMI 295977

Coal spoil material

Aspergillus spectabilis CABI-IMI 216611

Coal tip

Acrophialophora fusispora CABI-IMI 144749
Aspergillus fumigatus CABI-IMI 131020
Corynascus sepedonium CABI-IMI 131037
Malbranchea sulfurea CABI-IMI 131007
Talaromyces leycettanus CABI-IMI 178525
Thermoascus crustaceus CABI-IMI 131006
Thielavia terrestris CABI-IMI 131014
Trichophaea abundans CABI-IMI 128503

Coconut matting

Penicillium variabile CABI-IMI 040040

Conveyor belt (textile) in coal mine

Cephalotrichum purpureofuscum CABI-IMI 98505

Coprophilous organisms

Gymnascella confluens CABI-IMI 100873
Gymnascella confluens CABI-IMI 100873ii
Isaria felina CABI-IMI 054853
Kernia nitidia CABI-IMI 151090
Lophotrichus ampullus CABI-IMI 188081
Memnoniella echinata CABI-IMI 075133
Mycosphaerella CABI-IMI 163345
Oedocephalum glomerulosum CABI-IMI 074901
Syncephalis sphaerica CABI-IMI 212171

Coprophilous (animal dung)

Mortierella indohii CABI-IMI 242504
Mortierella indohii CABI-IMI 242505

Coprophilous (antelope dung)

Chaetomium indicum CABI-IMI 287277
Chaetomium strumarium CABI-IMI 291757
Neocosmospora vasinfecta CABI-IMI 170137

Coprophilous (bat dung)

Mortierella reticulata CABI-IMI 336529

Coprophilous (buffalo dung)

Geotrichum candidum CABI-IMI 190584
Lophotrichus ampullus CABI-IMI 171509
Coemansia erecta CABI-IMI 279145
Mortierella capitata CABI-IMI 280010
Mucor flavus CABI-IMI 280011

Coprophilous (burro dung)

Gelasinospora heterospora CABI-IMI 267933
Neocosmospora vasinfecta var. *africana* CABI-IMI 267939
Sporormiella minima CABI-IMI 267941

Coprophilous (camel dung)

Orbicula parietina CABI-IMI 268927

Coprophilous (carnivore dung)

Kernia nitidia CABI-IMI 151088
Kernia nitidia CABI-IMI 151082
Thelebolus caninus CABI-IMI 144397
Thelebolus caninus CABI-IMI 144378
Thelebolus crustaceus CABI-IMI 144389
Thelebolus crustaceus CABI-IMI 144390
Thelebolus crustaceus CABI-IMI 144395

Coprophilous (cat dung)

Crinella muscae CABI-IMI 040634

Coprophilous (chicken dung)

Cephaliophora tropica CABI-IMI 074900

Coprophilous (cockroach dung)

Heterocephalum aurantiacum CABI-IMI 236332

Coprophilous (cow dung)

Chaetomium succineum CABI-IMI 287289

Coniocheata tetraspora CABI-IMI 260594
Gelasinospora indica CABI-IMI 267935
Gelasinospora pseudoreticulata CABI-IMI 230600
Kernia hyalina CABI-IMI 151079
Kernia nitidia CABI-IMI 151081
Neocosmospora vasinfecta var. *africana* CABI-IMI 267940
Nigrosabulum globosum CABI-IMI 230591
Penicillium novae-zeelandiae CABI-IMI 204086
Preussia cylindrica CABI-IMI 166524
Sporormiella minima CABI-IMI 175093
Sporormiella schadospora CABI-IMI 230589
Thielavia terricola CABI-IMI 153731

Coprophilous (deer dung)

Cephalotrichum nanum CABI-IMI 068394
Cephalotrichum purpureofuscum CABI-IMI 071294
Chaetomium lentum CABI-IMI 291748
Chatomium cuniculorum CABI-IMI 172753
Pilaria anomala CABI-IMI 109387
Pilobolus umbonatus CABI-IMI 240956
Piptocephalis arrhiza CABI-IMI 196613
Thelebolus caninus CABI-IMI 144386
Thelebolus microsporus CABI-IMI 144391

Coprophilous (dog dung)

Circinella umbellata CABI-IMI 338439
Fimaria trochospora CABI-IMI 230598
Microascus cinereus CABI-IMI 086909
Mucor mucedo CABI-IMI 122485

Coprophilous (elephant dung)

Chaetomium lentum CABI-IMI 291749
Chaetomium subspirale CABI-IMI 287290
Humicola grisea var. *thermoidea* CABI-IMI 126329
Mycoarachis inversa CABI-IMI 148374
Sporormiella isomera CABI-IMI 170135
Sporormiella isomera CABI-IMI 170136

Coprophilous (field mouse dung)

Botryotrichum piluliferum CABI-IMI 092395
Mucor mucedo CABI-IMI 133976

Coprophilous (frog dung)

Thamnidium elegans CABI-IMI 337533

Coprophilous (giraffe dung)

Podosordaria australis CABI-IMI 080526

Coprophilous (goat dung)

Heterocephalum aurantiacum CABI-IMI 241265
Microascus cirosus CABI-IMI 075819
Pilobolus crystallinus CABI-IMI 240957
Thamnostylum piriforme CABI-IMI 337630
Veroneae botryose CABI-IMI 115127

Coprophilous (guinea pig dung)

Gymnascella hyalinospora CABI-IMI 099725
Gymnascella nodulosa CABI-IMI 168760

Coprophilous (hare dung)

Chaetomium convolutum CABI-IMI 073521
Coniocheata polysperma CABI-IMI 223876
Podosordaria jugoyasan CABI-IMI 223877

Coprophilous (herbovore dung)

Chaetocladium brefeldii CABI-IMI 336114
Chaetomium subspirale CABI-IMI 287291
Mucor hiemalis f. *silvaticus* CABI-IMI 309639
Mycoarachis inversa CABI-IMI 148375
Syncephalis coernui CABI-IMI 240688
Thelebolus crustaceus CABI-IMI 144387
Thelebolus crustaceus CABI-IMI 144388

Coprophilous (horse dung)

Chaetocladium brefeldii CABI-IMI 191243
Chaetocladium brefeldii CABI-IMI 266093
Chaetomium globosum CABI-IMI 287298
Chaetomium thermophilium var. *coprophilium* CABI-IMI 126331
Chaetomium unipiculatum CABI-IMI 293980
Coprinus hexagonosporus CABI-IMI 161417
Crinella muscae CABI-IMI 206881
Mycotypha microspora CABI-IMI 108621
Nigrosabulum globosum CABI-IMI 148373
Piptocephalis cylindrospora CABI-IMI 077593ii
Preussia isomera CABI-IMI 223878
Stibella thermophila CABI-IMI 295313

Coprophilous (human)

Mucor variabilis CABI-IMI 117670

Coprophilous (kangaroo dung)

Kernia bifurcotricha CABI-IMI 141564
Mucor circinelloides CABI-IMI 103844

Coprophilous (lizard dung)

Benjaminiella youngi CABI-IMI 336111
Benjaminiella youngi CABI-IMI 337532
Benjaminiella poitrasii CABI-IMI 377366
Circinella umbellata CABI-IMI 325631
Ellisomyces anomalus CABI-IMI 211277
Mortierella reticulata CABI-IMI 236323
Mucor mousanensis CABI-IMI 290247
Mucor mucedo CABI-IMI 337631
Radiomyces embreei CABI-IMI 338452
Radiomyces spectabilis CABI-IMI 338433
Radiomyces spectabilis CABI-IMI 338434
Radiomyces spectabilis CABI-IMI 142387
Thamnidium elegans CABI-IMI 325630
Thamnidium elegans CABI-IMI 336112
Thamnidium elegans CABI-IMI 337663
Thamnostylum nigricans CABI-IMI 208235
Thamnostylum nigricans CABI-IMI 208234
Thamnostylum piriforme CABI-IMI 290425
Thamnostylum piriforme CABI-IMI 290426
Thamnostylum piriforme CABI-IMI 336113
Thamnostylum piriforme CABI-IMI 337632
Thamnostylum piriforme CABI-IMI 337634

Coprophilous (Malayan squirrel dung)

Piptocephalis lepidula CABI-IMI 339043

Coprophilous (marmot dung)

Sporormiella similis CABI-IMI 230590

Coprophilous (millipede droppings)

Verticillium catenulatum CABI-IMI 113078

Coprophilous (mircocavia dung)

Pleuroascus nicholsonni CABI-IMI 230599

Coprophilous (moose dung)

Chaetomidium cephalothecoides CABI-IMI 180791
Chaetomium trilaterale var. *chiversii* CABI-IMI 387306
Cheatomium cupreum CABI-IMI 250968
Coemansia erecta CABI-IMI 281612
Crinella naumovii CABI-IMI 133977
Dichotomocladium elegans CABI-IMI 211271
Dichotomocladium robustum CABI-IMI 211274
Gymnascella nodulosa CABI-IMI 168761
Helicostylum pulchrum CABI-IMI 203381
Kickxella albastrina CABI-IMI 213918
Mortierella scerotiella CABI-IMI 133978
Mycotypha microspora CABI-IMI 282443
Phycomyces nitens CABI-IMI 281611
Pilaria caucasica CABI-IMI 134107
Piptocephalis tieghemia CABI-IMI 339051
Piptocephalis debaryana CABI-IMI 338753
Piptocephalis lepidula CABI-IMI 338765
Piptocephalis lepidula CABI-IMI 339044
Piptocephalis microcephala CABI-IMI 213565
Radiomyces embreei CABI-IMI 338449
Radiomyces embreei CABI-IMI 081586
Radiomyces spectabilis CABI-IMI 338435
Thamnidium elegans CABI-IMI 264665
Thamnidium elegans CABI-IMI 264662
Thamnostylum lucknowense CABI-IMI 211280
Thamnostylum repens CABI-IMI 208236
Thamnostylum repens CABI-IMI 211270

Coprophilous (mouse dung)

Actinomucor elegans CABI-IMI 171226
Piptocephalis brijmohanii CABI-IMI 339038

Coprophilous (opossum dung)

Pilaria anomala CABI-IMI 105546

Coprophilous (pack rat dung)

Chaetomium torulosum CABI-IMI 387304
Ellisomyces anomalus CABI-IMI 338443
Eremomyces bilateralis CABI-IMI 151076
Radiomyces embreei CABI-IMI 338451
Radiomyces embreei CABI-IMI 338450

Coprophilous (partridge dung)

Chaetomium trilaterale CABI-IMI 287305
Penicillium claviforme CABI-IMI 071357

Coprophilous (pony dung)

Poronia puntata CABI-IMI 245188

Coprophilous (porcupine dung)

Kernia hippocrepida CABI-IMI 151078

Coprophilous (rabbit dung)

Actinomucor elegans CABI-IMI 86138
Chaetocladium brefeldii CABI-IMI 234455
Chaetocladium brefeldii CABI-IMI 080480
Chaetocladium brefeldii CABI-IMI 266092
Chaetocladium brefeldii CABI-IMI 266091
Chaetocladium brefeldii CABI-IMI 266090
Chaetocladium brefeldii CABI-IMI 266088
Chaetomium bostrychodes CABI-IMI 039725
Chaetomium subspirale CABI-IMI 287292
Chatomium cuniculorum CABI-IMI 155487
Cheatomium gangligerum CABI-IMI 291744
Crinella muscae CABI-IMI 073461
Gliocladium roseum CABI-IMI 101020
Gymnascella hyalinospora CABI-IMI 099726
Gymnascella nodulosa CABI-IMI 168759
Isaria felina CABI-IMI 281451
Lophotrichus incarnatus CABI-IMI 135564
Mucor peasmaticus CABI-IMI 381705
Onychophora corpophila CABI-IMI 275663
Pilaria anomala CABI-IMI 043023
Pilaria anomala CABI-IMI 109386
Pilaria anomala CABI-IMI 101020
Piptocephalis tieghemia CABI-IMI 339474
Piptocephalis arrhiza CABI-IMI 196614
Piptocephalis brijmohanii CABI-IMI 164897
Piptocephalis debaryana CABI-IMI 205704
Piptocephalis debaryana CABI-IMI 205703
Piptocephalis freseniana CABI-IMI 199156
Piptocephalis freseniana CABI-IMI 199157
Piptocephalis indica CABI-IMI 339042
Piptocephalis indica CABI-IMI 199159
Piptocephalis lepidula CABI-IMI 196611
Piptocephalis lepidula CABI-IMI 196612
Piptocephalis repens CABI-IMI 149887
Piptocephalis repens CABI-IMI 149887ii
Sprorminula tenerifae CABI-IMI 311150
Sutravarana samala CABI-IMI 205842
Syncephalis nodosa CABI-IMI 203383
Syncephalis nodosa CABI-IMI 197709
Thamnidium elegans CABI-IMI 079897
Thamnidium elegans CABI-IMI 264663
Thelebolus caninus CABI-IMI 144379
Thelebolus crustaceus CABI-IMI 068073

Coprophilous (rabbit hutch dung)

Cephalotrichum nanum CABI-IMI 080254
Thelebolus crustaceus CABI-IMI 079997

Coprophilous (rat dung)

Benjaminiella poitrasii CABI-IMI 081585
Benjaminiella youngi CABI-IMI 337208
Cokeromyces recurvatus CABI-IMI 077585
Crinella umbellata CABI-IMI 054855
Piptocephalis tieghemia CABI-IMI 339621
Utharomyces epallocaulis CABI-IMI 211283
Utharomyces epallocaulis CABI-IMI 213005
Utharomyces epallocaulis CABI-IMI 211282
Utharomyces epallocaulis CABI-IMI 211588

Coprophilous (rock hyrax dung)

Podosordaria austro-americana CABI-IMI 174498

Coprophilous (rodent dung)

Dichotomocladium elegans CABI-IMI 211272
Dispira parvispora CABI-IMI 101210iii
Mucor faisalabadensis CABI-IMI 304471
Piptocephalis cylindrospora CABI-IMI 339466
Radiomyces embreei CABI-IMI 338453
Utharomyces epallocaulis CABI-IMI 213007
Utharomyces epallocaulis CABI-IMI 236319
Utharomyces epallocaulis CABI-IMI 236320
Utharomyces epallocaulis CABI-IMI 240373

Coprophilous (sheep dung)

Coprinus utrifer CABI-IMI 161422
Iodophanus carneus CABI-IMI 070912ii
Penicillium novae-zeelandiae CABI-IMI 204087
Piptocephalis lepidula CABI-IMI 211369

Coprophilous (squirrel dung)

Pirella circinans CABI-IMI 338448

Coprophilous (toad dung)

Coemansia erecta CABI-IMI 312319

Coprophilous (unspecified)

Basidiobolus microsporus CABI-IMI 093345
Chaetomidium fimeti CABI-IMI 116692
Chaetomium brasiliense CABI-IMI 338828
Chaetomium brasiliense CABI-IMI 042201
Piptocephalis cylindrospora CABI-IMI 339465

Coprophilous (weasel dung)

Thamnidium elegans CABI-IMI 236139

Coprophilous (wolf dung)

Thelebolus caninus CABI-IMI 144375
Thelebolus microsporus CABI-IMI 144392

Coprophilous (woodchuck dung)

Hapsidospora irregularis CABI-IMI 173727

Coprophilous (zebra dung)

Chaetomium strumarium CABI-IMI 291758
Hapsidospora irregularis CABI-IMI 291760

Coprophilous (*Mus formosanus* dung)

Utharomyces epallocaulis CABI-IMI 213006

Eelworms

Harposporium bysmatosporum CABI-IMI 087014
Harposporium helicoides CABI-IMI 087013

Fermenter system at elevated temperature

Aspergillus fumigatus CABI-IMI 293740

Foam sample

Porocladium aquaticum CABI-IMI 199900

Nematodes

Dactyella illequens CABI-IMI 372759
Fusarium oxysporum CABI-IMI 369179*
Fusarium sp. CABI-IMI 369180*

Nematophagous (*Hepialis* sp.)

Cordyceps gracilis CABI-IMI 016573

Nematophagous

Harposporium anguillulae CABI-IMI 232243
Harposporium arthrosporum CABI-IMI 232244
Harposporium crasum CABI-IMI 174854

Pickled cured fish (spoilage)

Polyschema sagarii CABI-IMI 291231

River foam

Articulospora tetracladia CABI-IMI 209374
Ceuthospora phacidioides CABI-IMI 209750
Lemonniera terrestris CABI-IMI 209372
Mariannaea camptospora CABI-IMI 133904
Tricladium brunneum CABI-IMI 177447
Tricladium malaysianum CABI-IMI 177449
Tricladium spendens CABI-IMI 209373
Triposermum campelopardus CABI-IMI 123907
Varicosporium helicosporum CABI-IMI 177446a
Varicosporium macrosporeum CABI-IMI 177446b

Rock

Ramalina decipeus CABI-IMI 400185*

Sacking

Preussia vulgaris CABI-IMI 089575

Saline soil

Phoma ostiolata CABI-IMI 113691
Phoma ostiolata var. *brunnea* CABI-IMI 113692

Sewage

Aspergillus stellatus CABI-IMI 146289
Neocosmospora tenuicristata CABI-IMI 277708
Penicillium crustosum CABI-IMI 146306

Sewage beds

Fusarium aquaeductuum CABI-IMI 205331

Symbiont (lichen mycobiont)

Dictyostelium mucoroides CABI-IMI 074707

Waste water

Fusarium merismoides CABI-IMI 297028
Fusarium merismoides CABI-IMI 297027
Fusarium merismoides CABI-IMI 297032

Biodeteriogens, Biodegraders/Utilizers

See also: Food spoilage organisms and Environmental isolates

Acetic preserve

Monilinia acetobutans CABI-IMI 180570

Aircraft fuel tank

Penicillium miniolutеum CABI-IMI 147406
Syncephalostrum racemosum CABI-IMI 362748*

Alkali pulp manufacture

Trichoderma harzianum CABI-IMI 297033
Trichoderma harzianum CABI-IMI 295979

Animal specimen in formalin

Penicillium janthinellum CABI-IMI 117655

Army socks

Mucor oblongisporus CABI-IMI 100706

Art

Gliomastix luzulae CABI-IMI 133983

Artificial leather

Penicillium viridicatum CABI-IMI 091952

Artificial leather cloth

Penicillium commune CABI-IMI 091957

Asbestos cord

Cephalotrichum nanum CABI-IMI 076251

Asbestos substitute building

Stachybotrys atra CABI-IMI 292096

Aviation fuel

Amorpotheca resinae CABI-IMI 342896
Amorpotheca resinae CABI-IMI 331223
Amorpotheca resinae CABI-IMI 325811
Amorpotheca resinae CABI-IMI 331941
Amorpotheca resinae CABI-IMI 135416
Amorpotheca resinae CABI-IMI 325704
Amorpotheca resinae CABI-IMI 325706
Amorpotheca resinae CABI-IMI 344050
Amorpotheca resinae CABI-IMI 325812
Amorpotheca resinae CABI-IMI 331939
Amorpotheca resinae CABI-IMI 331942
Amorpotheca resinae CABI-IMI 331973
Amorpotheca resinae CABI-IMI 325705
Amorpotheca resinae CABI-IMI 296266
Aspergillus fumigatus CABI-IMI 181419
Aspergillus fumigatus CABI-IMI 184361
Aspergillus fumigatus CABI-IMI 184363
Paecilomyces variotii CABI-IMI 184364
Penicillium brevicompactum CABI-IMI 329660
Penicillium canescens CABI-IMI 184360

Baby powder

Mucor racemosus CABI-IMI 316960

Balsa blocks in river

Zopfiella leucotricha CABI-IMI 153733

Balsa wood submerged in patuxent river

Trichocladium achrasporium CABI-IMI 132773

Batter pudding

Aspergillus wentii CABI-IMI 040641

BEA vanguard wing tank

Amorpotheca resinae CABI-IMI 138247

Betacyanin

Aspergillus foetidus CABI-IMI 041871

Betaxanthin

Aspergillus foetidus CABI-IMI 041871

Binocular lens

Aspergillus tonophilus CABI-IMI 108299

Bitumen treated cardboard

Amorpotheca resinae CABI-IMI 090126

Blue stain (wood spoilage)

Ceratocystis pilifera CABI-IMI 274362

Bromthymol blue indicator solution

Penicillium spinulosum CABI-IMI 091919

N-Butane

Graphium putredinis CABI-IMI 151810

Canteen cover

Myrothecium verrucaria CABI-IMI 140593

Cannery (walls and ceiling)

Exophiala pisciphila CABI-IMI 288044

Canvas

Memnoniella echinata CABI-IMI 061273ii
Memnoniella echinata CABI-IMI 061273

Canvas bag (Officer's)

Myrothecium verrucaria CABI-IMI 140592

Canvas in mine

Aspergillus ustus CABI-IMI 089359

Cellophane (buried in soil)

Oidiodendron rhodogenum CABI-IMI 068735

Cellophane (weathering treated)

Penicillium sclerotiorum CABI-IMI 092040

Cellophane in sand

Botryotrichum piluliferum CABI-IMI 081303
Botryotrichum piluliferum CABI-IMI 068736

Cellophane paper

Aspergillus versicolor CABI-IMI 045554

Cellulose

Trichoderma viride CABI-IMI 296458

Cellulose board

Chaetomium globosum CABI-IMI 107512
Scopulariopsis brevicaulis CABI-IMI 107513

Cellulose fibres

Penicillium kloeckeri CABI-IMI 202320

Cellulosic material

Penicillium albicans CABI-IMI 233074

Cellulosic material (buried in forest soil)

Fennellia flavipes CABI-IMI 321308

Cellulosic material (India)

Chaetomium cymbiforme CABI-IMI 252060
Cheatomium fungicola CABI-IMI 318396
Chaetomium cymbiforme CABI-IMI 318414

Cellulosic soil buried in forest material

Fennellia flavipes CABI-IMI 171883

Charcoal

Hypocenomyce australis CABI-IMI 401280

Cigar tobacco

Dictyoarthrinium rabaulense CABI-IMI 181940

Cloth

Cladosporium sphaerospermum CABI-IMI 094153
Aspergillus terreus CABI-IMI 089355
Aspergillus versicolor CABI-IMI 094152

Coal spoil

Calcarisporium thermophilium CABI-IMI 144750

Coal mine spoilage

Aspergillus spectabilis CABI-IMI 316611

Concrete wall

Arthrobotrys javinica CABI-IMI 255232

Contact lens

Verticillium lecanii CABI-IMI 176057

Copper electroplating solution

Penicillium crustosum CABI-IMI 166526

Copper sulphate solution

Penicillium ochrochloron CABI-IMI 039806
Penicillium spinulosum CABI-IMI 087416

Cotton

Chaetomium globosum CABI-IMI 045550
Chaetomium globosum CABI-IMI 045550ii
Chaetomium globosum CABI-IMI 045550iv
Chaetomium globosum CABI-IMI 045550iii

Cotton (American)

Aspergillus terreus CABI-IMI 089357
Aspergillus terreus CABI-IMI 044339

Cotton (Egyptian)

Aspergillus terreus CABI-IMI 089356

Cotton fabric

Stachybotrys atra CABI-IMI 082021
Memnoniella echinata CABI-IMI 016201

Cotton textile

Stachybotrys atra CABI-IMI 115289

Cotton yarn

Aspergillus versicolor CABI-IMI 094159
Aspergillus tamarii CABI-IMI 092019

Creosoted wood

Amorpotheca resinae CABI-IMI 089357
Amorpotheca resinae CABI-IMI 089838
Amorpotheca resinae CABI-IMI 149211

Crude oil

Exophiala dermatitidis CABI-IMI 294929

Cutting fluid

Geotrichum candidum CABI-IMI 329542
Penicillium expansum CABI-IMI 329544

Dextrin paste

Penicillium spinulosum CABI-IMI 091953

Diesel

Amorpotheca resinae CABI-IMI 179266
Penicillium corylophilum CABI-IMI 290817
Penicillium spinulosum CABI-IMI 320722
Trichoderma harzianum CABI-IMI 290819
Acremonium strictum CABI-IMI 299023

Diso naval fuel

Tritirachium oryzae CABI-IMI 320903

Distilled water container

Penicillium spinulosum CABI-IMI 091954
Penicillium spinulosum CABI-IMI 091950
Penicillium spinulosum CABI-IMI 091955

Earthenware plant pot

Oedocephalum glomerulosum CABI-IMI 140796

Electrical equipment

Cladosporium herbarum CABI-IMI 170353

Electrical fuse

Aspergillus versicolor CABI-IMI 091890
Penicillium commune CABI-IMI 091918

Electrical wire sleeving

Absidia californica CABI-IMI 227977

Explosives

Acremonium strictum CABI-IMI 077044

Film (deteriorated)

Corynascus sepedonium CABI-IMI 316334

Flannel bag

Aspergillus ustus CABI-IMI 089360

Formalin tanks (surface)

Penicillium simplicissium CABI-IMI 252091

Fuel filters

Penicillium corylophilum CABI-IMI 273248
Penicillium hirsutum CABI-IMI 273250

Gelatine factory effluent

Rhizopus stolonifera CABI-IMI 324540

Glycerine

Penicillium solitum CABI-IMI 039819

Glyphosphate solution

Verticillium lecanii CABI-IMI 293226

Hardwood

Perenniporia fraxinea CABI-IMI 081803

Heat seal foil material

Penicillium chrysogenum CABI-IMI 175765

Hydaulic oil

Aspergillus fumigatus CABI-IMI 160349

Ice cream carton

Trichoderma harzianum CABI-IMI 049268

Industrial waste

Chaetomium medusarum CABI-IMI 317647
Chaetomium jodhpurense CABI-IMI 317631
Monascus purpueus CABI-IMI 317626

Insect moulting hormones

Rhizopus arrhizus CABI-IMI 090340

Iron bark in cooling tower

Penicillium erubescens CABI-IMI 190275

Italian talcum powder

Cladosporium cladosporioides CABI-IMI 185487

Jet fuel

Acremonium strictum CABI-IMI 178506

Kerosene filter pad

Cheatomium fungicola CABI-IMI 087410

Kerosene fuel filter

Aspergillus fumigatus CABI-IMI 173460
Pestalotiopsis aquatica CABI-IMI 087402
Amorpotheca resinae CABI-IMI 088973
Amorpotheca resinae CABI-IMI 088972
Amorpotheca resinae CABI-IMI 088971
Amorpotheca resinae CABI-IMI 088968
Yarrowia lipolytica CABI-IMI 093743

Label (museum)

Crinella umbellata CABI-IMI 136605

Leather

Paecilomyces variotii CABI-IMI 114930
Penicillium diversum CABI-IMI 040579iii
Penicillium diversum CABI-IMI 040579

Leather boot

Penicillium emmonsii CABI-IMI 089713

Lignin

Phanerochaete chrysosporium CABI-IMI 232175

Linseed oil based red primer

Penicillium canescens CABI-IMI 166199

Lubricating oil

Fusarium oxysporum CABI-IMI 290814
Chaetomium globosum CABI-IMI 290815

Manilla fibre

Ulocladium chartarum CABI-IMI 116369

Marine diesel oil

Amorpotheca resinae CABI-IMI 322372

Metal working fluid

Amorpotheca resinae CABI-IMI 322170
Fusarium solani CABI-IMI 289157
Geotrichum candidum CABI-IMI 289156
Scopulariopsis brevicaulis CABI-IMI 321976
Chaetomium globosum CABI-IMI 314227

Military equipment (deterioratinng)

Penicillium implicatum CABI-IMI 040578ii

Mine timber

Penicillium duclauxii CABI-IMI 200309

Mines

Penicillium crustosum CABI-IMI 297891

Mould proofing

Aspergillus versicolor CABI-IMI 04554iv
Aspergillus versicolor CABI-IMI 045554

Mattress cover

Chaetomium atrobrunneum CABI-IMI 291742

Mussels in preserving fluid

Aspergillus versicolor CABI-IMI 188634
Scopulariopsis brevicaulis CABI-IMI 188635

Nylon

Tritirachium oryzae CABI-IMI 045535

Oil palm trunk

Thielaviopsis paradoxa CABI-IMI 378941
Thielaviopsis paradoxa CABI-IMI 378942

Ointment

Amorpotheca resinae CABI-IMI 049621

Optical instrument

Penicillium capsulatum CABI-IMI 040576

Paint

Penicillium kloeckeri CABI-IMI 210615
Cladosporium sphaerospermum CABI-IMI 210616
Cladosporium herbarum CABI-IMI 131128
Cladosporium sphaerospermum CABI-IMI 214669
Cladosporium sphaerospermum CABI-IMI 084420

Paint (emulsion with chromate)

Penicillium minioluteum CABI-IMI 178519
Penicillium spinulosum CABI-IMI 178523

Paint (phenolic tung-Zn oxide/chromate)

Aspergillus flavus CABI-IMI 178520

Paint (PVA emulsion)

Phoma epicoccina CABI-IMI 178513
Phoma violacea CABI-IMI 178508
Phoma violacea CABI-IMI 178507

Paint (red lead)

Penicillium expansum CABI-IMI 163739
Penicillium brevicompactum CABI-IMI 163740

Paint (bathroom)

Phoma violacea CABI-IMI 090179

Painted beach box

Penicillium corylophilum CABI-IMI 222984

Painted surface metal amplifier case

Aspergillus versicolor CABI-IMI 096330

Paint (white lead)

Phoma violacea CABI-IMI 049948
Phoma violacea CABI-IMI 049948ii

Painted wood

Aureobasidium pullans CABI-IMI 045533ii
Aureobasidium pullans CABI-IMI 045533
Cladosporium sphaerospermum CABI-IMI 061886
Curvularia lunata CABI-IMI 063375

Paintwork

Cladosporium sphaerospermum CABI-IMI 235764

Palm oil

Paecilomyces variotii CABI-IMI 227571

Paper - 17th century rag-paper document

Mycotypha microspora CABI-IMI 031239

Paper

Chaetomium globosum CABI-IMI 016203
Chaetomium nigricolor CABI-IMI 144976
Penicillium rubrum CABI-IMI 040036
Penicillium sclerotiorum CABI-IMI 146640
Penicillium viridicatum CABI-IMI 297902
Phoma violacea CABI-IMI 129267
Stachybotrys atra CABI-IMI 072155
Trichoderma harzianum CABI-IMI 061381

Paper (damp)

Ulocladium botrytis CABI-IMI 123533

Paper mill slime

Mucor plumbeus CABI-IMI 014781

Paper pulp

Penicillium funiculosum CABI-IMI 061385
Penicillium funiculosum CABI-IMI 061383
Penicillium variabile CABI-IMI 061384

Paracetamol tablet

Penicillium aurantiogriseum CABI-IMI 305618

Parachute nylon

Penicillium janthinellum CABI-IMI 075589

Paradise nuts

Crinella muscae CABI-IMI 200308

Paraffin wax

Aspergillus versicolor CABI-IMI 091859

p-digunidino-benzoyl-l-glutamic acid hydrochloride

Penicillium spinulosum CABI-IMI 029467

Penicillium powder for blueing cheese (Denmark)

Penicillium crustosum CABI-IMI 173028

Pilot plant usng kerosene for copper extraction

Penicillium spirillum CABI-IMI 152804

Pinus strobus boards

Cephaloascus fragrans CABI-IMI 068865

Plaster and wallpaper

Botryotrichum piluliferum CABI-IMI 051266
Rhinotrichum lanosum CABI-IMI 051266

Plastic

Aspergillus fumigatus CABI-IMI 095927
Aspergillus terreus CABI-IMI 095928
Aureobasidium pullans CABI-IMI 145194
Cunninghamella elegans CABI-IMI 313951
Penicillium pinophilum CABI-IMI 087160ii
Penicillium pinophilum CABI-IMI 087160

Plastic covered nylon sail cover

Drechslera dematioidea CABI-IMI 100720

Polycaprolactone polyurethane

Trichoderma harzianum CABI-IMI 286366
Trichoderma harzianum CABI-IMI 287340

Polyester based Polyurethane sheet

Alternaria alternata CABI-IMI 299022

Polyester polyurethane

Cunninghamella elegans CABI-IMI 313951
Gliocladium roseum CABI-IMI 313952
Penicillium simplicissium CABI-IMI 313950
Penicillium simplicissium CABI-IMI 313950
Trichoderma harzianum CABI-IMI 314381

Polyester polyurethane foam

Aspergillus ustus CABI-IMI 314382
Aspergillus versicolor CABI-IMI 314386
Fusarium culmorum CABI-IMI 314383
Fusarium solani CABI-IMI 314385
Penicillium chrysogenum CABI-IMI 314384
Penicillium chrysogenum CABI-IMI 314387

Polyether polyurethane

Penicillium simplicissium CABI-IMI 313953
Penicillium simplicissium CABI-IMI 313950
Cunninghamella elegans CABI-IMI 313951

Polypropylene diver's hose

Paecilomyces lilacinus CABI-IMI 319302

Polyurethane

Alternaria alternata CABI-IMI 166751
Aspergillus fischeri CABI-IMI 319453

Aspergillus fischeri var. *spinosus* CABI-IMI 301296
Trichoderma harzianum CABI-IMI 286366
Ulocladium chartarum CABI-IMI 127962

Polyurethane coated fabric

Chaetomium globosum CABI-IMI 184731
Chaetomium cochliodes CABI-IMI 183700

Polyurethane coated fabric

Mucor racemosus CABI-IMI 184727
Mucor mucedo CABI-IMI 184726
Penicillium chrysogenum CABI-IMI 184732

Polyurethane foam

Aspergillus ustus CABI-IMI 319457
Paecilomyces lilacinus CABI-IMI 174677
Aspergillus fischeri var. spinosus CABI-IMI 319455
Aspergillus fumigatus CABI-IMI 319454
Penicillium expansum CABI-IMI 319460
Penicillium chrysogenum CABI-IMI 319452
Penicillium simplicissium CABI-IMI 319456

Polyurethane footware

Trichoderma harzianum CABI-IMI 324751
Trichoderma harzianum CABI-IMI 324750

Polyurethane products

Aspergillus versicolor CABI-IMI 299019

Polyurethanes used in soil burial experiments

Gliocladium roseum CABI-IMI 304503
Gliocladium roseum CABI-IMI 304501
Gliocladium roseum CABI-IMI 304501

Poplin fabric

Fennellomyces linderi CABI-IMI 208237

Potassium thiocyanate containing mercuric nitrate

Paecilomyces farinosus CABI-IMI 307991

Preservative treated telegraph pole

Geomyces pannorum CABI-IMI 212118

PVA emulsion

Cladosporium cladosporioides CABI-IMI 178517
Cladosporium cladosporioides CABI-IMI 178510
Cladosporium cladosporioides CABI-IMI 178509
Penicillium crustosum CABI-IMI 178511

PVA emulsion paint

Penicillium chrysogenum CABI-IMI 178514

PVC

Aplosporella sp. CABI-IMI 361388*
Aureobasidium pullans CABI-IMI 361381*
Aureobasidium pullans CABI-IMI 361386*
Aureobasidium pullans CABI-IMI 361387*
Bacillus megaterium CABI-IMI 361382*
Cladosporium cladosporoides CABI-IMI 361389*
Coryneform sp. CABI-IMI 361383*
Cunninghamella elegans CABI-IMI 313951
Curtobacterium sp. CABI-IMI 361384
Penicillium pinophilum CABI-IMI 114933
Penicillium simplicissium CABI-IMI 313950
Phoma sorghina CABI-IMI 361393*
Rhodosporidium sp. CABI-IMI 361385*
Escherichia coli CABI-IMI 362054*

PVC covered wallpaper

Aspergillus ustus CABI-IMI 113731
Penicillium pinophilum CABI-IMI 113730
Penicillium rubrum CABI-IMI 113729
Aspergillus terreus CABI-IMI 113732
Penicillium janthinellum CABI-IMI 106294

Quinine containing medicine

Paecilomyces dactylethromorphus CABI-IMI 082348

Radio equipment

Aspergillus tamarii CABI-IMI 061268

Radio set

Aspergillus ustus CABI-IMI 089361
Penicillium pinophilum CABI-IMI 087160

Penicillium pinophilum CABI-IMI 087160ii
Rhizopus stolonifera CABI-IMI 061269

Rubber

Aspergillus fischeri var. *glaber* CABI-IMI 061447ii
Aspergillus fischeri var. *glaber* CABI-IMI 061447
Paecilomyces lilacinus CABI-IMI 117109

Sawdust

Thermoascus auranticus CABI-IMI 067936

Seed -Organo mercury treated

Syncephalastrum racemosum CABI-IMI 154276

Shoe leather

Cephaliophora tropica CABI-IMI 045555
Aspergillus unguis CABI-IMI 136526

Silcoset test panel

Cladosporium sphaerospermum CABI-IMI 135301

Silk screen of duplicating machine

Penicillium corylophilum CABI-IMI 095275

Starch

Amorpotheca resinae CABI-IMI 344050

Stone

Oidiodendron rhodogenum CABI-IMI 235256

Sugar sack

Aspergillus mangini CABI-IMI 072050

Synthetic rubber

Paecilomyces variotii CABI-IMI 108007

Talcum powder

Aspergillus versicolor CABI-IMI 180556
Penicillium chrysogenum CABI-IMI 185483
Penicillium chrysogenum CABI-IMI 180550
Penicillium griseofulvum CABI-IMI 180551
Stephanosporim cereale CABI-IMI 180559
Syncephalastrum racemosum CABI-IMI 180560

Tanning liquor

Penicillium spinulosum CABI-IMI 167427

Tent canvas

Chaetomium brasiliense CABI-IMI 045557
Chaetomium brasiliense CABI-IMI 044212

Tent canvas

Memnoniella echinata CABI-IMI 045547ii
Memnoniella echinata CABI-IMI 045547
Trichoderma longibrachiatum CABI-IMI 045548
Penicillium duclauxii CABI-IMI 040210
Myrothecium verrucaria CABI-IMI 140591

Termites

Paecilomyces variotii CABI-IMI 242891

Test block

Fusarium melanochlorum CABI-IMI 112906

Textile exposure test sample

Myrothecium verrucaria CABI-IMI 140594
Myrothecium gramineum CABI-IMI 140595

Textile in soil

Trichoderma viride CABI-IMI 110145
Trichoderma viride CABI-IMI 110138

Textiles

Penicillium aculeatum CABI-IMI 040588
Penicillium sclerotiorum CABI-IMI 040574

Vinyl pyridine latex

Graphium penicillioides CABI-IMI 280035

Vinyl wall covering

Penicillium spinulosum CABI-IMI 341998
Penicillium pupurogenum CABI-IMI 341997

Wallpaper

Aspergillus tamarii CABI-IMI 318157
Cladosporium sphaerospermum CABI-IMI 317350
Penicillium aurantiogriseum CABI-IMI 317752

Penicillium chrysogenum CABI-IMI 318158
Ulocladium chartarum CABI-IMI 183940

Water bag

Memnoniella echinata CABI-IMI 015377

Wine bottle cork

Oidiodendron rhodogenum CABI-IMI 291885

Wood

Amyloporia xantha NCWRF62C
Antrodia juniperina NCWRF136
Antrodia sinosa NCWRF 303A
Apoxona nitida NCWRF 232
Armillaria mellea NCWRF6H
Armillaria tabescens NCWRF 203A
Armillaria tabescens NCWRF 203
Bjerkandera adusta NCWRF35B
Bondarzewia montana NCWRF 356
Boreostereum radiatum NCWRF 414
Buglossoporous pulvinus NCWRF 313
Cephaloascus fragrans CABI-IMI 046868
Cereporiopsis NCWRF 127
Chondrostereum purpureum NCWRF88
Climacocystis borealis NCWRF 101
Coniophora puteana NCWRF 11A
Coriolopsis polyzina NCWRF 230
Coriolus hirstus NCWRF1
Coriolus versicolor NCWRF28C
Coriolus zonatus NCWRF199
Crepidotus mollis NCWRF400
Cyathus striatus NCWRF22
Cylindrocarpon destructans CABI-IMI 313222
Daedaleopsis confragosa NCWRF 235
Daldinia concentrica NCWRF 26C
Datronia mollis NCWRF 154
Dichomitus squalens NCWRF 70A
Dictyopanus rhipidium NCWRF 260
Ditiola radicata NCWRF 364
Donkioporia expansa NCWRF 178E
Eschinodontium tinctorium NCWRF 78
Exophiala mansonii CABI-IMI 170820
Exophiala mansonii CABI-IMI 173394
Fibroporia radiculosa NCWRF 346
Fibroporia vaillantii NCWRF 14
Flammulina veloptipes NCWRF 54
Flavodon flavus NCWRF 292
Fomes fomentarius NCWRF50C
Fomes fraxinophilus NCWRF 80
Fomes lividus NCWRF 293
Fomes melanosporus NCWRF 283
Fomes sclerodermeus NCWRF 223
Fomitopsis cajanderi NCWRF 106
Fomitopsis pinicola NCWRF 98
Fomitopsis rosea NCWRF 140
Funalia trogii NCWRF 247
Ganoderma lucideum NCWRF 89B
Ganoderma oregonense NCWRF 111
Ganoderma valesiacum NCWRF 189
Gleophyllum abietinum NCWRF 77C
Gleophyllum odoratum NCWRF 123B
Gleophyllum protractum NCWRF 314
Gleophyllum sepiarium NCWRF 10C
Gleophyllum trabeum NCWRF 108B
Gleosporus dichrous NCWRF 225
Grifola frondosa NCWRF 117
Gymnopilus spectabillis NCWRF 126
Gyrodontium versicolor NCWRF 339
Hapalopilus croceus NCWRF 145
Hericium coralloides NCWRF 60
Heteroporus biennis NCWRF 84C
Heteroporus biennis NCWRF 84D
Hirschioporus abietinus NCWRF 2
Hymenochaete rubiginosa NCWRF 243
Hypholoma fasciculare NCWRF 32
Hypoderma tenue NCWRF420
Hypoxylon fragiforme NCWRF 209A
Incrustoporia nivea NCWRF131B
Inonotus andersonnii NCWRF 288
Inonotus cuticularis NCWRF 278
Inonotus glomeratus NCWRF 284
Inonotus hispidus NCWRF43
Inonotus mikadoi NCWRF226
Inonotus nothofagi NCWRF332
Inonotus obliquus NCWRF298C
Inonotus radiatus var. *nodulosus* NCWRF 37B
Inonotus rickii NCWRF 289
Inonotus weirii NCWRF 286
Irpex lacteus NCWRF 191A
Ischnoderma benzonium NCWRF 137B
Kuehneromyces mutabalis NCWRF 49A
Laetiporus sulpureus NCWRF 29
Laricifomes officinalis (syn. *Fomes*) NCWRF81
Lentinellus cochleatus (syn. *Lentinus*) NCWRF171A
Lentinula deodes (syn. *Lentinus*) NCWRF335
Lentinus cyathiformis NCWRF 153B
Lentinus lepideus NCWRF 7
Lenzites betulina NCWRF 76
Lenzites palisotii NCWRF 231B
Lenzites subferruginea NCWRF 245
Leptoporus mollis NCWRF253

Leucogyrophana mollusca (syn. *Merulius*) NCWRF94A
Leucogyrophana pinastri NCWRF 141B
Lopharia spadicea NCWRF69A
Marasmius alliaceus NCWRF 275
Merulius tremellosus NCWRF 13
Merulopsis corium NCWRF30C
Onnia circinata NCWRF115
Oudemansiella radicata (syn. *Collybia*) NCWRF244
Oxysporus late-marginatus (syn. *Poria*) NCWRF375C
Oxysporus obduscens (syn. *Poria*) NCWRF51B
Panus tigrinus (syn. *Lentinus*) NCWRF68A
Panus torulosus NCWRF64A
Paxillus atromentosus NCWRF 211
Paxillus panuoides NCWRF8B
Peniphora gigantea NCWRF175A
Peniphora quercinia NCWRF 152A
Phellinus conchatus (syn. *Fomes*) NCWRF142
Phellinus contiguus (syn. *Poria*) NCWRF376
Phellinus demidoffii (syn. *Fomes*) NCWRF158
Phellinus fastuosus (syn. *Fomes*) NCWRF132
Phellinus ferruginosus (syn. *Fomes*) NCWRF47C
Phellinus gilvus (syn. *Polyporus*) NCWRF133a
Phellinus igniarius (syn. *Fomes*) NCWRF97B
Phellinus laevigatus (syn. *Poria*) NCWRF147
Phellinus pini (syn. *Trametes*) NCWRF45B
Phellinus pini (syn. *Trametes*) NCWRF45A
Phellinus senex (syn. *Fomes*) NCWRF389
Phellinus torulosus (syn. *Fomes*) NCWRF222
Phlebia radiata NCWRF19B
Phlebia rufa (syn. *Merulius*) NCWRF 330
Pholiota adiposa NCWRF52B
Pholiota aurivella NCWRF349
Piptoporus betulinus NCWRF 4B
Pleurotus eumosus NCWRF66
Pleurotus cystidiosus (syn. *P.corticats*) NCWRF212A
Pleurotus olearius NCWRF325
Pleurotus osteatus NCWRF40B
Polyporus rugulosus NCWRF 214
Polyporus anthracophilus NCWRF 327
Polyporus brumalis NCWRF174A
Polyporus galactinus (syn. *Spongipellis*) NCWRF355
Polyporus lentus NCWRF 337
Polyporus ochroleucus NCWRF 261
Polyporus scopulosus NCWRF 318
Polyporus spraguei (syn. *Tyromyces*) NCWRF227
Polyporus tumulosus (var. *westralensis*) NCWRF256A
Polyporus xanthopus (syn. *Microporus*) NCWRF365
Polystictus holsteii NCWRF366
Polystictus leoninus (syn. *Coriolus leoninus*) NCWRF251
Polystictus proteus NCWRF295
Poria aurea (syn. *Auriporia*) NCWRF367
Poria carbonacea NCWRF374
Poria clelandi (syn. *Fomes*) NCWRF259
Poria hypobrunnea NCWRF354
Poria incrassata (syn. *Merulioporia*) NCWRF71
Poria placenta NCWRF280
Poria sericeo-mollis (syn. *Leptoporus*) NCWRF358
Poria subacida (syn. *Perenniporia*) NCWRF104
Pulcherricium coeruleum (syn. *Corticum*) NCWRF371
Pycnoporus cinnabranus (syn. *Polystictus*) NCWRF 176
Pycnoporus sanguineus (syn. *Polystictus*) NCWRF150A
Rigidosporus lineatus (syn. *Polyporus*) NCWRF216
Rigidosporus microporus (syn. *Fomes*) NCWRF202
Rigidosporus ulmarius (syn. *Fomes*) NCWRF241
Rigidosporus vitreus (syn. *Polyporus adiposus*) NCWRF279
Scizopora (syn. *Irpex paradoxa*) NCWRF16B
Serpula himantoides NCWRF 73B
Serpula lacrymans NCWRF12
Spongipellis delectans (syn. *Polyporus*) NCWRF360
Spongipellis spumeus (syn. *Polyporus*) NCWRF166A
Spongipellis unicolor (syn. *Polyporus*) NCWRF198
Stereum lacteus NCWRF 418
Stereum lobatum NCWRF 263
Stereum necator NCWRF 268
Stereum ostrea NCWRF 105
Stereum rameale NCWRF 265
Stereum rugosum NCWRF 61A
Stereum sanguinolentum (syn. *Haematostereum*) NCWRF 27
Stereum vellereum NCWRF269
Sterum hirsutum NCWRF 3
Trametes feei NCWRF 201
Trametes lactinea NCWRF329
Trametes lilacino-gilva NCWRF 172A
Trametes ochro-leuca NCWRF 328
Trechispora mollusca (syn. *Poria*) NCWRF234
Tyromyces immitis (syn. *Polyporus*) NCWRF359
Tyromyces undosus (syn. *Polyporus*) NCWRF361
Wolfiporia extensa (syn. *Poria*) NCWRF344
Xylaria polymorpha NCWRF18
Xylobolus frustulatus (syn. *Stereum*) NCWRF83
Xylobolus subpileatus (syn. *Stereum*) NCWRF

Wood rotting fungi

Antrodia carbonica CABI-IMI 387930
Antrodia juniperina CABI-IMI 387542
Antrodia oleracea CABI-IMI 387544
Antrodia sinuosa CABI-IMI 387549

Coriolopsis polyzona CABI-IMI 387597
Coriolopsis polyzona CABI-IMI 387598
Fistulina hepatica CABI-IMI 387642
Flavodon flavus CABI-IMI 387647
Fomes fasciatus CABI-IMI 387658
Fomitopsis cajanderi CABI-IMI 387660
Fomitopsis feei CABI-IMI 388034
Fomitopsis lilacinogilva CABI-IMI 388039
Ganoderma pseudoferreum CABI-IMI 389522
Gloeophyllum trabeum CABI-IMI 387699
Gloeophyllum trabeum CABI-IMI 387700
Gloeophyllum trabeum CABI-IMI 387701
Gloeophyllum trabeum CABI-IMI 387705
Gloeophyllum trabeum CABI-IMI 387706
Gymnopilus sapineus CABI-IMI 387716
Hapalopilus croceus CABI-IMI 387719
Hericium erinaceus CABI-IMI 387725
Inonotus andersonii CABI-IMI 387750
Inonotus cuticularis CABI-IMI 387751
Inonotus rheades CABI-IMI 387766
Inonotus rickii CABI-IMI 387767
Irpex lacteus CABI-IMI 387769
Irpex lacteus CABI-IMI 387769
Laetiporus sulphurous CABI-IMI 389523
Loweporus lividus CABI-IMI 387654
Loweporus lividus CABI-IMI 387655
Nigroporus durus CABI-IMI 387906
Peniophora quercina CABI-IMI 387834
Phanerochaete sanguinea CABI-IMI 387848
Phellinus chrysoloma CABI-IMI 387849
Phellinus robiniae CABI-IMI 387878
Phellinus punctatus CABI-IMI 387881
Phellinus weirii CABI-IMI 387768
Pholiota populnea CABI-IMI 387886
Pleurotus cornucopiae CABI-IMI 387892
Polyporus arcularius CABI-IMI 387904
Perenniporia ochroleuca CABI-IMI 387911
Polyporus platensis CABI-IMI 387913
Polyporus rugulosus CABI-IMI 387914
Polyporus squamosus CABI-IMI 387918
Polystictus proteus CABI-IMI 387925
Postia placenta CABI-IMI 387942
Postia placenta CABI-IMI 387945
Postia placenta CABI-IMI 387948
Postia sericeomollis CABI-IMI 387952
Pycnoporus cinnabarinus CABI-IMI 387963
Pycnoporus sanguineus CABI-IMI 387964
Pycnoporus sanguineus CABI-IMI 387965
Pycnoporus sanguineus CABI-IMI 387966
Resinicium bicolor CABI-IMI 387970
Rhodotus palmatus CABI-IMI 387972
Rigidoporus microporus CABI-IMI 387975
Serpula himantioides CABI-IMI 389154
Sistotrema brinkmannii CABI-IMI 388904
Skeletocutis nivea CABI-IMI 387746
Skeletocutis nivea CABI-IMI 387747
Sphaerobolus stellatus CABI-IMI 388006
Spongipellis unicolor CABI-IMI 388009
Stereum necator CABI-IMI 388016
Stereum sanguinolentum CABI-IMI 389524
Postia balsamea CABI-IMI 388044
Xylobolus subpileatus CABI-IMI 388077
Xylobolus subpileatus CABI-IMI 388078

Wood (from mine)

Stigmina longispora CABI-IMI 139632
Tritirachium heimii CABI-IMI 149642

Wood pulp chips

Scytalidium album CABI-IMI 212120

Wood pulp chips of *Pinus sylvestris*

Xylogone sphaerospora CABI-IMI 212121

Wood (treated with preservatives)

Trichoderma viride CABI-IMI 153416

Wood rot (*Abies alba*)

Ischnoderma benzonium NCWRF 137A

Wood - dry rot

Serpula lacrymans NCWRF12C

Wood rot (*Abies balsamea*)

Gleophyllum trabeum NCWRF 108F

Wood rot (*Abies balsamea*)

Tyromyces guttulatus (syn. *Polyporus*) NCWRF185
Xeromphalina campanella (syn. *Omphalia*) NCWRF 319

Wood rot (*Abies concolor*)

Ganoderma oregonense NCWRF 111A

Wood rot (*Abies* sp.)

Columnocystis abietina NCWRF 177A
Gleophyllum sepiarium NCWRF 10E

Hericium abitis NCWRF 300A
Hericium abitis NCWRF 300B
Heterobasidion annosum NCWRF 41F
Poria placenta NCWRF 304G
Poria placenta NCWRF 304E
Trametes lilacino-gilva NCWRF 291

Wood rot (*Acer negundo*)

Pholiota formosa NCWRF342

Wood rot (*Acer* sp.)

Climacodon septentrionalis NCWRF 237B

Wood rot (*Albies balsamea*)

Poria subacida (syn. *Perenniporia*) NCWRF104A

Wood rot (*Alnus glutinosa*)

Daedaleopsis confragosa NCWRF 59A

Wood rot (*Alnus incana*)

Tyromyces chioneus (syn. *Polyporus*) NCWRF350C

Wood rot (*Alnus* sp.)

Inonotus radiatus var. *radiatus* NCWRF37C

Wood rot (*Alnus* sp.)

Plicatura sp. (syn. *Merulius*) NCWRF240

Wood rot (*Araucaria* sp.)

Climacocystis borealis NCWRF 101A

Wood rot (*Astronium balansae*)

Fomes sclerodermeus NCWRF 223A

Wood rot (*Betula lutea*)

Daedaleopsis confragosa NCWRF 82A

Wood rot (*Betula papyrifera*)

Cerrena unicolor NCWRF110A
Daedaleopsis confragosa NCWRF 82B
Hypochnicium vellereum NCWRF 321

Wood rot (*Betula* sp.)

Hypoxylon multiforme NCWRF 384
Lentinus lepideus NCWRF7F
Piptoporus betulinus NCWRF 4C
Tyromyces fissilis (syn. *Polyporus*) NCWRF85B

Wood rot (brickwork)

Coniophora marmorata NCWRF 410A
Coniophora marmorata NCWRF 410B

Wood rot (*Carya aquatica*)

Rigidosporus ulmarius (syn. *Fomes*) NCWRF241B

Wood rot (*Carya* sp.)

Globiformes graveolens NCWRF 118

Wood rot (*Cassia siamea*)

Phaeolus manihotis (syn. *Ganoderma*) NCWRF338

Wood rot (*Cassipourea elliottii*)

Coriolopsis polyzina NCWRF 230A

Wood rot (*Cassipourea elliottii*)

Trametes cingulate NCWRF250B

Wood rot (*Castanea sativa*)

Coriolus versicolor NCWRF28A
Fistulina hepatica NCWRF67
Laetiporus sulpureus NCWRF 29A

Wood rot (*Chamearops* sp.)

Podoscypha elegans (syn. *Stereum*) NCWRF341

Wood rot (chipboard)

Pleurotus osteatus NCWRF40C

Wood rot (*Citrus* sp.)

Trametes cervine NCWRF167

Wood rot (*Cocus mucifera* leaves)

Grammothele fuligo NCWRF 390

Wood rot (conifer)

Amyloporia crassa NCWRF373A
Amyloporia xantha NCWRF62D
Gymnopilus sapineus NCWRF 155
Junghuhnia collabens NCWRF 336
Leptoporus mollis NCWRF253B
Rigidosporus nigrescens (syn. *Poria*) NCWRF315
Tyromyces chioneus (syn. *Polyporus*) NCWRF350
Tyromyces stipticus (syn. *Polyporus*) NCWRF208

Wood rot (*Corylus avellana*)

Dichomitus campestris NCWRF 393
Incrustoporia nivea NCWRF131C

Wood rot (*Crataegus* sp.)

Phellinus ribis (syn. *Fomes*) NCWRF42

Wood rot (*Cupressus* sp.)

Coniophora puteana NCWRF 11E
Gleophyllum trabeum NCWRF 108E

Wood rot (*Dalbergia* sp.)

Pycnoporus sanguineus (syn. *Polystictus*) NCWRF150D

Wood rot (*Elaeis guineensis*)

Ganoderma lucideum NCWRF 89C

Wood rot (*Erigthrina crista-galli*)

Polyporus platensis NCWRF200

Wood rot (*Eucalyptus marginata*)

Fistulina hepatica NCWRF67C
Polyporus eucalyptorum NCWRF 257

Wood rot (*Eucalyptus paniculata*)

Fomes lividus NCWRF 293A

Wood rot (*Eugenia uniflora*)

Hexagonia variegata NCWRF 340

Wood rot (*Fagus* sp.)

Antrodia serpens NCWRF156
Aurantiporus alborobescens NCWRF368
Chondrostereum purpureum NCWRF88A
Ganoderma applanatum NCWRF 20C
Ganoderma applanatum NCWRF 50E
Ganoderma laccatum NCWRF 409
Grifola frondosa NCWRF36A
Grifola frondosa NCWRF117B
Hericium erinaceus NCWRF277
Incrustoporia nivea NCWRF131E
Kuehneromyces mutabalis NCWRF 49B
Meripilus giganteus NCWRF407A
Ossicaulis lignatilis (syn. *Pleurotus*) NCWRF128
Oudemansiella mucida (syn. *Armillaria*) NCWRF5
Oxysporus late-marginatus (syn. *Poria*) NCWRF375
Phaeolus schweinitzii NCWRF 102B
Podoscypha multizonata (syn. *Stereum*) NCWRF391
Pseudotrametes gibbosa (syn. *Tramete*) NCWRF25B
Stereum insignitum NCWRF369
Tyromyces floriformis (syn. *Polyporus*) NCWRF351

Wood rot (*Fagus sylvatica*)

Fomes fomentarius NCWRF50F

Wood rot (fibre board)

Amyloporia xantha NCWRF62G
Sistotrema brinkmannii NCWRF 386D

Wood rot (*Fragus* sp.)

Hypoxylon fragiforme NCWRF 209B

Wood rot (*Fraxinus excelsior*)

Daldinia concentrica NCWRF 26D

Wood rot (*Fraxinus* sp.)

Fomitopsis cytisina NCWRF 17A
Gleophyllum striatum NCWRF 204

Wood rot (hardwood)

Odontia nothofagi NCWRF403

Wood rot (*Hevea brasiliensis*)

Trametes incana NCWRF299

Wood rot (*Hevea* sp.)

Ganoderma philippii NCWRF 388A

Wood rot (*Hevea* sp.)

Phellinus fastuosus (syn. *Fomes*) NCWRF132A
Phellinus lamaensis (syn. *Fomes*) NCWRF383

Wood rot (*Hevea* sp.)

Phellinus noxius (syn. *Fomes*) NCWRF408
Polystictus vittas NCWRF348A
Polystictus vittas NCWRF348
Rigidosporus microporus NCWRF 202B
Trametes badia NCWRF345A
Trametes meyenii NCWRF353
Trametes scarbrosa NCWRF347

Wood rot (*Juniperus bermudiana*)

Fomitopsis cajanderi NCWRF 301

Wood rot (*Juniperus procera*)

Antrodia juniperina NCWRF136a
Phellinus demidoffii (syn. *Fomes*) NCWRF158A

Wood rot (*Khaya* sp.)

Poria placenta NCWRF 304H
Schizophyllum commune NCWRF 9

Wood rot (*Larix decidua*)

Stereum sanguinolentum (syn. *Haematostereum*) NCWRF27B

Wood rot (*Larix kaempferi*)

Stereum sanguinolentum (syn. *Haematostereum*) NCWRF27C

Wood rot (*Larix laricina*)

Fomitopsis pinicola NCWRF98D

Wood rot (*Larix* sp.)

Tyromyces lacteus (syn. *Polyporus*) NCWRF124C
Bjerkandera fumosa NCWRF 24B
Laricifomes officinalis (syn. *Fomes*) NCWRF81A
Laricifomes officinalis (syn. *Fomes*) NCWRF81B
Resinicium bicolor (syn. *Odontia*) NCWRF165A
Serpula himantoides NCWRF 233A

Wood rot (*Magnolia* sp.)

Rigidosporus ulmarius (syn. *Fomes*) NCWRF241A

Wood rot (*Malus* sp.)

Inonotus hispidus NCWRF43A
Pleurotus cornucopiae NCWRF 167B
Sarcodontia setosa (syn. *Acia*) NCWRF362
Tyromyces fissilis (syn. *Polyporus*) NCWRF85D
Tyromyces fissilis (syn. *Polyporus*) NCWRF85A
Tyromyces fissilis (syn. *Polyporus*) NCWRF85

Wood rot (*Mitragnya* sp.)

Gleophyllum striatum NCWRF 204A

Wood rot (*Ocotea rodiaeri*)

Pycnoporus sanguineus (syn. *Polystictus*) NCWRF150B

Wood rot (*Olex aquifolium*)

Stereum rugosum NCWRF61

Wood rot (*Peltophorum dubium*)

Fomes pseudosenex NCWRF190

Wood rot (*Picea excelsia*)

Pycnoporellus albo-luteus (syn. *Hapalopilus*) NCWRF 399

Wood rot (*Picea abies*)

Gleophyllum sepiarium NCWRF 10A

Wood rot (*Picea albeis*)

Resinicium bicolor (syn. *Odontia*) NCWRF165B

Wood rot (*Picea canadensis*)

Phellinus ferrugineo-fuscus (syn. *Poria*) NCWRF193

Wood rot (*Picea excelsa*)

Amyloporia crassa NCWRF373B
Hapalopilus ochraceo-lateritius NCWRF397
Incrustoporia stellae NCWRF 398
Phellinus ferrugineo-fuscus (syn. *Phellinus*) NCWRF193A

Wood rot (*Picea glauca*)

Peniphora septentrionalis NCWRF 372
Peniphora septentrionalis NCWRF 372A

Wood rot (*Picea mariana*)

Tyromyces balsameus (syn. *Leptoporus*) NCWRF114

Wood rot (*Picea sitchensis*)

Armillaria mellea NCWRF6F

Wood rot (*Picea sitchensis*)

Bondarzewia berkelyi NCWRF 100

Wood rot (*Picea sitchensis*)

Columnocystis abietina NCWRF 177
Heterobasidion annosum NCWRF 41E

Wood rot (*Picea sitchensis*)

Phaeolus schweinitzii NCWRF 102A

Wood rot (*Picea* sp.)

Amylostererеum areolatum NCWRF 404A
Amylostererеum chaillettii NCWRF 318D
Antrodia seriales NCWRF 107C
Coniophora marmorata NCWRF 410
Fomitopsis pinicola NCWRF98B
Gleophyllum abietinum NCWRF 77B
Gleophyllum odoratum NCWRF 123C
Lentinus kauffmanii NCWRF309A
Peniphora gigantea NCWRF175C
Phellinus chrysoloma (syn. *Fomes*) NCWRF246
Phellinus contiguus (syn. *Poria*) NCWRF376B
Poria sericeo-mollis (syn. *Leptoporus*) NCWRF296A
Stereum sanguinolentum (syn. *Haematostereum*) NCWRF27D
Tyromyces caesius (syn. *Polyporus*) NCWRF91F
Tyromyces revolutus (syn. *Polyporus*) NCWRF392
Phellinus nigrolimitatus (syn. *Fomes*) NCWRF320

Wood rot (*Pinus*)

Ceratocystis pilfera CABI-IMI 025316-

Wood rot (*Pinus caribaea*)

Lentinus pallidus NCWRF 406

Wood rot (*Pinus nigra* subsp. *larico*)

Sistotrema brinkmannii NCWRF 386A

Wood rot (*Pinus nigra* subsp. *larico*)

Sparassis crispa NCWRF312
Sphaerobolus stellatus NCWRF 169

Wood rot (*Pinus resinosa*)

Sctinostroma galactinum (syn. *Corticum*) NCWRF324B

Wood rot (*Pinus* sp.)

Tyromyces lacteus (syn. *Polyporus*) NCWRF124A
Tyromyces stipticus (syn. *Polyporus*) NCWRF208B
Tyromyces stipticus (syn. *Polyporus*) NCWRF208C
Armillaria mellea NCWRF 6G
Boletopsis subsquamosa NCWRF 402
Coniophora arida NCWRF411
Gleophyllum abietinum NCWRF 77D
Kuehneromyces mutabalis NCWRF 49
Lentinus lepideus NCWRF7E
Paxillus panuoides NCWRF8D
Phaeolus schweinitzii NCWRF 102
Phanerochaete sanguinea NCWRF 161A
Polyporus palustris NCWRF307A
Ptychogaster albus NCWRF333
Resinicium bicolor (syn. *Odontia*) NCWRF165D
Serpula himantoides NCWRF 233B
Stereum sanguinolentum (syn. *Haematostereum*) NCWRF27E
Tyromyces balsameus (syn. *Leptoporus*) NCWRF311B
Tyromyces fragilis (syn. *Polyporus*) NCWRF173A

Wood rot (*Pinus strobus*)

Amyloporia xantha NCWRF62F
Sctinostroma galactinum (syn. *Corticum*) NCWRF324
Wood rot (*Pinus sylvestris*)
Amylostererеum chaillettii NCWRF 318E
Cartilosoma subsinosa NCWRF 394
Dacromyces stillatus NCWRF 412A
Fibroporia vaillantii NCWRF 14F
Gleophyllum sepiarium NCWRF 10D
Gloeocystidiellum lactescens NCWRF379
Heterobasidion annosum NCWRF 41G
Hypoderma puberum NCWRF 419A
Merulopsis taxicola NCWRF401
Peniphora pinus NCWRF267A
Phaeolus schweinitzii NCWRF 102D
Phellinus contiguus (syn. *Poria*) NCWRF376A
Piloporia albobrunnea (syn. *Leptoporus*) NCWRF378

Resinicium bicolor (syn. *Odontia*) NCWRF165
Sistotrema brinkmannii NCWRF 368F
Sistotrema brinkmannii NCWRF 368E
Sistotrema brinkmannii NCWRF 386C
Steccherinium fimbriatum (syn. *Mycoleptodon*) NCWRF395
Tyromyces fragilis (syn. *Polyporus*) NCWRF173E
Poria placenta NCWRF 304B

Wood rot (*Podocarpus* sp.)

Gleophyllum trabeum NCWRF 108D

Wood rot (*Populus* sp.)

Inonotus rheades var. *rheades* NCWRF183
Lentinus cyathiformis NCWRF 153E

Wood rot (*Populus tremuloides*)

Wolfiporia extensa (syn. *Poria*) NCWRF344B

Wood rot (*Populus tremuloides*)

Wolfiporia extensa (syn. *Poria*) NCWRF344A

Wood rot (*Prunus avium*)

Pycnoporus sanguineus (syn. *Polystictus*) NCWRF150C

Wood rot (*Prunus* sp.)

Armillaria mellea NCWRF6E
Daedaleopsis confragosa NCWRF59C
Ganoderma applanatum NCWRF20F
Phellinus pomaceus (syn. *Fomes*) NCWRF33A
Phellinus pomaceus (syn. *Fomes*) NCWRF33B
Pholiota lucifera NCWRF113
Tyromyces fissilis (syn. *Polyporus*) NCWRF85C
Xylaria hypoxylon NCWRF65B

Wood rot (*Pseudotsuga menziesii*)

Poria carbonica NCWRF308B
Poria carbonica NCWRF308C
Poria carbonica NCWRF308D
Fomitopsis pinicola NCWRF98E
Poria placenta NCWRF304C
Poria placenta NCWRF304D

Wood rot (*Pseudotsuga* sp.)

Poria rivulosa NCWRF316

Wood rot (*Pycnanthus angolensis*)

Lentinus squarrosolus NCWRF 413

Wood rot (*Quercus albus*)

Phellinus gilvus (syn. *Polyporus*) NCWRF133

Wood rot (*Quercus ilex*)

Inonotus dryadeus NCWRF96A

Wood rot (*Quercus* sp.)

Antrodia oleracea NCWRF196
Bjerkandera fumosa NCWRF 24
Daedalea quercina NCWRF38D
Daedalea quercina NCWRF38
Daedalea quercina NCWRF38E
Fistulina hepatica NCWRF67B
Ganoderma resinaceum NCWRF 144A
Grifola frondosa NCWRF117A
Inonotus dryadeus NCWRF96
Lopharia spadicea NCWRF69C
Panus tigrinus (syn. *Lentinus*) NCWRF68
Peniphora quercinia NCWRF 152B
Polyporus durus (syn. *Fomes*) NCWRF205
Poria medulla-panis (syn. *Perenniporisa*) NCWRF258B
Tyromyces chioneus (syn. *Polyporus*) NCWRF350A
Xylobolus frustulatus (syn. *Stereum*) NCWRF83A
Xylobolus subpileatus (syn. *Stereum*) NCWRF
Phellinus ferrus (syn. *Poria*) NCWRF387

Wood rot (*Robinia pseudoacacia*)

Fomitopsis cytisina NCWRF17
Phellinus robinae (syn. *Fomes*) NCWRF 180

Wood rot (*Rubus odoratus*)

Peniphora incarnata NCWRF 86A

Wood rot (*Salix* sp.)

Oxysporus late-marginatus (syn. *Poria*) NCWRF375A
Phellinus igniarius (syn. *Fomes*) NCWRF97D
Phellinus igniarius (syn. *Fomes*) NCWRF97E
Pholiota destruens NCWRF159

Wood rot (*Sapindus* sp.)

Oxysporus late-marginatus (syn. *Poria*) NCWRF375B

Wood rot (*Schinopsis balansae*)

Fomes chaquensis NCWRF417

Wood rot (*Sequoia sempervirens*)

Poria sequoniae NCWRF 306

Wood rot (softwood)

Hygrophoropsis aurantiaca NCWRF207
Hypoderma puberum NCWRF 419

Wood rot (*Sorbus americana*)

Irpex lacteus NCWRF 191

Wood rot (*Swietenia*)

Trametes hydnoides NCWRF 377

Wood rot (*Thuja plicata*)

Poria sericeo-mollis (syn. *Leptoporus*) NCWRF296
Poria sericeo-mollis (syn. *Leptoporus*) NCWRF296B

Wood rot (*Thuja* sp.)

Amylosterereum chaillettii NCWRF 318

Wood rot (*Trachylobium verrucosum*)

Polystictus fulvo-cinereus NCWRF 385

Wood rot (tree)

Polystictus fimbriatus NCWRF 343

Wood rot (*Tsuga heterophylla*)

Coriolus versicolor NCWRF28B
Fomitopsis pinicola NCWRF98C
Phellinus contiguus (syn. *Poria*) NCWRF376C
Poria placenta NCWRF 304J
Haematostereum) NCWRF27F

Wood rot (*Tsuga* sp.)

Columnocystis abietina NCWRF 177C
Dacromyces stillatus NCWRF 412
Hericium abitis NCWRF 300C
Phellinus punctatus (syn. *Poria*) NCWRF 194
Poria placenta NCWRF304F

Wood rot (*Uinas* sp.)

Fomitopsis cytisina NCWRF 17B

Wood rot (*Ulmus* sp)

Agrocybe aererita NCWRF302
Flammulina veloptipes NCWRF 54A
Lentinus adhaerens NCWRF415
Pleurotus osteatus NCWRF40A
Phellinus ferruginosus (syn. *Fomes*) NCWRF47B
Pleurotus cornucopiae NCWRF 167A
Polyporus squamosus (syn. *Cerioporus*) NCWRF103B
Poronidulus conchifer (syn. *Polyporus*) NCWRF116B
Rhodotus palmatus (syn. *Pleurotus*) NCWRF170

Wood rot (*Ulmus americana*)

Climacodon septentrionalis NCWRF237A

Wood rot (yellow straw of *Eucalyptus*)

Poria healeyi NCWRF 380

Wood shavings

Mucor racemosus CABI-IMI 296421

Wooden storage bin

Malbranchea pulchella CABI-IMI 096741

Wood (chemically treated)

Penicillium arenicola CABI-IMI 367261*

Wood blocks in sea water

Piricauda articoceanorum CABI-IMI 081625

Wood blocks in sea water

Zalerion maritima CABI-IMI 081619
Zalerion maritima CABI-IMI 081620
Zalerion maritima CABI-IMI 081624

Wood destroying (patent strain)

Aspergillus candidus CABI-IMI 127260

Wood pulp

Penicillium roquefortii CABI-IMI 057201
Trichoderma harzianum CABI-IMI 055217

Wood pulp (Scandinavian)

Penicillium melinii CABI-IMI 060041

Zinc based oil paint

Penicillium chrysogenum CABI-IMI 178521

Food Spoilage Strains

See also: Biodeteriogens, Biodegraders and utilizers

Animal feed

Aspergillus repens CABI-IMI 321288

Animal feed nuts

Penicillium solitum CABI-IMI 291543

Apple

Phytohpthora cactorum CABI-IMI 381060*
Penicillium expansum CABI-IMI 378779*
Penicillium expansum CABI-IMI 378781*
Penicillium expansum CABI-IMI 378782*
Penicillium expansum CABI-IMI 378783*
Penicillium expansum CABI-IMI 378785*
Penicillium expansum CABI-IMI 378786*
Penicillium expansum CABI-IMI 378787*
Penicillium expansum CABI-IMI 379324*
Penicillium expansum CABI-IMI 379325*
Penicillium expansum CABI-IMI 379326*
Penicillium expansum CABI-IMI 379327*
Penicillium expansum CABI-IMI 379328*
Penicillium pinophilum CABI-IMI 378780
Penicillium pinophilum CABI-IMI 379323

Apricot

Penicillium ulaiense CABI-IMI 380701*

Australian apple

Monilinia fructicola CABI-IMI 158424

Bakers yeast (mouldy)

Penicillium echinulatum CABI-IMI 293180
Penicillium roquefortii CABI-IMI 285518

Bird feed

Penicillium griseofulvum CABI-IMI 293195

Blackcurrent

Penicillium crustosum CABI-IMI 113137

Blue mould (food)

Penicillium spp. CABI-IMI Many

Blue peas (dried)

Penicillium phoeniceum CABI-IMI 089286

Bran (Toxic)

Aspergillus flavus CABI-IMI 092875

Bread

Penicillium crustosum CABI-IMI 191240

Bread (India)

Aspergillus terricola var. *indicus* CABI-IMI 172295

Butter

Paecilomyces fumosoroseus CABI-IMI 058413

Cake

Penicillium crustosum CABI-IMI 028040
Wallemia sebi CABI-IMI 037338

Canned blueberries

Penicillium lapidosum CABI-IMI 039743

Canned peaches

Penicillium viridicatum CABI-IMI 297971

Cashew nuts

Penicillium miczynskii CABI-IMI 191323

Cattle feed

Cladosporium tenuissimum CABI-IMI 224833

Cheese

Mucor racemosus f. *sphaerosporus* CABI-IMI 318800
Penicillium atramentosum CABI-IMI 039752ii
Penicillium commune CABI-IMI 039812ii
Penicillium chrysogenum CABI-IMI 024314
Penicillium crustosum CABI-IMI 089385
Penicillium crustosum CABI-IMI 214071
Penicillium palitans CABI-IMI 092276
Penicillium palitans CABI-IMI 089378
Penicillium palitans CABI-IMI 089379
*Penicillium solitum*CABI-IMI 293210

Cheese (blue)

Penicillium roquefortii CABI-IMI 337464

Cheese (Brie)

Penicillium commune CABI-IMI 092215

Cheese (Camembert)

Penicillium atramentosum CABI-IMI 039752

Cheese (Danish blue)

Penicillium roquefortii CABI-IMI 129207
Penicillium roquefortii CABI-IMI 173224

Cheese (Ellischauer)

Penicillium jensenii CABI-IMI 039804
Cheese (Gorgonzola)
Penicillium roquefortii CABI-IMI 092261

Cheese (mouldy Danbo)

Penicillium verrucosum CABI-IMI 285523

Cheese (Parmesan)

Penicillium solitum CABI-IMI 297969

Cheese (Roquefort)

Penicillium roquefortii CABI-IMI 024313

Cheese (Stilton blue)

Penicillium roquefortii CABI-IMI 148775

Chocolate

Aspergillus mangini CABI-IMI 068766

Chocolate covered sponge biscuit

Xeromyces bisporus CABI-IMI 179743

Chocolate sponge biscuit

Penicillium viridicatum CABI-IMI 154731
Xeromyces bisporus CABI-IMI 154732

Chopped chinese dates

Xeromyces bisporus CABI-IMI 317902

Citrus sinensis

Penicillium ulaiense CABI-IMI 380875*
Penicillium ulaiense CABI-IMI 380876*
Penicillium ulaiense CABI-IMI 380877*
Penicillium italicum CABI-IMI 380880*

Citrus x paradisi

Penicillium ulaiense CABI-IMI 380878*

Citrus x paradisi

Penicillium ulaiense CABI-IMI 380879*

Coconut (desiccated)

Aspergillus repens CABI-IMI 298307

Cooked meat

Penicillium aurantiogriseum CABI-IMI 089375

Corn (mouldy field)

Penicillium rubrum CABI-IMI 136127
Penicillium rubrum CABI-IMI 136128
Penicillium rubrum CABI-IMI 136126

Corn and rice cake

Penicillium herquei CABI-IMI 007651

Corn, barley and sorghum grain

Penicillium islandicum CABI-IMI 105272

Cottage cheese

Geotrichum candidum CABI-IMI 29023

Crumpets

Penicillium solitum CABI-IMI 297997

Dairy products

Fennellia flavipes CABI-IMI 135422

Dried fish

Penicillium chalybeum CABI-IMI 288722
Penicillium chalybeum CABI-IMI 288721
Polypaecilium pisce CABI-IMI 288727ii
Polypaecilium pisce CABI-IMI 288720ii

Dried prunes

Bettsia alvei CABI-IMI 151276
Bettsia alvei CABI-IMI 126288

Fermented coccoa

Aspergillus tamarii CABI-IMI 016129

Fermented rice grains

Monascus purpueus CABI-IMI 210765

Foodstuffs

Aspergillus tonophilus CABI-IMI 226077
Bettsia alvei CABI-IMI 226074

Fruit based gel (baby food)

Penicillium oblatum CABI-IMI 288718

Fruit sauce (sweeet)

Monilinia acetobutans CABI-IMI 159918

Gelatine

Geomyces pannorum CABI-IMI 059628
Penicillium atramentosum CABI-IMI 089373

Grape

Penicillium crustosum CABI-IMI 293182

Grapes (stored)

Penicillium crustosum CABI-IMI 174717

Groundnuts

Penicillium aurantiogriseum CABI-IMI 091020

Hazelnut

Penicillium solitum CABI-IMI 321495
Aspergillus ruber CABI-IMI 071762
Aspergillus repens CABI-IMI 298305

Hops (stored)

Penicillium carneolutescens CABI-IMI 040218

Irn Bru drink

Brettanomyces sp. CABI-IMI 357902

Jam

Wallemia sebi CABI-IMI 037338

Lemon ice tea

Paecilomyces variotii CABI-IMI 371343*

Lemon juice

Penicillium brevicompactum CABI-IMI 194127

Lime flavour carton drink

Trichoderma viride CABI-IMI 296843

Liquorice

Xeromyces bisporus CABI-IMI 317904

Madeira cake

Penicillium expansum CABI-IMI 297965

Maize (stored)

Penicillium brevicompactum CABI-IMI 092034

Malt extract

Geotrichum candidum CABI-IMI 042074

Margarine

Penicillium sclerotiorum CABI-IMI 304282

Penicillium sclerotiorum CABI-IMI 304281

Meat

Cladosporium sphaerospermum CABI-IMI 091964
Geomyces pannorum CABI-IMI 023281i
Mucor racemosus CABI-IMI 017313

Milk

Geotrichum candidum CABI-IMI 283163
Penicillium crustosum CABI-IMI 089386

Milk (pasteurised)

Dichotomomycetes cepji CABI-IMI 151074

Milled rice

Microascus trigonosorus CABI-IMI 086912

Millet alcoholic beverage

Mucor circinelloides forma *circinelloides* CABI-IMI 375454*

Nigerian groundnuts

Aspergillus flavus CABI-IMI 091020

Orange

Penicillium digitatum CABI-IMI 091956
Penicillium ulaiense CABI-IMI 380702*

Pecans (weevil damaged)

Penicillium solitum CABI-IMI 321496

Pepper

Stemphylium majusculum CABI-IMI 191067

Pistachio nuts

Aspergillus flavus CABI-IMI 190443

Plum juice

Aspergillus itaconicus CABI-IMI 016119

Poultry feed

Microdochium dimerum CABI-IMI 117087
Cheatomium cupreum CABI-IMI 299545

Rye bread

Penicillium roquefortii CABI-IMI 293204
Penicillium roquefortii CABI-IMI 293189

Salami sausage

Paecilomyces variotii CABI-IMI 131711
Penicillium nagliovense CABI-IMI 376118*

Salt fish

Wallemia sebi CABI-IMI 074824
Wallemia sebi CABI-IMI 076804

Salted anchovy paste

Wallemia sebi CABI-IMI 086292

Sausage product

Penicillium aurantiogriseum CABI-IMI 300381

Savoury preserve

Aspergillus amstelodami CABI-IMI 212944

Soya products

Aspergillus tamarii CABI-IMI 321319
Aspergillus tamarii CABI-IMI 091888

Stick liquorice

Xeromyces bisporus CABI-IMI 063718

Stored food

Penicillium purpurescens CABI-IMI 141658
Curvularia borreriae CABI-IMI 141697

Sugar

Aspergillus terreus CABI-IMI 016044

Table jelly

Xeromyces bisporus CABI-IMI 076589
Bettsia alvei CABI-IMI 307744

Tobacco

Aspergillus versicolor CABI-IMI 091883
Aspergillus versicolor CABI-IMI 016041
Myrothecium leucotrichum CABI-IMI 001495ii
Myrothecium leucotrichum CABI-IMI 001495
Paecilomyces victoriae CABI-IMI 058412
Penicillium crustosum CABI-IMI 295697
Penicillium crustosum CABI-IMI 295696

Tobacco (cured)

Penicillium pseudostromaticum CABI-IMI 197491

Tobacco (pipe)

Fusarium culmorum CABI-IMI 291777

Walnut kernal

Penicillium spinulosum CABI-IMI 113139

Wheat bread

Penicillium italicum CABI-IMI 293196

Wine

Monascus pilosus CABI-IMI 296424

Tolerant Strains

See also: Environmental strains;
Strains tolerant to chemical and environmental parameters

Acidophile

Scytalidium acidophilum CABI-IMI 183518
Scytalidium acidophilum CABI-IMI 173066ii

Acidophile (acetic acid)

Paecilomyces dactylethromorphus CABI-IMI 065752
Paecilomyces dactylethromorphus CABI-IMI 065752

Acidophile (acetic preserve)

Moniliella acetoabutans CABI-IMI 180570

Acidophile (grows on acetic acid media)

Penicillium solitum CABI-IMI 300728
Penicillium simplicissium CABI-IMI 296073

Alkali (from pulp manufacture)

Trichoderma viride CABI-IMI 295977

Alkalophile

Acremonium restrictum CABI-IMI 206958
Acremonium implicatum CABI-IMI 59628
Acremonium curvulum CABI-IMI 297016
Geomyces pannorum CABI-IMI 059628
Trichoderma harzianum CABI-IMI 297033
Trichoderma harzianum CABI-IMI 295979
Coprinus amphibius CABI-IMI 133855
Fusarium merismoides CABI-IMI 297028
Fusarium merismoides CABI-IMI 297032
Fusarium merismoides CABI-IMI 297027
Gliocladium virens CABI-IMI 295989
Trichoderma viride CABI-IMI 295977

Benzimidazole

Venturia inaequalis CABI-IMI 223845

Copper (tolerant to high concentrations)

Scytalidium acidophilum CABI-IMI 173066ii

High altitude

Colletotrichum kahawe CABI-IMI
High temperature: see thermophiles
Low temperature: see psychrophiles

Psychrophile

Cladosporium sphaerospermum CABI-IMI 091964
Mucor hiemalis CABI-IMI 289880
Mucor racemosus CABI-IMI 017313
Alternaria alternaria CABI-IMI 378423*
Unamed CABI-IMI 378656*
Unamed CABI-IMI 378657*
Leptodontidium elatus CABI-IMI 369788*
Cladosporum sp. CABI-IMI 377828*

Psychrophile (Alpine tundra)

Thielavia peruviana CABI-IMI 135024

Psychrophile (Artic isolate (soil))

Phlyctochytrium CABI-IMI 143636

Thermophile

Calcarisporium thermophilium CABI-IMI 144750
Chaetomium thermophilium CABI-IMI 289313
Humicola grisea var. *themoidea* CABI-IMI 369788
Penicillium lapidosum CABI-IMI 039743

Thermophile (45°C and above)

Aspergllu fumigatus CABI-IMI 361368

Thermophile (from fermenter system at elevated temperature)

Aspergillus fumigatus CABI-IMI 293740

Resistant Strains

Strains resistant to Antibiotics, chemicals, radiation, environmental parameters etc.

Benomyl fungicide

Botrytis cinerea CABI-IMI 313860
Cercospora api CABI-IMI 186563
Colletotrichum musae CABI-IMI 165753
Penicillium hirsutum CABI-IMI 170486

Heat resistant

Penicillium lapidosum CABI-IMI 039743

Phenol mercury

Cladosporium cladosporioides CABI-IMI 154279
Cladosporium cladosporioides CABI-IMI 154278

Phenol mercury 2.5ppm

Penicillium crustosum CABI-IMI 154282
Penicillium crustosum CABI-IMI 154283

Thia-bendazole

Colletotrichum musae CABI-IMI 165753

Benzimidazole

Venturia inaequalis CABI-IMI 223844

Test Strains

For cited specifications

Aerosols

Chaetomium globosum CABI-IMI 045550
Chaetomium globosum CABI-IMI 045550iv
Chaetomium globosum CABI-IMI 045550iii
Chaetomium globosum CABI-IMI 045550ii

Antimicrobial preservative testing

Aspergillus niger CABI-IMI 149007

ASTM G21

Aureobasidium pullans CABI-IMI 145194
Aspergillus niger CABI-IMI 091855
Penicillium pinophilum CABI-IMI 087160
Chaetomium globosum CABI-IMI 045550ii
Trichoderma viride CABI-IMI 045553ii

Australian test mould proofing strain

Alternaria alternata CABI-IMI 071749

Fungal toxicity of alcohols and fatty acids

Mucor circinelloides CABI-IMI 078379

Fungus resistance test BS2011 PART 2J

Aspergillus niger CABI-IMI 017454
Aspergillus terreus CABI-IMI 045543
Aureobasidium pullans CABI-IMI 045533
Paecilomyces variotii CABI-IMI 108007
Penicillium pinophilum CABI-IMI 114933
Penicillium ochrochloron CABI-IMI 061271
Scopulariopsis brevicaulis CABI-IMI 049528
Trichoderma viride CABI-IMI 045553ii

Fungus resistance test DEF 133

Aspergillus amstelodami CABI-IMI 017455
Aspergillus niger CABI-IMI 017454
Chaetomium globosum CABI-IMI 016203
Paecilomyces variotii CABI-IMI 108007
Penicillium brevicompactum CABI-IMI 017456

Penicillium cyclopium CABI-IMI 019759
Stachybotrys atra CABI-IMI082021

Fungus resistance test MIL 810D (MIL C 62122b)

Aspergillus niger CABI-IMI 091855
Aspergillus flavus CABI-IMI 091856
Aspergillus versicolor CABI-IMI 045554ii
Penicillium funiculosum CABI-IMI 211742
Penicillium pinophilum CABI-IMI 087160
Chaetomium globosum CABI-IMI 045550ii

Fungus resistance testing

Aureobasidium pullans CABI-IMI 145194
Paecilomyces lilacinus CABI-IMI 117109
Paecilomyces variotii CABI-IMI 108007

Mould proof

Aspergillus amstelodami CABI-IMI 017455
Aspergillus clavatus CABI-IMI 155703
Aspergillus flavus CABI-IMI 114929
Aspergillus flavus CABI-IMI 0901856
Aspergillus flavus CABI-IMI 091856
Aspergillus flavus CABI-IMI 091856iii
Aspergillus niger CABI-IMI 070102
Aspergillus niger CABI-IMI 017454
Aspergillus niger CABI-IMI 114927
Aspergillus niger CABI-IMI 059374
Aspergillus niger CABI-IMI 091855iii
Aspergillus ustus CABI-IMI 045552
Aureobasidium pullans CABI-IMI 070103
Chaetomium globosum CABI-IMI 045550iv
Chaetomium globosum CABI-IMI 045550ii
Chaetomium globosum CABI-IMI 016203
Chaetomium globosum CABI-IMI 045550
Chaetomium globosum CABI-IMI 045550iii
Fusarium moniliforme CABI-IMI 061274
Paecilomyces variotii CABI-IMI 017457
Penicillium brevicompactum CABI-IMI 114932
Penicillium brevicompactum CABI-IMI 017456
Penicillium citrinum CABI-IMI 061272
Penicillium commune CABI-IMI 039812ii
Penicillium ochrochloron CABI-IMI 061271ii
Penicillium ochrochloron CABI-IMI 061271
Penicillium pinophilum CABI-IMI 087160
Penicillium pinophilum CABI-IMI 114933
Penicillium pinophilum CABI-IMI 087160ii
Rhizopus stolonifera CABI-IMI 061269
Scopulariopsis brevicaulis CABI-IMI 049528

Mould proofing strain (Australia)

Aspergillus niger CABI-IMI 091855
Aspergillus niger CABI-IMI 091855ii
Pestalotiopsis gracilis CABI-IMI 089307

Mold proofing (USA)

Aspergillus niger CABI-IMI 045551
Aspergillus tamarii CABI-IMI 061268
Trichoderma viride CABI-IMI 045553

Mould testing resistance

Aspergillus niger CABI-IMI 091855iii

Pullan 4-glucanhydrolase

Aspergillus niger CABI-IMI 091855iii

Wall covering adhesive

Aspergillus niger CABI-IMI 017454
Chaetomium globosum CABI-IMI 016203
Penicillium pinophilum CABI-IMI 113730
Penicillium rubrum CABI-IMI 113729
Stachybotrys atra CABI-IMI 082021

Wood preservatives BS1982: Part 1 of contructional

Amyloporia xantha NCWRF62G
Coniophora puteana NCWRF 11E
Coriolus versicolor NCWRF28A

Wood preservatives BS1982:Part 1 of contructional

Gleophyllum trabeum NCWRF 108N
Pleurotus ostreatus NCWRF40C
Serpula lacrymans NCWRF12C

Wood preservatives BS6009:1982 Additional

Coniophora puteana NCWRF 11E
Gleophyllum abietinum NCWRF 77C
Gleophyllum sepiarium NCWRF 10E
Paxillus panuoides NCWRF8D
Poria placenta NCWRF304K
Serpula himantiodes NCWRF 233C
Serpula lacrymans NCWRF12E
Stereum hisutum NCWRF3B

Wood preservatives BS6009:1982 Creosotes etc.

Lantinus lepideus NCWRF7H
Lentinus cyathiformis NCWRF 153E

Wood preservatives BS6009: 1982 General

Coniophora puteana NCWRF 11R
Coriolus versicolor NCWRF28G
Wood preservatives BS6009:1982 Other
Gleophyllum trabeum NCWRF 108N
Poria placenta NCWRF 280

Wood CEN test strain

Gloeophyllum trabeum CABI-IMI 387707 (NCWRMF)

Assay Strains Including Assay of Antibiotics and Vitamins

B317 -9-9a Argenine 10

Neurospora crassa CABI-IMI 147008

B46004-Argenine 1

Neurospora crassa CABI-IMI 147007

Biotin

Saccharomyces cerevisiae CABI-IMI 019391

Choline

Neurospora crassa CABI-IMI 019419

Copper Mulders strain in soils

Aspergillus niger CABI-IMI 031283

Inositol

Neurospora crassa CABI-IMI 147006
Neurospora crassa CABI-IMI 017836

Leucine A

Neurospora crassa CABI-IMI 019420ii

Magnesium in soils

Aspergillus niger (Mulders strain) CABI-IMI 031283

Molybdenum in soils

Aspergillus niger (Mulders strain) CABI-IMI 031283

Phosphorus in soils

Aspergillus niger CABI-IMI 094145

Potassium in soils

Aspergillus niger CABI-IMI 094145

Proline

Neurospora crassa CABI-IMI 147005
Neurospora crassa CABI-IMI 044244

Zinc in soils

Aspergillus niger (Mulders strain) CABI-IMI 031283

Special Features: Anatomical and Morphological

Adiaspore

Chrysosporium merardium CABI-IMI 201979

Appresoria

Colletotrichium coccodes CABI-IMI 078581
Ventura inaequalis CABI-IMI 273208

Arthoconidia

Crinipellis periciosa CABI-IMI 367909

Ascospore (dark discharge)

Daldinia concentrica CABI-IMI 193348

Ascospore (enanescent ascii)

Ceratocystis CABI-IMI 159574

Ascospore (perithecial)

Sordaria fimicola CABI-IMI 146886

Ascospores

Aspergillus alliaceus CABI-IMI 051982

Ballistospore

Itersonilia perplexans CABI-IMI 264396

Beige spored mating type A

Sordaria brevicollis CABI-IMI 210764

Black yeasts

Aureobasidium pullans CABI-IMI 269216

Chlamydospores (teaching)

Fusarium oxysporum CABI-IMI 244439
Phoma glomerata CABI-IMI 327750
Phoma jolgana CABI-IMI 361393

Clamps

Crinipellis periciosa CABI-IMI 367909*
Serpula lacrymans NCWRF12C

Macroconidia

Fusarium oxysporum CABI-IMI 244439

Macroconidia (5 septate)

Nectridiscophora sp. CABI-IMI 370949

Microconida

Fusarium oxysporum CABI-IMI 244439

Micoconidia (aseptate)

Periconiella sapienfumicola CABI-IMI 380798

Mitichondrial plasmid containing

Neurospora crassa CABI-IMI 317528
Neurospora intermedia CABI-IMI 317527

Non-sporulating in glass bottles

Aspergillus spathulatus CABI-IMI 308593

Strand formation (basidiomycete)

Serpula lacrymans NCWRF12C

Special Features: Physiological

Including nutrient requiring strains, fixators, teaching strains and strains exhibiting specific physiological processess, and mutant strains.

Adenine

Neurospora crassa CABI-IMI 075724

Albino strain

Neurospora crassa CABI-IMI 147004

Amoboidophagous

Arthrobotrys pyriformis CABI-IMI 056803

Armillaria mellea (inhibits)

Ascomycotina sp. CABI-IMI 381608*

Autoclaved wool (attacks)

Trichoderma longibrachiatum CABI-IMI 045548

Autotrophic mutant

Verticillium lamellicola CABI-IMI 093439
Acremonium fusidioides CABI-IMI 087758

Chemotype 1

Penicillium aurantiogriseum CABI-IMI 293186

Cinnamon coloured mutant

Aspergillus cinnamomeus CABI-IMI 016148

Colourless strain

Humicola hyalothermophila CABI-IMI 204250

Cysteine requiring

Acremonium fusidioides CABI-IMI 087758

Endophyte

Diploidia salicorniar CABI-IMI 295925

Fungistatic (on *Heterobasidion annosum*)

Trichoderma harzianum CABI-IMI 206040
Trichoderma polysporum CABI-IMI 206039

Haploid isolate stable

Penicillium viridicatum CABI-IMI 039758ii

High galactosamine content

Aspergillus niger CABI-IMI 146891

Homothallic

Pythium splendens CABI-IMI 323145
Pythium splendens CABI-IMI 323148
Rhizomucor pusillus CABI-IMI 226166
Rhizomucor miehei CABI-IMI 089882
Rhizomucor pusillus CABI-IMI 226172

Hygrophobic

Aspergillus fumigatus CABI-IMI 051984
Aspergillus glaucus CABI-IMI 053242

Hypoxanthine

Neurospora crassa CABI-IMI 075724

Isolated from casein hydrolysate

Mucor racemosus CABI-IMI 114891

Keratinolytic

Chrysosporium tropicum CABI-IMI 094288

Methionine requiring

Acremonium fusidioides CABI-IMI 087758

Monokaryon culture (SS)

Ganoderma sp. CABI-IMI 380170*
Ganoderma sp. CABI-IMI 380171*
Ganoderma sp. CABI-IMI 380172*
Ganoderma sp. CABI-IMI 380173*
Ganoderma sp. CABI-IMI 380174*
Ganoderma sp. CABI-IMI 380175*

Morphological mutant

Fusarium coccineum CABI-IMI 087759

Nematophagous

Arthrobotrys amerospora CABI-IMI 232238

Niacin requiring

Neurospora crassa CABI-IMI 075727ii

Nicotinic acid requiring

Neurospora crassa CABI-IMI 075726ii

Non-aggressive waxy strain from *Ulmus* sp.

Ceratocystis ulmi CABI-IMI 177907

Non-cyclopiazonic acid producer

Aspergillus oryzae CABI-IMI 309454

Non-nitrate utilizer

Trichoderma reesei CABI-IMI 192654

Phytophthora (antagonist)

Penicillium expansum CABI-IMI 028619

Potential challenge organism

Scopulariopsis brevicaulis CABI-IMI 321976

Produces teleomorph crossed with IMI 43707 or 288002

Cochliobolus tuberculatus CABI-IMI 278848

Produces teleomorph 288003 crossed with IMI 43707

Cochliobolus tuberculatus CABI-IMI 286597

Produces teleomorph 288004 when crossed with IMI 43707

Cochliobolus tuberculatus CABI-IMI 287394

Soil analyis

Cunninghamella blakesleena CABI-IMI 063877

Tape ketan production

Amylomycs rouxii CABI-IMI 283878

Toxic strain

Aspergillus versicolor CABI-IMI 173208

Toxic to termites

Aspergillus flavus CABI-IMI 102171

Tryptophan requiring

Neurospora crassa CABI-IMI 075727ii

Virus infected

Penicillium chrysogenum CABI-IMI 026211
Penicillium brevicompactum CABI-IMI 143520
Penicillium brevicompactum CABI-IMI 039824
Aspergillus niger CABI-IMI 146891
Penicillium chrysogenum CABI-IMI 026210
Aspergillus niger CABI-IMI 050566
Acremonium chrysogenium CABI-IMI 049137
Aspergillus foetidus CABI-IMI 104688

White mutant

Aspergillus sp. CABI-IMI 378801*

White spores

Neurospora crassa CABI-IMI 142820

Wild strain

Penicillium funiculosum CABI-IMI 134755

Wild-type A

Neurospora crassa CABI-IMI 053240
Neurospora crassa CABI-IMI 053239ii
Neurospora crassa CABI-IMI 053238ii
Neurospora crassa CABI-IMI 147002

Mating Strains

Achyla bisexualis CABI-IMI 141473

Female mates with CABI-IMI 141474

Achyla bisexualis CABI-IMI 141474

Male mates with CABI-IMI 141473

Arthroderma multifidium CABI-IMI 094206

Mates with CABI-IMI 094205

Arthroderma multifidium CABI-IMI 094205

Mates with CABI-IMI 094206

Arthroderma cuniculi CABI-IMI 096245

Mates with CABI-IMI 096244

Arthroderma tuberculatum CABI-IMI 086177

Mates with CABI-IMI 086178

Arthroderma tuberculatum CABI-IMI 086178

Mates with CABI-IMI 086177

Arthroderma cuniculi CABI-IMI 096244

Mates with CABI-IMI 096245

Aspergillus heterothallicus CABI-IMI 139278

Mating type A

Aspergillus heterothallicus CABI-IMI 139277

Mating type A

Blakeslea trispora CABI-IMI 179034

Mates(+) with CABI-IMI 179035

Blakeslea trispora CABI-IMI 179035

Mating strain (–)Mates with CABI-IMI 179034

Cochliobolus tuberculatus CABI-IMI 287394

Produces teleomorph CABI-IMI 288004 when crossed with IMI 43707

Cochliobolus sativus CABI-IMI 166173

Mating type A

Cochliobolus sativus CABI-IMI 166172

Mating type A

Cochliobolus hawaiiensis CABI-IMI 208337

Mates with CABI-IMI 208336

Cochliobolus hawaiiensis CABI-IMI 208336

Mates with CABI-IMI 208337

Coprinus cinereus CABI-IMI 140506

Type A5B6

Coprinus cinereus CABI-IMI 140505

Type A5B5

Crinella muscae CABI-IMI 224150

Mating strain (+)

Crinella muscae CABI-IMI 224149

Mating strain (–)

Crinella umbellata CABI-IMI 325631

Female mating strain

Cunninghamella blakesleena CABI-IMI 053586

(–)Mates with CABI-IMI 053585

Cunninghamella blakesleena CABI-IMI 053585

(+)Mates with CABI-IMI 053586

Cunninghamella elegans CABI-IMI 021198

(+)Mates with CABI-IMI 021199

Cunninghamella elegans CABI-IMI 021199

(–)Mates with CABI-IMI 021198

Dichotomocladium elegans CABI-IMI 211272

(–)Mates with CABI-IMI 211271

Dichotomocladium elegans CABI-IMI 211271

(+)Mates with CABI-IMI 211272

Dichotomocladium robustum CABI-IMI 211273

(+)Mates with CABI-IMI 211274

Dichotomocladium robustum CABI-IMI 211274

(–)Mates with CABI-IMI 211273

Fusarium solani f.sp. *cucurbitae* CABI-IMI 325431

Hermaphrodite(+)

Fusarium solani f.sp. *cucurbitae* CABI-IMI 325432

Hermaphrodite(–)

Fusarium solani f.sp. *cucurbitae* CABI-IMI 325434

Hermaphrodite(+)

Fusarium solani f.sp. *cucurbitae* CABI-IMI 325433

Female(+)

Gongronella butleri CABI-IMI 038498

Mates with CABI-IMI 38499

Gongronella butleri CABI-IMI 038499

Mates with CABI-IMI 38498

Mucor circinelloides CABI-IMI 039478

Mates with CABI-IMI (–)25330

Mucor circinelloides CABI-IMI 025330

(+)Mates with CABI-IMI 39748

Mucor hiemalis CABI-IMI 021216

(+)Mates with CABI-IMI 21217

Mucor hiemalis CABI-IMI 138261

(+)Mates with CABI-IMI 138262

Mucor hiemalis CABI-IMI 089291

(+)Mates with CABI-IMI 089292

Mucor hiemalis f. _corticola_ CABI-IMI 129976

Mating strain (–) strain

Mucor hiemalis f. _corticola_ CABI-IMI 192541

(+) strain Mates with CABI-IMI 138261

Mucor hiemalis CABI-IMI 021217

(–)Mates with CABI-IMI 21216

Mucor hiemalis CABI-IMI 089292

(–)Mates with CABI-IMI 089291

Mucor hiemalis CABI-IMI 138262

(–)Mates with CABI-IMI 138261

Mucor mucedo CABI-IMI 133298

(–) Strain mates with CABI-IMI 133299 and 78407

Mucor mucedo CABI-IMI 078407

(+) Strain mates with CABI-IMI 133298

Mucor mucedo CABI-IMI 133299

(+) Strain mates with CABI-IMI 133298

Nectria fuckeliana CABI-IMI 277828

Mates with CABI-IMI 27827

Nectria penicillioides CABI-IMI 063372

Mates with CABI-IMI 063372ii

Neurospora crassa CABI-IMI 075721

Mates with CABI-IMI 075722

Neurospora crassa CABI-IMI 142820

Mates with CABI-IMI 142819

Neurospora sitophila CABI-IMI 063919

Mates with CABI-IMI 063920

Parasitella parasitica CABI-IMI 320579

Mating strain (–)

Parasitella parasitica CABI-IMI 041058(+)

Mates with CABI-IMI 014057(–)

Parasitella parasitica CABI-IMI 014058ii

Mating strain(–)

Phycomyces blakesleeanus CABI-IMI 118796(–)

Mates with CABI-IMI 118497(+)

Phycomyces blakesleeanus CABI-IMI 200162

Mating strain(+)

Phycomyces blakesleeanus CABI-IMI 200163

Mating strain(+)

Phycomyces blakesleeanus CABI-IMI 200164

Mating strain(+)

Phycomyces blakesleeanus CABI-IMI 200165

Mating strain(+)

Phycomyces nitens CABI-IMI 063923

Mating strain(+)

Phycomyces nitens CABI-IMI 044751

Mating strain(–)

Phycomyces nitens CABI-IMI 051077

Mating strain(+)

Phytophora meadii CABI-IMI 325862

Mating strain A1

Phytophthora cryptogea CABI-IMI 325908

Mating strain A2

Phytophthora cryptogea CABI-IMI 325907

Mating strain A1

Phytophthora botrysa CABI-IMI 136915

Mating strain(–)

Phytophthora botrysa CABI-IMI 136916

Compatibility(+) type A2

Phytophthora cambivora CABI-IMI 162672

Compatibility type A2

Phytophthora cambivora CABI-IMI 059156

Compatibility type A1

Phytophthora cambivora CABI-IMI 040505

Compatibility type A2

Phytophthora cambivora CABI-IMI 223987

Compatibility type A2

Phytophthora cambivora CABI-IMI 162671

Compatibility type A2

Phytophthora cinnamomi CABI-IMI 022938

Mates with CABI-IMI 21278

Phytophthora cinnamomi CABI-IMI 158786

Mating strain A1

Phytophthora cinnamomi CABI-IMI 136395

Mating strain A2

Phytophthora cinnamomi CABI-IMI 230381

Mating strain A1

Phytophthora cinnamomi CABI-IMI 283248

Mating strain A1

Phytophthora cinnamomi CABI-IMI 077377

Mating strain A2

Phytophthora cinnamomi CABI-IMI 077375

Mating strain A2

Phytophthora nicotianae CABI-IMI 035087

(+)Mates with CABI-IMI 22176

Phytophthora nicotianae CABI-IMI 329726

Compatibility A1

Phytophthora nicotianae CABI-IMI 136411

Mates with CABI-IMI 35087

Phytophthora nicotianae CABI-IMI 022176

(–)Mates with CABI-IMI 35087

Phytophthora nicotianae CABI-IMI 325902

Mating strain A2

Pilaria anomala CABI-IMI 228216

(+)Mates with CABI-IMI 228217

Pilaria anomala CABI-IMI 228217

(–)Mates with 228216

Pirella circinans CABI-IMI 085609

Mates with CABI-IMI 085608

Pirella circinans CABI-IMI 085608

Mates with CABI-IMI 085609

Poitrasia circinans CABI-IMI 078521ii

Mates with CABI-IMI 078522ii

Protomyces macrosporus CABI-IMI 102385

(–)Mates with CABI-IMI 102384

Protomyces macrosporus CABI-IMI 115296(–)

Mates with CABI-IMI 115295(+)

Protomyces macrosporus CABI-IMI 102384

(+)Mates with CABI-IMI 102385

Pythium flevoense CABI-IMI 323128

Female

Pythium flevoense CABI-IMI 323126

Female

Pythium flevoense CABI-IMI 323127

Female

Pythium flevoense CABI-IMI 176046

Female

Pythium flevoense CABI-IMI 176045

Male

Pythium flevoense CABI-IMI 176044

Male

Pythium intermedium CABI-IMI 323126

Mating strain(+)

Pythium intermedium CABI-IMI 323124

Mating strain(+)

Pythium intermedium CABI-IMI 308149

Mating strain(+)

Pythium macrosporum CABI-IMI 323141

Mating strain(+)

Pythium macrosporum CABI-IMI 323139

Mating strain(+)

Pythium splendens CABI-IMI 323150

Mating strain(+)

Pythium splendens CABI-IMI 323149

Mating strain(+)

Pythium intermedium CABI-IMI 323134

Mating strain(–)

Pythium intermedium CABI-IMI 323132

Mating strain(–)

Pythium intermedium CABI-IMI 323131

Mating strain(–)

Pythium intermedium CABI-IMI 323130

Mating strain(–)

Pythium intermedium CABI-IMI 308150

Mating strain(–)

Pythium intermedium CABI-IMI 323133

Mating strain(–)

Pythium macrosporum CABI-IMI 323140

Mating strain(–)

Pythium macrosporum CABI-IMI 323144

Mating strain(–)

Pythium macrosporum CABI-IMI 323142

Mating strain(–)

Pythium splendens CABI-IMI 323147

Mating strain(–)

Pythium splendens CABI-IMI 323349

Mating strain(–)

Pythium splendens CABI-IMI 323350

Mating strain(–)

Pythium splendens CABI-IMI 323146

Mating strain(–)

Pythium macrosporum CABI-IMI 323143

Mating strain

Pythium sylvaticum CABI-IMI 323156

Male strain

Pythium sylvaticum CABI-IMI 323151

Male strain

Pythium sylvaticum CABI-IMI 308298

Mates with CABI-IMI 308299

Pythium sylvaticum CABI-IMI 248395

Mates with CABI-IMI 248394

Pythium sylvaticum CABI-IMI 248394

Mates with CABI-IMI 248395

Pythium sylvaticum CABI-IMI 323153

Female strain

Pythium sylvaticum CABI-IMI 323155

Female strain

Pythium sylvaticum CABI-IMI 323154

Female strain

Pythium sylvaticum CABI-IMI 323352

Male strain

Pythium sylvaticum CABI-IMI 323152

Male strain

Pythium sylvaticum CABI-IMI 323353

Female strain

Renispora flavissima CABI-IMI 241795

Mating strain

Renispora flavissima CABI-IMI 241796

Mating strain

Thamnostylum nigricans CABI-IMI 208234

Mates with CABI-IMI 208235

Thamnostylum nigricans CABI-IMI 208235

Mates with CABI-IMI 208234

Tapesia acuformis CABI-IMI 369188

MAT1 mating type

Tapesia acuformis CABI-IMI 369189

MAT2 Mating type

Thanatephorus cucumeris CABI-IMI 303158

Anastomosis group 3

Thanatephorus cucumeris CABI-IMI 305033

Anastomosis group 8

Thanatephorus cucumeris CABI-IMI 305038

Anastomosis group 8

Thanatephorus cucumeris CABI-IMI 305037

Anastomosis group 8

Thanatephorus cucumeris CABI-IMI 305034

Anastomosis group 8

Thanatephorus cucumeris CABI-IMI 305036

Anastomosis group 8

Thanatephorus cucumeris CABI-IMI 303155

Anastomosis group AG2 type1

Thanatephorus cucumeris CABI-IMI 303157

Anastomosis group AG2 type2

Thanatephorus cucumeris CABI-IMI 303156

Anastomosis group AG2 type2

Thanatephorus cucumeris CABI-IMI 303162

Anastomosis group AG4

Thanatephorus cucumeris CABI-IMI 303159

Anastomosis group AG5

Thanatephorus cucumeris CABI-IMI 303161

Anastomosis group AG7

Thanatephorus cucumeris CABI-IMI 303163

Anastomosis group AGB1

Thanatephorus cucumeris CABI-IMI 303152

Anastomosis group AGI culture type 1A

Thanatephorus cucumeris CABI-IMI 303153

Anastomosis group AGI cuture type 1B

Thanatephorus cucumeris CABI-IMI 303154

Anastomosis group AGI cuture type 1C

Thanatephorus cucumeris CABI-IMI 303160

Anastomosis group 6

Special Features: Chemical Transformation, Bioconversion and Bioaccumulation

11-beta (hydroxylation) of steroids

Cochliobolus lunatus CABI-IMI 312658

3-4 dichloroproprionanilide (metabolises)

Fusarium solani CABI-IMI 129056

6-alphapragane steroids to 11 alpha hydroxy analogues

Aspergillus nidulans CABI-IMI 227976

6-alphapragane steroids to 11 alpha hydroxy analogues

Aspergillus nidulans CABI-IMI 227976

Acronycine to 9-hydroxyacronycine (Bioconv.)

Aspergillus niger CABI-IMI 041874

Amylo (process)

Rhizopus oryzae CABI-IMI 309922
Rhizopus oryzae CABI-IMI 309923

Ang-kak (production)

Monascus purpueus CABI-IMI 283870

Anti tumour agents (transformation)

*Penicillium brevicompactum*CABI-IMI 040225

Citric acid (accumulation of)

Aspergillus niger CABI-IMI 309921

5-anilino 1,2,3,4-thiatriazole (Bioconv/Hydroxylates)

Rhizopus stolonifera CABI-IMI 061269

Steroid (hydroxylation)

Rhizopus arrhizus CABI-IMI 090340
Aspergillus niger CABI-IMI 041874

Ion exchange / complexation of the uranyl ion

Rhizopus oryzae CABI-IMI 309924
Rhizopus oryzae CABI-IMI 309922

Orotic acids (accumulation)

Neurospora crassa CABI-IMI 075730ii

Orotidine (accumulation)

Neurospora crassa CABI-IMI 075730ii

Penicillic acid (production deficient)

Penicillium viridicatum CABI-IMI 039758ii

Sesquiterpene lactone costunolide (Patent)

Penicillium spinulosum CABI-IMI 167427

Sesquiterpene lactone costunolide (transformation)

Rhizopus stolonifera CABI-IMI 057762
Geotrichum suaveolens CABI-IMI 096837
Cochliobolus lunatus CABI-IMI 061535
Rhizopus arrhizus CABI-IMI 090340
Penicillium chrysogenum CABI-IMI 037767
Penicillium glabrum CABI-IMI 040234
Mucor circinelloides f. *griseocyanus* CABI-IMI 116946
Fennellia flavipes CABI-IMI 135424
Rhizopus stolonifera CABI-IMI 057761

Steroid (oxygenation)

Rhizoctonia muneratii CABI-IMI 236679

Steroid (research)

Cunninghamella blakesleena CABI-IMI 063877

Steroid (transformation)

Cunninghamella elegans CABI-IMI 314507

Tannin-gallic acid (fermentation)

Aspergillus niger CABI-IMI 050566ii

Zinc oxide (dissociation)

Mortierella elongata CABI-IMI 375268*

Zinc phophate (dissociation)

Mortierella elongata CABI-IMI 375268*

Vectors, Phages, Transposons, Genetically Modified Organisms (Auxotrophs, Resistant, Sensitive, Producers), Mutants

The property and map (where available) is given for each organism with its name, collection acronym and number.

Bronze mutant

Aspergillus parasiticus CABI-IMI 091019bii

Brown mutant

Aspergillus fumigatus CABI-IMI 131705

Buff mutant

Aspergillus aurantiobrunneus CABI-IMI 201980

Chemically induced mutant (patent)

Penicillium chrysogenum CABI-IMI 142387
Penicillium chrysogenum CABI-IMI 142384
Penicillium chrysogenum CABI-IMI 142385
Penicillium chrysogenum CABI-IMI 142386
Penicillium chrysogenum CABI-IMI 142383

Dark mutant

Glomerella cingulata CABI-IMI 085091ii

Hyaline mutant

Cochliobolus sativus CABI-IMI 166173
Fulvia fulva CABI-IMI 296520

Light coloured mutant

Aspergillus fumigatus var. *helvolus* CABI-IMI 016153ii

Strontium chromate fungicide resistant

Acremonium strictum CABI-IMI 178506

Appendix B
Media recipes

Media for Bacteria

There are numerous recipes available for media which can either be made from scratch, created from commercially produced dehydrated products, or purchased as pre-prepared agar plates. Growth media can perform a number of functions, such as the initial isolation of organisms, for general growth and maintenance, for identification purposes or to selectively grow specific microbes whilst suppressing others.

The media listed in this section have been chosen because they support the growth of most of the common organisms encountered in a plant diagnostic laboratory, and also isolates from environmental sources. The main purpose of the media that we use is to grow and maintain pure cultures, which may then be analysed further using molecular, biochemical or serological tests. Although an extensive range of diagnostic media is available commercially, we make much of our media in-house, which allows modifications to be made to recipes and gives greater flexibility for the use of different containers and quantities.

Many labs, including ours, used diagnostic media for identification purposes in the past, although in recent years there has been a shift towards more reliable, quicker and less expensive methods of identification and analysis. Therefore, the list of media that we currently use on a regular basis is quite short, and has been cut down to a handful of recipes which will support the growth of most of the bacteria and yeasts that we encounter in the diagnostic laboratory.

Several specialized media recipes for plant pathogenic bacteria have been included in this appendix, although this list is by no means extensive and a more comprehensive list of growth media for specific organisms can be found online or in publications such as:

- *The Preservation and Maintenance of Living Fungi* (Smith and Onions, 1994)
- *The UK National Culture Collection Biological Resource: Properties, maintenance and management* (Smith *et al.*, 2001)

Detailed formulae are given for selected growth media, and where appropriate alternative commercially available dehydrated media have been suggested. Other reputable suppliers/brands of dehydrated media may be substituted as necessary.

Once made, the liquid media should be poured into bottles and autoclaved at 121°C for 15min unless otherwise specified. Media can be poured immediately after autoclaving when 'hand-hot', or placed in a water bath to maintain the correct temperature until it can be poured.

Alternatively, stocks of media may be produced, allowed to cool with the lids tightened, and 'melted' as required at a later date. It is important to label all bottles and to keep track of

DOI: 10.1079/9781800622135.appb

batches to ensure that they are used within a specified time frame.

General Purpose Media

Nutrient Agar (NA)

This is a commonly used non-selective general purpose medium suitable for a wide range of organisms. NA is the first choice in our lab for the growth and maintenance of bacteria, although we would select an alternative if the target organism had particular growth requirements.

Most plant pathogenic bacteria prefer a neutral or slightly alkaline pH and will grow sufficiently on NA, however some sugarcane pathogens, *Xanthomonas* spp., *Clavibacter* spp. and *Erwinia* spp. may require an alternative medium such as GP, YDC, SNA or KB.

NB. Oxoid and Difco dehydrated NA have different formulations and suitability should be checked prior to use.

Nutrient Agar (NA) - Oxoid

Lab-Lemco Powder (Oxoid; L29)	1.0g
Yeast extract (Oxoid; L21)	2.0g
Peptone Bacteriological (Oxoid; L34)	5.0g
Sodium chloride	5.0g
Agar No 3 (Oxoid; L13)	15.0g
Distilled water	1000ml
pH	7.2-7.4

Or use Oxoid CM0003 dehydrated formulation. 28g in 1 litre of distilled water.

We add an additional 3g agar/litre for a firmer medium as this is quite soft if made as instructed.

Nutrient Agar (NA) - Difco

Beef extract (Difco)	3.0g
Bacteriological Peptone	5.0g
Agar No. 3	15g
Distilled water	1000ml

Or use Difco NA dehydrated formulation. Suspend 31g dehydrated medium in 1 litre of distilled water.

Nutrient Broth

Liquid media can be prepared using the recipes above without the addition of agar.

Or use Oxoid CM001 or Oxoid CM0067 (NB No.2) dehydrated formulation.

Nutrient Broth Yeast Extract agar (NBY medium)

An alternative general-purpose medium to Nutrient Agar which is suitable for plant pathogenic bacteria.

Nutrient broth (Oxoid CM001)	8.0g
Yeast extract	2.0g
Di-potassium hydrogen phosphate	2.0g
Potassium dihydrogen phosphate	0.5g
Agar	15g
Distilled water	974ml

Autoclave media, then add sterile solutions:

Glucose (20% solution)	25ml
Magnesium sulphate heptahydrate (1M solution)	1ml

Autoclave glucose and Magnesium sulphate solutions separately and add aseptically.

Tryptone soya agar (TSA)

A general purpose medium for the growth of a wide variety of organisms. This medium is our second choice and a good all-rounder. It is commonly used for environmental monitoring purposes which means that it may be readily available in the laboratory.

Pancreatic digest of casein	15.0g
Enzymatic digest of soya bean	5.0g
Sodium chloride	5.0g
Agar	15.0g
Distilled water	1000ml
pH 7.3.	

Or use Oxoid CM0131 dehydrated formulation. Suspend 40.0g dehydrated medium in 1 litre of distilled water

Tryptone soya broth (TSB)

A highly nutritious general purpose medium without agar which can be used to grow bacterial cultures in liquid culture.

Pancreatic digest of casein	17.0g
Enzymatic digest of soya bean	3.0g
Sodium chloride	5.0g
Di-potassium hydrogen phosphate	2.5g
Glucose	2.5g
Distilled water	1000ml

Or use Oxoid CM0129 dehydrated formulation. Suspend 30.0g dehydrated medium in 1 litre distilled water.

Trypticase soya broth agar (TSBA)
A nutritious general purpose medium for the qualitative growth of organisms. We used TSBA in our lab to grow cultures for identification using Gas Chromatography although it is rarely used today.

Trypticase soya broth (BBL 11768)	30.0g
Agar	15.0g
Distilled water	1000ml

Other Bacteriological Media

Casamino-acid Peptone Glucose Agar (CPG)
Recommended for the growth and isolation of *Ralstonia solanacearum*, and can be used to differentiate colony types (fluidal and non-fluidal).

Casamino acid (casein hydolysate)	1.0g
Peptone	10.0g
Glucose	5.0g
Agar	17.0g

Adjust pH to 6.5-7.0 if necessary.
Autoclave at 121°C for 20 minutes.

Tetrazolium Chloride Medium (TTC or TZC)
Used as a diagnostic medium for *R. solanacearum*, and can be used to differentiate virulent and non-virulent strains.

Add 5ml of a sterile 1% stock of 2,3,5-triphenyl tetrazolium chloride to 1 litre CPG which has been cooled to 55°C.

Glucose Peptone Agar (GP)
Suitable for some species of *Xanthomonas*.

Peptone Bacteriological (Oxoid L34)	5.0g
Glucose	20.0g
Di-Potassium hydrogen phosphate (anhydrous)	0.5g
Magnesium sulphate heptahydrate	0.25g
Agar	15.0g
Distilled	1000ml
pH	7.2-7.4 (adjust with potassium hydroxide if necessary)

King's Medium B (KB)
A non-selective media which is suitable for Pseudomonads and *Erwinia* spp.

KB can also be used as a diagnostic medium to detect fluorescent Pseudomonads.

Proteose peptone (Difco No.3/ Oxoid; L46)	20.0g
Glycerol	10.0g
Di-Potassium hydrogen phosphate (anhydrous)	1.5g
Magnesium sulphate heptahydrate	1.5g
Agar	15.0g
Distilled water	1000ml
pH	7.2

Suspend ingredients, heat to dissolve, dispense into bottles and sterilize at 121°C for 15 minutes.

MacConkey agar
This medium is slightly selective and used for the detection and isolation of Gram-negative organisms. It is also used to differentiate lactose fermenting and non-fermenting organisms. We do not use this in our lab, but it is commonly used in many labs as a preliminary step in screening bacteria.

Supplied as dehydrated formulation by Oxoid and Difco.

Semi-selective medium for the isolation of *Ralstonia solanacearum*
Prepare 1 litre of TTC/TZC medium, but substitute 5ml glycerol for 5g glucose.

Add the following (dissolved in 5ml of 70% ethanol 30 minutes prior to use):

Crystal violet	5mg
Polymyxin ß sulphate	100mg
Bacitracin	25mg
Chloromycetin	5mg
Penicillin	0.5mg

For the inhibition of fungal contaminants add:

Cycloheximide	100mg

MRS Agar
For the enrichment, cultivation and isolation of *Lactobacillus* spp. We do occasionally use this medium although it is usually made up for specific pieces of work.

Peptone	10.0g
Meat extract	5.0g
Yeast extract	5.0g
D(+) glucose	20.0g

Di-potassium hydrogen phosphate	2.0g
Di-ammonium hydrogen citrate	2.0g
Sodium acetate	5.0g
Magnesium sulphate	0.1g
Manganous sulphate	0.05g
Agar	12.0g
Polysorbate 80 (Tween)	1.0ml
pH	6.5

Or use Sigma 69964 dehydrated media. Add 61.15g to 1litre of water with 1ml polysorbate 80 (Tween).

Nutrient Dextrose Agar (NDA)
For isolation of *Agrobacterium* spp., some *Clavibacter* spp. and some non-soft rot *Erwinia* spp.

Nutrient Agar (Oxoid CM3)	28.0g
D-glucose	10.0g
Distilled water	1000ml

Pseudomonas Agar F (PsF)
Used as a diagnostic medium to detect fluorescent Pseudomonads (can be substituted for King's Medium B). Although many of the diagnostic media for bacteria are no longer used in our labs, we do occasionally use KB or PsF.

NB. When checking for fluorescence (production of fluorescein) it is important to use ultraviolet light as some non-fluorescent species of *Pseudomonas* may produce yellowish diffusible pigments which should not be confused with fluorescence.

Bacto tryptone (Difco)	10.0g
Bacto Proteose peptone (Difco)	10.0g
Di-potassium hydrogen phosphate (anhydrous)	1.5g
Magnesium sulphate heptahydrate	1.5g
Agar	15.0g
Distilled water	1000ml
pH	7.0

Suspend in 1 litre of distilled water and add 10g glycerol.

Or use Difco dehydrated formulation. Suspend 38g of the medium in 1litre of distilled water.

R2A medium
Used for heterotrophic bacteria and ideal for slow-growing organisms. We used to use this medium for bacteria which wouldn't grow on TSBA when we were identifying strains using Gas Chromatography.

Yeast extract	0.5g
Proteose peptone	0.5g
Casamino acids	0.5g
Glucose	0.5g
Soluble starch	0.5g
Sodium pyruvate	0.3g
Di-potassium hydrogen phosphate	0.3g
Magnesium sulphate heptahydrate	0.05g
Noble agar	15.0g
Distilled water	1000ml

Adjust pH to 7.2 with potassium dihydrogen phosphate or di-Potassium hydrogen phosphate.

Or use Oxoid CM906 dehydrated media. Add 18.1g in 1 litre of distilled water.

Sucrose Nutrient Agar (SNA) or Levan medium (LEV)
This media is especially suited to green-fluorescent Pseudomonads and *Erwinia amylovora*

Nutrient Agar (Oxoid CM3)	28.0g
Sucrose	50.0g
Distilled water	1000ml
pH	7.2

Tap Water Agar (TWA)
This is a minimal medium recommended for the growth of *Streptomyces* spp.

Agar	20.0g
Tap water	1000ml
Autoclave at 121°C for 15min.	

V8 media
Suitable for the growth of Actinomycetes, although when we do encounter these they usually grow sufficiently well on our standard growth media.

V8 juice	355ml
Calcium carbonate	3.7g
Agar	35.0g
Tap water	1400ml
pH	7.3

Yeast Dextrose Chalk Agar (YDC)
We do use this medium on a regular basis as it is non-selective which is recommended for

isolation and maintenance of *Xanthomonas* spp. The calcium carbonate (chalk) buffers the medium and ensures the pH does not become acidic which would cause the organism to die as *Xanthomonas* spp. are not tolerant of acidic pH.

This medium can be used as diagnostic medium for selected plant pathogens.

Yeast extract (Oxoid; L21)	10.0g
D-Glucose	20.0g
Calcium carbonate	20.0g
Agar No 3 (Oxoid; L13)	12.0g
Distilled water	1000ml

Mix all the ingredients except the calcium carbonate and heat to dissolve.

Add the calcium carbonate and adjust pH to 7.2 if required.

The chalk in this medium will settle to the bottom of the plate or slant and be visible as a white layer.

Yeast Malt Extract Agar (YMEA)
For the growth of Streptomycetes

Yeast extract	4.0g
Malt extract	10.0g
Dextrose	4.0g
Agar	20.0g
Adjust pH to 7.0-7.2	

Media for Yeasts

Although we don't often encounter yeasts in the diagnostic laboratory, there are occasions when we do wish to isolate or maintain yeast cultures. Although some strains thrive in the natural environment or in association with plants, they are more commonly associated with food and beverages, especially as spoilage organisms.

The following list of media can be used to grow most of the species that we would want to maintain, although there are many others available. When present, yeasts often grow on our standard isolation media for fungi or bacteria (e.g., TSA, NA, TWA).

Diagnostic test media is still used in some labs although it has not been included in this section due to the vast range and complexity of the recipes available.

Yeast Malt Agar (YM)
A general purpose medium which supports the growth of a wide variety of organisms.

Formulae Difco™ YM Agar Approximate Formula per litre

Yeast extract	3.0 g
Malt extract	3.0 g
Peptone	5.0 g
Dextrose	10.0 g
Agar	20.0g

Several variations of this medium are available and dehydrated products may vary from the recipe given above.

Media may be made without agar as a broth if required.

Sabouraud Dextrose Agar (SDA)
A general purpose medium for fungi and aciduric microorganisms.

Mycological peptone (Oxoid)	10g
Glucose	40g
Agar No 3 (Oxoid)	15g
Distilled water	1000ml

Or use Oxoid CM0041 SDA dehydrated media. Add 65g to 1 litre distilled water.

Yeast extract Peptone Dextrose Agar (YPD)
A general purpose culture medium for most yeast species, especially *Saccharomyces cerevisiae*.

Yeast extract	10.0 g
Peptone	20.0 g
Dextrose	20.0 g
Agar	20.0 g
Distilled water	1000 ml
pH	5.6

Or use commercial dehydrated media from Difco or Sigma. Add 65g to 1 litre distilled water.

Media may be made without agar as a broth.

Media for Fungi

Preparation of pure cultures, including their characteristic fungal features still remains relevant for the culturing of rare live strains for distribution from our collection and for some aspects of research. Morphological verification will almost certainly be necessary for small-scale storage systems and always for teaching and learning purposes, therefore a need remains for traditional culturing methods which enable the development of optimal growth under artificial conditions.

This short media guide offers information developed over decades of culturing fungi in our laboratories at CABI. The table below shows a list of cultures frequently examined at CABI including the media and conditions required to grow and exhibit recognizable fungal features.

The media and conditions required for organisms frequently examined at CABI

Organism name	Recommended media for sporulation	UV- Y/N	Temp	Growth (days)
Curvularia lunata	PCA	Y	RT	14
Didymella glomerata	PCA	Y	RT	10
Lichtheimia corymbifera	PCA	N	37°C	5
Mucoraceous fungi			RT	
Mucor circineloides	PCA	N	RT	4
Rhizopus stolonifer	PCA	N	RT	5
Synchephalastrum racemosum	PCA	N	RT	5
Dematiatious hyphomycetes				
Alternaria alternata	PCA	y	RT	15
Alternaria botrytis	PCA	Y	RT	10
Arthrinium phaeospermum	PCA	Y	RT	21
Cadophora fastigiata	MA	Y	RT	12
Cephalototrichum gorgonifer	PCA	N	RT	10
Cladosporium herbarum	PCA	Y	RT	9
Epicoccum nigrum	PCA	Y	RT	14
Scopulariopsis brevicaulis	OA	Y	RT	9
Stachybotrys chartarum	PCA	Y	RT	7
Hyaline hyphomycetes				
Akanthomyces muscarium	PCA	Y	RT	10
Beauveria bassiana	MA/PCA	Y	RT	9
Clonostachys rosea	PCA	Y	RT	7
Geotrichum candidum	MA	N	RT	7
Sarocladium strictum	PCA	Y	RT	8
Trichoderma harzianum	PCA	N	RT	4
Black yeasts				
Aureobasidium pullulans	MA	Y	RT	12
Exophiala castellani	MA	Y	RT	16
***Penicillium* and *Aspergillus*-like organisms**				
Paecilomyces variotii	PCA	Y	RT	7
Penicillium spp.	CYA/MEA	N	25°C	7
Aspergillus spp.	CYA/MEA	N	25°C	7
Aspergillus herbariorum	CYA/MEA	N	25°C	21
Aspergillus herbariorum teleomorph	M20	Y	RT	21
Fusarium				
Fusarium spp.	SNA + FP /PSA	Y	RT	8
Ascomycetes and coelomycetes				
Chaetomium globosum	PCA	N	RT	13
Monascus ruber	M40	Y	RT	21
Phoma fimeti	PCA	Y	RT	12
Pyronema domestica	PCA	Y	RT	21
Sordaria humana	MA	N	RT	21

RT= Room Temp. 23-25°C

The table below provides an alphabetical list of media that can be used to grow and support fungal genera held within the CABI collection, with a few helpful comments.

Media abbreviation	Name	Fungus	Comments
AB	Antibiotic	N/A	This can be added to agar to remove/reduce bacterial contamination. This is not suitable for growing fungus for molecular work
CMA	Corn Meal Agar	Oomycetes/water moulds, *Phytophthora*, *Pythium*, *Achlya*, *Saprolegnia*, wood inhabiting fungi	Simple clear media, nutrient poor, encourages chlamydospore development
CZ	Czapek	*Penicillium* and *Aspergillus*	No longer used at CABI
CYAT	Czapek Yeast Autolysate	*Penicillium* and *Aspergillus*	Use with published keys
GYM	Glucose Yeast Media	Various fungal species	Liquid media, biomass growth for DNA extraction
LBA	Lima Bean Agar	*Phytophthora*	Clear nutrient poor agar
MA	Malt Extract Agar	Lots of Ascomycetes	Nutrient rich media encourages mycelial growth and fruit body formation, used for growth of black yeasts and molecular work
MEA	Blakeslee's Formula Malt Extract Agar	*Penicillium* and *Aspergillus*	For use with published keys e.g. Klich, (2002), Samson *et al.* (2019).
MB	Malt Broth	Numerous fungi	Creation of biomass for DNA extraction
MCZ	Malt Czapek	*Penicillium* and *Aspergillus*	No longer used at CABI, we now use MEA in line with published keys
M20, M40	MEA + Sucrose e.g. 200g/l for M20, 400g/l for M40	Sexual state of *Aspergillus*, M40 -*Monascus*	Xerotolerant and osmotolerant fungi
OA	Oat Agar	*Phytophthora*, *Trichoderma*, *Phoma*	Nutrient poor media, useful for fruit-body formation, ascomycetes and coelomycetes
PCA	Potato Carrot Agar	Mitosporic fungi to encourage sporulation	This is our 'first choice' agar, (always available). Relatively nutrient poor, contains small amounts of carbohydrates, minerals and cellulose, encourages sporulation
PDA	Potato Dextrose Agar	Numerous fungal strains	Nutrient rich agar, high in carbohydrates e.g., starch and sugars. Encourages vegetative growth. PDA made from fresh potatoes is difficult to see through, we do not currently select this media routinely
Pitt	Pitt Media	*Aspergillus* and *Penicillium*	Used to create a spore suspension for three-point inoculation and use with diagnostic keys

Continued

Continued.

Media abbreviation	Name	Fungus	Comments
PSA	Potato Sucrose Agar	*Fusarium*	Used to differentiate between *Fusarium* species by growth rate and colony colour
RDA	Rabbit Dung Agar	*Pilobolus*, *Ascobolus*	*Ascobolus* forms apothecia in culture
SNA + FP	Sucrose Nurient Agar + Sterile Filter paper (FP). Very similar to Spezieller Nährstoffarmer Agar + Sterile Filter paper, Carnation Leaf Agar	*Fusarium*	SNA is a low nutrient agar and the addition of cellulose (FP) induces the formation of macroconidia
TWA	Tap Water Agar	Numerous fungal strains	Nutrient poor media, fungal growth will be slow and relatively sparse, used for isolation and separation of fungal colonies. Addition of wheat straw will encourage sporulation of some fungi
V8	V8 Juice Agar	*Phytophthora*, *Pythium* and other Oomycetes	Encourages the development of spores
YES	Yeast Extract Sucrose		Broth used for cultivation and maintenance of fungi
YPSS	Yeast Phosphate Soluble Starch	*Thermomyces* growth and Actinomycete isolation	

Alphabetical list of Media Recipes

Antibiotic Solutions (AB)

Chloramphenicol	150mg
Penicillin G	150 mg
Distilled water	10ml
Ethanol	10ml

Chloramphenicol

Dissolve 150 mg in 10 ml of ethanol.

Penicillin G

Dissolve 150 mg in 10 ml of distilled water.

To prepare antibiotic media, add 1.25ml of each antibiotic per 250ml of hand-hot agar (50-60°C) after autoclaving.

Corn Meal Agar (CMA)

Maize	30g
Agar Oxoid No.3	20g
Tap water	1 litre

Place the maize and water in the saucepan. Simmer for 1 hour.

Filter through muslin, add the agar, and boil to dissolve.

Autoclave at 121°C for 20 minutes

For *Phytophthora* and *Pythium* spp. and similar sensitive species, 0.5g wheat germ oil may be added. Distilled water is preferable.

Cornmeal Agar (CMA OX)

Cornmeal Agar (oxoid)	17g
Distilled water	1 litre

Add agar to distilled water in a double saucepan. Heat until dissolved. Dispense into bottles.

Autoclave at 121°C for 15 min.

Final pH 6.0±0.2

Czapek Dox Agar (CZ)

Made with stock Czapek solutions

Stock solution A	50ml
Stock solution C	50ml
Distilled water	900ml

Sucrose Analar 30g
Agar Oxoid No. 3 20g

Add the stock solutions and the sucrose last.

To each litre add 1ml of the following trace elements.

Zinc sulphate $ZnSO_4.7H_2O$ 1g/100ml distilled water

Cupric sulphate $CuSO_4.5H_2O$ 0.5g/100ml distilled water

Autoclave at 121°C for 20 minutes

Stock solution A

Sodium nitrate $NaNO_3$	40g
Potassium chloride KCl	10g
Magnesium sulphate $MgSO_4.7H_2O$	10g
Ferrous sulphate $FeSO_4.7H_2O$	0.2g

Dissolve in 1 litre of distilled water and refrigerate.

Stock solution C

Di-potassium hydrogen phosphate K_2HPO_4 20g

Dissolve in 1 litre of distilled water and refrigerate.

Czapek Yeast Autolysate Agar (CYAT)

Make up Trace Metal Solution (TMS) as follows:

Zinc sulphate heptahydrate $ZnSO_4.7H_2O$	1g
Copper sulphate pentahydrate $CuSO_4.5H_2O$	0.5g
Distilled water	100ml

Make up CYA as follows:

Sodium nitrate $NaNO_3$	3g
Di-potassium hydrogen phosphate K_2HPO_4	1g
Potassium chloride KCl	0.5g
Magnesium sulphate $MgSO_4.7H_2O$	0.5g
Iron sulphate $FeSO_4.7H_2O$	0.01g
Yeast extract (or Autolysate) Oxoid	5g
Sucrose Analar	30g
Agar Oxoid No.3	15g
Trace Metal Solution	1ml
Distilled water	1 litre

Dissolve ingredients in double saucepan adding sucrose last when agar has dissolved.

Autoclave at 121°C for 15 minutes. pH before and after autoclaving should be 6.0±0.5. For *Penicillium* identifications.

Penicillium and *Aspergillus* identification (Pitt, 1973)

Liquid Culture (GYM)

Glucose	10g
Ammonium di-hydrogen phosphate $NH_4H_2PO_4$	1g
Potassium chloride KCl	0.2g
Magnesium sulphate hydrate $MgSO_4.7H_2O$	0.2g
Yeast extract (Oxoid)	5g
1% Zinc sulphate $ZnSO_4.7H_2O$ (cz soln)	1ml
0.5% Copper sulphate $CuSO_4.5H_2O$ (cz soln)	1ml
Distilled water	1 litre

Can be dispensed as;
10ml in universals
60ml in medical flats
60ml in flasks
Autoclave at 115°C for 15 minutes

Lima Bean Agar (LBA)

Lima beans (butter beans)	100g
Agar Oxoid No. 3	15g
Distilled water	1 litre

Soak beans overnight in 500ml distilled water. Simmer for 30mins. Strain beans and retain liquid. Add agar to 500ml distilled water and dissolve using a double saucepan. Add bean liquid and make up to 1 litre. Autoclave at 121°C for 20mins.

Malt Extract (MA)

Toffee barley malt extract	20g
Agar Oxoid No.3	20g
Distilled water	1 litre

Adjust pH of MA before autoclaving to 6.0±0.5. If outside this range after autoclaving, discard.Autoclave at 121°C for 15 minutes. Malt extract is bought in tins, opened and decanted into a clean plastic container with lid and stored in the fridge for up to 3 months.

Malt Extract Agar – Blakeslee's formulation (MEA)

Malt extract (powdered, Difco or Oxoid)	20g
Bacteriological peptone	1g
Glucose Analar	20g
Agar Oxoid No.3	15g
Distilled water	1 litre

Autoclave at 121°C for 15 minutes

For *Penicillium* identification, Raper & Thom (1949); Pitt (1973)

Malt Extract Agar plus sucrose (M20, M40, M60)

Prepare ingredients as for MA with the addition, before autoclaving, of the following quantitiesof sucrose per litre of medium. Add sugar last to avoid caramelization.

M20	200g
M40	400g
M60	600g

Autoclave at 121°C for 15 minutes

Oat Agar (OA)

Oat Meal (ground)	30g
Agar Oxoid No.3	20g
Distilled water	1 litre

Add oat meal to 500ml of water in a saucepan. Heat for 1 hour. To other 500ml of water add agar and dissolve in double saucepan. Pass cooked oat meal through fine strainer and add to agar mixture. Stir thoroughly. Autoclave at 121°C for 20 minutes

Pitt's Media

Pitt's method for inoculating *Penicillium* and *Aspergillus*

Agar Oxoid No. 3	0.3g
Distilled water	100ml
Washing up liquid	1 drop

Add agar and detergent to distilled water in a glass vessel and heat briefly in a saucepan until the agar dissolves. Pipette small amount of molten agar and detergent into 5 ml glass vials. Sterilize in a pressure cooker for 15 minutes at 15 psi.

Potato and Carrot Agar (PCA)

Avoid new potatoes, as they do not make a good medium.Wash and grate vegetables. Rinse the potato with water to remove starch

Grated potato	20g
Grated carrot	20g
Agar Oxoid No. 3	20g
Tap water or Distilled water	1 litre

Boil vegetables for 1 hour in 500ml tap water. Pass through a fine sieve keeping the liquid. The agar is dissolved add, strained liquid and stir. Autoclave at 121°C for 20 minutes

Malt Extract agar (MEAOX)

Malt Extract agar (Oxoid)	50g
Distilled water	1 litre

Suspend agar in distilled water and heat to dissolve.

Sterilize by autoclaving at 121°C for 15 min.

Potato Dextrose Agar (PDA)

Avoid new potatoes, as these do not make a satisfactory medium

Potatoes	200g
Agar	20g
Dextrose/Glucose	15g
Distilled water	1 litre

Soak 300g of washed, peeled and finely grated potato in 900ml of water overnight in a refrigerator. Filter through cheesecloth into bottles and autoclave at 110°C for 1 hour. Dissolve 15g of dextrose and 20g agar in 230ml of potato infusion and add 770ml distilled water. Autoclave at 121°C for 20 mins.

Potato Dextrose Agar (PDA OX)

Potato Dextrose Agar (Oxoid)	39.0g
Distilled water	1 litre

Add agar to distilled water in double saucepan.

Heat until dissolved. Dispense into bottles. Autoclave at 121°C for 15min. Shelf life: 6 months.

Final pH 5.6±0.2

Potato Sucrose Agar (PSA)

To make 2 litres of potato water:

Distilled water	1125ml
Potato	450g

Use old potatoes. Peel and dice potatoes, suspend in double cheesecloth and boil in the water until almost cooked. Retain the water and discard the cooked potatoes.

Potato water	500ml
Sucrose	20g
Agar Oxoid No.3	20g
Distilled water	500ml

Heat in double saucepan until agar is dissolved. pH 6.5 adjusted with calcium carbonate if necessary.
Autoclave at 121°C for 15 minutes.

Rabbit Dung Agar (RDA)

5 sterilized pellets in medical flats for plates
3 sterilized pellets in universals for slopes

Agar Oxoid No.3	15g
Tap water	1 litre

Heat agar to dissolve in a double saucepan.

Pour into medical flats or universals with the added pellets.

Autoclave at 126°C for 20 minutes

Dung pellets should be collected from wild rabbits. These are dried for 5 days before use.

The pellets can be stored once dried.

Sucrose Nutrient Agar (SNA) / SNA + FP

Potassium di-hydrogen phosphate KH_2PO_4	1g
Potassium nitrate $KN0_3$	1g
Magnesium sulphate $MgSO_4.7H_2O$	0.5g
Potassium chloride KCl	0.5g
Glucose Analar	0.2g
Sucrose Analar	0.2g
Agar Oxoid No.3	20g
Distilled water	1 litre

Adjust to pH 6 - 6.5
Autoclave at 121°C for 20 minutes
Add filter paper to the centre of each freshly poured plate prior to setting.

Tap Water Agar (TWA)

Agar Oxoid No.3	15gw
Tap water	1 litre

Autoclave at 121°C for 20 minutes

V8 Agar (V8)

V8 Vegetables juice	200ml
Agar	20g
Distilled water	800ml

Adjust pH to 6.0 with 10% sodium hydroxide.
[adjust final pH to 5.8]
See Deiner (1955)

V8 Agar for Phytophthora (V8PH)

V8 vegetable juice	200ml
$CaCO_3$	3g
Agar Oxoid No. 3	20g
Distilled water	800ml

Dissolve agar in water and add V8 juice and $CaCO_3$. Stir well to ensure the calcium carbonate is evenly distributed. Clarify the medium by passing through several layers of muslin before dispensing into bottles. Autoclave at 121°C for 20 min.

Yeast Extract-Potassium Phosphate Soluble Starch Agar (YPSS)

Yeast extract (DIFCO)	4.0g
Di-potassium hydrogen phosphate KH_2PO_4	1.0g
Magnesium sulphate $MgSO_4.7H_2O$	0.5g
Soluble starch	15.0 g
Agar Oxoid No 3	20.0g
Distilled water	1 litre

Separately dissolve agar and starch in ca 400 ml water. Heat to boiling. Dissolve the inorganic salts and yeast extract separately in ca 500 ml water. Mix both solutions thoroughly and adjust volume to 1000 ml before autoclaving.

References

Deiner, U.L. (1955) Sporulation in pure culture by *Stemphylium solani*. *Phytopathology* 45, p 141.

Klich, M.A. (2002). Identification of Common Aspergillus species. CBS-KNAW Fungal Biodiversity Centre. *ISBN*: 9789070351465. https://www.nhbs.com/identification-of-common-aspergillus-species-book

Pitt (1973) An Appraisal of Identification Methods for Penicillium Species: Novel Taxonomic Criteria Based on Temperature and Water Relations. *Mycologia* 65, 1135-1157. https://doi.org/10.2307/3758294

Raper, K.B. and Thom, C. (1949) *A Manual of the Penicillia*. The Williams & Wilkins Company, Baltimore.

Samson, R. A., Houbraken, J., Thrane, U., Frisvad, J. C., & Andersen, B. (2019) Food and Indoor Fungi. . CBS Laboratory Manual Series, Volume 2, CBS-KNAW Fungal Biodiversity Centre. ISBN: 9789491751189. https://www.nhbs.com/food-and-indoor-fungi-book

Smith, D. and Onions, A.H.S. (1994) *The Preservation and Maintenance of Living Fungi*. 2nd edn. *IMI Technical Handbooks* No. 2, pp 122. CAB International, Wallingford, UK. https://www.cabi.org/bookshop/book/9780851989020/

Smith, D., Ryan, M.J. and Day, J.G. (eds) (2001) *The UK National Culture Collection Biological Resource: Properties, maintenance and management*. pp 382. UK National Culture Collection, Egham. https://www.ukbrcn.org/media/1025/ukncc-biological-resource.pdf

Appendix C

Culture Collection contacts

UKBRCN – United Kingdom Biological Resource Centre Network https://www.ukbrcn.org/

The United Kingdom Biological Resource Centre Network (UKBRCN) was established to coordinate the activities of the UK national service collections of microbial organisms. The members collections hold over 100,000 microbial cultures and cell lines services and products include:

- the supply of cultures, including microbial resistance test strains, reference and Type strains and cell lines;
- DNA and extracts derived from authenticated strains;
- patent and safe deposits;
- cell line authentication;
- training;
- contract freeze-drying and invasive species management;

The member catalogues of organism strains held can be accessed via the UKBRCN website or via the direct links given below.

UKBRCN Member Collections

Animal and Plant Health Agency (APHA), Department of Food and Rural Affairs for Animal and plant microbes: http://apha.defra.gov.uk/apha-scientific/services/culture/index.htm

Centre for Environment, Fisheries and Aquaculture Science (Cefas), Department of Food and Rural Affairs for Bacteria: https://www.cefas.co.uk/

Culture Collection of Algae and Protozoa (CCAP), Scottish Association for Marine Science campus, Dunstaffnage, near Oban for Cyanobacteria, Algae, Protozoa: https://www.ukbrcn.org/who-we-are/culture-collection-of-algae-and-protozoa-ccap/

European Collection of Authenticated Cell Cultures (ECACC), Public Health England, Porton Down, Salisbury for Cell lines, Stem cells: https://www.ukbrcn.org/who-we-are/european-collection-of-authenticated-cell-cultures-ecacc/

Kew Scientific Collections, Royal Botanic Gardens Kew for Seed bank, endophytic microbes: https://www.kew.org/science/collections-and-resources/collections

National Collection of Fungus Cultures: CABI Genetic Resource Collection CABI UK Centre, Egham; Fungi and bacteria from agriculture and the environment: https://www.cabi.org/products-and-services/bioscience-services/microorganism-supply-services/

National Collection of Pathogenic Fungi (NCPF), PHE: https://www.ukbrcn.org/who-we-are/national-collection-of-pathogenic-fungi-ncpf/

DOI: 10.1079/9781800622135.appc

National Collection of Pathogenic Viruses (NCPV), Public Health England, Porton Down Salisbury for Viruses: https://www.ukbrcn.org/who-we-are/national-collection-of-pathogenic-viruses-ncpv/

National Collection of Yeast Cultures (NCYC) at the Quadram Institute, Norwich: https://www.ncyc.co.uk/

National Collection of Industrial and Marine Bacteria (NCIMB Ltd.), Aberdeen https://www.ncimb.com/

The National Collection of Plant Pathogenic Bacteria (NCPPB), Food and Environment Research Agency: https://www.fera.co.uk/ncppb

National Collection of Type Cultures (NCTC), PHE, for Bacteria pathogenic to man and animals https://www.ukbrcn.org/who-we-are/national-collection-of-type-cultures-nctc/

National Institute for Biological Standards and Controls (NIBSC) - UK Stem Cell Bank, Medicines and Healthcare products Regulatory Agency for Stem Cells: https://www.nibsc.org/

Natural History Museum, Molecular Collections, Natural History Museum, London: https://www.ukbrcn.org/who-we-are/natural-history-museum/

UK based Culture Collections not in the National Service Collections' Network

Marine Biology Association for genera of marine phytoplankton https://www.mba.ac.uk/facilities/culture-collection

University of Portsmouth Culture Collection for isolates of marine fungi https://www.bodc.ac.uk/resources/inventories/edmed/report/2035/

European based belonging to the European Culture Collection Organisation (ECCO)

The European Culture Collections' Organisation (ECCO) was established in 1981. The aim of the organisation is to promote collaboration and exchange of ideas and information about all aspects of culture collection activity. ECCO meetings are held annually and are a valuable forum for discussion and innovation on the future development of member collection activities. Together the member collections hold over 350,000 strains representing yeasts, filamentous fungi, bacteria and archaea, phages, plasmids including plasmid bearing strains and other recombinant DNA constructs, animal cells including human and hybridoma cell lines, animal and plant viruses, plant cells, algae and protozoa. The member collections are listed and contactable via the ECCO Website at https://www.eccosite.org/ or directly via the links given below.

Belgium

BCCM Belgian Coordinated Collections of Microorganisms; together the member collections hold over 250,000 biological resources https://bccm.belspo.be/catalogues

- BCCM/IHEM Fungi Collection: Human & Animal Health https://bccm.belspo.be/about-us/bccm-ihem
- BCCM/GeneCorner Plasmid Collection https://bccm.belspo.be/about-us/bccm-genecorner
- BCCM/LMG Bacteria Collection https://bccm.belspo.be/about-us/bccm-lmg
- BCCM/MUCL Agro-food & Environmental Fungal Collection https://bccm.belspo.be/about-us/bccm-mucl

Bulgaria

NBIMCC National Bank for Industrial Microorganisms and Cell Cultures; holds over 7000 strains including bacteria, actinomycetes, yeasts, filamentous fungi, plasmid-bearing microorganisms, animal and plant viruses, and animal cell cultures https://www.nbimcc.org/www_2020/en/

Czech Republic

CAPM Collection of Animal Pathogenic Microorganisms representing bacteria and viruses https://capm.vri.cz/

CCF Culture Collection of Fungi https://www.natur.cuni.cz/biology/botany/structure/culture-collection-of-fungi-ccf

CCM Czech Collection of Micro-organisms representing bacteria and archaea https://ccm.sci.muni.cz/

CNCTC Czech National Collection of Type Cultures representing bacteria, fungi and yeasts www.szu.cz/cnctc/a

Denmark

IBT Culture Collection of Fungi, Technical University of Denmark https://www.bioengineering.dtu.dk/english/researchny/strain-collections/ibt-culture-collection-of-fungi

Finland

VTT Biotechnology and Food Research, Culture Collection for bacteria, fungi, and plant cell cultures http://culturecollection.vtt.fi

France

CFBP Collection Francaise des Bacteries Phytopathogenes, INRA for plant associated bacteria https://cirm-cfbp.fr/page/CFBP_query

CIP Collection des Bactéries de l'Institut Pasteur for bacteria and viruses https://catalogue-crbip.pasteur.fr/recherche_catalogue.xhtml

CNCM Collection Nationale de Cultures de Micro-organismes, Institut Pasteur; for patent strains https://www.pasteur.fr/en/public-health/biobanks-and-collections/collection-nationale-cultures-microorganismes-cncm

Germany

DSMZ Deutsche Sammlung von Mikroorganismen und Zellkulturen GmbH for microorganisms (mainly bacteria), human and animal cell lines, plant viruses and antisera https://www.dsmz.de/

Greece

ACA-DC Dairy Collection, Agricultural University of Athens for microbial diversity of the Greek traditional foods http://www.aca-dc.gr/

ATHUM Collection of Fungi, University of Athens http://mycetotheca.biol.uoa.gr/

BPIC Collections of Phytopathogenic Fungi and Bacteria, Benaki Phylopathological https://en.bpi.gr/section.aspx?id=1&subid=21&subsubid=77

Hungary

HNCMB Hungarian National Collection of Medical Bacteria http://www.oek.hu/oek.web?to=1808&nid=911&pid=1&lang=eng

NCAIM National Collection of Agricultural and Industrial Micro-organisms representing filamentous fungi, yeasts and bacteria http://ncaim.etk.szie.hu/

Italy

ICLC Interlab Cell Line Collection http://wwwsql.iclc.it/

DBVPG Collezione dei Lieviti Industriali, Dipartimento di Biologia Vegetale, Università di Perugia for yeast cultures http://www.dbvpg.unipg.it/index.php/it/la-collezione-dbvpg

MUT Mycotheca Universitatis Taurinensis (MUT) fungal collection https://www.tucc-database.unito.it/mut/en.html

ITEM Agro-Food Microbial Culture Collection for fungi http://www.ispa.cnr.it/Collection

UMCC UNIMORE Microbial Culture Collection for bacteria and yeasts https://umcc.bio-aware.com/

BUSCOP Bologna University Scardovi Collection of *Bifidobacteria* https://site.unibo.it/buscob-microbial-collection/en

Latvia

MSCL Microbial Strain Collection of Latvia for bacteria, filamentous fungi and yeasts http://mikro.daba.lv/EN/

The Netherlands

Westerdijk Fungal Biodiversity Institute https://wi.knaw.nl/

Norway

NIVA Culture Collection of Algae, Norwegian Institute for Water Research https://niva-cca.no/

Poland

IPF Collection of Industrial Micro-organisms for bacteria, filamentous fungi and yeasts https://cim.ibprs.pl/

KOS Collection of Salmonella Micro-organisms, National Salmonella Centre, Medical University of Gdansk http://salmonella.gumed.edu.pl/index2.php

PCM Polish Collection of Micro-organisms for bacteria https://pcm.org.pl/home

Portugal

MUM Micoteca da Universidade do Minho for filamentous fungi http://www.micoteca.deb.uminho.pt/

PYCC Portuguese Yeast Culture Collection https://pycc.bio-aware.com/

Russia

IPPAS Collection of Microalgae of the Institute of Plant Physiology http://www.cellreg.org

SRC CCM Collection of Cultures of Micro-organisms for bacteria and archaea, filamentous fungi and yeasts, bacteriophages https://ccm.sci.muni.cz/

VKM Russian Collection of Micro-organisms for bacteria (including actinomycetes) and micro fungi (including yeasts) http://www.vkm.ru

VKPM Russian National Collection of Industrial Micro-organisms for bacteria, yeasts, mycelial fungi, phages and genetically modified strains http://www.vkm.ru/collecti.htm

Slovakia

CCY Culture Collection of Yeasts, Slovak Academy of Sciences http://ccy.sk/index.php/en/13-ccy

Spain

CECT Coleccion Espanola de Cultivos Tipo for bacteria, archaea, yeast and filamentous fungi https://www.uv.es/uvweb/spanish-type-culture-collection/en/spanish-type-culture-collection-1285872233521.html

Sweden

CCUG Culture Collection University of Göteborg for bacteria and some yeast and fungi https://www.ccug.se/

FCUG Fungal Cultures University of Göteborg http://www2.dpes.gu.se/database/FCUG.html

Switzerland

CCOS The Culture Collection of Switzerland for bacteria, fungi and cell lines of animals http://www.ccos.ch

Turkey

HÜKÜK Culture Collection of Animal Cells, Foot and Mouth Disease Institute https://vetkontrol.tarimorman.gov.tr/sap/Menu/24/Hucre-Kultur-Koleksiyonu

Some public service collections based in the rest of the world

The collections above and those of the global network can be found via the World Data Centre for Microorganisms (WDCM) http://www.wdcm.org/. There are 818 registered collections from 78 countries, holding a total of 3,338,809 microorganisms, comprising 1,469,385 bacteria, 885,898 fungi, 39,492 viruses and 38,023 cell lines. The WDCM have also established the Global Collection of Microorganisms (http://gcm.wdcm.org/) which amalgamates the holdings of 146 collections in 51 countries and to date holds 56,672 species represented by 526,637 strains. Examples of some of the individual collections in the WDCM are given below.

ANACC Australian National Algae Culture Collection https://www.csiro.au/en/about/facilities-collections/collections/anacc

ARS Agricultural Research Service Culture Collection, USA for bacteria and fungi https://nrrl.ncaur.usda.gov/

ATCC American Type Culture Collection, USA for bacteria, viruses, yeast, fungi, protozoa, algae, nucleic acids, cell lines and molecular tools https://www.atcc.org/

BIOTEC Culture Collection, Thailand for bacteria, yeasts, and filamentous fungi https://www.biotec.or.th/bcc/

CCAMF Canadian Collection of Arbuscular Mycorrhizal Fungi https://agriculture.canada.ca/en/agricultural-science-and-innovation/agriculture-and-agri-food-research-centres-and-collections/canadian-collection-arbuscular-mycorrhizal-fungi-ccamf

DAOMC Canadian Collection of Fungal Cultures https://agriculture.canada.ca/en/agricultural-science-and-innovation/agriculture-and-agri-food-research-centres-and-collections/canadian-collection-fungal-cultures-daomc

Fiocruz Biological Collections for bacteria, fungi, protozoa, helminths, arthropods, molluscs of medical and environmental importance https://portal.fiocruz.br/en/biological-collections

FGSC Fungal Genetic Stock Center, USA for fungi and yeast http://www.fgsc.net/

ICMP International Collection of Microorganisms from Plants, Landcare, Research, New Zealand for Oceania bacteria and fungi https://www.landcareresearch.co.nz/tools-and-resources/collections/icmp-culture-collection/

KCTC Korean Collection for Type Cultures, Korea for Asia bacteria, fungi, yeasts, plasmids, cell lines, archaea and microalgae https://kctc.kribb.re.kr/En/Kctc

NCMR National Centre for Microbial Resource, India for bacteria and fungi http://210.212.161.138/

NMRC National Mycology Reference Centre, SA Pathology Australia for Oceania fungi https://www.adelaide.edu.au/mycology/national-mycology-reference-centre

Phaff Yeast Collection (UCD-FST) USA for yeast https://phaffcollection.ucdavis.edu/

JCM Japan Collection of Micro-organisms for aerobic and anaerobic bacteria including actinomycetes, archaea and fungi including yeasts https://jcm.brc.riken.jp/en/?doing_wp_cron=1645807377.4885210990905761718750

UAMH Center for Global Microfungal Biodiversity, Canada for bacteria, fungi and yeast https://www.uamh.ca/

NBRC NITE Biological Resource Center, Japan for bacteria including actinomycetes and archaea, yeasts, fungi, algae, bacteriophages and DNA resources https://www.nite.go.jp/en/nbrc/index.html

Appendix D

Compliance with controls on the distribution of dangerous organisms

1. This Appendix is written from the perspective of UK legislation and represents best practice. Other countries have different regulations and lists of restricted organisms, collections should ensure, that the export of organisms adheres to local legislation and, as far as reasonably practicable, does not contravene the legislation of the requestor's country.

2. Anyone handling microorganisms is committed to preventing dangerous or hazardous pathogens from falling into the hands of non-legitimate users. This document describes a means by which collections can comply. It is based on procedures that are used by collections and the legislation governing the distribution of organisms that may be a biosecurity threat (Rohde *et al.*, 2013; Smith *et al.*, 2017). An example Biosecurity Risk Assessment form is presented below. There are several pieces of legislation that require permission before distribution and these have been discussed in Chapter 7. In this volume the focus is biosecurity (threat of misuse and terrorism). In the UK this is governed by the Counter-Terrorism and Security Act 2015 (https://www.gov.uk/government/collections/counter-terrorism-and-security-bill). Globally, this is covered by the Biological Weapons Convention 2019 (https://www.un.org/disarmament/biological-weapons/). The Biological Weapons Convention (BWC) effectively prohibits the development, production, acquisition, transfer, stockpiling and use of biological and toxin weapons. It was the first multilateral disarmament treaty banning an entire category of weapons of mass destruction (WMD). The Australia Group (AG) is an informal forum of 43 countries which, through the harmonization of export controls, seeks to ensure that exports do not contribute to the development of chemical or biological weapons (https://www.dfat.gov.au/publications/minisite/theaustraliagroupnet/site/en/index.html). The 'Australia Group' have provided a list of the organisms that they wish to control distribution of both inside and outside of this group of countries (see: https://www.dfat.gov.au/publications/minisite/theaustralia-groupnet/site/en/human_animal_pathogens.html). Import and export permits must be sought through Animal and Plant Health Agency (http://apha.defra.gov.uk/vet-gateway/index.htm) and Department for Environment, Food & Rural Affairs (Defra) (https://www.gov.uk/guidance/moving-prohibited-plants-plant-pests-pathogens-and-soil) as appropriate.

3. The organisms covered by this best practice are placed in three categories; the current UK ACDP Approved list of Biological Agents (https://www.hse.gov.uk/pubns/misc208.pdf) must be consulted to ensure the correct assignment of human pathogen hazard group of the organism before it is supplied. In addition, collections staff (microbiologists shipping the organism) must assess the dangers presented by an organism and subsequently restrict distribution accordingly.

DOI: 10.1079/9781800622135.appd

Category 1. Hazardous organisms, including Advisory Committee on Dangerous Pathogens (ACDP) category 3 & 4 pathogens, included in the 'Australia Group' regulations.
Category 2. Hazardous organisms including ACDP category 3 & 4 pathogens and others not included in the Australia Group regulations.
Category 3. Animal and plant pathogens controlled by Defra, and other, legislation.
4. The best practice governing the distribution of organisms in each of these categories is outlined below.

Category 1: Hazardous organisms, including ACDP category 3 & 4 pathogens, included in the 'Australia Group' regulations.

5. The export of organisms specified by the 'Australia Group' of countries is governed by legislation, with export licences being required for all of these organisms. Guidance Moving Dangerous Goods (https://www.gov.uk/guidance/moving-dangerous-goods) and the Export Control Order 2008 (https://www.legislation.gov.uk/uksi/2008/3231/contents/made). Exports of the listed organisms to countries outside the Australia Group require Standard individual export licences (SIELs) see https://www.gov.uk/guidance/standard-individual-export-licences. Exports to countries within the 'Australia Group' require an Open General Export Licence (OGEL) which removes the need for an individual export licence provided there are no grounds for knowing or suspecting that goods are going to be used for biological weapons purposes. Note: Guidance on the transport of infectious substances is given in Guidance Note number: 17/2012[REV.7] https://assets.publishing.service.gov.uk/government/uploads/system/uploads/attachment_data/file/917954/dangerous-goods-guidance-note-17-document.pdf.
Licence requirements for goods can be checked using the 'OGEL and Goods Checker Tools' https://www.ecochecker.trade.gov.uk/spirefox5live/fox/spire/OGEL_GOODS_CHECKER_LANDING_PAGE/new
A code of practice for export is also available and should be consulted,
https://assets.publishing.service.gov.uk/government/uploads/system/uploads/attachment_data/file/341998/10-668-codepractice-compliance.pdf
6. Failure to comply with these requirements is a criminal offence. Any enquiries from the UK should be directed to the Export Control Joint Unit which is part of the Department for International Trade (https://www.gov.uk/government/organisations/export-control-organisation).
7. In addition to possessing the appropriate licence, collections supplying organisms in this category should take all practicable steps to ensure that these organisms are going to legitimate users. Supply of listed organisms may continue to existing users however, the request should be signed by the Head of Department/Division (or the safety officer or other person authorized by the Head of Department) and the registered user. The signatures should be matched against those held on record in the collection on the appropriate form.
8. The Biological Weapons Convention 2019 includes the clause that 'No person shall develop, produce, stockpile, acquire or retain any biological agent or toxin of a type and in a quantity that has no justification for prophylactic, protective or other peaceful purposes'. Contravention of this act is a criminal offence with a maximum penalty of life imprisonment.

Category 2: Hazardous organisms including ACDP category 3 & 4 pathogens and others not included in the Australia Group regulations.

9. All other ACDP group 3 & 4 organisms, and selected other organisms offering potential dangers to humans, animals or plants, are also included in this best practice. Supply of listed hazardous organisms may continue to existing users but the request should be signed by the Head of Department/Division or the registered user. The signatures are matched against those held on record in the collection on the appropriate form.
10. ACDP hazard group 4 viruses are only supplied to known established scientists with a track record of legitimate work with such organisms and who are known to have appropriate containment facilities.

Category 3: Animal and plant pathogens controlled by Defra

11. The regulations governing the distribution of animal and plant pathogens are shown below. Some animal viral pathogens and some plant bacterial and fungal pathogens are also included in the 'Australia Group' of organisms. Distribution arrangements for these organisms conforms to both sets of regulations.

Animal Pathogens

12. The Specified Animal Pathogens Order 2008 states that 'No person may deliberately introduce into any animal any animal pathogen specified in either Part 1 or Part 2 of that Schedule, except under the authority of a licence issued by the Secretary of State' (https://www.legislation.gov.uk/uksi/2008/944/contents/made). It is an offence to possess or spread a listed animal pathogen within the UK without a licence. It is supplemented by the Importation of Animal Pathogens Order 1980 (https://www.legislation.gov.uk/uksi/1980/1212/contents/made) which makes it an offence to import any animal pathogen, or potential or actual carrier, of an animal pathogen from a non-EC country, except under licence (https://www.gov.uk/government/publications/animal-pathogens-guidance-on-controls). Enquiries should be made to the Animal and Plant Health Agency (https://www.gov.uk/government/organisations/animal-and-plant-health-agency). The culture collection and the customer must hold the appropriate licences to hold these organisms. Orders will be refused where the customer is unable to produce a copy of the appropriate licence.

Plant Pathogens

13. The Plant Health Order (England) 2015 (https://www.legislation.gov.uk/uksi/2015/610/contents) regulates the import, movement and keeping of plant pests including plant pathogens. The order notes that licensed pathogens may be provided to persons or organizations that hold a relevant and current Defra licence. Licensed pathogens may also be sent to persons or organizations overseas that have authority from their national plant health service to receive such material. However, material must not be released to other persons or organizations without permission from Defra (https://www.gov.uk/guidance/plant-health-controls) who are responsible for plant health controls, imports and exports, certification schemes, plant passporting and listed quarantine plant pests.

14. Collections include in this category any other pathogen that may not be indigenous to, and which may cause damage to plants within, the UK. If there is any uncertainty over the status of any organism the matter is referred to Defra.

Record keeping

15. Collections maintain records of all requests for all controlled organisms including those requests which are refused for any reason. Records are kept for a period of 25 years.

This would include records of Standard individual export licences (SIELs)https://www.gov.uk/guidance/standard-individual-export-licences, exports to countries within the 'Australia Group' with an Open General Export Licence (OGEL). Separate records of the supply of 'Australia Group' organisms are required whether under an SIEL or an OGEL.

Conditions of supply

16. All controlled organisms are supplied on the basis that they are not passed on to third parties. Recipients of these cultures are cautioned that some third-party transfers, such as the unlicensed export of 'Australia Group' organisms and the transfer of some Defra controlled pathogens, may be a criminal offence. Users may also be liable under civil law for disease or damage following third party transfer. Collection users receiving requests for third party transfer of controlled organisms should refer such requests back to the collection.

17. Collection staff assess user registrations and reserve the right to refuse any order which does not show clear evidence that it is for the safe and legitimate use of an organism. A collection accepts no obligation to give reasons for this refusal.

References

Rohde, C., Smith, D., Martin, D., Fritze, D. and Stalpers, J. (2013) Code of Conduct on Biosecurity for Biological Resource Centres: procedural implementation. *International Journal of Systematic and Evolutionary Microbiology* 63, 2374-2382. https://www.microbiologyresearch.org/content/journal/ijsem/10.1099/ijs.0.051961-0

Smith, D., Martin, D. and Novossiolova, T. (2017) Microorganisms: good or evil, MIRRI provides biosecurity awareness. *Current Microbiology* 74, 299. https://doi.org/10.1007/s00284-016-1181-y

BIOSECURITY RISK ASSESSMENT

Business Area: _____ Location:__________ Assessment undertaken by: _____ Checked by:.

Notes for completion

Biosecurity risk may include theft or breach of security, equipment or infrastructure failure or compromise, inadvertent supply due to human error or non-compliance with legislation. This covers organisms, data, 'know how' and information concerning plant pathogens, biopesticides, toxin producing organisms and includes mechanisms to store, distribute and 'mass-produce.'

Specific risks are included for guidance only, while some will be generic - please amend to your specific business areas

Risk should be scored as: Impact (I) 1 low to 5 High / Likelihood (L) 1 low to 5 High / Total (T) [L x I] 1 low to High 25

If the score is more than 4, then a management strategy should be assessed and residual risk calculated

Residual risk is the assessment after all the measures have been documented, so unless the likelihood and impact have been eliminated some risk will remain.

The impact score should reflect the risk to Company profile and ability to operate the business

The impact is difficult to reduce much, as if it got out (material or info) it could still be used to about the same extent. The likelihood can be made lowest by ensuring robust enough procedures to prevent it happening – as much as possible.

Specific risks and implications	I	L	T	Management Strategy for dealing with risk	Total Residual Risk I x L = T	Required Action	Responsible Individual (s)
Breach of high security containment/ escape or other internal problem, e.g. HEPA containment, autoclave or LN tank leakage.	4	2	8	Maintain integrity of facilities and procedures; maintain correct waste disposal methods and procedures. Implement and maintain Standard Operating Procedures for quarantine and laboratory procedures as appropriate. Regular preventive maintenance of equipment.	4 x 1 = 4	Continue monitoring, annual testing and certification. Regularly review SOPs. Notify police/security services/national authorities. In the event of an issue that may affect Company profile, marketing should be contacted to help draft a public statement	
Breach of high security containment, caused by exceptional circumstance such as storm, violence, tree fall, fire or flood including accidental release of strains into the environment.	3	2	6	Where possible, take preventive action to mitigate risks for predicted events. Maintain integrity of facilities, ensuring fit for purpose as part of ongoing maintenance regime. Disaster plan strategy in place and reviewed	3 x 2 = 6	Ensure back up measures are in place to ensure security to the resource. Notify police/security services/national authorities In the event of an issue that may affect company profile, marketing should be contacted to help draft a public statement	

Continued

Continued.

Specific risks and implications	I	L	T	Management Strategy for dealing with risk	Total Residual Risk I x L = T	Required Action	Responsible Individual (s)
Theft of data or information related to high biosecurity risk organisms from high security area	4	2	8	Ensure data subject to password access. Servers kept in secure location. Procedures to ensure that data are correctly assigned to risk group to ensure they are 'flagged up' on the database (full data sets visible to key staff only).	3 x 2 = 6	Yearly biosecurity audit. Notify Police in the event of theft. Ensure staff are familiar with SOPs. Staff accessing high security areas to be checked via the Disclosure and Barring Service 'DBS' checked In the event of an issue that may affect company profile, marketing should be contacted to help draft a public statement	
Theft of materials related to high biosecurity risk organisms from high security area	4	2	8	High security areas 'point of entry' access controlled. Regular security patrols, contact with Police Force (alarm contact) and security.	3 x 2 = 6	Yearly biosecurity audit. Inform Police in the event of theft. Ensure staff are familiar with SOPs. In the event of an issue that may affect company profile, marketing should be contacted to help draft a public statement	
Lack of screening procedure for visitors to restricted areas. Visitors / Staff accessing restricted areas when not authorized to do so	3	3	9	All visitors to sign in and to be accompanied at all times. Suggested staff ratio of 1 to 8, to be increased when individuals from 'countries of concern' are likely to be present. Access to high security areas restricted by point of access card entry to key staff only	3 x 2 = 6	Yearly biosecurity audit. Ensure staff are familiar with SOPs. Key staff 'DBS' checked In the event of an issue that may affect company profile, marketing should be contacted to help draft a public statement	
Inadvertent supply of 'Know how' to individuals, countries or others listed by the Australia group or other national legislation	4	2	8	Staff aware of Biosecurity policy and risks. Consultancy arrangements agreed by unit Directors	3 x 2 = 6	Review and adherence to SOPs Staff fully trained In the event of an issue that may affect company profile, marketing should be contacted to help draft a public statement	

Continued

Continued.

Specific risks and implications	I	L	T	Management Strategy for dealing with risk	Total Residual Risk I x L = T	Required Action	Responsible Individual (s)
Inadvertent supply of 'Data' to individuals, countries or others listed by the Australia group or other national legislation	3	2	6	Staff training, risk data are correctly assigned to risk category and 'flagged up' on the database (visible to key staff only).Data restricted to trained staff only	3 x 2 = 6	Review and adherence to SOPs Staff fully trained In the event of an issue that may affect company profile, marketing should be contacted to help draft a public statement	
Inadvertent supply of restricted organism or supply of technology.	4	2	8	Secondary check before dispatch in place. Staff fully aware and signed up to biosecurity policy. Training reviewed annually	4 x 2 = 8	Review and adherence to SOPs Staff fully trained In the event of an issue that may affect company profile, marketing should be contacted to help draft a public statement	
Legal liability as a result of non-compliance with national, regional or international legislation	4	2	8	Staff aware of legislation though biosecurity policy and procedures and the obligations under it.	3 x 2 = 6	Review and adherence to SOPs Staff fully trained In the event of an issue that may affect company profile, marketing should be contacted to help draft a public statement	
Release or transport of organisms in incorrect packaging or otherwise non-compliant with IATA guidelines	4	3	12	Release of living organisms from company site to be undertaken by IATA trained staff only. Staff advised not to carry living organisms by hand unless it is legally permitted, with additional permission granted by senior manager in exceptional circumstances	3 x 2 = 6	Review and adherence to SOPs Staff fully trained In the event of an issue that may affect company profile, marketing should be contacted to help draft a public statement	

Inadvertent supply covers human error and/or non-compliance with operating procedures.

Appendix E

Examples of Forms/Data Sheets for Culture Collection Activities

Form 1: Culture Order Form

Customer details

Name:
Company / organisation:
Address:
Country:
Post/zip code:
Contact number(s):
Fax:
E-mail:
Customer order/reference id:

Shipping address if different from above

Name:
Company/organisation:
Address:
Country:
Post/zip code:
Contact number(s):

Continued overleaf....................

DOI: 10.1079/9781800622135.appe

Authorised signature:
Print name:
Position held:
Date: / /

	CULTURE NAME	Collection number (i.e. IMI, NCIMB)	UKNCC number (not necessary)	Quantity
1				
2				
3				
4				
5				
6				
7				
8				
9				
10				
11				
12				
13				
14				
15				

NOTES

CABI Bioscience was formerly known as the International Mycological Institute (IMI). Cultures obtained from CABI retain their IMI prefix.

NCWRF cultures are obtained from CABI; NCFB cultures are obtained from NCIMB.

CABI Genetic Resources Collection: Culture Order Form

Please only use this form if you couldn't find what you were looking for in our electronic catalogue

www.cabi.org/services/microbial-services/culture-collection-microorganism-supply/grc/

All orders are subject to our conditions of sale, which are included with this form. Please e-mail this form to: cultures@cabi.org

CUSTOMER DETAILS:

Name		Position held	
Organisation			
Address			
Country		Post or Zip Code	
Telephone		Fax	
Email Address			

SHIPPING ADDRESS (If different from above):

Contact Name			
Organisation			
Address			
Country		Post or Zip Code	
Telephone			

Authorised Signature		Date	
Print Name		Position Held	

PLEASE SUPPLY:

Name of Organism	**IMI Number**	**Quantity**	**Form of supply** (Please state A, B, C, D or E) A: Freeze-dried culture B: Culture growing on agar C: Genomic DNA D: Culture Extract E: Other

Method of payment	Bank transfer		Credit/debit card	
Do you need an import permit?	Yes		No	
Do you need a licence to work with the samples?	Yes		No	
Do you have Hazard Group 2 facilities?	Yes		No	
PO number				

On occasion, microorganisms can be difficult to recover after preservation. In the event of this happening with any of the cultures you have requested, please indicate whether you would like us to recommend alternatives:

Yes: ☐ **No:** ☐

CABI aims to supply high quality resources. So that we can ensure that the cultures we provide meet your expectations, please indicate the intended use for the microorganisms requested from CABI:

Taxonomic study: ☐ **Molecular study:** ☐ **Teaching:** ☐ **Research:** ☐
Assay/Testing: ☐ **Other (please state):** ☐

Additional Information about your order

Cultures supplied by CABI are provided with either their originally-deposited name or with a subsequently-amended name so please check any synonyms.

Our cultures are supplied either as freeze-dried ampoules or live on agar. Please be aware that some cultures may need to be provided live. We strongly recommend that unless you have knowledge or experience of growing fungal cultures from freeze-dried ampoules, you order the live culture.

Plant Pathogens

Plant pathogens ordered from outside England and Wales require paperwork to accompany the cultures in transit; either a Letter of Authority (LoA) **or** a letter from your National Plant Protection Organisation (NPPO) stating that no licence is required. If you are ordering a plant pathogen please check with your authority promptly - obtaining the necessary approval often takes time.

Payment

We are now requesting payment prior to culture dispatch. To satisfy this, we require a purchase order number or you can pay via credit card by calling our finance department directly. A pro-forma invoice can be raised for you if you would prefer. Payment terms are 30 days from the invoice date and the amount shown on the invoice must be the amount received by us in full after all bank charges have been settled. **Once payment is agreed, you will be liable for the cost of culture preparation should you have to cancel your order.**

Our prices are listed below:

Service	Price (excluding VAT)
1x freeze-dried ampoule	£155
1x live culture	£210
UK postage and packaging (up to multiples of 2 cultures)	£15
European postage and packaging	£150
Worldwide postage and packaging	From £150 (exact price dependent on country)

Please note, additional charges may apply for supplementary administrative tasks, including (but not limited to) the following: the drafting of Pest Risk Forms, new supplier forms, liaison with NPPO's directly or the provision of supplementary information about our cultures

Biosafety
We will only send cultures to laboratories with the appropriate facilities, not private addresses. *You are responsible for ensuring that your facilities comply with biosafety and biosecurity regulations for your own country and that appropriate risk assessments are undertaken.* Cultures supplied by CABI are classified using the United Kingdom ACDP and COSHH guidelines. If ordering cultures assigned by ACDP to Hazard Group 2, your establishment must be equipped to this level.

Processing Time
The dispatch of your order is subject to the completion of any permit or licensing requirements (explained above). Working with over 30,000 specimens, providing cultures is a bespoke service tailored to our customers. We always endeavour to dispatch our cultures in a timely fashion, however please be aware that some organisms may present challenges which hinder this.

CABI may use the information provided in this form, along with scientific results arising from analysis of the sample, for scientific research purposes. Your personal data such as contact details will not be included in any scientific reports, may be shared with partners or donors but not be shared with third parties. Our Privacy Policy is available at https://www.cabi.org/privacy-policy/. We would like to retain the contact details you have provided indefinitely in order to contact you in the future with regard to the samples you wish to purchase if there are further scientific developments we think may be of interest to you. Please indicate if you are willing for CABI to contact you for this purpose:

Email ☐ Phone ☐ Post ☐ SMS ☐ No ☐

CABI would like to contact you in the future about other services we offer that we think may be of interest to you. Please indicate below if you are willing for CABI to contact you for this purpose:

Email ☐ Phone ☐ Post ☐ SMS ☐ No ☐

Supply of Genetic Resources
MATERIAL TRANSFER AGREEMENT

This Material Transfer Agreement ("MTA") governs the transfer of the Original Material from CAB International for research, teaching and identification purposes only.

This MTA constitutes a contract between you (the “Customer”) and CAB International, an international governmental not-for-profit organisation (“CABI”).

CABI and the Customer may be individually referred to as a Party or jointly, as the Parties.

1. Definitions

“Commercial Purposes” means the use or exploitation of the Material, with the object of, or resulting in, financial gain, and includes but is not limited to the following activities: sale, use of the Material to provide a service, applying for, obtaining or transferring intellectual property rights or other tangible or intangible rights by sale or license or in any other manner, commencement of product or service development, conducting market research, and seeking regulatory approval.

“Effective Date” means the date of the last signature of this MTA by a Party.

“Force Majeure” means an event or sequence of events beyond a party's reasonable control preventing or delaying it from performing its obligations under this MTA, including but not limited to war, civil commotion or riots, war, threat of or preparation for war, armed conflict, natural disaster, transportation difficulties, labour or trade dispute, strikes, industrial action interruption or failure of utility service.

“Material” means Original Material, Progeny and Unmodified Derivatives. The Material shall not include Modifications.

“Modifications” means substances created by the Recipient which contain or incorporate all or any part of the Original Material, Progeny or Unmodified Derivatives.

"Original Material" shall mean any biological material transferred from CABI to Customer identified in Schedule 1.

"Prior Informed Consent" has the meaning set out in the Nagoya Protocol to the Convention on Biological Diversity. It refers to the requirement that a country which has ratified the Nagoya Protocol and which is providing access to a genetic resource provide its consent on the basis of mutually agreed terms before access to the said genetic resources is granted.

"Progeny" means all unmodified descendant from the Original Material, such as virus from virus, cell from cell, or organism from organism.

"Unmodified Derivatives" means any substance created by the Customer which constitutes (a) a substantially unmodified copy of the Original Material or Progeny or any of their components or (b) a product expressed or produced by the Original Material or Progeny. Some examples include: material produced by growth of a microorganism, purified or fractionated subsets of the Original Material, and proteins expressed by DNA/RNA supplied by CABI.

"Third Party" shall mean any person other than the Customer and CABI.

2. Transfer of Material

2.1 Subject to Clause 2.4 below, CABI agrees to transfer the Original Material to the Customer. CABI shall label, package, and arrange for transport of the Original Material in accordance with applicable laws and regulations.

2.2 The Customer shall be responsible for all duties, tax(es), tariffs and fees arising in connection with this MTA and the transfer of the Original Material. The Customer shall, upon demand, pay to CABI an amount equal to any such tax(es), duties, tariffs and permit fees actually paid or required to be collected or paid by CABI.

2.3 If the Original Material is not viable on receipt by the Customer, and provided the Customer has handled and continues to handle the Original Material in accordance with instructions provided by CABI, CABI will provide (a) replacement of the Original Material or (b) a refund of the fee paid to CABI (if any).

2.4 Nothing in this MTA obliges CABI to supply the Original Material if (a) CABI is prohibited from doing so under any applicable legislation, including legislation relating to terrorism or threat of biological weapons or (b) CABI has reasonable grounds to suspect that the Customer intends to utilise the Material in any manner which contravenes the terms of this MTA or for any other prohibited or unethical purpose.

2.5 At CABI's request the Customer shall promptly and at its cost return or destroy (at CABI's option) all Material on termination or expiry of this MTA.

3. Conditions of Use

3.1 The Customer may use the Material solely for research, teaching and identification purposes and shall not use the Material for Commercial Purposes.

3.2 The Material is not intended for human in vivo use. The Customer shall not use the Material in human subjects, in clinical trials, or for diagnostic purposes involving human subjects.

3.3 The Customer shall not distribute, release, sell, lend or otherwise transfer the Material to a Third Party without CABI's prior written consent.

3.4 The Customer agrees to use the Material in compliance with all applicable laws, regulations, codes of practice and guidelines such as, for example, those relating to research involving the use of animals or recombinant DNA.

4. Compliance with the Convention on Biological Diversity

4.1 The Customer shall use the Material in a legitimate, safe, legal, ethical and sustainable way, and in respect of the principles laid down in the Convention on Biological Diversity ("CBD").

Nothing in this MTA shall be construed as changing or affecting the rights and obligations of the Parties under the CBD.

4.2 This MTA fully complies with Article 15 of the CBD, which recognises the sovereign rights of States over their natural resources.

4.3 The Original Material is provided under this MTA on the understanding that it was collected either before the CBD came into force or if collected after the CBD came into force with Prior Informed Consent, where procedures are in place and recognised authorities exist.

4.4 The Customer shall uphold the mutually agreed terms of the Prior Informed Consent and any other terms that CABI notifies to the Customer in Schedule 1.

4.5 To the extent any terms and conditions set out in Schedule 1 conflict with the terms of this MTA, the terms and conditions set out in Schedule 1 shall prevail.

5. Ownership and Intellectual Property

5.1 The transfer of the Original Material pursuant to this MTA does not affect the ownership of the Original Material or any intellectual property rights in or relating to it or its use.

5.2 In accordance with the CBD which recognizes that nations have sovereign rights over their own biological resources, the ownership of Material, including any Material incorporated in Modifications, or derivatives remains with the Provider Country, as defined in the CBD.

Form 2: Compliance with Convention on Biological Diversity (CBD)

CABI Bioscience, in the spirit of the terms of the Convention on Biological Diversity (CBD) and along with other major international culture collections, require that customers sign the declaration below.

Declaration under the terms of the Convention on Biological Diversity

*I/we agree not to claim ownership over the micro-organisms or cell lines received from *..*, nor to seek intellectual property rights over them or related information. If we wish to utilise or exploit such organisms commercially, suitable and adequate sharing of benefits in the spirit of the Convention on Biological Diversity will first be discussed with the relevant UKNCC collection.*

I/we also agree to ensure that any subsequent person or institution to whom I/we make samples of the micro-organism available, is bound by the same provision.

* Enter name of culture collection*

Authorised signature:	**Date: / /**
Print name:	
Position held:	
Address:	

Form 3: Accessions Form to Accompany Deposits

	Name & authority:	CABI (IMI) collection accession Nº:
	Principal Synonyms/Name Change:	collection accession date:
	Identified by:	Type of deposit:

Depositor: please complete as much as possible:

Name of isolate:	Date sent to CABI:
Name & address of depositor:	
Designation:	
Previous history: (other collection/owners/isolate designation):	
Other collections where held: (give collection number and acronym):	
Isolated from (substratum/genus & species of organism):	
Anatomical part/substratum part:	
Geographical location:	
Isolated by:	
Date of isolation:	
Isolation method (soil plate, damp chamber, surface sterilisation, micromanipulation etc.):	
Special features & usage (metabolic products, culture derived from type, etc.): References (journal, volume, page, year) please attach copies/reprints if available:	
Recommendations for Maintenance and Preservation: i) Growth medium: ii) Incubation temperature: iii) Incubation time: iv) Light requirements/pH etc.: v) Period between transfer: **Please tick appropriate method(s)** i) Freeze-drying ○ ii) Liquid nitrogen storage ○ iii) Water storage ○ iv) Soil storage ○ v) Oil storage ○ vi) Silica gel storage ○ vii) Other - please specify ○	For CABI use only:

Permission to deposit in CABI Genetic Resource Collection and distribute to third parties under the Convention on Biological Diversity (CBD).

Was the organism collected after December 1993?............................Yes ☐ or No ☐

If the answer to the above question is YES please answer the following questions and provide the requested information.

From whom did you receive prior informed consent to collect the material?

Land owner:

National Authority:

To the best of your knowledge do the above have the authority to grant this permission?...... . Yes ☐ or No ☐

Do you have the authority to deposit them in living and dead dried collections for further use and distribution?...Yes ☐ or No ☐

If so under what terms:

For official use
Enquiry Number:

Form 4 CABI DAS Service

Email: diagnostic.service@cabi.org
Website: www.plantwise.org

Please complete Part 1 and Part 2 of this form using the guidelines provided below, and include the form in the package with your samples. Please use a separate form for each country of origin. Note that an incomplete form may delay a diagnosis.

Part 1. Request for diagnostic service:

Name: Enter here the name of the person requesting the service. This may not be the same as the person submitting the samples, but is the contact person for diagnostic service staff.
Country: Enter the country from which the samples originated.
Date: Enter the date on which this submission form was completed (dd/mm/yy).

NAME	
ORGANISATION/DEPARTMENT	
COUNTRY	
HOW DID YOU HEAR ABOUT THE DIAGNOSTIC AND ADVISORY SERVICE	
DATE	

Part 2. Information on samples:

Your sample number: Provide your own reference number for each individual sample. We will refer to these numbers when sending our report to you.

Collection site/location: Specify the geographic or administrative area where sample was collected (e.g. province, division, town, village, parish).

Date collected: Enter the date when the sample was collected (dd/mm/yy)

Host plant: Enter the common name as well as the scientific name and variety/ cultivar if known.

Sample type: Specify the type of sample e.g. whole plant, leaves, roots, stem, seeds, fruit, soil, culture or other (specify type).

Symptoms/problem: Provide a short description of the problem and particularly the symptoms that you are most concerned about (e.g. spots on leaves). Include symptoms that you have observed but may not be visible on the sample you send to us.

Collector: Provide the name of the person(s) who provided the sample if this is different from the person specified in Part 1 above (we do not need to know the name of individuals who collected specimens in the field, unless the requestor thinks this is important).

Your Sample Number	Collection Site / Location	Date Collected	Host Plant	Sample Type	Describe symptoms and/or nature of problem	Collector	For official use W no.

Requirements under the Convention on Biological Diversity:

The UK is a signatory to the Convention on Biological Diversity (CBD) and party to the Nagoya Protocol on Access and Benefit Sharing (ABS). CABI operates in accordance with European legislation to implement these requirements. Those wishing to submit samples to CABI for identification or deposit should check if their country is party to the CBD and the Nagoya Protocol and if so ensure compliance. All customers are required to read this form & sign the Declaration below.

For applicants from Universities, the Declaration must be signed by the relevant person in Authority.

Declaration:

I understand the requirements and my responsibilities relating to the Convention on Biological Diversity and the Nagoya Protocol. I confirm that the samples in this consignment have been submitted to CABI in compliance with requirements.

Signature:

Date:

Print Name:

Position:

Samples for deposit:
Requirements under the **Convention on Biological Diversity** and the **Nagoya Protocol** relating to **Mutually Agreed Terms, Access and Benefit Sharing** and **Material Transfer Agreements** apply to all samples to be deposited in the CABI Genetic Resource Collection. If you wish to deposit samples, please state the reasons why this is required on the front of this form. We will then consider your request. To be considered for free deposit, samples must be fully identified. Charges will apply if you require identification prior to deposit.

Alternatively, CABI may wish to place samples of scientific interest in its herbarium or culture collection e.g. new or uncommon species. **To maintain confidentiality in the case of samples from commercial clients,** no information about the company or its employees, or any information about a sample is disclosed that may enable the company to be identified as the source of a sample.

Please indicate if you are **willing to permit deposit of samples** by ticking the box ☐

The UK is a signatory to the Convention on Biological Diversity (CBD) and a party to the Nagoya Protocol on Access and Benefit Sharing (ABS). CABI operates in accordance with European legislation to implement these requirements. Those wishing to submit samples to CABI for identification or deposit from outside the UK, from countries that are signatory to the Convention on Biological Diversity, are required to sign a declaration on Page 1 of this form, stating that the requirements of the country of origin, relating to the CBD, have been followed. If Prior Informed Consent (PIC) is required to collect the material and send outside the country of origin, the relevant national authority, if PIC is not required by law permissions may still be required from the land owner and/or stakeholders or rights holders, according to national law of the originating country. Permission to send the material outside the country of origin should be obtained from the relevant national authority. Where utilization of genetic resources is involved, Mutually Agreed Terms (MAT) are required between the collector and relevant parties covering Access and Benefit Sharing (ABS) relating to the genetic resources. Under the requirements of the CBD, CABI is unable to accept samples for deposit in its collection without such confirmation.

Details of the CBD and the relevant national authority are available at
http://www.cbd.int/convention/text/default.shtml

(See article 15.5 relating to PIC).

For details of countries that are signatories to the CBD and relevant policies see:
http://www.cbd.int/information/parties.shtml

For a list of countries that are party to the Nagoya Protocol see:
http://www.cbd.int/abs/nagoya-protocol/signatories/default.shtml

MATERIAL TRANSFER AGREEMENT

This Material Transfer Agreement ("MTA") governs the transfer of the Original Material from CAB International for research, teaching and identification purposes only.

This MTA constitutes a contract between you (the "Customer") and CAB International, an international governmental not-for-profit organisation ("CABI"). CABI and the Customer may be individually referred to as a Party or jointly, as the Parties.

1. Definitions

"Commercial Purposes" means the use or exploitation of the Material, with the object of, or resulting in, financial gain, and includes but is not limited to the following activities: sale, use of the Material to provide a service, applying for, obtaining or transferring intellectual property rights or other tangible or intangible rights by sale or licence or in any other manner, commencement of product or service development, conducting market research, and seeking regulatory approval.

"Effective Date" means the date of the last signature of this MTA by a Party.

"Force Majeure" means an event or sequence of events beyond a party's reasonable control preventing or delaying it from performing its obligations under this MTA, including but not limited to war, civil commotion or riots, war, threat of or preparation for war, armed conflict, natural disaster, transportation difficulties, labour or trade dispute, strikes, industrial action interruption or failure of utility service.

"Material" means Original Material, Progeny and Unmodified Derivatives. The Material shall not include Modifications.

"Modifications" means substances created by the Recipient which contain or incorporate all or any part of the Original Material, Progeny or Unmodified Derivatives.

"Original Material" shall mean any biological material transferred from CABI to Customer identified in Schedule 1.

"Prior Informed Consent" has the meaning set out in the Nagoya Protocol to the Convention on Biological Diversity. It refers to the requirement that a country which has ratified the Nagoya Protocol and which is providing access to a genetic resource provide its consent on the basis of mutually agreed terms before access to the said genetic resources is granted.

"Progeny" means all unmodified descendant from the Original Material, such as virus from virus, cell from cell, or organism from organism.

"Unmodified Derivatives" means any substance created by the Customer which constitutes (a) a substantially unmodified copy of the Original Material or Progeny or any of their components or (b) a product expressed or produced by the Original Material or Progeny. Some examples include: material produced by growth of a microorganism, purified or fractionated subsets of the Original Material, and proteins expressed by DNA/RNA supplied by CABI.

"Third Party" shall mean any person other than the Customer and CABI.

2. Transfer of Material

2.1 Subject to Clause 2.4 below, CABI agrees to transfer the Original Material to the Customer. CABI shall label, package, and arrange for transport of the Original Material in accordance with applicable laws and regulations. The Customer shall be responsible for all duties, tax(es), tariffs and fees arising in connection with this

2.2 The Customer shall be responsible for all duties, tax(es), tariffs and fees arising in connection with this MTA and the transfer of the Original Material. The Customer shall, upon demand, pay to CABI an amount equal to any such tax(es), duties, tariffs and permit fees actually paid or required to be collected or paid by CABI.

2.3 If the Original Material is not viable on receipt by the Customer, and provided the Customer has handled and continues to handle the Original Material in accordance with instructions provided by CABI, CABI will provide (a) replacement of the Original Material or (b) a refund of the fee paid to CABI (if any).

2.4 Nothing in this MTA obliges CABI to supply the Original Material if (a) CABI is prohibited from doing so under any applicable legislation, including legislation relating to terrorism or threat of biological weapons or (b) CABI has reasonable grounds to suspect that the Customer intends to utilise the Material in any manner which contravenes the terms of this MTA or for any other prohibited or unethical purpose.

2.5 At CABI's request the Customer shall promptly and at its cost return or destroy (at CABI's option) all Material on termination or expiry of this MTA.

3. Conditions of Use

3.1 The Customer may use the Material solely for research, teaching and identification purposes and shall not use the Material for Commercial Purposes.

3.2 The Material is not intended for human in vivo use. The Customer shall not use the Material in human subjects, in clinical trials, or for diagnostic purposes involving human subjects.

3.3 The Customer shall not distribute, release, sell, lend or otherwise transfer the Material to a Third Party without CABI's prior written consent.

3.4 The Customer agrees to use the Material in compliance with all applicable laws, regulations, codes of practice and guidelines such as, for example, those relating to research involving the use of animals or recombinant DNA.

4. Compliance with the Convention on Biological Diversity

4.1 The Customer shall use the Material in a legitimate, safe, legal, ethical and sustainable way, and in respect of the principles laid down in the Convention on Biological Diversity ("CBD"). Nothing in this MTA shall be construed as changing or affecting the rights and obligations of the Parties under the CBD.

4.2 This MTA fully complies with Article 15 of the CBD, which recognises the sovereign rights of States over their natural resources.

4.3 The Original Material is provided under this MTA on the understanding that it was collected either before the CBD came into force or if collected after the CBD came into force with Prior Informed Consent, where procedures are in place and recognised authorities exist.

4.4 The Customer shall uphold the mutually agreed terms of the Prior Informed Consent and any other terms that CABI notifies to the Customer in Schedule 1.

4.5 To the extent any terms and conditions set out in Schedule 1 conflict with the terms of this MTA, the terms and conditions set out in Schedule 1 shall prevail.

5. Ownership and Intellectual Property

5.1 The transfer of the Original Material pursuant to this MTA does not affect the ownership of the Original Material or any intellectual property rights in or relating to it or its use.

5.2 In accordance with the CBD which recognizes that nations have sovereign rights over their own biological resources, the ownership of Material, including any Material incorporated in Modifications, or derivatives remains with the Provider Country, as defined in the CBD.

5.3 The Customer agrees to acknowledge CABI and any contributor indicated by CABI as the source of the Original Material in all publications. The Customer may not seek to patent the Material or protect it by any other intellectual property right.

5.4 Nothing in this MTA shall be interpreted as granting a licence to the Customer to use CABI's name or any of CABI's intellectual property rights, including its trade marks.

6. Liability

6.1 The Customer acknowledges that any Original Material designated as Hazard group 2 or 3 constitutes known human pathogens. The Material may be toxic, carcinogenic, mutagenic or pathogenic under certain conditions. The Customer assumes all risk and responsibility to ensure that the Material is used in a safe and appropriate manner. This includes the receipt, handling, (containment), storage and disposal.

6.2 CABI makes no representations and gives no warranties or conditions of any kind, either express or implied. There are no express or implied warranties or conditions of satisfactory quality or fitness for a particular purpose, or that the use of the Material will not infringe any intellectual property rights or other proprietary rights of any other person.

6.3 CABI shall not be liable for any failure of the Material to meet the objectives or requirements of the Customer. CABI shall have no liability under this MTA except as provided under clause 2.3. Under no circumstances will CABI be liable for any indirect, incidental or consequential loss or damage or for loss of profits or revenue (whether direct or indirect) in connection with or arising out of the use of the Material or this MTA, however caused (including as a result of negligence) including for any delays whether or not caused by Force Majeure, even if CABI has been advised of the possibility of such loss or damage.

6.4 [The Customer shall indemnify and hold harmless CABI, its officers, agents, employees and any person acting on their behalf from and against: (1) any Third Party claims or demands for loss, liability or damage, including, but not limited to, claims for personal injury or death, arising from any act or omission of the Customer's officers, employees, agents, subcontractors or any person acting on the Customer's behalf in the performance of this MTA; and (2) all claims, suits, and damages, brought by any Third Party, by reason of the non-observance or non-performance by the Customer, or its officers, employees, agents, subcontractors, or any person acting on the Customer's behalf, of any of the terms and conditions of this MTA or any applicable laws, regulations, codes of practice or guidelines. The Customer shall reimburse CABI its officers, employees, agents and any person acting on CABI's behalf for all legal fees, costs, and expenses incurred in connection with the defence of any such Third-Party claim.]

7. Governing Law

7.1 This MTA is governed by, and is to be construed in accordance with, English law.

7.2 Any proceedings relating to any dispute or claim arising out of or in connection with this MTA or its subject matter or formation (including non-contractual disputes or claims) instituted against CABI by the Customer shall be brought in the courts of England and any such proceeding against the Customer by CABI shall be brought in the courts of the jurisdiction where the Customer is domiciled. Each party agrees that the specified courts shall have exclusive jurisdiction over such disputes or claims, save that any counterclaim may be brought in any proceedings already commenced.

8. Term and Termination

8.1 This MTA will commence on the Effective Date and unless terminated earlier under clause 8.2 will terminate on the earliest of (a) completion of the Customer's use of the Material in accordance with Clause 2, and (b) [NUMBER] year(s) after the Effective Date.

8.2 CABI may terminate this MTA with immediate effect by giving notice to the other Party if:

8.2.1 The Customer is in breach of any provision of this Agreement and (if it is capable of remedy) the breach has not been remedied within thirty (30) days after the Customer receives a written notice specifying the breach and requiring its remedy, or

8.2.2 CABI has reasonable grounds to suspect that the Customer intends or is utilizing the Material for any illegal or unethical purpose.

9. Miscellaneous

9.1 This MTA constitutes the entire agreement between CABI and the Customer and supersedes and extinguishes all previous drafts, agreements, arrangements and understandings between them, whether written or oral, relating to its subject matter.

9.2 Nothing in this MTA creates, implies or evidences any partnership or joint venture between the Parties, or the relationship between them of principal and agent. Neither Party has any authority to make any representation or commitment, or to incur any liability, on behalf of the other.

9.3 No variation or amendment of this MTA (including the Schedule) will be effective unless it is made in writing and signed by each Party's representative.

9.4 Failure to exercise, or any delay in exercising, or a single or partial exercise of any right or remedy provided under this MTA or by law shall not constitute a waiver of that or any other right or remedy, nor shall it preclude or restrict any further exercise of that or any other right or remedy.

9.5 A person who is not a Party to this MTA shall not have any rights under or in connection with it.

9.6 Each of the provisions contained within this MTA shall be enforceable independently of each of the others and its validity shall not be affected if any of the others is invalid. If any of these provisions is void and would be valid if some part of the provision were deleted, the provision in question shall apply with such modification as may be necessary to make it valid.

10. Notices

Any notice to be given under this MTA must be in writing and sent to the following addresses: To CABI: The Chief Executive Officer Head Office CAB International Nosworthy Way Wallingford Oxon, OX10 8DE U

CABI BRC Audit Procedure - Guidance Notes

Documentation
Ensure that the Quality Manual is available and staff aware of content and amendments
Ensure that staff have up to date and assigned Laboratory Procedures Manuals and that staff can give evidence that they follow the protocols

Supply
Select representative order(s) from the order receipt book
Compare details in order book (and computer version) with customer order documents
Check customer supply agreement

Review all customer correspondence
Check that any conditions of supply requested by customer have been followed
e.g. specific properties, delivery conditions or supply form (freeze-dried, active)
Examine freeze-dried batch details (within shelf-life*) and stock levels
Check growth conditions were followed
Ensure incubator temperature was compared with calibrated thermometer
Check supply date – freeze-dried material supplied within 3 working days, active within 3 weeks

Accession
Select strain(s) from CABI Bioscience Reference Collection Accessions Book (select entries 6 months, 3 months and 1 month after entry)
Check BRC Accession Book details against BRC Accession Book
Check preservation records
Ensure post preservation checks have been completed before release into open collection
Check that seed stock and distribution stock are marked and that samples have been placed in the duplicate (back up) collection
Check freeze dry batch details
Equipment maintenance records
Ensure equipment is operating within specified parameters*
Ensure representative ampoules pass the vacuum spark test
Check liquid nitrogen storage run
Check equipment maintenance records
Ensure equipment operates within stated parameters*
Ensure correct cooling protocol has been used
Check sample temperature against second resistance thermometer
Check cryo-storage tank location against record
Check level of nitrogen and operation of automatic fill system

Database
Check selected data records entered on BRC database (a minimum of 10) against depositor details and data on the BRC Accession Forms, cross check field validation as defined in Joining CABRI Database and Catalogues Review

Training records
Check collection staff have appropriate training, experience and that records are up to date
Check that they have appropriate authorisation for the use of BRC equipment
Equipment and Laboratory maintenance
Check maintenance and calibration (where appropriate) records
Ensure maintenance programmes are in place
Assess laboratory safety and cleanliness
Enquiries and complaints records
Check that adequate responses have been made within 3 working days.

External support services
Check suppliers lists for traceability to standards, responsibility of Technical Service Manager.
* See relevant technical procedure in this manual

CABI BRC Audit Procedure Forms

Name of Assessor	BRC staff member	Date

Documentation

Quality Manual		
	Yes	No
Does each member of staff have a numbered version assigned? (Check against central record of issue)		
Does the amendment summary table match the changes made to the manual text?		
Comments:		

Corrective action	Date completed	Signature

Signature of Assessor	Signature of staff member

CABI BRC Audit Procedure Form

Name of Assessor	BRC staff member	Date

Documentation

Laboratory Procedures Manual (LPM)		
	Yes	No
Are all aspects defined in the Quality Manual adequately addressed?		
Are protocols clear and do they match current procedures?		
Does each member of staff have a numbered version assigned? (Check against central record of issue)		
Does the amendment summary table match the changes made to the manual text?		
Comments:		

Corrective action	Date completed	Signature

Signature of Assessor	Signature of staff member

CABI BRC Audit Procedure Form

Name of Assessor	BRC staff member	Date

Documentation

Laboratory Books (records)		
	Yes	No
Does each member of staff maintain an up to date Lab Book(s)?		
Are entries signed and appropriately dated?		
Comments:		

Corrective action	Date completed	Signature

Signature of Assessor	Signature of staff member

CABI BRC Audit Procedure Form

Name of Assessor	BRC staff member	Date

Documentation

Review and audit reports		
	Yes	No
Is there an appropriate schedule for reviews and audits?		
Is there an appropriate standard agenda for review meetings?		
Have issues raised in reviews and audits been adequately addressed in a timely manner?		
Comments:		

Corrective action	Date completed	Signature

Signature of Assessor	Signature of staff member

CABI BRC Audit Procedure Form

Name of Assessor	BRC staff member	Date

Accession		
Select a minimum of 3 strain(s) from CABI Bioscience Reference Collection Accessions Book (select entries 6 months, 3 months and 1 month after entry)	Yes	No
Do BRC Accession Book details agree with BRC Accession Book		
Do strains deposited meet the acquisition criteria*(Sections 1.2, 1.3 in LPM)?		
Are terms of deposit and further distribution in place?		
Has the strain's hazard status been ascertained?		
Has the strain had viability, purity and identity checks?		
Have appropriate preservation techniques been selected (see Section 2.6.1 in LPM)?		
Do preservation records correspond with preserved materials?		
Have post preservation checks been completed before release into open collection?		
Are seed and distribution stocks clearly marked?		
Have samples been placed in the duplicate (back up) collection		
Has cryo-storage tank location been correctly recorded?		
Comments:		

Corrective action	Date completed	Signature

Signature of Assessor	Signature of staff member

CABI BRC Audit Procedure Form

Name of Assessor	BRC staff member	Date

Supply		
Select a minimum of 2 representative order(s) from the order receipt book	Yes	No
Do details in order book (and computer version) agree with customer order documents?		
Is there a customer supply agreement in place?		
Have customer requirements outlined in correspondence been followed?		
Is the freeze-dried batch within the expected shelf-life* and are stock levels adequate?		
Were specified growth conditions (BRC database) followed?		
Was the BRC incubator temperature compared with a calibrated thermometer?		
Were strains supplied within specified periods? – freeze-dried material supplied within 3 working days, active within 3 weeks		
Were restrictions (if any) on supply observed?		
Is the information provided on supply comprehensive and up to date?		
Comments:		

Corrective action	Date completed	Signature

Signature of Assessor	Signature of staff member

CABI BRC Audit Procedure Form

Name of Assessor	BRC staff member	Date

Database		
Select a minimum of 10 data records	Yes	No
Does data entered on BRC database correspond to depositor supplied details on accession forms?		
Are procedures for validation of data in place and adequate?		
Are CABRI definitions of input, authenticity and format followed?		
Comments:		

Corrective action	Date completed	Signature

Signature of Assessor	Signature of staff member

CABI BRC Audit Procedure Form

Name of Assessor	BRC staff member	Date

Staff and their training records		
	Yes	No
Are assigned staff aware of their duties in relation to quality management?		
Are deputies assigned and aware of their responsibilities?		
Do collection staff have appropriate training and experience?		
Do staff have appropriate authorisation for the use of BRC equipment?		
Are records up to date?		
Comments:		

Corrective action	Date completed	Signature

Signature of Assessor	Signature of staff member

CABI BRC Audit Procedure Form

Name of Assessor	BRC staff member	Date

Equipment and Laboratory maintenance		
	Yes	No
Is BRC equipment maintained to specification* (Section 1.10 in LPM)?		
Are appropriate maintenance programmes in place?		
Does laboratory safety and cleanliness meet appropriate standards?		
Freeze driers (to be checked on the batch records)		
Is equipment is operating within specified parameters*?		
Do representative ampoules pass the vacuum spark test?		
Liquid nitrogen storage (to be checked on the cooler printout records)		
Is equipment is operating within specified parameters*?		
Has controlled rate cooler temperature been checked against a resistance thermometer standard?		
Is the automatic fill system operating satisfactorily?		
Is the oxygen monitoring equipment maintained adequately?		
Culture store		
Is the room temperature within the set parameters (15-18°C)?		
Comments:		

Corrective action	Date completed	Signature

Signature of Assessor	Signature of staff member

CABI BRC Audit Procedure Form

Name of Assessor	BRC staff member	Date

Enquiries and complaints records		
	Yes	No
Have responses been made within 3 working days?		
Are responses appropriate?		
Comments:		

Corrective action	Date completed	Signature

Signature of Assessor	Signature of staff member

CABI BRC Audit Procedure Form

Name of Assessor	BRC staff member	Date

External support services		
	Yes	No
Do suppliers have mechanisms for traceability to standards for the products supplied?		
Comments:		

Corrective action	Date completed	Signature

Signature of Assessor	Signature of staff member

CABI BRC Audit Procedure Form

Name of Assessor	BRC staff member	Date

Compliance with legislation

Procedures		
Does the collection comply, as far as reasonably practicable, with legislation and conventions governing the handling, shipping and access to organisms supplied?	Yes	No
Health and Safety e.g. Biological agents' directive		
Packaging and shipping: e.g. IATA Dangerous Goods Regulations		
Convention on Biological Agents		
Control of Dangerous Pathogens		
Quarantine Regulations		
Comments:		

Corrective action	Date completed	Signature

Signature of Assessor	Signature of staff member

Appendix F

Cryopreservation Protocols

DOI: 10.1079/9781800622135.appf

Cryopreservation Protocols for Microalgae and Cyanobacteria

Species	Growth conditions or pretreatment before freezing	Cryoprotective Additives	Cooling protocol/ rate	Thaw(°C min^{-1})	Recovery	Storage temp. (°C)	Reference
Actinocyclus subtilis	–	Methanol (5% v/v)	Two-step 15 min (-40)	Rapid warming	2%	–196	McLellan, 1989
Anabaena cylindrica	–	Me_2SO (5% w/v)	Two-step 20min (-30)	Rapid warming	100%	–196	Day, 1998
Ankistrodesmus angustus	–	Me_2SO (5% w/v)	Two-step 15min (-30)	Rapid warming	93%	–196	Day, 1998
Attheya decora	–	Me_2SO (10% v/v)	Two-step 15 min (-40)	Rapid warming	54%	–196	McLellan, 1989
Brachiomonas submarina	Encapsulation in calcium alginate with 0.5M sucrose, desiccated at 30°C	–	Rapid cooling	Rapid warming	33%	–196	Hirata *et al.*, 1996
Calothrix brevissima	Encapsulation in calcium alginate with 0.5M sucrose, desiccated at 30°C	–	Rapid cooling	Rapid warming	12%	–196	Hirata *et al.*, 1996
Chaetoceros gracilis	–	Me_2SO (15% v/v)	Two-step min 0.5-4 (-35)	Rapid warming	9-29%	–196	Canavate and Lubian, 1995a, b
Chaetoceros neogracile	–	Me_2SO (10% v/v)	Two-step 15 min (-40)	Rapid warming	56%	–196	McLellan, 1989
Chlamydomonas	Encapsulation in calcium alginate with 0.5M sucrose, desiccated at 30°C	–	Rapid cooling	Rapid warming	34%	–196	Hirata *et al.*, 1996
C. reinhardtii	18h in Me_2SO (1% v/v)	Me_2SO (5% v/v)	Two-step Step 1 (-20)	Rapid warming	0.1-10%	–196	Gresshoff, 1977
C. reinhardtii	–	Me_2SO (10% v/v)	Two-step Step 1 (-30)	Rapid warming	+	<–180	Hwang and Hudock, 1971
C. reinhardtii	–	Me_2SO (5% v/v)	Two-step Step 1 (-55)	Rapid warming	<10%	–150	McGarth and Daggett, 1977
Chlorella	–	Me_2SO (5% v/v)	200 (-196)	Rapid warming	<30%	–196	Morris, 1976a
C. emersonii	–	Me_2SO (5% v/v)	Plunge into liquid nitrogen	Rapid warming	70%	–196	Day *et al.*, 1997

C. emersonii	–	Methanol (1.5M)	One step 10 (-196)	Rapid warming	87%	–196	Morris *et al.*, 1980
C. protothecoides	–	Me_2SO (5% v/v)	Plunge into liquid nitrogen	Rapid warming	85%	–196	Day *et al.*, 1997
C. pyrenoidosa	Encapsulation in calcium alginate with 0.5M sucrose, desiccated at 30°C	–	Rapid cooling	Rapid warming	30%	–196	Hirata *et al.*, 1996
C. vulgaris	Cultivated in 1 M Tris buffered f/2 medium, stirred at 300 rpm and aerated with 1% CO_2 mixed with air; incubated at room temperature under continuous light	5% (v/v) methanol in the f/2 medium	Cooled –1 °C min^{-1} to –40 °C, held at –40 °C for 15 min and plunged into liquid nitrogen	Rapidly thawed in a 37 °C water bath	Maintained near 100% viability for 4 months	–196	Kapoore *et al.*, 2019
Chlorocloster engadinensis	–	Me_2SO (5% w/v)	Two-step 20min (-30)	Rapid warming	70%	–196	Day, 1998
Chlorococcales (250 strains)	–	Glycerol (10% v/v) Me_2SO (5% v/v)	Two-step 15min (-30)	Rapid warming	>50%	–196	Morris, 1977, 1978
Chroomonas placoidea	–	Me_2SO (5% w/v)	Two-step 15min (–30)	Rapid warming	<10%	–196	Day, 1998
Cricosphaera roscoffensis	Encapsulation in calcium alginate with 0.5M sucrose, desiccated at 30°C	–	Rapid cooling	Rapid warming	8%	–196	Hirata *et al.*, 1996
Cyanobacteria 150 strains	Cultured in cryotube on agar	Me_2SO (8% w/v)	Two-step ~1 (-70)	Rapid warming	+	–196	Bodas *et al.*, 1995
Cyclotella cryptica	–	Me_2SO (10% v/v)	Two-step 15 (-40)	Rapid warming	52%	–196	McLellan, 1989
C. cryptica	–	Me_2SO (5% w/v)	Two-step 15 (-40)	Rapid warming	57%	–196	Day, 1998
Cylindrocystis brebissonii	–	Me_2SO (5% v/v)	Two-step 1 (-60)	Rapid warming	55%	–196	Morris *et al.*, 1986

Continued

Species	Growth conditions or pretreatment before freezing	Cryoprotective Additives	Cooling protocol/ rate	Thaw(°C min^{-1})	Recovery	Storage temp. (°C)	Reference
Dunaliella salina	3 days at 4°C in dark	Glycerol added at 20μL min^{-1} until concentration of 3.5 M. Then 1 hour at 4° C	Two-step 1 (-40)	Rapid warming	78%	–196	Mortain-Bertrand *et al.*, 1996
D. tertiolecta	Encapsulation in calcium alginate with 0.5M sucrose, desiccated at 30°C	–	Rapid cooling	Rapid warming	30%	–196	Hirata *et al.*, 1996
Eisenia bicyclis		Proline (10% w/v) + ethylene glycol (10% v/v)	Two-step 1 (-40)	Rapid warming	52%	–196	Kono, *et al.*, 1998
Enteromorpha intestinalis	–	Me_2SO (15% v/v) + 5% proline	Two-step 1 (-40)	Rapid warming	71%	–196	Kono *et al.*, 1997
Enteromorpha intestinalis	–	Me_2SO (5% v/v)	Two-step 1 (-35)	Rapid warming	100%	–196	Day *et al.*, 1998
Enteromorpha intestinalis	–	PVS 2Vitrification solution	Plunge	Rapid warming	100%	–196	Day *et al.*, 1998
Euglena gracilis	–	Methanol 10% (v/v)	Two-step 0.3 (-60)	Rapid warming	34%	–196	Day, 1998
Euglena gracilis	Encapsulation in calcium alginate 3h air drying	Methanol 10% (v/v)	1 step plunge	1min room temperature then rapid warming	37%	–196	Day *et al.*, 2000
Euglena gracilis	Pre-culture 48h in 0.75 M sucrose Encapsulation in calcium alginate 2h air drying	Methanol 10% (v/v)	Two-step 0.5 (-60)	1min room temperature then rapid warming	24%	–196	Day *et al.*, 2000

Organism	Pre-treatment	Cryoprotectant	Cooling	Thawing	Viability	Storage temp.	Reference
Euglena gracilis	Pre-culture 24h in 0.5 M sucrose and then Pre-culture 48h in 0.75 M sucrose Pre-culture 24h in 0.75 M sucrose Encapsulation in calcium alginate	Methanol 10% (v/v)	Two-step 0.5 (-60)	1min room temperature then rapid warming	48%	–196	Day *et al.*, 2000
Euglena gracilis (26 strains)	–	Methanol (10% v/v)	Two-step 1 (-30)	Rapid warming	~30%	–196	Morris and Canning, 1978
Eukaryotic algae (16 strains)	Cultured in cryotube on agar	Polyethylene glycol 10% (w/v) in agar Me_2SO (5% w/v) + methanol (5% v/v)	Two-step ~1 (-70)	Rapid warming	+	–196	Bodas *et al.*, 1995
Eukaryotic algae (365 strains)	Low light, 1% agar media, 20°C	Me_2SO (5% v/v)	Two-step 1 (-40), then -40 for 15 min	Rapid warming	+	–196	Beaty and Parker, 1992
Fragilaria pinnata	–	Me_2SO (10% v/v)	Two-step 15 (-40)	Rapid warming	57%	–196	McLellan, 1989
Geitlerinema amphibium	Stationary phase culture	Me_2SO (10% v/v)	Rapid cooling	Rapid warming	+	<80	Romo and Bécares, 1992
Heterococcus caespitosus	–	Me_2SO (5% w/v)	Two-step 20min at –30	Rapid warming	65%	–196	Day, 1998
H. protonematoides	–	Me_2SO (5% w/v)	Two-step 30min at –30	Rapid warming	34%	–196	Day *et al.*, 1997
Isochrysis sp.	–	Me_2SO (5% w/v)	Two-step 15min (-30)	Rapid warming	<10%	–196	Day, 1998
Marine microalgae (12 strains)	–	Me_2SO (5% v/v)	Two-step 15min (-30)	Rapid warming	20-87%	–196	Ben Amotz and Gilboa, 1980
Microcystis aeruginosa	–	Me_2SO (5% v/v)	DI	Rapid warming	60%	–196	Box, 1988
Microcystis aeruginosa	–	Me_2SO (5% v/v)	DI	Rapid warming	+	–196	Box, 1988
Microcystis aeruginosa f. *aeruginosa*	Stationary phase culture	Me_2SO (3% v/v), 15min at room temp.	Two-step 1 (-30) then –30 for 15 min	Rapid warming	>50%	–196	Watanabe and Sawaguchi, 1995

Continued

Species	Growth conditions or pretreatment before freezing	Cryoprotective Additives	Cooling protocol/ rate	Thaw(°C min⁻¹)	Recovery	Storage temp. (°C)	Reference
Microcystis auruginosa	–	Me_2SO (5% w/v)	Two-step 20 min (-30)	Rapid warming	100%	–196	Day, 1998
Monodopsis subtteranea	–	Me_2SO (5% w/v)	Two-step 15min (-30)	Rapid warming	100%	–196	Day, 1998
Nannochloropsis	Encapsulation in calcium alginate with 0.5M sucrose, desiccated at 30°C	–	Rapid cooling	Rapid warming	42%	–196	Hirata *et al.*, 1996
Nitzschia ovalis (axenic)	–	Me_2SO (10% v/v)	Two-step 15 (-40)	Rapid warming	82%	–196	McLellan, 1989
Nitzschia ovalis (non-axenic)	–	Me_2SO (10% v/v)	Two-step 15 (-40)	Rapid warming	59%	–196	McLellan, 1989
Nostoc commune	Encapsulation in calcium alginate with 0.5M sucrose, desiccated at 30°C	–	Rapid cooling	Rapid warming	76%	–196	Hirata *et al.*, 1996
N. muscorum	–	None	Rapid cooling	Rapid warming	Good	–196	Holm-Hansen, 1963
N. muscorum	–	Glycerol (10% v/v) Me_2SO (5% v/v)	Rapid cooling	Rapid warming	60-70%	–196	Tsuru, 1973
N. spongiaeforme	Encapsulation in calcium alginate with 0.5M sucrose, desiccated at 30°C	–	Rapid cooling	Rapid warming	21%	–196	Hirata *et al.*, 1996
Pediastrum duplex	–	Me_2SO (10% v/v)	Two-step 1 (-30)	Rapid warming	96%		Watanabe *et al.*, 1992 Day *et al.*, 1997
Phaeodactylum tricornutum	–	Me_2SO (10% v/v)	Two-step 15 (-40)	Rapid warming	68%	–196	McLellan, 1989
Phaeodactylum tricornutum	–	Me_2SO (5% w/v)	Two-step 15 (-40)	Rapid warming	65%	–196	Day, 1998
Phormidium foveolarum	Encapsulation in calcium alginate with 0.5M sucrose, desiccated at 30°C	–	Rapid cooling	Rapid warming	44%	–196	Hirata *et al.*, 1996

Porphyra minata	–	Me_2SO (5% w/v)	Two-step 15min at –30	Rapid warming	+	–196	Day, 1998
Prorocentrum micans	–	Me_2SO (5% w/v)	Two-step 15min at –30	Rapid warming	<10%	–196	Day, 1998
Prototheca (5 species)	Precultured for 14 days at 4°C	-	Two-step 0.3 (-60)	Rapid warming	34-89%	–196	Morris, 1976b
Pseudanabaena galeata	Stationary phase culture	Me_2SO (10% v/v)	Rapid cooling	Rapid warming	+	<80	Romo and Bécares, 1992
Scenedesmus acutus	–	Me_2SO (10% v/v)	Two-step 1 (-30)	Rapid warming	75%	–196	Watanabe *et al.*, 1992; Day *et al.*, 1997
S. basiliensis	–	Me_2SO (5% v/v)	Two-step 15min at -30	Rapid warming	97%	–196	Morris, 1978 Day *et al.*, 1997
Selenastrum capricornutum	–	Me_2SO (5% w/v)	Two-step 15min at –30	Rapid warming	90%	–196	Day, 1998
Tetraselmis (4 strains)	Grown at room temp.	Glycerol (10% v/v)	Two-step1 (-30)	Rapid warming	>50%	–196	Day and Fenwick, 1993
T. suecica	–	Glycerol 10% (v/v)	Two-step 1 (-30)	Rapid warming	70%	–196	Day, 1998
T. suecica	Room temp.	Glycerol (5% v/v)	Two-step 5 (-30)	Rapid warming	>65%	–196	Fenwick and Day, 1992

DI, direct immersion in liquid coolants; Two/ multi-step cooling protocols, intermediate temperatures given in parenthesis. Unless otherwise stated final step is a plunge into liquid nitrogen. See also: Day and Deville (1995); Taylor and Fletcher (1999); Day (2007).

Cryopreservation Protocols for Bacteria

Species	Growth conditions or pretreatment before freezing	Cryoprotective Additives	Cooling (°C min^{-1})	Thaw (°C min^{-1})	Recovery	Storage period (years) and temp. (°C)	Reference
Actinomyces	–	Blood serum	Rapid	Not specified	100%	Immediate thaw	Bradshaw *et al.*, 1989
Aegyptianella pullmorum	Blood serum	Glycerol (10% v/v)	30min (4°C) –0.4-5°C min^{-1}	Rapid	Infectious for vector	At least 7 years	Gothe and Hartman, 1979
Alcalgenes eutrophus	Not specified	Glycerol (10%v/v) Polyvinyl ethanol (10%v/v)	Rapid	Rapid warming	Good viability; plasmids retained	Immediate thaw	Beyersdorf-Radeck *et al.*, 1993
Azotobacter vinelandii	–	Glycerol (10% v/v)	Rapid	Rapid	60-99%	Immediate thaw	Thompson, 1987
Bacteria not amenable to freeze-drying	Mid to late log phase to a concentration of (2-6) x10^6 cells/ml	Glycerol (5% v/v) Me_2SO (5% w/v)	–1 to –3°C min^{-1} to –30°C, -15 to –30°C min^{-1} to –100°C or lower, store at –196°C	Water bath 30-37°C	95%	10 years	
Bacterioides intermedius	Not specified	Blood serum	Rapid	Rapid warming	100%	Immediate thaw	Bradshaw *et al.*, 1989
Borrelia burgdorferi	Infected dog, bull, ram semen, citrate – saccharose – egg yolk in BSK medium	Growth medium	Rapid	Rapid warming	79-91%	12 weeks	Kumi-Diaka and Harris, 1995
Clostridium chauvoei	Beef liver medium	Beef liver medium + glucose (1%w/v) + peptone (1%w/v)	Rapid	Rapid warming	No loss in infectivity	Immediate thaw	McCrindle, 1979

Cytophaga	–	No additive	Rapid	Not specified	Not provided		Sanfilippo and Lewin, 1970
Flexibacter	–	No additive	Rapid	Not specified	Not provided		Sanfilippo and Lewin, 1970
Lactobacillus casei	Collected on Millipore membrane	No additive	Ultra-rapid on membrane	Rapid warming	100%	Immediate thaw	Albrecht *et al.*, 1973
Lactobacillus delbrueckii ssp. *bulgaricus*	Cells were exposed to osmotic stress in solutions of cryoprotective additives (0.5 h incubation time at 25°C)	Dimethyl sulfoxide, DMSO 4.5% w/v	5°C min^{-1} from 4°C to -5°C for 15 minutes, then cooled at 2°C min^{-1} from -5°C to -80°C then transferred to a -80°C freezer	Rapidly warmed during 120 s in a 42°C water bath	0.5 log CFU. mL^{-1} loss in viability	Immediate thaw	Fonseca *et al.*, 2016
Leptospira	Grown and preserved in Ellinghausen and McCulloughs medium	Ellinghausen and McCulloughs medium	Placed in the vapour phase of nitrogen tank	Rapid warming	10-20%	2 years	
Medical bacteria	Growth on non-selective solid media, e.g. nutrient or blood agar; $10^8 – 10^{10}$ cells/mL in glass beads (Microbank Pro-Lab Diagnostics)	2.5g nutrient broth powder N°2, 15mL glycerol, 85mL distilled water	Placed in a –70°C freezer or direct into liquid nitrogen vapour	Streak a bead directly onto solid medium	Not indicated	Not provided	Perry, 1998
Methanococcoides, Methanosarcina, Methanophilus, Methanobolus	Growth in a H_2 + CO_2 gas mixture	Glycerol (10% v/v) Me_2SO (5% w/v)	-50°C min^{-1} gas phase of liquid nitrogen tank	Water bath 30-37°C	Not indicated	Not provided	Hippe, 1991

Continued

Species	Growth conditions or pretreatment before freezing	Cryoprotective Additives	Cooling (°C min^{-1})	Thaw (°C min^{-1})	Recovery	Storage period (years) and temp. (°C)	Reference
Methanotrophic bacteria	Growth in a H_2 + CO_2 gas mixture; in glass capillary tubes	Glycerol (10% v/v) Me_2SO (5% w/v)	Ice bath and directly into storage	Warm water: +1000°C min^{-1}	Not indicated	15 years	Hippe, 1991
Microsilla	-	No additive	Rapid	Not specified	Not provided	Immediate thaw	Sanfilippo and Lewin, 1970
Mycoplasma	Hayflick medium + horse serum and yeast extract	Growth medium	Rapid	Rapid warming	Decreased by 1 log value	Not provided	Ritter and Ulrich, 1986
Obligate methanotrophic bacteria	Normal growing conditions for these bacteria	PVP dialysed molecular weight 40,000	plunge	Rapid warming	Not quantified	Not provided	Green and
Obligate methanotrophic bacteria	Normal growing conditions for these bacteria	Methanol	Plunge	Rapid warming	Not quantified	Not provided	Green and
Phototrophic bacteria	Anaerobic conditions, under a nitrogen blanket	Me_2SO (5% w/v)	Plunge	Rapid warming	Not quantified	Not provided	Malik, 1984
Streptococcus	Blood serum	Blood serum	Rapid	Rapid warming	100%	Immediate thaw	Bradshaw *et al.*, 1989
Streptomyces griseus	Nutrient broth	Growth medium	Rapid	Rapid warming	100%	5 years	Fortney and Thoma, 1977

Cryopreservation Protocols for Fungi

Species	Growth conditions before freezing	Cryoprotective Additives	Cooling (°C min^{-1})	Thaw (°C min^{-1})	Recovery	Storage period (years) and temp. (°C)	Reference
Achyla spp. (9 strains)	Mature cultures washed from agar slopes or mature liquid cultures	Glycerol (10% v/v)	1	c. 400	+	1-3 at –196	Hwang, 1966
Achyla spp. (7 strains)	14d culture mycelium washed from agar surface	Glycerol (10% v/v)	1	Rapid warming	+	1-8 at –196	Smith, 1982
Agaricus spp.	Agar blocks cut from growing colony	Glycerol (10% v/v) Me_2SO (5% v/v)	Slow cooling	Rapid warming	+	9 at –196	Elliot and Challen, 1981 Challen and Elliot, 1986
Allomyces spp. (5 strains)	14d culture mycelium washed from agar surface	Glycerol (10% v/v)	1	Rapid warming	+	3-8 at –196	Smith, 1982
Albugo occidentalis	On infected spinach leaves	None	Slow cooling	Rapid warming	+	0.5 at –23	O'Brien and Webb, 1958
Aphanomyces spp. (6 strains)	14d culture mycelium washed from agar surface	Glycerol (10% v/v)	1	Rapid warming	+	4-7 at –196	Smith, 1982
Aphanomyces spp. (2 strains)	Mycelium washed from agar or liquid culture	Glycerol (10% v/v)	1	c. 400	+	3 at –196	Hwang, 1966
Ascomycota (62 strains)	Spores or mycelium washed from agar surface	Glycerol (10% v/v) Me_2SO (5% v/v)	1	Rapid warming	+	2-8 at –196	Butterfield *et al.*, 1974

Continued

Species	Growth conditions before freezing	Cryoprotective Additives	Cooling (°C min^{-1})	Thaw (°C min^{-1})	Recovery	Storage period (years) and temp. (°C)	Reference
Ascomycota (44 strains)	Spores or mycelium washed from agar surface	Glycerol (10% v/v) Me_2SO (5% v/v)	1	Rapid warming	+	0.5-3 at –196	Hwang, 1966
Basidiobolus (239 strains)	Spores or mycelium washed from agar surface	Glycerol (10% v/v) Me_2SO (5% v/v)	1	Rapid warming	+	1-8 at –196	Butterfield *et al.*, 1974
Basidomycota (5 strains)	Spores or mycelium washed from agar surface	Glycerol (10% v/v) Me_2SO (5% v/v)	1	Rapid warming	+	1-3 at –196	Hwang, 1966
Bremia lactucae	Spores washed from infected plant tissue	Skimmed milk (8.5% w/v) + Glycerol (10% v/v)	1	Rapid warming	80%*	5-9 at –196	Dahmen *et al.*, 1983
Bremia lactucae	Spores washed from infected plant tissue	Me_2SO (15% v/v)	1	Rapid warming	80%*	5-9 at –196	Dahmen *et al.*, 1983
Brettanomyces anomalus	–	Me_2SO (10% v/v) + 6% Ficoll400 + 0.24% yeast extract + 0.5% glucose	1	Rapid warming	79%	–196	Smentek and Windisch, 1982
Bullera alba	–	Me_2SO (10% v/v) + 6% Ficoll400 + 0.24% yeast extract + 0.5% glucose	1	Rapid warming	24%	–196	Smentek and Windisch, 1982
Candida albicans	–	Me_2SO (10% v/v) + 6% Ficoll400 + 0.24% yeast extract + 0.5% glucose	1	Rapid warming	97%	–196	Smentek and Windisch, 1982
Coprinus cinereus	Spores or mycelium washed from agar surface	Glycerol (10% v/v) Me_2SO (5% v/v)	1	Rapid warming	+	2 at –196	Butterfield *et al.*, 1974

Entomophthorales (19 strains)	Spores or mycelium washed from agar surface	Glycerol (10% v/v) Me_2SO (5% v/v)	1	Rapid warming	+	1-5 at –196	Hwang, 1968
Entophlyctis cinfervaeglomerata	14d culture mycelium washed from agar surface	Glycerol (10% v/v)	1	Rapid warming	+	4-8 at –196	Smith, 1982
Ganoderma oregonense	Mycelium	Glycerol, 10% v/v)	1	Rapid warming	+	–196	Butterfield *et al.*, 1974
Ganoderma spp (36 strains)	Cultures grown on millet grains	Glycerol (5-10% v/v)	Slow cooling	Rapid warming	+	1 in nitrogen vapour	Wang *et al.*, 1990
Helminthosporium (12 strains)	Spores or mycelium washed from agar surface	None	DI	Rapid warming	+	2 at 189	Joshi *et al.*, 1974
Hyphochytrium catenoides	14d culture mycelium washed from agar surface	Glycerol (10% v/v) Me_2SO (5% v/v)	1	Rapid warming	+	4-10 at –196	Smith, 1982
Mitotic fungi** (33 strains)	Spores or mycelium washed from agar surface	Glycerol (10% v/v) Me_2SO (5% v/v)	1	c. 400	+	1-3 at –196	Hwang, 1966
Mitotic fungi** (239 strains)	Spores or mycelium washed from agar surface	Glycerol (10% v/v) Me_2SO (5% v/v)	1	Rapid warming	+	2-8 at –196	Butterfield *et al.*, 1974
Neocallimastix patricarum	10^5/ml zoospores	Me_2SO (0.68 M)	Placed at -80	Rapid warming	40%	1 at –80	Yarlett *et al.*, 1986
Peronospora effusa	On infected spinach leaves	None	Slow cooling	Rapid warming	+	0.5 at –23	O'Brien and Webb, 1958
Peronospora tabacina	Harvested and filter concentrated conidia	Me_2SO (15% v/v)	Two-step	Rapid warming	61%	1 at –196	Bromfield and Schmitt, 1967

Continued

Species	Growth conditions before freezing	Cryoprotective Additives	Cooling (°C min^{-1})	Thaw (°C min^{-1})	Recovery	Storage period (years) and temp. (°C)	Reference
Phlyctochytrium acuminatum	14d culture mycelium washed from agar surface	Glycerol (10% v/v)	1	Rapid warming	+	4-10 at –196	Smith, 1982
Phytophthora boehmeriae	Agar discs	Glycerol (10% v/v) Me_2SO (5% v/v)	Two-step 24 min –20	Rapid warming	75-100%	–	Tooley, 1988
Phytophthora palmivora	7-14d culture on agar plugs	Glycerol (10% w/v)	1	Rapid warming	100%	0.5 at –196	Smith, 1982
Phytophthora phaseoli	Infected lima bean epicotyl	None	DI	Air	+	1 at –196	Antonio and Blount, 1973
Phytophthora spp. (3 strains)	14-21d spores or mycelium washed from agar surface	Skimmed milk (8.5% w/v) and glycerol (10% v/v)	1	Rapid warming	100%*	5.9 at –196	Dahmen *et al.*, 1983
Phytophthora spp. (104 strains)	14d culture mycelium washed from agar surface	Glycerol (10% v/v)	1	Rapid warming	+	2-11 at –196	Smith, 1982
Plasmopara viticola	Spores harvested from infected leaves	Agar powder	Two-step	Rapid warming	c. 20%	1 at –196	Tetsuka and Katsuya, 1983
Plasmopara viticola	Spores harvested from infected leaves	Polyvinyl alcohol	Two-step	Rapid warming	c. 20%	1 at –196	Tetsuka and Katsuya, 1983
Plasmopara viticola	Spores harvested from infected plant tissue	Skimmed milk (8.5% w/v) and glycerol (10% v/v)	1	Rapid warming	51%	5-9 at –196	Dahmen *et al.*, 1983
Plasmopara viticola	Spores harvested from infected plant tissue	Me_2SO (15% v/v)	1	Rapid warming	50% and 96% infection	5-9 at –196	Dahmen *et al.*, 1983

Pseudoperonospora cubensis	Spores washed from infected plant tissue	Skimmed milk (8.5% w/v) and glycerol (10% v/v)	1	Rapid warming	50% and 100% infection	5-9 at –196	Dahmen *et al.*, 1983
Pseudoperonospora humuli	Spores harvested from infected leaves	Agar powder	Two-step	Rapid warming	c. 22% (control 37.4%)	1 at –196	Tetsuka and Katsuya, 1983
Pseudoperonospora humuli	Spores harvested from infected leaves	Polyvinyl alcohol	Two-step	Rapid warming	c. 29% (control 37.4%)	1 at –196	Tetsuka and Katsuya, 1983
Puccinia graminis	Uredospores	None	Rapid warming	Rapid warming	62-74%	Immediate	Kilpatrick *et al.*, 1971
Puccinia (12 strains)	Uredospores from infected plants	None	DI	Rapid warming	75-90%	2 at –189	Joshi *et al.*, 1974
Pythium sylvaticum (2 strains)	7-14d culture on agar plugs	Glucose (8% w/v) and Me_2SO (10% v/v)	1	Rapid warming	100%	0.5 at –196	Smith, 1983b
P. ultimum	14-21d spores or mycelium washed from agar surface	Skimmed milk (8.5% w/v) + glycerol (10% v/v)	1	Rapid warming	100%**	5-9 at –196	Dahmen *et al.*, 1983
Pythium spp. (10 strains)	Culture washed from agar or mature liquid culture	Glycerol (10% v/v)	1	c. 400	+	3 at –196	Hwang, 1966
Pythium spp. (28 strains)	14d culture; mycelium washed from agar surface	Glycerol (10% v/v)	1	Rapid warming	+	2-9 at –196	Smith, 1982
Rhizophydium bisporum	14d culture; mycelium washed from agar surface	Glycerol (10% v/v)	1	Rapid warming	+	4 at –196	Smith, 1982

Continued

Species	Growth conditions before freezing	Cryoprotective Additives	Cooling (°C min^{-1})	Thaw (°C min^{-1})	Recovery	Storage period (years) and temp. (°C)	Reference
Saprolegnia spp. (4 strains)	14d culture; mycelium washed from agar surface	Glycerol (10% v/v)	1	Rapid warming	+	1-6 at −196	Smith, 1982
Sclerospora philippinensis	Conidia harvested from infected host	Me_2SO (10% v/v)	1	Rapid warming	76%	2 at −196	Long *et al.*, 1978
S. sacchari	Conidia harvested from infected host	Me_2SO (10% v/v)	1	Rapid warming	10%	2 at −196	Long *et al.*, 1978
S. sorghi	Conidia harvested from infected host	Me_2SO (10% v/v)	1	Rapid warming	14-20%	2 at −196	Long *et al.*, 1978
S. sorghi	Conidia washed from leaves of *Sorghum bicolor*	Me_2SO (15% v/v)	1	Rapid warming	100%	7d at −196	Gale *et al.*, 1975
S. sorghi (9 strains)	Infected tissue taken from host	Glycerol (10% v/v)	1	Rapid warming	+	1 at −196	Smith, 1982
S. sorghi (9 strains)	Infected tissue taken from host	None	1	Rapid warming	+	1 at −196	Smith, 1982
Sterile mycelia (20 strains)	Spores or mycelium washed from agar surface	Glycerol (10% v/v) Me_2SO (5% v/v)	1	Rapid warming	+	0.5-4 at −196	Hwang, 1966
Thraustotheca clavata	14d culture; mycelium washed from agar surface	Glycerol (10% v/v)	1	Rapid warming	+	1 at −196	Smith, 1982
Ustilago nuda tritici	Chlamydospores from wheat	None	DI	Rapid warming	+	2 at −189	Joshi *et al.*, 1974

* , Germination and infection;
**, fungi not linked to a perfect state; DI, direct immersion in liquid coolants; Two-step, cooling at -22°C for 2 h and stored at -196°C.
Also see: Ryan and Smith (2007); Homolka (2013); Ryan *et al.* (2014)

Cryopreservation Protocols for Nematodes

Species	Growth conditions before freezing	Cryoprotective Additives	Cooling (°C min^{-1})	Thaw (°C min^{-1})	Recovery	Storage period (years) and temp. (°C)	Reference
Dictyocaulus viviparus	Exsheathment of third stage larvae with sodium hypochlorite	None	10-70	6800	30%	Immediate thaw	James and Peacock, 1986
Dictyocaulus viviparus	Exsheathment of third stage larvae with sodium hypochlorite	None	1°C min^{-1} to -10°C plunged into liquid nitrogen	6800	50-75%	Immediate thaw	James and Peacock, 1986
Heterorhabditis bacteriophora	Preincubation in 15% glycerol	10min in cold 70% methanol	Rapid	Rapid	+	24h	Popiel and Vasquez, 1991
Heterorhabditis megidis	Preincubation for 6-8 days in Me_2SO	10min in cold 70% methanol	Plunge into liquid nitrogen	Rapid	81%	24h	Nugent *et al.*, 1994
Heterorhabditis megidis	Preincubation for 4 days in 11% glycerol	10min in cold 70% methanol on filter paper strips	Plunge into liquid nitrogen	Rapid	75%	24h	Nugent *et al.*, 1994
Heterorhabditis zealandica	Preincubation for 4 days in 15% glycerol	10min in cold 70% methanol on filter paper strips	Plunge into liquid nitrogen	Rapid	28%	24h	Nugent *et al.*, 1994
Heterorhabditis spp.	Preincubation for 6 days in 15% glycerol	10min in cold 70% methanol on filter paper strips	Plunge into liquid nitrogen	Rapid	88%	24h	Nugent *et al.*, 1994
Heterorhabditis spp.	72h at 23°C in 17% glycerol	Rinse with ice cold 70% methanol, incubate for 10 min, absorb onto filter paper strips	Plunge into liquid nitrogen	Rapid	+	6 years	CABI Bioscience Collaborating Research Group Method
Steinernema carpocapsae	Preincubation in Me_2SO	10min in 0.5% methanol	Rapid	Rapid	+	24h	Popiel and Vasquez, 1991

Continued

Steinernema feltiae	No pretreatment	60% methanol for 45 sec at 0°C	5100	10000	12.3% motility	Immediate thaw	Smith *et al.*, 1990
Steinernema feltiae	Surface sterilized in 0.5% hyamine for 30min, desiccated to 97%RH at 25°C over 24h	70% methanol for 10min at 0°C	330-4800	600	84% motility infectivity not quantified	Immediate thaw	James, 1990
Steinernema spp.	72h at 23°C in 22% glycerol	Rinse with ice cold 70% methanol, incubate for 10 min, absorb onto filter paper strips	Plunge into liquid nitrogen	Rapid	+	6 years	CABI Bioscience Collaborating Research Group Method

DI, direct immersion in liquid coolants; Two-step, cooling at -22°C for 2 h and stored at -196°C (unless cited otherwise).

Cryopreservation Protocols for Protozoa

Species	Growth conditions or pretreatment before freezing	Cryoprotective Additives	Cooling (°C min^{-1})	Thaw (°C min^{-1})	Recovery	Storage temp. (°C)	Reference
Free living pathogenic and nonpathogenic amoeba	–	Me_2SO (10% v/v)	Three step –20 (60 min), –70 (60 min)	Hold at room temp for 3 min, then rapid warming	8 - 49%	–196	Kilvington, 1995
Pathogenic protozoa	Chill harvested cultures on ice, centrifuge 3min resuspend pellet in fresh medium, harvest resuspend pellet in cryoprotectant equilibrate at RT 15 min	Basal broth 6ml; Inactivated adult bovine serum 2ml; glucose (2.5M) 0.8ml; DMSO 1ml; cystein/ ascorbic acid sol. (Cystein monohydrochloride 1g; ascorbic acid 0.1g distilled water 9.3ml; 10N NaOH 0.7ml) 0.2ml	Four step 8-10 (~–8); 1.5 (–10); 1–2 (–40); plunge	Rapid warming	+	–160/ –196	Diamond, 1995
Aegypian pullorum	30min 4°C	Glycerol (10% w/v)	Plunge into LN_2	Rapid warming	+	–196	Raether and Seidenath, 1977
Anaplasma marginale	–	Me_2SO (2M)	Plunge into LN_2	Rapid warming	+	–196	Love, 1972
Babsia bigemina	30min 0°C	Me_2SO (14% v/v)	Two-step 82 (–60)	Rapid warming	+	–196	Dalgliesh and Mellors, 1974
B. bovis	–	PVP (15% v/v)	Two-step 20 (–70), slow (–196)	Rapid warming	+	–196	Vega *et al.*, 1985
B. rodhaini	0°C	Me_2SO (10% v/v)	Two-step 100 (–100), fast (–196)	Rapid warming	+	–196	Dalgliesh *et al.*, 1980
Cryptosporidium	–	Me_2SO (5% v/v)	Two-step 0.3 (–80), fast (–196)	Rapid warming	+	–196	Rossi *et al.*, 1990

Continued

Species	Growth conditions or pretreatment before freezing	Cryoprotective Additives	Cooling (°C min^{-1})	Thaw (°C min^{-1})	Recovery	Storage temp. (°C)	Reference
Cryptothecodinium cohnii (16 strains)	–	Glycerol (7.5%)	Two-step 1 (–40), fast (–196)	Rapid warming	20-76	–150	Simione and Daggett, 1977
Cytilophosis mucicola	–	Me_2SO (5% v/v)	2-3	Rapid warming	+	–196	Simon and Schneller, 1973
Dientamoeba fragilis	–	Me_2SO (2.75% v/v)	Three-step 5.5 (–6), 3.5 (–60), fast (–196)	Rapid warming	+	–196	Dwyer and Honigberb, 1971
Eimeria	–	Me_2SO (2.75% v/v)	Two-step 1 (–70), fast (–196)	Rapid warming	+	–196	Norton *et al.*, 1968
Entamoeba histolytica	–	Me_2SO (5% w/v)	Two-step 1 (–100), fast (–196)	Rapid warming	+	–196	Farri *et al.*, 1983
Entodinium caudatum	–	Me_2SO (4% w/v)	Two-step 45min at -25, fast (–196)	Rapid warming	36%	–196	Kisidayova, 1995
Giardia intestinalis	–	Me_2SO (7.5-10% v/v)	Two-step 1 (–25), 5 (–196)	Rapid warming	+	–196	Phillips *et al.*, 1984
G. lamblia	–	Me_2SO (6.5% w/v)	Two-step 1.2 (–40), fast (–196)	Rapid warming	+	–196	Lyman and Marchin, 1984
Leishmania tropica	–	Me_2SO (10.6% w/v)	Two-step 1.9 (–65), fast (–196)	Rapid warming	+	–196	Callow and Farrant, 1973
Naegleria	–	Me_2SO (5% v/v)	Two-step 1.3 (–55), fast (–196)	Rapid warming	+	–196	Simione and Daggett, 1976

N. gruberi	–	Me_2SO (5% w/v)	Three-step 2 h at –29, 2 h at –70 plunge	Rapid warming	>40%	–196	Brown and Day, 1993
Paramecium aurelia	–	Me_2SO (5% v/v)	2-3	Rapid warming	1.3-16	–196	Simon and Scheller, 1973
Plasmodium chabaudi	–	Glycerol (10% v/v)	Plunge to LN_2	Rapid warming	+	–196	Mutetwa and James, 1984
P. falciparum		Glycerol (10% v/v)	One step 300 to –196	Rapid warming	+	–196	Mutetwa and James, 1985
P. galinaceum	–	Glycerol (10% v/v)	Two-step 1 (–70), fast (–196)	Rapid warming	+	–196	Mutetwa and James, 1985
P. berghei	–	Glycerol (10% v/v)	Two-step 50 (–80), fast (–196)	Rapid warming	+	–196	Hollingdale *et al.*, 1985
Tetrahymena pyriformis	–	Me_2SO (5% v/v)	2-3	Rapid warming	+	–196	Simon and Scheller, 1973
T. pyriformis	–	Me_2SO (1.4 M)	Two-step 1 (–53), fast (–196)	Rapid warming	0.4	–196	Osbourne and Lee, 1975
Theileria parva	–	Glycerol (7.5% v/v)	Two-step 1 (–80), fast (–196)	Rapid warming	+	–196	Cunningham *et al.*, 1973
Toxoplasma gondii	–	Glycerol (10% v/v)	1 (–160)	Rapid warming	+	–196	Bollinger *et al.*, 1974
T. gondii	–	Glycerol (10% v/v)	0.3 (–79)	Rapid warming	+	–79	Eyles *et al.*, 1956
Trichomonas vaginalis	–	Me_2SO (5% v/v)	Two-step 1 (–35), fast (–196)	Rapid warming	+	–196	Ivey, 1975
Trypanosoma brucei spp.	–	Glycerol (7.5 v/v)	Two-step 0.7 (–60), fast (–196)	Warming in air 20°C	+	–196	Lumsden *et al.*, 1973

Continued

Species	Growth conditions or pretreatment before freezing	Cryoprotective Additives	Cooling (°C min^{-1})	Thaw (°C min^{-1})	Recovery	Storage temp. (°C)	Reference
Trypanosoma brucei congolense	–	Glycerol (12 v/v)	Two-step 1-2 (–59), fast (–196)	Rapid warming	+	–196	Diffley *et al*., 1976
T. cruzi	–	Me_2SO (10% v/v)	Two-step 3 (–60), fast (–196)	Rapid warming	+	–196	Ribiero dos Santos *et al*., 1978
T. cruzi	–	Glycerol (7% v/v)	Two-step 2 (–60), fast (–196)	Rapid warming	+	–196	Engel *et al*., 1980
Zoa entodinium simplex	–	Me_2SO (38% w/v)	Two-step 0.5 –2.2 (–100), fast (–196)	Rapid warming	15%	–196	Marcin *et al*., 1989

DI, direct immersion in liquid coolants; Two/ multi-step cooling protocols, intermediate temperatures given in parenthesis. Unless otherwise stated final step is a plunge into liquid nitrogen.
See also Lee and Soldo (1992).

References

Albrecht, R.M., Orndorff, G.R. and MacKenzie, A.P. (1973) Survival of certain microorganisms subjected to rapid and very rapid freezing on membrane filters. *Cryobiology* 10, 233–239. https://doi.org/10.1016/0011-2240(73)90036-9

Antonio, J.P. san and Blount, V. (1973) Use of liquid nitrogen to preserve downy mildew (*Phytophthora phaseoli*) inoculum. *Plant Disease Reporter* 57, 724. https://doi.org/10.1016/S0953-7562(09)80247-5

Beaty, M.H. and Parker, B.C. (1992) Cryopreservation of eukaryotic algae. *Virginia Journal of Science* 43, 403–410. http://www.vacadsci.org/vjsArchives/V43/43-4/43-403.pdf

Ben-Amotz, A. and Gilboa, A. (1980) Cryopreservation of marine unicellular algae. II. Induction of freezing tolerance. *Marine Ecology Progress Series* 2, 221–224. https://www.int-res.com/articles/meps/2/m002p221.pdf

Beyersdorf-Radeck, B., Schmid, R.D. and Malik, K.A. (1993) Long-term storage of bacterial inoculum for direct use in a microbial biosensor. *Journal of Microbiological Methods* 18, 36–39.

Bodas, K., Brennig, C., Diller, K.R. and Brand, J.J. (1995) Cryopreservation of blue-green and eukaryotic algae in the culture collection at the University of Texas at Austin. *Cryo-Letters* 16, 267–274. https://www.semanticscholar.org/paper/Cryopreservation-of-blue-green-and-eukaryotic-algae-Bodas-Brennig/74b93a58a3576d0bef9bcbb7b2b7e1900e01e854

Bollinger, R.O., Musallam, N. and Stulberg, C.S. (1974) Freeze preparation of tissue and propagated *Toxoplasma gondii*. *Journal of Parasitology* 60, 368–369.

Box, J.D. (1988) Cryopreservation of the blue-green alga *Microcystis aeruginosa*. *British Phycology Journal* 23, 385-386. https://doi.org/10.1080/00071618800650411

Bradshaw, D.J., McKee, A.S. and Marsh, P.D. (1989) The use of defined inocula stored in liquid nitrogen for mixed culture chemostat studies. *Journal of Microbiological Methods* 9, 123–128. DOI: 10.1016/0167-7012(89)90062-6. https://eurekamag.com/research/007/942/007942573.php

Bromfield, K.R. and Schmitt, C.G. (1967) Cryogenic storage of conidia of *Peronospora tabacinae*. *Phytopathology* 57, 1133–1134.

Brown, S. and Day, J.G. (1993) An improved method for the long-term preservation of *Naegleria gruberi*. *Cryo-Letters* 14, 347–352.

Butterfield, W., Jong, S.C. and Alexander, M.J. (1974) Preservation of living fungi pathogenic for man and animals. *Canadian Journal of Microbiology* 20, 1665–1673. https://doi.org/10.1139/m74-258

Callow, L.L. and Farrant, J. (1973) Cryopreservation of the promastigote form of *Leishmania tropica* var. *major* at different cooling rates. *International Journal of Parasitology* 3, 77–88. https://doi.org/10.1016/0020-7519(73)90011-8

Canavate, J.P. and Lubian, L.M. (1995a) Some aspects on the cryopreservation of microalgae used as food for marine species. *Aquaculture* 136, 277–290. https://doi.org/10.1016/0044-8486(95)01056-4

Canavate, J.P. and Lubian, L.M. (1995b) Relationship between cooling rates, cryoprotectant concentrations and salinates in the cryopreservation of marine microalgae. *Marine Biology* 124, 325–334. https://link.springer.com/article/10.1007/BF00347136

Challen, M.P. and Elliott, T.J. (1986) Polypropylene straw ampoules for the storage of micro-organisms in liquid nitrogen. *Journal of Microbiological Methods* 5, 11–23. https://doi.org/10.1016/0167-7012(86)90019-9

Cunningham, M.P., Brown, C.G.D., Burridge, M.J. and Purnell, R.E. (1973) Cryopreservation of infective particles of *Theileria parva*. *International Journal of Parasitology* 3, 583–587. https://doi.org/10.1016/0020-7519(73)90082-9

Dahmen, H., Staub, T. and Schwinn, F.T. (1983) Technique for long-term preservation of phytopathogenic fungi in liquid nitrogen. *Phytopathology* 73, 241–246. https://www.apsnet.org/publications/phytopathology/backissues/Documents/1983Articles/Phyto73n02_241.PDF

Dalgliesh, R.J. and Mellors, L.T. (1974) Survival of the parasitic protozoan, *Babesia bigemina*, in blood cooled at widely different rates to -196°C. *International Journal of Parasitology* 4, 169–172.

Dalgliesh, R.J., Mellors, L.T. and Blight, G.W. (1980) Comparisons of glucose, sucrose and dimethyl sulfoxide as cryoprotective agents for *Babesia rodhaini*, with estimates of survival rates. *Cryobiology* 17, 410-417. https://doi.org/10.1016/0011-2240(80)90048-6

Day, J.G. (1998) Cryoconservation of microalgae and cyanobacteria. *Cryo-Letters supplement* 1, 7–14.

Day J.G. (2007) Cryopreservation of Microalgae and Cyanobacteria. In: Day J.G., Stacey G.N. (eds) *Cryopreservation and Freeze-Drying Protocols. Methods in Molecular Biology™*, vol 368. Humana Press. https://doi.org/10.1007/978-1-59745-362-2_10

Day, J.G. and DeVille, M.M. (1995) Cryopreservation of algae. *Methods in Molecular Biology* 38, 81-89.

Day, J.G. and Fenwick, C. (1993) Cryopreservation of members of the genus *Tetraselmis* used in Aquaculture. *Aquaculture* 118, 151-160.

Day, J.G., Watanabe, M.M., Morris, G.J., Fleck, R.A. and McLellan, M.R. (1997) Long-term viability of preserved eukaryotic algae. *Journal of Applied Phycology* 9, 121-127. https://link.springer.com/article/10.1023/A:1007991507314

Day, J.G., Fleck, R.A. and Benson, E.E. (1998) Cryopreservation of multicellular algae: Problems and perspectives. *Cryo-Letters* 19, 205-206.

Day, J.G., Fleck, R.A. and Benson, E.E. (2000) Cryopreservation-recalcitrance in microalgae: novel approaches to identify and avoid cryo-injury. *Journal of Applied Phycology* 12, 369-377. https://www.academia.edu/61210314/Cryopreservation_recalcitrance_in_microalgae_novel_approaches_to_identify_and_avoid_cryo_injury

Diamond, L.S. (1995) Cryopreservation and storage of parasitic protozoa in liquid nitrogen. *Journal of Eukaryotic Microbiology* 42, 585-590. https://doi.org/10.1111/j.1550-7408.1995.tb05911.x

Diffley, P., Honigberg, B.M. and Mohn, F.A. (1976) An improved method of cryopreservation *Trypanosoma (Nannomonas) congolense* Borden in liquid nitrogen. *Journal of Parasitology* 62, 136-137. https://pubmed.ncbi.nlm.nih.gov/1255362/

Dwyer, D.M. and Honigberg, B.M. (1971) Freezing and maintenance of *Dientamoeba fragilis* in liquid nitrogen. *Journal of Parasitology* 57, 190-191. https://doi.org/10.2307/3277781

Elliott, T.J. and Challen, M.P. (1981) The storage of mushroom strains in liquid nitrogen. *Annual Report of the Glasshouse Crops Research Institute* 1979, 194-204.

Engel, J.C., Perez, A.C. and Wynne de Martini, G. (1980) Cryopreservation of *Trypanosoma cruzi*. *Medicina* 40, 103-108.

Eyles, D.E., Coleman, N. and Cavanaugh, D.J. (1956) Preservation of *Toxoplasma gondii* by freezing. *Journal of Parasitology* 42, 408-413. https://pubmed.ncbi.nlm.nih.gov/13346430/

Farri, T.A., Warhurst, D.C. and Marshall, T.F.D. (1983) The use of infectivity titrations for measurement of viability of *Entamoeba histolytica* after cryopreservation. *Transactions of the Royal Society of Tropical Medicine and Hygiene* 77, 259-266. https://doi.org/10.1016/0035-9203(83)90085-8

Fenwick, C. and Day, J.G. (1992) Cryopreservation of *Tetraselmis suecica* cultured under different nutrients regimes. *Journal of Applied Phycology* 4, 105-109. https://www.academia.edu/61210306/Cryopreservation_ofTetraselmis_suecica_cultured_under_different_nutrients_regimes

Fonseca, F., Meneghel, J., Cenard, S., Passot, S. and Morris, G.J. (2016) Determination of intracellular vitrification temperatures for unicellular micro organisms under conditions relevant for cryopreservation. *PLoS ONE* 11(4): e0152939. https://doi.org/10.1371/journal.pone.0152939

Fortney, K.F. and Thoma, R.W. (1977) Stabilization of culture productivity. *Developmental and Industrial Microbiology* 18, 319-325.

Gale, A.W., Schmitt, C.G. and Bromfield, K.R. (1975) Cryogenic storage of *Sclerospora sorghi*. *Phytopathology* 65, 828-829. https://www.apsnet.org/publications/phytopathology/backissues/Documents/1975Articles/Phyto65n07_828.PDF

Gothe, R. and Hartman, S. (1979) The viability of cryopreserved *Aegyptianella pullorum* Carpano, 1928 in the vector *Argas* (*Persicargas*) *walkerae* Kaiser and Hoogstraal, 1969. *Z. Parasitenk.* 58, 189-190.

Gresshoff, P.M. (1977) *Chlamydomonas reinhardi* – A model plant system. *II Cryopreservation. Plant Science Letters* 9, 23-25.

Hippe, H.H. (1991) Maintenance of methanogenic bacteria. In: Kirsop, B.E. and Doyle, A. (eds) *Maintenance of Microorganisms and Cell Cultures*, 2nd ed. Academic Press, London, pp. 101–114.

Hirata, K., Phunchindawan, M., Tukamoto, J., Goda, S. and Miyamoto, K. (1996) Cryopreservation of microalgae using encapsulation/dehydration. *Cryo-Letters* 17, 321-328.

Hollingdale, M.R., Leland, P., Singer, C.I. and Leef, J.L. (1985) *In vitro* infectivity of cryopreserved *Plasmodium berghei* sporozoites to cultured cells. *Transactions of the Royal Society of Tropical Medicine and Hygiene* 79, 206-208.

Holm-Hansen, O. (1963). Viability of blue-green and green algae after freezing. *Physiology of Plants* 16, 530-539.

Homolka, L. (2013) Methods of Cryopreservation in Fungi. In: Gupta, V., Tuohy, M., Ayyachamy, M., Turner, K. and O'Donovan, A. (eds) *Laboratory Protocols in Fungal Biology. Fungal Biology*. Springer, New York, NY. https://doi.org/10.1007/978-1-4614-2356-0_2

Hwang, S-W. (1966) Long term preservation of fungus cultures with liquid nitrogen refrigeration. *Applied Microbiology* 14, 784-788. https://doi.org/10.1128/am.14.5.784-788.1966

Hwang, S-W. (1968) Investigation of ultra-low temperature for fungal cultures. I. An evaluation of liquid nitrogen storage for preservation of selected fungal cultures. *Mycologia* 60, 613-621. https://doi.org/10.1080/00275514.1968.12018610

Hwang, S-E. and Hudock, G.A. (1971) Stability of *Chlamydomonas reinhardi* in liquid nitrogen storage. *Journal of Phycology* 7, 300-303.

Ivey, M.H. (1975) The use of solid medium techniques to evaluate factors affecting the ability of *Trichomonas vaginalis* to survive freezing. *Journal of Parasitology* 61, 1101-1103.

James, E.R. (1990) Cryopreservation of helminth parasites. In: Smyth, J.D. (ed.). *In vitro Cultivation of Parasitic Helminths*. CRC Press Inc., Boca Raton, FL. https://doi.org/10.1201/9780203711798

James, E.R. and Peacock, R. (1986) Studies on the cryopreservation of *Dictyocaulus viviparus* (Nematoda) third stage larvae. *Journal of Helminthology* 60, 65-73. DOI: 10.1017/s0022149x00008269

Joshi, L.M., Wilcoxson, R.D., Gera, S.D. and Chatterjee, S.C. (1974) Preservation of fungal cultures in liquid air. *Indian Journal of Experimental Biology* 12, 598-599.

Kapoore, R.V., Huete-Ortega, M., Day, J.G. *et al.* (2019) Effects of cryopreservation on viability and functional stability of an industrially relevant alga. *Sci Rep* 9, 2093. https://doi.org/10.1038/s41598-019-38588-6

Kilpatrick, R.A., Harmon, D.L., Loegering, W.Q. and Clark, W.A. (1971) Viability of uredospores of *Puccinia graminis* f. sp. *tritici* stored in liquid nitrogen. *Plant Disease Reporter* 55, 871-873.

Kilvington, S. (1995) Cryopreservation of pathogenic and non-pathogenic free-living amoebae. *Methods in Molecular Biology* 38, 63-70.

Kisidayova, S. (1995) 2-step freezing of the rumen ciliate protozoan *Entodinium caudatum. Journal of Microbiological Methods* 22, 185-192.

Kono, S., Kuwano, K., Ninomiya, M., Onishi, J. and Saga, N. (1997) Cryopreservation of *Enteromorpha intestinalis* in liquid nitrogen. *Phycologia* 36, 76-78.

Kono, S., Kuwano, K. and Saga, N. (1998) Cryopreservation of *Eisenia bicyclis* in liquid nitrogen. *Journal of Marine Biotechnology* 6, 220-223.

Kumi-Diaka, J. and Harris, O. (1995) Viability of *Borrelia burgdorferi* in stored semen. *British Veterinary Journal* 151, 221-224.

Lee, J.J. and Soldo, A.T. (1992) *Protocols in Protozoology*. Society of Protozoologists, Kansas. https://protistologists.org/wp-content/uploads/2016/06/Protocols-in-protozoology.pdf

Long, R.A., Wood, J.M. and Schmitt, G.C. (1978) Recovery of viable conidia of *Sclerospora philippinensis, S. sacchari* and *S. sorghi* after cryogenic storage. *Plant Disease Reporter* 62, 479-481.

Love, J.N. (1972) Cryogenic preservation of anaplasma marginale with dimethyl sulfoxide. *American Journal of Veterinary Research* 33, 2557-2560.

Lumsden, W.H.R., Herbert, W.J. and McNeillage, G.J.C. (1973) *Techniques with Trypanosomes*. Churchill Livingstone, Edinburgh.

Lyman, J.R. and Marchin, G.L. (1984) Cryopreservation of *Giardia lamblia* with dimethylsulfoxide using a dewer flask. *Cryobiology* 21, 170-176. https://doi.org/10.1016/0011-2240(84)90208-6

Malik, K.A. (1984) A new method for liquid nitrogen storage of phototrophic bacteria under anerobic conditions. *Journal of Microbiological Methods* 2, 4147.

Marcin, A., Gyulai, F., Vardy, M. and Sorokova, M. (1989) A simple technique for cryopreservation of the rumen protozoa *Entodinium simplex*. *Cryo-Letters* 10, 89-104.

McCrindle, C.M.E. (1979) The preservation of *Clostridium chauvoei* in liquid nitrogen. *Journal of the South African Veterinary Association* 50, 221-222.

McGrath, M.S. and Daggett, P.M. (1977) Cryopreservation of flagellar mutants of *Chlamydomonas reinhardii. Canadian Journal of Botany* 55, 1794-1796.

McLellan, M.R. (1989) Cryopreservation of diatoms. *Diatom Research* 4, 301-318. https://doi.org/10.1080/0269249X.1989.9705078

Morris, G.J. (1976a) The cryopreservation of *Chlorella* 2. Effect of growth temperature on freezing tolerance. *Archives of Microbiology* 107, 309-312. https://link.springer.com/article/10.1007/BF00425345

Morris, G.J. (1976b) Effect of growth temperature on the cryopreservation of *Prototheca. Journal of General Microbiology* 94, 395-399. https://www.microbiologyresearch.org/docserver/fulltext/micro/94/2/mic-94-2-395.pdf?expires=1645288195&id=id&accname=guest&checksum=303271B772B07A270D6030EED6414414

Morris, G.J. (1977) Preservation of algae by the method of Two-step cooling. *Cryobiology* 14, 691-692.

Morris, G.J. (1978) Cryopreservation of 250 strains of Chlorococcales by the method of Two-step cooling. *British Phycology Journal* 13, 15-24. https://doi.org/10.1080/00071617800650031

Morris, G.J. and Canning, C.E. (1978) The cryopreservation of *Euglena gracilis*. *Journal of General Microbiology* 108, 27-31. https://www.microbiologyresearch.org/docserver/fulltext/micro/108/1/mic-108-1-27.pdf?expires=1645288383&id=id&accname=guest&checksum=EE9150543763551689727C78B912F692

Morris, G.J., Coulson, G.E. and Engels, M. (1986) A cryomicroscopic study of *Cylindrocystis brebissonii* and two species of *Micrasterias conjugatophyceae* chlorophyta during freezing and thawing. *Journal of Experimental Botany* 37, 842-856. https://www.jstor.org/stable/23691552

Morris, G.J., Clarke, A. and Fuller, B.J. (1980) Methanol as a cryoprotective additive for *Chlorella*. *Cryo-Letters* 1, 121-128. http://nora.nerc.ac.uk/id/eprint/524985/

Mortain-Bertrand, A., Etchart, F. and de Boucayd, M-T. (1996) A method for the cryoconservation of *Dunaliella salina* (Chlorophyceae): Effect of glycerol and cold adaptation. *Journal of Phycology* 32, 346-352. https://doi.org/10.1111/j.0022-3646.1996.00346.x

Mutetwa, S.M. and James, E.R. (1984) Cryopreservation of *Plasmodium chabaudi*. 1. Protection by glycerol and dimethylsulphoxide during cooling and by glucose following thawing. *Cryobiology* 21, 329-339. https://www.meta.org/papers/cryopreservation-of-plasmodium-chabaudi-ii/6499502

Mutetwa, S.M. and James, E.R. (1985) Low-temperature preservation of *Plasmodium* spp. *Parasitology* 90, 589-603. https://doi.org/10.1017/S003118200005558X

Norton, C.C., Pout, D.D. and Joyner, L.P. (1968) Freeze preservation of *Eimeria acervulina* Tyzzer, 1929. *Folia Parasitology* 15, 203-211. https://folia.paru.cas.cz/pdfs/fol/1968/03/02.pdf

Nugent, M.J., O'Leary, S.A. and Burnell, A.M. (1994) Optimised procedures for the cryopreservation of different species of *Heterorhabditis*. *Eclair Grant 151*. EU Commission, Brussels. https://mural.maynoothuniversity.ie/186/

O'Brien, M.J. and Webb, R.E. (1958) Preservation of conidia of *Albugo occidentalis* and *Peronospora effusa*, obligate parasites of spinach. *Plant Disease Reporter* 42, 1312-1315.

Osbourne, J.A. and Lee, D. (1975) Studies on the conditions required to optimum recovery to *Tetrahymena pyriformis* strain S (Phenoset A) after freezing to and thawing from −196°C. *Journal of Protozoology* 22, 233-237. https://doi.org/10.1111/j.1550-7408.1975.tb05857.x

Perry, S.F. (1998) Freeze-drying and cryopreservation of bacteria. *Molecular Biotechnology* 9, 59–64. https://doi.org/10.1007/BF02752697

Phillips, R.E., Boreham, P.F.L. and Shepherd, R.W. (1984) Cryopreservation of viable *Giardia intestinalis* trophozoites. *Transactions of the Royal Society of Tropical Medicine and Hygiene* 78, 604-606. https://doi.org/10.1016/0035-9203(84)90218-9

Popiel, I. and Vasquez, E.M. (1991) Cryopreservation of *Steinernema carpocapsae* and *Heterorhabditis bacteriophora*. *Journal of Nematology* 23, 432-437. https://www.ncbi.nlm.nih.gov/pmc/articles/PMC2619178/

Raether, W. and Seidenath, H. (1977) Survival of *Aegyptianella pullorum*, *Anaplasma marginale* and various parasitic protozoa following prolonged storage in liquid nitrogen. *Zeitschrift fur Parasitenkunde* 53(1), 41–46. https://doi.org/10.1007/BF00383113

Ribeiro dos Santos, R., von Gal Furtado, C.C., Martins, J.B. and Martins, A.C.P. (1978) *Brazilian Journal of Medical and Biological Research* 11, 99-104.

Ritter, E. and Ulrich, U. (1986) Long-term preservation of mycoplasma in liquid nitrogen. *Total Hygiene* 32, 618-620.

Romo, S. and Becares, E. (1992) Preservation of filamentous cyanobacteria cultures. *Journal of Microbiological Methods* 16, 85-89. https://www.academia.edu/32756121/Preservation_of_filamentous_cyanobacteria_cultures_Pseudanabaena_galeata_B%C3%B6cher_and_Geitlerinema_amphibium_Ag_ex_Gom_Anagn_under_low_temperatures

Rossi, P., Pozio, E. and Besse, M.G. (1990) Cryopreservation of Cryptosporidium sp. *Oocysts. Transactions of the Royal Society of Tropical Medicine and Hygiene* 84, 68. https://doi.org/10.1016/0035-9203(90)90387-T

Ryan, M.J. and Smith, D. (2007) Cryopreservation and Freeze-Drying of Fungi Employing Centrifugal and Shelf Freeze-Drying. In: Day, J.G. and Stacey, G.N. (eds) *Cryopreservation and Freeze-Drying Protocols. Methods in Molecular Biology*™, vol 368. Humana Press. https://doi.org/10.1007/978-1-59745-362-2_9

Ryan, M.J., Kasulyte-Creasey, D., Kermode, A., San, S.P. and Buddie, A.G. (2014) Controlled rate cooling of fungi using a Stirling cycle freezer. *Cryoletters* 35, 63-69. https://www.ingentaconnect.com/content/cryo/cryo/2014/00000035/00000001/art00009#expand/collapse

Sanfilippo, A. and Lewin, R.A. (1970) Preservation of viable flexibacteria at low temperatures. *Canadian Journal of Microbiology* 16, 441-444. https://doi.org/10.1139/m70-075

Simione, F.P. and Daggett, P.-M. (1976) Freeze preservation of pathogenic and nonpathogenic *Naegleria* species. *Journal of Parasitology* 62, 49. https://pubmed.ncbi.nlm.nih.gov/1255382/

Simione, F.P. and Daggett, P-M. (1977) Recovery of a marine dinoflagellate following controlled and uncontrolled freezing. *Cryobiology* 14, 362-366. https://doi.org/10.1016/0011-2240(77)90183-3

Simon, E.M. (1972) Freezing and storage in liquid nitrogen of axenically and monoxenically cultivated *Tetrahymena pyriformis. Cryobiology* 9, 75-81.

Simon, E.M. and Schneller, M.V. (1973) The preservation of ciliated protozoa at low temperature. *Cryobiology* 10, 421-426. https://doi.org/10.1016/0011-2240(73)90069-2

Smentek, P. and Windisch, S. (1982). Studies on the survival of yeast strains under liquid nitrogen. *Central Journal for Bacteriology, Microbiology and Hygiene: I. Dept. Originals C: General, applied and ecological microbiology* 3, 432-439. https://doi.org/10.1016/S0721-9571(82)80027-6

Smith, B.S., Hodgson-Smith, A., Popiel, I., Minter, D.M. and James, E.R. (1990) Cryopreservation of the entomogenous nematode parasite *Steinernema feltiae* (=*Neoaplectana carpocapsae*). *Cryobiology* 27, 319. https://doi.org/10.1016/0306-4565(95)00033-X

Smith, D. (1982) Liquid nitrogen storage of fungi. *Transactions of the British Mycological Society* 79, 415-421. https://doi.org/10.1016/S0007-1536(82)80035-1

Smith, D. (1983) Cryoprotectants and the cryopreservation of fungi. *Transactions of the British Mycological Society* 80, 360–363.

Taylor, R. and Fletcher, R.L. (1998) Cryopreservation of eukaryotic algae - a review of methodologies. *Journal of Applied Phycology*10, 481-501. https://link.springer.com/article/10.1023/A:1008094622412

Tetsuka, Y. and Katsuya, K. (1983) Storage of sporangia of hop and vine downy mildews in liquid nitrogen. *Annals of the Phytopathological Society of Japan* 49, 731-735.

Thompson, J.P. (1987) Cryopreservation of Azotobacteriaceae in liquid nitrogen. *MIRCEN Journal – Journal of Applied Microbiology and Biotechnology* 3, 323-336.

Tooley, P.W. (1988) Use of uncontrolled freezing for liquid nitrogen storage of *Phytophthora* spp. *Plant Diseases* 72, 680-682. https://www.apsnet.org/publications/plantdisease/backissues/Documents/1988Articles/PlantDisease72n08_680.PDF

Tsuru, S. (1973) Preservation of marine and freshwater algae by means of freezing and freeze-drying. *Cryobiology* 10, 445-452. https://doi.org/10.1016/0011-2240(73)90074-6

Vega, C.A., Buening, G.M., Rodriguez, S.D., Carson, C.A. and McLaughlin, K. (1985) Cryopreservation of *Babesia bigemina* for *in vitro* cultivation. *American Journal of Veterinary Research* 46, 421-423. https://pubmed.ncbi.nlm.nih.gov/3922260/

Wang, G.T., Lin, C.T. and Hua, J. (1990) Long-term preservation of *Ganoderma* mycelia. *Journal of Chinese Agriculture and Chemistry Society* 28, 86-93.

Watanabe, M.M. and Sawaguchi, T. (1995) Cryopreservation of a water bloom forming cyanobacterium *Microcystis aeruginosa. Japanese Journal of Phycology* 43, 111-118.

Watanabe, M.M., Shimizu, A. and Satake, K. (1992) NIES-Microbial Culture Collection at the National Institute of Environmental Studies: Cryopreservation and database of culture strains of microalgae. In: Watanabe M.M. (ed.), *Proceedings of Symposium on Culture Collection of Algae.* NIES, Tsukuba, Japan, 33-41.

Yarlett, N.C., Yarlett, N., Orpin, C.G. and Lloyd, D. (1986) Cryopreservation of the anaerobic rumen fungus *Neocallimastixpatriciarum.LettersonAppliedMicrobiology*3,1-3.https://doi.org/10.1111/j.1472-765X.1986.tb01533.x

Appendix G

Example risk assessments from CABI

Contents

Microbiological Assessment 1 Covering microorganisms of ACDP Hazard Group 3

CABI holds only one species under this category: *Penicillium marnefei*, two strains are preserved in the collection by freeze-drying and cryopreservation. These organisms are of Risk Group 3 under EU Directive 2000/54/EC Classification of Biological Agents, and implemented in the UK through The Advisory Committee on Dangerous Pathogens. These are biological agents that can cause severe human disease and may be a serious hazard to employees. It may spread to the community but there is usually effective prophylaxis treatment available. At CABI there is no appropriate laboratory (CL3) that currently meets the required standards to safely handle this organism. Therefore, it must not be removed from the culture store under any circumstance.

Entry in Risk Assessment field on GRC Database
Hazard Group 3: Microbiological Assessment 1

Special instruction
Avoid all contact with the organism, growth media or materials on which they have grown.

Containment Level 3
Containment level 3 is suitable for work with pathogens in Hazard Group 3. Laboratory personnel

DOI: 10.1079/9781800622135.appg

must have training in handling pathogenic and potentially lethal organisms and in the use of safety equipment and controls. A high standard of supervision of the work must be maintained. See COSHH Regulation 7(10) - SCHEDULE 3 (Control of Substances hazardous to Health 2002)

1. The laboratory must be easy to clean. Bench surfaces and the floor must be impervious to water and resistant to acids, alkalis, solvents and disinfectants.
2. The laboratory must be sealable to permit fumigation.
3. There should be adequate space ($24m^3$) in the laboratory for each worker.
4. The laboratory must be sited in an area away from general circulation. Access to the laboratory should be limited to laboratory personnel and other specified persons. The laboratory door must be locked when the room is unoccupied.
5. A specific biohazard sign must be posted at the entry to the laboratory and the door must contain a glass panel so that the occupants can be seen.
6. A continuous airflow into the laboratory must be maintained during work with pathogens by one of the following means:

a. extracting the laboratory air through independent ducting to the outside air through a HEPA filter;
b. extracting the laboratory air to the outside air with a fan and HEPA filter sited in a wall or window of the laboratory;
c. ducting the exhaust air from a microbiological safety cabinet to the outside air through a HEPA filter; or
d. a safe variation of these provisions (provisions should also be made for comfort factors e.g. fresh air, temperature control).

In laboratories that have a mechanical air supply system, the supply and extract airflow must be interlocked to prevent positive pressurization of the room in the event of failure of the extract fan. The ventilation system must also incorporate a means of preventing reverse airflows.
7. A wash hand basin must be provided near the laboratory exit. Taps must be of a type which can be operated without being touched by hand.
8. An autoclave for sterilization of waste materials should be situated within the laboratory, if not, one must be readily accessible in the laboratory suite.
9. The laboratory door must be kept closed when work is in progress.
10. Side or back fastening gowns must be worn in the laboratory and they must be autoclaved before being sent for laundering. These gowns must not be used outside the laboratory suite.
11. Gloves must be worn for all work with infective materials and hands washed before leaving the laboratory.
12. Eating, chewing, drinking, smoking, storing of food and applying cosmetics must not take place in the laboratory.
13. Mouth pipetting must not take place.
14.
a. All laboratory procedures with infective materials must be conducted in a microbiological safety cabinet (Class I or Class III, BS 5726: 1979, or unit with equivalent protection factor or performance) except where the equipment to be used provides containment of the potential aerosol.
b. The cabinet must exhaust through a HEPA filter to the outside air or to the laboratory air extract system, and in other respects such as siting, performance, protection factor and air filtration, it must comply with the specifications detailed in BS 5726: 1979. When laboratories are faced with a major problem because of difficulties in arranging for the cabinet to exhaust to the open air, recirculation of the exhaust through two HEPA filters in series may, in exceptional circumstances, be considered as an alternative. In this case the maintenance of a continuous airflow into the laboratory during work with pathogens (see 6a and 6b above) will be of particular importance and such an option must not be adopted without prior consultation with the HSE.
15. The laboratory should contain its own equipment, e.g. centrifuge (in which sealed buckets must be used), incubator, refrigerator, deep freeze, vapour phase liquid nitrogen chest etc, so that all infective Hazard Group 3 pathogenic materials are held within the laboratory and nowhere else. Where this is not reasonably practicable material must be transported and stored without spillage in properly labelled robust containers which must be opened only in Containment level 3 accommodation.
16. Effective disinfectants must be available for routine disinfection and immediate use in the event of spillage.

17. Materials for autoclaving must be transported to the autoclave in robust containers, without spillage.
18. All waste materials must be made safe before disposal or removal to the incinerator.
19. All accidents and incidents, spills and exposures to infective materials must be immediately reported to, and recorded by, the person responsible for the work.

Accidents to personnel

If an accident occurs in the Bisby Parcel Reception Room (PRR), immediately contact someone outside e.g. request switchboard to page a first aider, the person responsible for the work, or the Safety Officer or designate, who will assess the situation and initiate appropriate action.

Spillages

If damage occurs to equipment/containers, cover with 'Hycolin' or other suitable disinfectant, in cotton wool, followed by absorbent spill granules, then remove to a container for autoclaving. The room should then be cleaned before further use. First Aiders, person responsible for the work and designates should be made aware of the particular dangers of spillages in the PRR.

Procedures for decontamination

In all procedures the Hycolin disinfectant solution must be made up on the day it is to be used.

Decontamination of the safety cabinet

All surfaces within the cabinet must be wiped with concentrated disinfectant (Hycolin) and 20min contact time allowed before any items are removed from the cabinet. All items to be retained should be packaged for autoclaving and containers of viable cultures or cultures to be saved should be carefully wiped with disinfectant. Appropriate gloves must be worn when handling disinfectant. If a high-risk organism (Category 3) has been released into the cabinet, fumigation may be necessary.

Decontamination of the PRR in the event of spillage or breakage

Airborne spores can be precipitated by spraying a dilution of Hycolin into the atmosphere. This work must be done wearing protective overalls, mask and goggles. The room must be left for 30min to allow the disinfectant to settle. All outer clothing worn inside the room must be removed and placed in an autoclave bag for sterilization. After this procedure decontamination of the surfaces and floor of the room can be carried out. The broken tube or Petri dish must be covered with cotton wool or tissue and soaked in disinfectant (this prevents further release of spores into the room) and then left for a minimum of 24 h. Work surfaces and the floor can be cleaned as for low risk decontamination within the cabinet. Once the broken material has had time to soak (minimum of 24h) it must be swept up into an autoclavable dustpan, using the covering tissue rather than a brush, placed into an autoclave hazard bag and autoclaved along with the dustpan.

Opening cultures and ampoules All parcels containing microorganisms must be opened in a CL3 laboratory by trained personnel, in a cabinet that will prevent inhalation of aerosols. As CABI does not have appropriate laboratory (CL3) that currently meets the required standards to safely handle this organism, work with this fungus is forbidden and the following information is for information only.

The microorganism is supplied either growing on a suitable agar slope or as a freeze-dried preparation. Both may be stored in the laboratory prior to use, the first in a 20-25°C microbiological incubator for 2-6 weeks and the latter in a refrigerator for microbiological use for 5-30 years. Storage must be in a secure area.

Details of suitable media, incubation temperatures for the growth of the strains, and any known special hazards, are supplied with the strains.

The freeze-dried preparation is opened and transferred to growth media as follows:

Care should be taken in opening freeze-dried or lyophilized cultures. The ampoule should be marked with a glass-cutter/file near the middle of the cotton wool plug and a hot glass rod used to crack the glass. Time is allowed for air to filter through the plug into the ampoule and the pointed tip should be snapped off.

Failure to allow air to gently enter will result in a rush of air generating an aerosol and releasing the microorganism. The plug may be impregnated with dried culture and should not be handled. This plug should be replaced with a fresh sterile plug. The addition of water/medium should be a gradual process avoiding squirting the liquid onto the dried material which will generate an aerosol. Transfer of the now liquid culture after a period for absorption of water (at least 30min) should be carried out carefully as previously described.

Transport This microorganism **must not** be forwarded to third parties outside CABI as we currently have no trained shippers certified to dispatch this material. Further guidance can be found at http://www.eccosite.org/ (transport regulations)

Disposal All fungi, media and containers should be sterilized by autoclaving at 121°C for 15min before disposal by suitable means such as incineration.The autoclave must be situated in the laboratory suite where the organism was handled.

Procedures in case of spillage If the fungus is spilt or its container broken, thoroughly wet with a disinfectant, such as 4% sodium hypochlorite, and allow 30min before swabbing up and transferring into a container for autoclaving.

Microbiological Assessment 2 Covering Microorganisms of ACDP Hazard Group 2

The microorganisms subject to this Microbiological Assessment are of Risk Group 2 under EU Directive 2000/54/EC Classification of Biological Agents, and implemented in the UK through The Advisory Committee on Dangerous Pathogens. These are biological agents that can cause human disease and may be a hazard to laboratory workers but are unlikely to spread to the community. Laboratory exposure rarely produces infection and effective prophylaxis or effective treatment is available.

Entry in Risk Assessment field on GRC Database

Hazard Group 2: Microbiological Assessment 2

Key precautions are:

- Avoid all contact with the organism, growth media or materials on which they have grown. Many fungi produce extracellular metabolites which may be toxic and although these have not been detected for the organisms under this risk assessment precautions to avoid contact must be taken.
- Avoid inhalation of spores as they may cause infection or allergic reaction. Such allergies may be caused by the spores of fungi including *Alternaria, Aspergillus, Cladosporium, Doratomyces, Sporobolomyces* and many more.

To avoid these possible hazards, and reduce the risk in handling, normal aseptic microbiological techniques should be employed. Further guidance can be found at http://www.hse.gov.uk/biosafety/biologagents.pdf. Outside the UK, countries may have different requirements and it is strongly advised that customers should adhere to the regulations in their own countries. Any work that may result in the creation of an aerosol containing the organism must be carried out in an appropriate microbiological safety cabinet.

Containment Level 2

Containment level 2 is suitable for use with agents in Hazard Group 2. Laboratory personnel must receive instruction and training in handling pathogens and an appropriate standard of supervision of the work must be maintained. See COSHH Regulation 7(10) - SCHEDULE 3 (Control of Substances hazardous to Health 2002)

1. The laboratory should be easy to clean. Bench surfaces should be impervious to water and resistant to acids, alkalis, solvents and disinfectants.
2. Access to the laboratory should be limited to laboratory personnel and other specified persons.
3. There should be adequate space ($24m^3$) in the laboratory for each worker.
4. If the laboratory is mechanically ventilated, an inward airflow into the laboratory must be maintained by extracting room air to atmosphere.
5. The laboratory must contain a wash hand basin which should be located near the laboratory exit. Taps should be of a type which can be operated without being touched by hand.
6. An autoclave for the sterilization of waste materials must be readily accessible, normally in the same building as the laboratory.
7. The laboratory door should be closed when work is in progress.
8. Laboratory coats or gowns, which should be side or back fastening, must be worn in the laboratory and removed when leaving the laboratory suite. Separate storage (pegs) must be provided in the laboratory suite for this clothing.
9. Eating, drinking, smoking, storing of food and applying cosmetics must not take place in the laboratory.

10. Mouth pipetting must not take place.
11. Hands must be disinfected or washed immediately when contamination is suspected, after handling infective materials, and also before leaving the laboratory.
12. In general, work can be conducted on the open bench***. If aerosols are produced and for manipulations such as vigorous shaking or mixing and ultrasonic disruption etc., a microbiological safety cabinet (Class 1 or Class II) or equipment which is designed to contain the aerosol must be used. The cabinet must exhaust to the outside air or to the laboratory air extract system (for further guidance see www.hse.gov.uk/biosafety)
13. Effective disinfectants must be available for routine disinfection and immediate use in the event of spillage.
14. Bench tops should be disinfected after use.
15. Used laboratory glassware and other materials awaiting sterilization must be stored in a safe manner. Pipettes if placed in disinfectant must be totally immersed.
16. Material for autoclaving must be transported to the autoclave in robust containers, without spillage.
17. All waste materials must be made safe before disposal or removal to the incinerator.
18. All accidents and incidents must be immediately reported to, and recorded by, the person responsible for the work.

***CABI recommends that all work involving Hazard group II organisms is undertaken in a Class II microbiological safety cabinet

Opening cultures and ampoules All parcels containing microorganisms must be opened in a laboratory by trained personnel, in a cabinet that will prevent inhalation of aerosols.

The microorganism is supplied either growing on a suitable agar slope or as a freeze-dried preparation. Both may be stored in the laboratory prior to use, the first in a 20-25°C microbiological incubator for 2-6 weeks and the latter in a refrigerator for microbiological use for 5-30 years.

Details of suitable media, incubation temperatures for the growth of the strains, and any known special hazards, are supplied with the strains.

The freeze-dried preparation is opened and transferred to growth media as follows:

Care should be taken in opening freeze-dried or lyophilized cultures. The ampoule should be marked with a glass-cutter/file near the middle of the cotton wool plug and a hot glass rod used to crack the glass. Time is allowed for air to filter through the plug into the ampoule and the pointed tip should be snapped off.

Failure to allow air to gently enter will result in a rush of air generating an aerosol and releasing the microorganism. The plug may be impregnated with dried culture and should not be handled. This plug should be replaced with a fresh sterile plug. The addition of water/medium should be a gradual process avoiding squirting the liquid onto the dried material which will generate an aerosol. Transfer of the now liquid culture after a period for absorption of water (at least 30min) should be carried out carefully as previously described.

Transport
All hazard group 2 organisms must be placed in appropriate packaging meeting UN 3373, IATA packaging instruction 650 and should be labelled as containing a 'Biological Substance Category B'. Specialized couriers should be used and all postal regulations of the recipient country must be followed. Microorganisms should not be forwarded to third parties outside your company. Further guidance can be found at http://www.eccosite.org/ (transport regulations)

Disposal All fungi, media and containers should be sterilized by autoclaving at 121°C for 15min before disposal by suitable means such as incineration.

Procedures in case of spillage If the fungus is spilt or its container broken, thoroughly wet with a disinfectant, such as 4% sodium hypochlorite, and allow 30min before swabbing up and transferring into a container for autoclaving.

Microbiological Assessment 3 Covering microorganisms of ACDP Hazard Group 1 that have been isolated from man as opportunistic pathogens

The fungi supplied (as listed above) are not categorized as Risk Group 2, 3 or 4 under EU Directive 2000/54/EC Classification of Biological Agents, and implemented in the

UK through The Advisory Committee on Dangerous Pathogens. These microorganisms have been isolated from man causing secondary infections and are particularly associated with immunocompromised patients.

Entry in Risk Assessment field on GRC Database

Hazard Group 1 - opportunistic pathogens of man: Microbiological Assessment 3

- Avoid all contact with the organism, growth media or materials on which they have grown. Many fungi produce extracellular metabolites which may be toxic.
- Avoid inhalation of spores as they may cause allergic reactions. Such allergies can be caused by the spores of many fungi including *Alternaria, Aspergillus, Cladosporium, Doratomyces, Sporobolomyces.*

To avoid these possible hazards and reduce the risk in handling, normal aseptic microbiological techniques should be employed.

All parcels containing fungi should be opened in a laboratory with Containment Level 1 as summarized below. Further guidance can be found at http://www.hse.gov.uk/biosafety/biologagents.pdf. Outside the UK, countries may have different requirements and it is strongly advised that customers should adhere to the regulations in their own countries.

Containment Level 1

Containment level 1 is suitable for work with organisms of hazard group 1. Laboratory personnel must have received instruction in the procedures conducted in the laboratory.

1. The laboratory should be easy to clean. Bench surfaces should be impervious to water and resistant to acids, alkalis, solvents and disinfectants.
2. If the laboratory is mechanically ventilated, it is preferable to maintain an inward airflow into the laboratory by extracting room air to the atmosphere.
3. The laboratory must contain a wash-basin or sink that can be used for hand washing.
4. The laboratory door should be closed when work is in progress.
5. Laboratory coats or gowns should be worn in the laboratory and removed when leaving the laboratory suite.
6. Eating, chewing, drinking, smoking, storing of food and applying cosmetics must not take place in the laboratory.
7. Mouth pipetting must not take place.
8. Hands must be disinfected or washed immediately when contamination is suspected, after handling viable materials, and also before leaving the laboratory.
9. All procedures must be performed so as to minimize the production of aerosols.
10. Effective disinfectants must be available for immediate use in the event of spillage.
11. Bench tops should be cleaned regularly after use.
12. Used laboratory glassware and other materials awaiting sterilization must be stored in a safe manner. Pipettes, if placed in disinfectant, must be totally immersed.
13. All waste material which is not to be incinerated should be rendered non-infective before disposal.
14. Materials for disposal must be transported in robust containers without spillage.
15. All accidents and incidents must be reported.

Opening cultures and ampoules All parcels containing microorganisms must be opened in a laboratory by trained personnel and, ideally, in a cabinet that will prevent inhalation of aerosols.

The fungus is supplied either growing on a suitable agar slope or as a freeze-dried preparation. Both may be stored in the laboratory prior to use, the first in a 20-25°C microbiological incubator for 2-6 weeks and the latter in a refrigerator for microbiological use for 5-30 years.

Details of suitable media, incubation temperatures for the growth of the strains and any known special hazard are given with the strain(s) supplied.

The freeze-dried preparation is opened and transferred to growth media as follows:

Care should be taken in opening freeze-dried or lyophilized cultures. The ampoule should be marked with a glass cutter/file near the middle of the cotton wool plug and a hot glass rod used to crack the glass. Time is allowed for air to filter through the plug into the ampoule and the pointed tip should be snapped off.

Failure to allow air to gently enter will result in a rush of air generating an aerosol and releasing the microorganism. The plug may be impregnated with dried culture and should not be handled. This plug should be replaced with a fresh sterile plug. The addition of water/medium should be a gradual process avoiding squirting the liquid onto the dried material which will generate an aerosol. Transfer of the now liquid culture after a period for absorption of water (at least 30min) should be carried out carefully.

Transport
Although cultures may be sent by postal letter mail in the UK and abroad, a minimum standard of packaging is required. If the materials are to be transported to another laboratory they should be packaged with enough absorptive material to absorb all contents of the containers in case of breakage. They should be placed in containers that will prevent breakage. If they are to be sent abroad they must be placed in appropriate packaging, ideally to UN 3373, IATA packaging instruction 650 and should be labelled as containing a 'Biological Substance Category B'. All postal regulations of the recipient country must be followed. Microorganisms should not be forwarded to third parties outside your company. Further guidance can be found at http://www.eccosite.org/ (transport regulations)

Disposal
All fungi, media and containers should be sterilized by autoclaving at 121°C for 15min before disposal by suitable means such as incineration.

Procedures in case of spillage
If the fungus is spilt or its container broken, thoroughly wet with a disinfectant, such as 4% sodium hypochlorite, and allow 30min before swabbing up and transferring into a container for autoclaving.

Microbiological Assessment 4 Covering microorganisms of ACDP Hazard Group 2 that produce known toxins

The microorganisms subject to this Microbiological Assessment are of Risk Group 2 under EU Directive 2000/54/EC Classification of Biological Agents, and implemented in the UK through The Advisory Committee on Dangerous Pathogens. These are biological agents that can cause human disease and may be a hazard to laboratory workers but are unlikely to spread to the community. Laboratory exposure rarely produces infection and effective prophylaxis or effective treatment is available. They are also known to produce toxins. The precautions described here that protect the user from infection will also protect from contact with the known toxins. However, extracts from the organism or its growth media that would not normally be considered as infectious may contain dangerous toxins and must be handled as a toxic chemical.

Entry in Risk Assessment field on GRC Database
Hazard Group 2 toxin producer: Microbiological Risk Assessment 4

Key precautions are:

- Avoid all contact with the organism, growth media or materials on which they have grown.
- Avoid inhalation of spores as they may cause infection or allergic reaction. Such allergies may be caused by the spores of fungi including *Alternaria, Aspergillus, Cladosporium, Doratomyces, Sporobolomyces* and many more.

The toxins produced may be carcinogenic, nephrotoxic, hepatotoxic, haemorrhagic, oestrogenic or cause inflammatory effects. The most commonly known is aflatoxin which is considered to be carcinogenic, hepatotoxic and potentially mutagenic and is produced by strains of *Aspergillus flavus* and *A. parasiticus*. Chapter 5 lists some mycotoxins that may be present in growth media and present additional problems in both use and disposal.

Mycotoxicoses are poisonings caused by the ingestion of food contaminated (and sometimes rendered carcinogenic) by toxin producing microfungi. Toxins are also produced by many other fungi, for example, citreoviridin, citrinin, islanditoxin and patulin by species of *Penicillium*, ochratoxin by *Aspergillus* and trichothecenes and zearelenone by species of *Fusarium*, and various other compounds including cochliodinol by *Chaetomium*. It should always be remembered that many fungi have not been studied chemically and because mycotoxins are not reported for a species does not mean it does not produce them. The handling of materials contaminated by these toxins can lead to their ingestion and subsequent poisoning. Inhalation of mycotoxins can also be dangerous.

Microbial toxins – Bacterial toxins

As infection patterns caused by bacterial pathogens are so different and depend on the bacterial pathogen and the individual host, every infection is an extremely individual process.

Diseases caused by bacteria may be grouped as follows (Anon, 1998):

- Local infections.
  Manifestation of the pathogen in a localized tissue.
  Examples: *Staphylococcus aureus, Neisseria gonorrhoeae.*
- Local infections with production of a potent toxin.
  Low invasiveness as above, but general diffusion of the toxin via the lymphatic and bloodstream.
  Examples: *Clostridium tetani, Corynebacterium diphtheriae.*
- Acute generalized infections.
  Usually highly invasive distribution of the pathogens after infection leading to possible septic-toxic shock. Sometimes tissue specific (organotroph) manifestation. Examples: plague, typhoid, brucellosis, some types of tuberculosis.

The virulence of strains of pathogenic bacterial species is determined by their invasiveness, production of aggressins and toxigenicity. Most bacterial toxins are capable of damaging or killing normal host cells and are effective upon infection. In contrast, most mycotoxins are effective without invasion or infection. The role played by bacterial invasiveness in damaging the host varies widely: sometimes infection can be extremely localized (e.g. *Corynebacterium diphtheriae*), the toxin diffuses and reaches almost all tissues. Alternatively, pathogens may invade and need to multiply to large numbers in order to generate enough toxin to cause damage to the host (e.g. *Bacillus anthracis*). Two classes of bacterial toxins have been designated which can be distinguished by their chemical nature. The first are protein-like exotoxins (examples are diphtheria, tetanus, botulinum toxins and enterotoxins) and the second are endotoxins which are molecular complexes containing protein, lipid and polysaccharide components. Generally, endotoxins are relatively non-specific, are derived from the outer layers of cell walls of Gram-negative bacteria and released after bacterial lysis. Cells of nearly all Gram-negative pathogenic bacteria are intrinsically toxic. The best-known endotoxins exhibiting pyrogenicity and toxicity are those of the enteric bacteria of the genera *Escherichia, Salmonella* and *Shigella*. Endotoxins are also inflammatory agents increasing capillary permeability.

Aggressins are enzyme-like substances e.g. proteases, collagenases, lipases, phospholipases or neuraminidases which usually support the invasion of a pathogen by damaging host tissue.

A complete list of all known bacterial toxins cannot be given here, some examples are given in Chapter 5 and further toxin producers can be found in Annexe III, Community Classification of the EU Directive 2000/54/EC

To avoid these possible hazards, and reduce the risk in handling, normal aseptic microbiological techniques should be employed. All parcels containing microorganisms should be opened and handled in a laboratory with containment level 2. Further guidance can be found at http://www.hse.gov.uk/biosafety/biologagents.pdf. Outside the UK, countries may have different requirements and it is strongly advised that customers should adhere to the regulations in their own countries. Any work that may result in the creation of an aerosol containing the organism must be carried out in an appropriate microbiological safety cabinet.

Containment Level 2

Containment level 2 is suitable for use with agents of Hazard Group 2. Laboratory personnel must receive instruction and training in handling pathogens and an appropriate standard of supervision of the work must be maintained. See COSHH Regulation 7(10) - SCHEDULE 3 (to Control of Substances hazardous to Health 2002)

1. The laboratory should be easy to clean. Bench surfaces should be impervious to water and resistant to acids, alkalis, solvents and disinfectants.
2. Access to the laboratory should be limited to laboratory personnel and other specified persons.
3. There should be adequate space ($24m^3$) in the laboratory for each worker.
4. If the laboratory is mechanically ventilated, an inward airflow into the laboratory must be maintained by extracting room air to atmosphere.
5. The laboratory must contain a wash hand basin which should be located near the laboratory

exit. Taps should be of a type which can be operated without being touched by hand.

6. An autoclave for the sterilization of waste materials must be readily accessible, normally in the same building as the laboratory.

7. The laboratory door should be closed when work is in progress.

8. Laboratory coats or gowns, which should be side or back fastening, must be worn in the laboratory and removed when leaving the laboratory suite. Separate storage (pegs) must be provided in the laboratory suite for this clothing.

9. Eating, drinking, smoking, storing of food and applying cosmetics must not take place in the laboratory.

10. Mouth pipetting must not take place.

11. Hands must be disinfected or washed immediately when contamination is suspected, after handling infective materials, and also before leaving the laboratory.

12. In general, work can be conducted on the open bench***, If aerosols are produced and for manipulations such as vigorous shaking or mixing and ultrasonic disruption etc., a microbiological safety cabinet (Class 1 or Class II) or equipment which is designed to contain the aerosol must be used. The cabinet must exhaust to the outside air or to the laboratory air extract system (for further guidance see www.hse.gov.uk/biosafety)

13. Effective disinfectants must be available for routine disinfection and immediate use in the event of spillage.

14. Bench tops should be disinfected after use.

15. Used laboratory glassware and other materials awaiting sterilization must be stored in a safe manner. Pipettes if placed in disinfectant must be totally immersed.

16. Material for autoclaving must be transported to the autoclave in robust containers, without spillage.

17. All waste materials must be made safe before disposal or removal to the incinerator.

18. All accidents and incidents must be immediately reported to, and recorded by, the person responsible for the work.

*** CABI recommends that all work involving Hazard group II organisms is undertaken in a Class II microbiological safety cabinet.

Opening cultures and ampoules All parcels containing microorganisms must be opened in a laboratory by trained personnel, in a cabinet that will prevent inhalation of aerosols.

The microorganism is supplied either growing on a suitable agar slope or as a freeze-dried preparation. Both may be stored in the laboratory prior to use, the first in a 20-25°C microbiological incubator for 2-6 weeks and the latter in a refrigerator for microbiological use for 5-30 years.

Details of suitable media, incubation temperatures for the growth of the strains, and any known special hazards, are supplied with the strains.

The freeze-dried preparation is opened and transferred to growth media as follows:

Care should be taken in opening freeze-dried or lyophilized cultures. The ampoule should be marked with a glass-cutter/file near the middle of the cotton wool plug and a hot glass rod used to crack the glass. Time is allowed for air to filter through the plug into the ampoule and the pointed tip should be snapped off.

Failure to allow air to gently enter will result in a rush of air generating an aerosol and releasing the microorganism. The plug may be impregnated with dried culture and should not be handled. This plug should be replaced with a fresh sterile plug. The addition of water/medium should be a gradual process avoiding squirting the liquid onto the dried material which will generate an aerosol. Transfer of the now liquid culture after a period for absorption of water (at least 30min) should be carried out carefully as previously described.

Transport

All hazard group 2 organisms must be placed in appropriate packaging meeting UN 3373, IATA packaging instruction 650 and should be labelled as containing a 'Biological Substance Category B'. Specialized couriers should be used and all postal regulations of the recipient country must be followed. Microorganisms should not be forwarded to third parties outside your company. Further guidance can be found at http://www.eccosite.org/ (transport regulations)

Disposal All fungi, media and containers should be sterilized by autoclaving at 121°C for 15min before disposal by suitable means such as incineration.

Procedures in case of spillage If the fungus is spilt or its container broken, thoroughly wet with a disinfectant, such as 4% sodium hypochlorite, and allow 30min before swabbing up and transferring into a container for autoclaving.

Microbiological Assessment 5 Covering microorganisms of ACDP Hazard Group 1 that produce known toxins

The microorganisms under this Microbiological Assessment are not categorized as Risk Group 2, 3 or 4 under EU Directive 2000/54/EC Classification of Biological Agents, and implemented in the UK through The Advisory Committee on Dangerous Pathogens, and have been assessed by CABI as Risk Group 1, i.e. a biological agent that is most unlikely to cause human disease. However, these fungi are known to produce toxins presenting a chemical hazard.

Entry in Risk Assessment field on GRC Database
Hazard Group 1 toxin producer: Microbiological Risk Assessment 5

- Avoid all contact with the organism, growth media or materials on which they have grown.
- Avoid inhalation of spores as they may cause allergic reactions or poisoning.

The toxins produced may be carcinogenic, nephrotoxic, hepatotoxic, haemorrhagic, oestrogenic or cause inflammatory effects. The most commonly known is aflatoxin which is considered to be carcinogenic, hepatotoxic and potentially mutagenic and is produced by strains of *Aspergillus flavus* and *A. parasiticus*. Customers must be aware that toxins may be present in growth media and present additional problems in both use and disposal.

Mycotoxicoses are poisonings caused by the ingestion of food contaminated (and sometimes rendered carcinogenic) by toxin producing microfungi. Toxins are also produced by many other fungi, for example, citreoviridin, citrinin, islanditoxin and patulin by species of *Penicillium*, ochratoxin by *Aspergillus* and trichothecenes and zearelenone by species of *Fusarium*, and various other compounds including cochliodinol by *Chaetomium*. It should always be remembered that many fungi have not been studied chemically and because mycotoxins are not reported for a species does not mean it does not produce them. The handling of materials contaminated by these toxins can lead to their ingestion and subsequent poisoning. Inhalation of mycotoxins can also be dangerous.

Microbial toxins – Bacterial toxins

In addition to those bacteria that produce toxins during infection there are also those that are non-infectious toxin producers. The most important group of the latter is the Cyanobacteria of which there are *ca.* 2000 species and are currently considered as hazard group 1. A complete list of all known bacterial toxins cannot be given here, some examples are given in the attached Table and further toxin producers can be found in Annexe III, Community Classification of the EU Directive 2000/54/EC. To avoid these possible hazards and reduce the risk in handling, normal aseptic microbiological techniques should be employed.

All parcels containing fungi should be opened in a laboratory with Containment Level 1 as described by the Advisory Committee on Dangerous Pathogens and summarized below. Further guidance can be found at http://www.hse.gov.uk/biosafety/biologagents.pdf/. Outside the UK, countries may have different requirements and it is strongly advised that customers should adhere to the regulations in their own countries.

Any activity that is likely to cause the generation of an aerosol must be carried out in a suitable microbiological safety cabinet. Some toxins are volatile at ambient temperature and in such cases, work should be carried out in a fume cupboard.

Containment Level 1
Containment level 1 is suitable for work with organisms of hazard group 1. Laboratory personnel must have received instruction in the procedures conducted in the laboratory.

1. The laboratory should be easy to clean. Bench surfaces should be impervious to water and resistant to acids, alkalis, solvents and disinfectants.
2. If the laboratory is mechanically ventilated, it is preferable to maintain an inward airflow into the laboratory by extracting room air to the atmosphere.
3. The laboratory must contain a wash-basin or sink that can be used for hand washing.

4. The laboratory door should be closed when work is in progress.
5. Laboratory coats or gowns should be worn in the laboratory and removed when leaving the laboratory suite.
6. Eating, chewing, drinking, smoking, storing of food and applying cosmetics must not take place in the laboratory.
7. Mouth pipetting must not take place.
8. Hands must be disinfected or washed immediately when contamination is suspected, after handling viable materials, and also before leaving the laboratory.
9. All procedures must be performed so as to minimize the production of aerosols.
10. Effective disinfectants must be available for immediate use in the event of spillage.
11. Bench tops should be cleaned regularly after use.
12. Used laboratory glassware and other materials awaiting sterilization must be stored in a safe manner. Pipettes if placed in disinfectant, must be totally immersed.
13. All waste material which is not to be incinerated should be rendered non-infective before disposal.
14. Materials for disposal must be transported in robust containers without spillage.
15. All accidents and incidents must be reported.

Opening cultures and ampoules: All parcels containing microorganisms must be opened in a laboratory by trained personnel and, ideally, in a cabinet that will prevent inhalation of aerosols.

The fungus is supplied either growing on a suitable agar slope or as a freeze-dried preparation. Both may be stored in the laboratory prior to use, the first in a 20-25°C microbiological incubator for 2-6 weeks and the latter in a refrigerator for microbiological use for 5-30 years.

Details of suitable media, incubation temperatures for the growth of the strains and any known special hazard are given with the strain(s) supplied.

The freeze-dried preparation is opened and transferred to growth media as follows: Care should be taken in opening freeze-dried or lyophilized cultures. The ampoule should be marked with a glass cutter/file near the middle of the cotton wool plug and a hot glass rod used to crack the glass. Time is allowed for air to filter through the plug into the ampoule and the pointed tip should be snapped off. Failure to allow air to gently enter will result in a rush of air generating an aerosol and releasing the microorganism. The plug may be impregnated with dried culture and should not be handled. This plug should be replaced with a fresh sterile plug. The addition of water/medium should be a gradual process avoiding squirting the liquid onto the dried material which will generate an aerosol. Transfer of the now liquid culture after a period for absorption of water (at least 30min) should be carried out carefully.

Transport

Although cultures may be sent by postal letter mail in the UK and abroad, a minimum standard of packaging is required. If the materials are to be transported to another laboratory they should be packaged with enough absorptive material to absorb all contents of the containers in case of breakage. They should be placed in containers that will prevent breakage. If they are to be sent abroad they must be placed in appropriate packaging, ideally to UN 3373, IATA packaging instruction 650 and should be labelled as containing a 'Biological Substance Category B'. All postal regulations of the recipient country must be followed. Microorganisms should not be forwarded to third parties outside your company. Further guidance can be found at http://www.eccosite.org/ (transport regulations)

Disposal

All fungi, media and containers should be sterilized by autoclaving at 121°C for 15min before disposal by suitable means such as incineration.

Procedures in case of spillage

If the fungus is spilt or its container broken, thoroughly wet with a disinfectant, such as 4% sodium hypochlorite, and allow 30min before swabbing up and transferring into a container for autoclaving.

Microbiological Assessment 6 Covering Biological agents of Hazard Group 1 with no identified toxic metabolite

The microorganisms under this Microbiological Assessment are not categorized as Risk Group 2,

3 or 4 under EU Directive 2000/54/EC Classification of Biological Agents, and implemented in the UK through The Advisory Committee on Dangerous Pathogens, and have been assessed by CABI as Risk Group 1, i.e. a biological agent that is most unlikely to cause human disease. However, these fungi are known to produce toxins presenting a chemical hazard.

Entry in Risk Assessment field on GRC Database

Hazard Group 1 toxin producer: Microbiological Risk Assessment 5

- Avoid all contact with the organism, growth media or materials on which they have grown.
- Avoid inhalation of spores as they may cause allergic reactions or poisoning.

The toxins produced may be carcinogenic, nephrotoxic, hepatotoxic, haemorrhagic, oestrogenic or cause inflammatory effects. The most commonly known is aflatoxin which is considered to be carcinogenic, hepatotoxic and potentially mutagenic and is produced by strains of *Aspergillus flavus* and *A. parasiticus*. Customers must be aware that toxins may be present in growth media and present additional problems in both use and disposal.

Mycotoxicoses are poisonings caused by the ingestion of food contaminated (and sometimes rendered carcinogenic) by toxin producing microfungi. Toxins are also produced by many other fungi, for example, citreoviridin, citrinin, islanditoxin and patulin by species of *Penicillium*, ochratoxin by *Aspergillus* and trichothecenes and zearelenone by species of *Fusarium*, and various other compounds including cochliodinol by *Chaetomium*. It should always be remembered that many fungi have not been studied chemically and because mycotoxins are not reported for a species does not mean it does not produce them. The handling of materials contaminated by these toxins can lead to their ingestion and subsequent poisoning. Inhalation of mycotoxins can also be dangerous.

Microbial toxins – Bacterial toxins

In addition to those bacteria that produce toxins during infection there are also those that are non-infectious toxin producers. The most important group of the latter is the Cyanobacteria of which there are *ca.* 2000 species and are currently considered as hazard group 1. A complete list of all known bacterial toxins cannot be given here, some examples are given in the attached Chapter 5 and further toxin producers can be found in Annexe III, Community Classification of the EU Directive 2000/54/EC.

To avoid these possible hazards and reduce the risk in handling, normal aseptic microbiological techniques should be employed.

All parcels containing fungi should be opened in a laboratory with Containment Level 1 as described by the Advisory Committee on Dangerous Pathogens and summarized below. Further guidance can be found at http://www.hse.gov.uk/biosafety/biologagents.pdf/. Outside the UK, countries may have different requirements and it is strongly advised that customers should adhere to the regulations in their own countries.

Any activity that is likely to cause the generation of an aerosol must be carried out in a suitable microbiological safety cabinet. Some toxins are volatile at ambient temperature and in such cases, work should be carried out in a fume cupboard.

Containment Level 1

Containment level 1 is suitable for work with organisms of hazard group 1. Laboratory personnel must have received instruction in the procedures conducted in the laboratory.

1. The laboratory should be easy to clean. Bench surfaces should be impervious to water and resistant to acids, alkalis, solvents and disinfectants.
2. If the laboratory is mechanically ventilated, it is preferable to maintain an inward airflow into the laboratory by extracting room air to the atmosphere.
3. The laboratory must contain a wash-basin or sink that can be used for hand washing.
4. The laboratory door should be closed when work is in progress.
5. Laboratory coats or gowns should be worn in the laboratory and removed when leaving the laboratory suite.
6. Eating, chewing, drinking, smoking, storing of food and applying cosmetics must not take place in the laboratory.
7. Mouth pipetting must not take place.
8. Hands must be disinfected or washed immediately when contamination is suspected, after

handling viable materials, and also before leaving the laboratory.
9. All procedures must be performed so as to minimize the production of aerosols.
10. Effective disinfectants must be available for immediate use in the event of spillage.
11. Bench tops should be cleaned regularly after use.
12. Used laboratory glassware and other materials awaiting sterilization must be stored in a safe manner. Pipettes if placed in disinfectant, must be totally immersed.
13. All waste material which is not to be incinerated should be rendered non-infective before disposal.
14. Materials for disposal must be transported in robust containers without spillage.
15. All accidents and incidents must be reported.

Opening cultures and ampoules: All parcels containing microorganisms must be opened in a laboratory by trained personnel and, ideally, in a cabinet that will prevent inhalation of aerosols.

The fungus is supplied either growing on a suitable agar slope or as a freeze-dried preparation. Both may be stored in the laboratory prior to use, the first in a 20-25°C microbiological incubator for 2-6 weeks and the latter in a refrigerator for microbiological use for 5-30 years.

Details of suitable media, incubation temperatures for the growth of the strains and any known special hazard are given with the strain(s) supplied.

The freeze-dried preparation is opened and transferred to growth media as follows: Care should be taken in opening freeze-dried or lyophilized cultures. The ampoule should be marked with a glass cutter/file near the middle of the cotton wool plug and a hot glass rod used to crack the glass. Time is allowed for air to filter through the plug into the ampoule and the pointed tip should be snapped off. Failure to allow air to gently enter will result in a rush of air generating an aerosol and releasing the microorganism. The plug may be impregnated with dried culture and should not be handled. This plug should be replaced with a fresh sterile plug. The addition of water/medium should be a gradual process avoiding squirting the liquid onto the dried material which will generate an aerosol. Transfer of the now liquid culture after a period for absorption of water (at least 30min) should be carried out carefully.

Transport

Although cultures may be sent by postal letter mail in the UK and abroad, a minimum standard of packaging is required. If the materials are to be transported to another laboratory they should be packaged with enough absorptive material to absorb all contents of the containers in case of breakage. They should be placed in containers that will prevent breakage. If they are to be sent abroad they must be placed in appropriate packaging, ideally to UN 3373, IATA packaging instruction 650 and should be labelled as containing a 'Biological Substance Category B'. All postal regulations of the recipient country must be followed. Microorganisms should not be forwarded to third parties outside your company. Further guidance can be found at http://www.eccosite.org/ (transport regulations)

Disposal

All fungi, media and containers should be sterilized by autoclaving at 121°C for 15min before disposal by suitable means such as incineration.

Procedures in case of spillage

If the fungus is spilt or its container broken, thoroughly wet with a disinfectant, such as 4% sodium hypochlorite, and allow 30min before swabbing up and transferring into a container for autoclaving.

CABI Standard Protocol SP20 Procedures for Handling Potential Human Pathogens

1. Introduction

This document should be read in conjunction with SP21, SP25, SP29 and SP37

Microorganisms are divided into four hazard groups as designated by the UK Health and Safety Executive (HSE) Advisory Committee on Dangerous Pathogens (ACDP). Full details can be found in the HSE publication: *The Approved List of Biological Agents* (Fourth edition, published 2021) which is available from the HSE website at http://www.hse.gov.uk/pubns/misc208.pdf. The four groups are categorized as:

Group 1	A biological agent that is unlikely to cause human disease.
Group 2	A biological agent that can cause human disease and which may be a hazard to employees. It is unlikely to spread to the community and there is usually effective prophylaxis or treatment available.
Group 3	A biological agent that can cause severe human disease and may be a serious hazard to employees. It may spread to the community but there is usually effective prophylaxis or treatment available.
Group 4	A biological agent that causes severe human disease and is a serious hazard to employees. It is likely to spread to the community and there is usually no effective prophylaxis or treatment available.

Microorganisms handled at CABI are mostly of hazard group 1 or 2. CABI Bioscience does not intentionally handle cultures above hazard group 2 but it is always possible that one may be received as an unnamed culture. It is therefore essential that all parcels containing microbial samples are opened by trained personnel in the Class III/I microbiological safety cabinet in the Parcel Reception Laboratory (PRL), Room C1.01. (see SP 25, Operation of the Safety Cabinet in Parcel Reception Laboratory).

A list of bacteria and fungi categorized by ACDP in hazard groups 2 and 3 is shown in Appendices 1 and 2 of this SP. Those of hazard group 2 which are accepted for handling by CABI Bioscience are emboldened in the list and marked with an asterisk. Other listed genera/species arriving as named samples of hazard group 2 organisms, are **not** accepted for processing. When samples are rejected, the reason for rejection should be recorded on LIMS also (see paragraphs 4 and 5 for procedure).

Organisms received for processing named as hazard group 3 organisms are to be destroyed on arrival following procedures outlined in paragraphs 6 and 7.

All samples for CABI Bioscience are to be logged into LIMS. (See SP 21 and SP 29, paragraph 1).

2. Containment requirements

Most samples from clinical sources require containment level 2 [Refer to: Part I of Schedule 3 paragraph 3(4) in Control of Substances Hazardous to Health (HSE, 2002)], unless a higher degree of containment is specified. Genetically Modified Organisms (GMOs) also require containment level 2 for handling.

3. Samples to be handled as potential human pathogens

These are as follows:

i. Samples provisionally identified by the sender as organisms of hazard group 3 (See SP30)

ii. Samples received with a provisional identification from the sender as a hazard group 2 organism not handled by CABI (listed in Appendix 1). These are to be dealt with following procedures set out in paragraphs 4 and 5.

iii. Samples isolated from other potentially hazardous sources as listed in Appendix 3.

4. Procedure for batches containing named samples of hazard group 2 not handled by CABI

If during parcel reception processing (refer to SP 19) named samples of hazard group 2 are received that are not among taxa to be handled by CABI, the relevant samples should be stored temporarily in refrigerator 6 in the Parcel Reception Laboratory. The samples are to be rejected and reasons for the rejection should be recorded in the Biosafety check comments field in the QC check test in LIMS. Place the samples in a plastic bag and enclose a label giving the name of the sender, date of arrival of the batch and LIMS number. Place the bag and samples in a plastic container fitted with a secure lid and store in refrigerator 6. The Delivery team and Customer Liaison Coordinator are to be informed. The sender is to be contacted and agreement reached on whether to discard the samples (refer to SP37) or forward to PHE. The former is the preferred course of action. The customer may be advised to submit a subculture of the sample to the PHE laboratory themselves and sent the following link for bacteria (https://www.gov.uk/government/

publications/microbial-identification-request-form) for filamentous fungi and yeasts (https://www.gov.uk/guidance/mycology-reference-laboratory-mrl-reference-and-diagnostic-services#samples-request-form)

If the latter is agreed, PHE is to be contacted (see details in paragraph 5) to obtain their agreement to receive the samples and to obtain details of their submission procedure. The samples are then to be forwarded to PHE following procedures set out in paragraph 5. The customer is be advised that as the sample(s) are being sent to PHE any requested turnaround time will no longer be met.

5. Procedures for forwarding samples to Public Health England (PHE)

Contact the relevant Public Health Laboratory service and establish whether or not they are willing to accept the samples for processing and, if so, the current prices. Also check prices of packaging and courier services for sending hazard group 2 samples to PHE. Notify the customer accordingly. If PHE are willing to accept the samples and the customer agrees for them to be forwarded and to pay all relevant charges, trained persons are to safely repackage the isolates according to procedures outlined in SP 23 (Postal and Packaging regulations) and send the samples and documentation by courier service to the relevant PHE Laboratory. Invoice the customer accordingly for packaging and courier services. A note is to be made in the sample metadata if individual samples are to be sent to PHE. If a whole batch is to be sent, a note should be made in the folder metadata. The documentation for submitting sample can be found via the hyperlinks in section 4.

A copy of any documentation sent to PHE is to be scanned (if necessary) and attached to the customer's LIMS folder.

If only part of the batch contains isolates of hazard group 2 normally handled by PHE and remaining isolates are acceptable for processing by CABI, the latter can be processed following standard procedures. Those to be forwarded to PHE should be noted in LIMS.

6. Receipt of potential hazard group 2 organisms and samples from potentially hazardous sources

If samples from potentially hazardous sources are received, log the batch into LIMS as per SP21 but do not scan any documentation and complete details on the QC checks on LIMS (see SP37). Any samples from a hazardous source are to be destroyed by autoclaving.

Inform the delivery team and a member of facilities that potentially hazardous samples have been received.

Inform the customer liaison team, who will inform the customer the sample(s) have been destroyed and to request an electronic copy of the request form.

7. Procedure for batches containing named samples of hazard group 2, handled by CABI

Log the batch in following the normal procedure, except once LIMS labels have been produced, mark these with a red spot to serve as a hazard warning. Samples can then be distributed to technical staff.

APPENDIX 1 to CABI Standard Protocol SP20

HAZARD GROUP 2 ORGANISMS

****Note** Emboldened species in the list are accepted for handling by the Microbial Identification Service.

All other named samples received, of genera or species listed below, are **not** accepted for processing by the Microbial Identification Service but may be accepted for deposit or preservation contracts at the discretion of the curator and/or the biosafety officer.

HAZARD GROUP 2 BACTERIA

Actinobacillus actinomycetemcomitans
Actinomadura madurae**
Actinomadura pelletieri
Actinomyces gerencseriae
Actinomyces israelii
Actinomyces pyogenes
Actinomyces* spp*
Alcaligenes* spp*
Arcanobacterium haemolyticum (formerly *Corynebacterium haemolyticum*)
Arcanobacterium pyogenes (formerly *Actinomyces pyogenes*)
Arcobacter butzleri (formerly *Campylobacter butzleri*)
Bacillus cereus**
Bacteroides fragilis
Bacteroides* spp*
Bartonella bacilliformis
Bartonella quintana (*Rochalimaea quintana*)
Bartonella spp (*Rochalimaea* spp)
Bordetella bronchiseptica
Bordetella parapertussis
Bordetella pertussis
Bordetella spp
Borrelia burgdorferi
Borrelia duttonii
Borrelia recurrentis
Borrelia spp
Brachispira spp (formerly *Serpulina* spp)
Burkholderia cepacia**
Campylobacter fetus
Campylobacter jejuni
Campylobacter spp
Cardiobacterium hominis
Chlamydophila pneumoniae
Chlamydophila psittaci (avian strains)
Chlamydophila psittaci (non avian strains)
Chlamydophila trachomatis
Clostridium botulinum
Clostridium perfringens
Clostridium tetani
Clostridium* spp*
Corynebacterium diphtheriae
Corynebacterium minutissimum
Corynebacterium pseudotuberculosis
Corynebacterium ulcerans
Corynebacterium* spp*
Edwardsiella tarda
Ehrlichia spp
Eikenella corrodens
Elizabethkingia meningoseptica (formerly *Flavobacterium meningosepticum*)
Enterobacter aerogenes/cloacae**
Enterobacter* spp*
Enterococcus* spp*
Erysipelothrix rhusiopathiae
Escherichia coli (except non-pathogenic strains)
Fluoribacter bozemanae (formerly *Legionella*)
Francisella tularensis (Type B)
Fusobacterium necrophorum
Fusobacterium spp
Gardnerella vaginalis
Haemophilus ducreyi
Haemophilus influenzae
Haemophilus spp
Helicobacter pylori
Klebsiella oxytoca
Klebsiella pneumoniae
Klebsiella* spp*
Legionella pneumophila
Legionella spp
Leptospira interrogans (all serovars)
Listeria ivanovii
Listeria monocytogenes
Moraxella catarrhalis
Morganella morganii
Mycobacterium avium/intracellulare
Mycobacterium bovis (BCG strain)
Mycobacterium chelonae
Mycobacterium fortuitum
Mycobacterium kansasii
Mycobacterium marinum
Mycobacterium paratuberculosis
Mycobacterium scrofulaceum
Mycobacterium simiae
Mycobacterium xenopi
Mycoplasma caviae
Mycoplasma hominis
Mycoplasma pneumoniae

Neisseria gonorrhoeae
Neisseria meningitidis
Nocardia asteroides**
Nocardia brasiliensis
Nocardia farcinica
Nocardia nova
Nocardia otitidiscaviarum
Pasteurella multocida
Pasteurella spp
Peptostreptococcus anaerobius
Peptostreptococcus spp
Plesiomonas shigelloides
Porphyromonas spp
Prevotella spp
Proteus mirabilis
Proteus penneri
Proteus vulgaris
Providencia alcalifaciens
Providencia rettgeri
Providencia spp
Pseudomonas aeruginosa**
Rhodococcus equi **
Rochalimaea quintana –(Bartonella quintana)
Rochalimaea spp
Salmonella arizonae
Salmonella enteritidis
Salmonella (other serovars)
Salmonella typhimurium
Shigella boydii
Shigella dysenteriae (other than Type 1)
Shigella flexneri
Shigella sonnei
Staphylococcus aureus
Streptobacillus moniliformis
Streptococcus agalactiae (Strep group B)
Streptococcus dysgalactiae equisimilis
Streptococcus pneumoniae
Streptococcus pyogenes (Strep group A)
Streptococcus suis
Streptococcus* spp*
Treponema carateum
Treponema pallidum
Treponema pertenue
Treponema spp
Ureaplasma parvum
Ureaplasma urealyticum
Vibrio cholerae (including El Tor)
Vibrio parahaemolyticus
Vibrio spp
Yersinia enterocolitica
Yersinia pseudotuberculosis
Yersinia* spp*

HAZARD GROUP 2 FUNGI

Arthroderma spp
Aspergillus fumigatus**
Aspergillus* spp*
Candida albicans**
Candida tropicalis**
Candida* spp*
Cryptococcus neoformans
(Filobasidiella neoformans var *neoformans)*
Cryptococcus neoformans var *gattii*
(Filobasidiella bacillispora)
Emmonsia crescens (Ajellomyces crescens)
Emmonsia parva**
Epidermophyton floccosum
Exophiala* spp*
Fonsecaea compacta**
Fonsecaea pedrosoi
Malassezia spp
Microsporum spp
Neotestudina rosatii
Rhizomucor pusillus**
Rhizopus microsporus**
Saksenaea vasiformis
Scedosporium apiospermum (Pseudallescheria boydii)**
Scedosporium proliferans
Scopulariopsis brevicaulis**
Sporothrix schenkii**
Trichophyton rubrum
Trichophyton spp
Trichosporon spp

Genera not Listed in The Approved List of Biological Agents (2021) but may be used due to changes in names of fungi
Lophophyton
Nannizza
Paraphyton

APPENDIX 2 to CABI Standard Protocol SP20

HAZARD GROUP 3 ORGANISMS

- Note: Organisms in this hazard group are to be destroyed on arrival at CABI

HAZARD GROUP 3 BACTERIA

Bacillus anthracis
Brucella abortus
Brucella canis
Brucella melitensis
Brucella suis
Burkholderia mallei (Pseudomonas mallei)
Burkholderia pseudomallei (Pseudomonas pseudomallei)
Coxiella burnetii
Ehrlichia sennetsu (Rikettsia sennetsu)
Escherichia coli (verocytotoxigenic strains O157 and O103)
Francisella tularensis (Type A)
Mycobacterium africanum
Mycobacterium bovis
Mycobacterium leprae
Mycobacterium malmoense
Mycobacterium microti
Mycobacterium szulgai
Mycobacterium tuberculosis
Mycobacterium ulcerans
Rickettsia akari
Rickettsia canada
Rickettsia conorii
Rickettsia montana
Rickettsia prowazekii
Rickettsia rickettsii
Rickettsia tstsugamushi
Rickettsia typhi (*Rickettsia mooseri*)
Rickettsia spp
Salmonella paratyphi A, B (java), C (Choleraesuis)
Salmonella typhi
Shigella dysenteriae (Type 1)
Yersinia pestis

HAZARD GROUP 3 FUNGI

Blastomyces dermatitidis (Ajellomyces dermatitidis)
Cladophialophora bantiana (formerly *Xylohypha bantiana*)
Coccidioides immitis
Coccidioides posadasii
Histoplasma capsulatum
(Histoplasma capsulatum var. *capsulatum)*
(Histoplasma capsulatum var. *farcinimosum)*
(Ajellomyces capsulatus)
Histoplasma duboisii
(Histoplasma capsulatum var. *duboisii)*
Paracoccidioides brasiliensis
Talaromyces marneffei (formerly *Penicillium marneffei*)
Rhinocladiella mackenziei
(formerly *Ramichloridium*)

APPENDIX 3 to CABI Standard Protocol SP20

HAZARDOUS SOURCES OF BACTERIA AND FUNGI

Organisms isolated from the following substrates are to be handled as potentially hazardous samples. The sample is to be rejected and destroyed.
Hazardous substrates:

Tissues, organs and body-fluids of man and other vertebrates.

Medical and veterinary fabrics and swabs taken from humans/animals.

Hair, nails and feathers of vertebrates.

Bait provided as skin, hair, nails, or feathers of vertebrates.

Vertebrate dung/excreta.

Eggs and nesting material of vertebrates.

Meat products.

Animal feed (potentially containing animal products).

Soil treated with animal products and materials from burrows or cavities inhabited by vertebrates.

Sewage.

Sludge.

Water: effluent and waste water from all countries; river water from developing countries.

CABI Standard Protocol SP30 Procedures for receipt of Hazard Group 3 Human Pathogens

1. Introduction

A description of hazard groups 3 organisms as designated by the UK Health and Safety Executive (HSE) Advisory Committee on Dangerous Pathogens (ACDP) is given in SP20.

Customers are notified that CABI Bioscience does not handle hazard group 3 organisms, nevertheless it is always possible that one may be received either as a named or unnamed culture. It is therefore essential that all parcels containing microbial samples are opened by trained personnel in the Class III/I microbiological safety cabinet in the Parcel Reception Laboratory (PRL), Room C1.01. (see SP 25, Operation of the Safety Cabinet in Parcel Reception Laboratory).

A list of bacteria and fungi currently categorized by ACDP in hazard group 3 is given in Appendix 1 of this document.

Organisms received for processing, named as hazard group 3 organisms, are to be destroyed on arrival following procedures outlined below. Reasons for the rejection should be recorded in the Biosafety check comments field in the QC check test in LIMS.

2. Procedure for disposal of hazard group 3 organisms

On opening a batch of samples, if documentation indicates that any of the enclosed samples may be dangerous human pathogens, classified as hazard group 3, the entire batch is be destroyed by autoclaving. If it is not possible to do so immediately, the entire batch is to be retained inside the safety cabinet with the cabinet operating in Class I mode as outlined in SP 25. Checks should be made to ensure that the velocity is set between 0.7-1.0 m/s. This may be adjusted, if necessary, by turning the velocity dial on the control panel. With the cabinet operating in Class I mode, samples of hazard group 3 can be safely contained. Inform the Delivery team and Safety Officer that a dangerous sample has been received and instruct all relevant staff that no access will be permitted to the laboratory until further notice. A sign, to this effect, must be placed on the outside of the C1.01 laboratory door.

If it is necessary to leave the laboratory unattended for a short period, it is recommended to operate the cabinet in Class III mode:

The right-hand side panel of the cabinet should be removed, with the airflow running between 0.7-1.0 m/s and the front of the cabinet sealed by fitting the closure panel. Air will then be drawn into the cabinet through the side filter.

The documentation should be destroyed by autoclaving immediately. A comment should be made in the folder metadata.

The sender is to be contacted, to inform them of the disposal of the samples and paperwork.

The customer shall also be asked to supply an electronic copy of the request form which can be attached to the LIMS folder. If this is not possible a comment should be made in the folder metadata stating why there is no request form.

3. Autoclaving of hazard group 3 organisms

Potential hazard group 3 organisms should be retained in the safety cabinet in the PRL until the batch can be autoclaved in autoclave 8. Whilst wearing a lab coat and gloves, an autoclave bag should be placed into the safety cabinet in the PRL. Samples, packaging and documentation, should be transferred into the autoclave bag, and then transferred to Autoclave 8 (see LF-109). Once the cycle is complete and the load has been deemed safe, the autoclaved material is to be placed in yellow incineration bags and transferred to the Autoclave Room (A1.10) in the Ainsworth Building for incineration.

4. Decontamination of the safety cabinet

Following disposal of Class 3 pathogens, the safety cabinet should be decontaminated. The procedure is outlined in SP43 and can be carried out only by trained personnel. Inform the Safety Officer and a member of the Delivery team or deputies.

5. Handling spillage of hazard group 3 organisms

If spillage or breakage of a sample of a hazard group 3 organism has occurred, or if there has been any spillage or breakage of a sample suspected to be hazard group 3, decontamination of the laboratory must be carried out.

- Leave the laboratory immediately and switch off the air exchange system to the laboratory at the control panel in the adjoining room (C1.04)
- Inform all relevant staff that no access will be permitted to the laboratory until further notice
- Place a sign to this effect on the outside of the laboratory door and inform the safety officer who will decide appropriate decontamination procedures. If any staff member has been exposed to a hazard group 3 organism, inform the safety officer and seek medical advice.

APPENDIX 1 to CABI Standard Protocol SP30

HAZARD GROUP 3 ORGANISMS

Note: Organisms in this hazard group are to be destroyed on arrival at CABI

HAZARD GROUP 3 BACTERIA

Bacillus anthracis
Brucella abortus
Brucella canis
Brucella melitensis
Brucella suis
Burkholderia mallei (formerly *Pseudomonas mallei)*
*Burkholderia pseudomallei (*formerly *Pseudomonaspseudomallei)*
Chlamydia psittaci (Avian strains)
Coxiella burnetii
Ehrlichia sennetsu (Rikettsia sennetsu)
Escherichia coli (verocytotoxigenic strains)
Francisella tularensis (Type A)
Mycobacterium africanum
Mycobacterium avium/intracellulare
Mycobacterium bovis
Mycobacterium leprae
Rickettsia akari
Rickettsia canada
Rickettsia conorii
Rickettsia montana
Rickettsia prowazekii
Rickettsia rickettsii
Rickettsia tsutsugamushi
Rickettsia typhi (Rickettsia mooseri)
Rickettsia spp.
Salmonella paratyphi A, B, C
Salmonella typhi
Shigella dysenteriae (Type 1)
Yersinia pestis

Mycobacterium malmoense
Mycobacterium microti
Mycobacterium szulgai
Mycobacterium tuberculosis
Mycobacterium ulcerans

HAZARD GROUP 3 FUNGI

Blastomyces dermatitidis (*Ajellomyces dermatitidis*)
Cladophialophora bantiana (*Xylohypha bantiana*)
Coccidioides immitis
Coccidioides posadasii
Histoplasma capsulatum (*Ajellomyces capsulatus*) (formerly *Histoplasma capsulatum* var. *capsulatum*, *Histoplasma capsulatum* var. *farcinimosum*)
Histoplasma duboisii (formerly *Histoplasma capsulatum* var. *duboisii*)
Paracoccidioides brasiliensis
Penicillium marneffei
Rhinocladiella mackenziei (formerly *Ramichloridium*)

Reference

Anon (1998) Safe Biotechnology, Classification of Biological Agents: Bacteria. Leaflet B006. BG Chemie, Germany.

Appendix H

CABI Culture collection (Genetic Resource Collection) database structure and field content

The CABI Genetic Resources Collection (GRC) is tasked with the collection of organisms to provide a resource for the scientific programmes of CABI and incorporates the UK National Collection of Fungus Cultures. CABI GRC is formerly the collection of the Commonwealth Mycological Institute and subsequently the International Mycological Institute (IMI); it has always been part of CABI, formerly known as the Commonwealth Agricultural Bureau (CAB). The GRC accesses filamentous fungi, yeasts and plant pathogenic bacteria with a focus on biocontrol agents, biostimulants and other microorganisms used in agriculture belonging to these groups. It endeavours to store the data recommended by the OECD Best Practice for BRCs and has collaborated in various European projects to try to make its data systems interoperable with the culture collection community. Some examples of these projects are:

- MINE - Development of Microbial Information Network in Europe to a centralized European culture collections database system, 1991-1994 https://cordis.europa.eu/project/id/BIOT0280
- CABRI - Common Access to Biotechnological Resources and Information; FP4-BIOTECH 2 - Specific research, technological development and demonstration programme in the field of biotechnology, 1994-1998 https://cordis.europa.eu/project/id/BIO4960231
- EBRCN - European Biological Resource Centres Network, 2004–2006 http://www.cabri.org/FAQ/faq_ebrcn.html
- EMbaRC - European Consortium of Microbial Resources Centres; To improve, coordinate and validate microbial resource delivery to European and International researchers from both public and private sectors, http://www.embarc.eu/
- MIRRI - Microbial Resource Research Infrastructure; An EU preparatory phase project for a European Strategy Forum Research Infrastructure, 2009-2012 https://www.mirri.org/
- EMBRIC - European Marine Biological Research Infrastructure Cluster; designed to accelerate the pace of scientific discovery and innovation from marine bio-resources. To promote new applications derived from marine organisms in fields such as drug discovery, novel foods and food ingredients, aquaculture selective breeding, bioremediation, cosmetics and bioenergy, 2015-2019, http://www.embric.ugent.be/

Through these projects common data sets were established and published; CABI's data fields are described below. These data fields and content formats were taken up by subsequent database projects and form the basis for the minimal, recommended and full data-sets of the OECD Best Practice for BRCs. The data fields and information

DOI: 10.1079/9781800622135.apph

systems for the collections involved in the Microbial Resources Research Infrastructure have also influenced CABI, these were analysed and reported (Smith *et al.*, 2017). CABI has used a variety of software to manage the data and currently utilizes Biolomics see https://www.bio-aware.com/.

References

Gams, W., Hennebert, G.L., Stalpers, J.A., Jansens, D., Schipper, M.A.A., Smith, J., Yarrow, D. and Hawksworth, D.L. (1988) Structuring strain data for the storage and retrieval of information on fungi and yeasts in MINE, Microbial Information Network Europe. *Journal of General Microbiology* 134, 1667-1689. https://www.microbiologyresearch.org/docserver/fulltext/micro/134/6/mic-134-6-1667.pdf?expires=1635696846&id=id&accname=guest&checksum=6F2AF496667C155D6F9068A71EC2CE57

Stalpers, J.A., Kracht, M., Jansens, D., De Ley, J., Van der Toorn, J., Smith, J., Claus, D. and Hippe, H. (1990) Structuring strain data for storage and retrieval of information on bacteria in MINE, Microbial Information Network Europe. *Systematic and Applied Microbiology* 13, 92-103. https://www.sciencedirect.com/science/article/abs/pii/S0723202011801852

Smith, D., Stackebrandt, E., Casaregola, S., Romano, P. and Glöckner, F.O. (2017) MIRRI Recommendations for Exploiting the Full Potential of Microorganism Data. *Annals of Biometrics and Biostatistics* 4, 1027. http://www.jscimedcentral.com/Biometrics/biometrics-4-1027.pdf

Fungi data fields analysis

The CABI Genetic Resources Collection (GRC) holds over 28,000 filamentous fungi

Catalogue availability

The catalogue is available on CABI Website: https://www.cabi.org/services/microbial-services/culture-collection-microorganism-supply/grc/

The last printed individual catalogue was the tenth edition published in 1992. A combined catalogue of fungi was published as part of the integrated UK National Culture Collection (UKNCC). The IMI produced catalogues every 3-4 years but now makes its data available via its online catalogue and through the World Data Centre for Microorganisms Global Catalogue of Microorganisms (http://gcm.wdcm.org/).

Sample of records

CABI - Collection of Filamentous Fungi, Yeasts and Plant Pathogenic Bacteria

IMI	303389
NAME	ASPERGILLUS AWAMORI Nakazawa
PRESERVATION	FD 15MAY86 G REFD 15AUG89 T Oil
TAXONOMY	hyphomycetes
MYCOLOGIST	R
DATE OF PROCESSING	860515
METABOLITES	citric acid < (C1) I;
LITERATURE	Eswaran, Sarasa, 1978, Ph.D. thesis, Botany Department, National University of Singapore. MIRCEN J., 1985, 1: 55-61.
ENZYMES	protease < (C1) I; amylase < (B4) I; cellulase (E2, B4) I;
ORIGIN	Japan
IDENTIFIED BY	A.H.S. Onions

HISTORY (FOR MINE)	G. Lim > 1986 IMI
PRESERVATION	LY (10% sk. milk + 5% inositol) MO
ORGANISM TYPE	F
DATE OF INPUT	860710
DATE OF UPDATE	960711
SECURITY CODE	1
CONDITIONS	CZ, MCZ, 23
SUPPLY FORM	Freeze-dried

CABI - Collection of Filamentous Fungi, Yeasts and Plant Pathogenic Bacteria

IMI	024314*
NAME	PENICILLIUM CHRYSOGENUM Thom
PRESERVATION	FD 16NOV77 4G FT 07JAN81 4S REFD 05FEB81 4S REFD 22FEB85 (EF6) 3S (M) 3S OIL 26NOV57
TAXONOMY	hyphomycetes
MYCOLOGIST	R
DATE OF PROCESSING	850222
METABOLITES	chrysogenin < D;
SYNONOMY	*Penicillium harmonense* Baghd.; *Penicillium camerunense* Heim; *Penicillium aromaticum* var. *microsporum* Romankova; *Penicillium notatum* Westl.; *Penicillium meleagrinum* Biourge; *Penicillium aromaticum* f. *microsporum* Romankova;
OTHER COLLECTIONS	FRR 807; LSHB P19; NCTC 589; CBS 306.48; ATCC 10106; QM 7500; Thom 26; NRRL 808; LSHB Ad3;
STATUS	T
SUBSTRATE	cheese
ORIGIN	USA
IDENTIFIED BY	A.H.S. Onions
DEPOSITED BY	NCTC
HISTORY (FOR MINE)	1904 Thom 26 > NRRL 807 > LSHB Ad3 > 1949 IMI
PRESERVATION	LY (10% sk. milk + 5% inositol), MO
ORGANISM TYPE	F
DATE OF INPUT	850705
DATE OF UPDATE	950524
SECURITY CODE	1
CONDITIONS	CZ, MCZ, 23
PIGMENTS	Yellow < diffusible into oil.
SUPPLY FORM	Freeze-dried

Database information

The GRC database uses Biolomics software available from BioAware https://www.bio-aware.com/. All new entries are given security code 2 or 3 (SEC field). Data are subsequently edited by the curator of the collection. The data specification used is a modification of that published in Gams *et al.*, (1988).

Update:

database	constantly
catalogue	hard copy catalogues are no longer produced
WEB	Currently twice a year on the web but software is being installed to update constantly

Data fields

Field name	Brief description	Catalogue Output (indicated by x)	Authentication
Internal Administration			
ACCN	Accession number		CABI
DOI	Date of input		CABI
DOU	Date of update		CABI
SEC	Security code		CABI
OT	Organisms type	x (implicit)	CABI
Name			
Name	Species name, authors	x	Depositor, CABI
RACE	Race name & authors	x	Depositor, CABI
TAX	Taxonomy	x	CABI
TAXREM	Taxonomic remarks		CABI
SYN	Synonymous names	x	CABI
NAMCH	Name changes / previous name	x	CABI
LIT	Taxonomic & other literature related to a species or a strain		Depositor, CABI
Strain administration			
STN	Strain number	x	CABI
OCC	Other collections	x	CABI, Depositor
DOA	Date of accession		CABI
DOP	Date of processing		CABI
NOTES	General remarks		CABI
CHK	Internal check of identity at collection		CABI
PRE	Preservation	x	CABI by usage
PREREM	Remarks on preservation		CABI by usage
RESTR	Precautions and restrictions		Depositor, CABI
Environment & History			
SSTR	Original substratum	x	CABI, Depositor
HAB	Habitat (original biotype)		CABI, Depositor
ORI	Geographic location	x	Depositor
COLL	Collector		Depositor
ISOL	Isolated by		Depositor
ISOM	Source & method of isolation		Depositor
DET	Identified by		Depositor, CABI
DEP	Depositor	x	Depositor
HIST	History for MINE	x	Depositor
Biological interactions			
MPAR	Mycoparasitism		Depositor, CABI
PATH	Pathogenicity and virulence		Depositor, CABI
PATHREM	Remarks on pathogenicity		Depositor, CABI
TOX	Toxicity to other organisms		Depositor, CABI
Sexuality			
SEX	Sexual behaviour		CABI
SEXST	Sexual state		Depositor, CABI
Properties			
PIGMENT	Pigment production & autofluorescence		Depositor, CABI
Genotype and genetics			
GEN	Genotype		Depositor
RNAD	Hybridization rRNA-DNA with other strains		CABI (literature)
MUT	Mutant	x	Depositor
MUTMET	Method of mutation		Depositor

Continued

Continued.

Field name	Brief description	Catalogue Output (indicated by x)	Authentication
Growth conditions			
CONDS	Conditions for growth & maintenance on solid media	x	CABI
CONDSP	Conditions for fruiting or sporulation		CABI
CSOR	Carbon sources that have been tested for growth		CABI
NSOR	Nitrogen sources that have been tested for growth		CABI
TOL	Tolerances & sensitivities		CABI
TEMPR	Temperature relationships for growth	x	CABI
Chemistry & enzymes			
PHY - ENZ	Enzymes produced		Depositor, CABI
DEC	Decomposition & deteriorating capacities		Depositor, CABI
MET	Metabolites produced		Depositor, CABI
Practical applications			
APLI	Industrial & general applications		Depositor, CABI
TEST	Application in testing and assays		Depositor, CABI
PATENT	Patents		Depositor
Status			
STAT	Status, indicating type or authentic strains		Depositor, CABI

Retention: records are maintained so long as an organism remains viable. On the loss of a strain the database record is printed and stored on file before the entry is removed from the database.

Yeasts data fields analysis

The CABI Genetic Resources Collection (GRC) holds ca. 300 yeasts.

Catalogue availability

The catalogue is available on CABI Website: https://www.cabi.org/services/microbial-services/culture-collection-microorganism-supply/grc/. The last printed individual catalogue was the tenth edition published in 1992. A combined catalogue of fungi was published as part of the integrated UK National Culture Collection (UKNCC). The IMI produced catalogues every 3-4 years but now makes its data available via its online catalogue and through the World Data Centre for Microorganisms Global Catalogue of Microorganisms (http://gcm.wdcm.org/).

Sample of records

CABI - Collection of Filamentous Fungi, Yeasts and Plant Pathogenic Bacteria

IMI	369526
NAME	HANSENIASPORA VALBYENSIS Klocker
PRESERVATION	FD 07DEC95 P+V

DATE OF ACCESSION	951122
DATE OF PROCESSING	951207
OTHER COLLECTIONS	NCYC 17;
STATUS	T of *Hanseniaspora valbyensis* Klocker
SUBSTRATE	soil
ORIGIN	Denmark
COLLECTOR	A. Klocker
HISTORY (FOR MINE)	1920 A. Klocker > NCYC > 1995 IMI
PRESERVATION	LY (10% sk. milk + 5% inositol)
ORGANISM TYPE	Y
DATE OF INPUT	960325
DATE OF UPDATE	960920
SECURITY CODE	1
CONDITIONS	YM, 25
SUPPLY FORM	Freeze-dried

CABI - Collection of Filamentous Fungi, Yeasts and Plant Pathogenic Bacteria

IMI	298055
NAME	DEKKERA ANOMALA M. Th. Smith & v. Grinsven
PRESERVATION	LN 29NOV85 (FR 4I1 a(iii)) T FD 29OCT85 REFD 05JAN90 G
TAXONOMY	Endomycota
MYCOLOGIST	F
DATE OF PROCESSING	851129
SUBSTRATE	soft drink
ORIGIN	Germany
ISOLATED BY	A.M.U. Grinsuen, 30.6.81
IDENTIFIED BY	D.W. Minter
DEPOSITED BY	A.P. Williams
HISTORY (FOR MINE)	IMI 1985
PRESERVATION	LN (10% glycerol) LY (10% sk. milk + 5% inositol),
ORGANISM TYPE	Y
DATE OF INPUT	890808
DATE OF UPDATE	961108
SECURITY CODE	1
CONDITIONS	MA, 25
SUPPLY FORM	Freeze-dried

Database information

The GRC database uses Biolomics software available from BioAware https://www.bio-aware.com/. All new entries are given security code 2 or 3 (SEC field). Data are subsequently edited by the curator of the collection. The data specification used is a modification of that published in Gams, W. (1988) *Journal of General Microbiology* 134:1667-1689.

Update:

database	constantly
catalogue	hard copy 3 yearly moving to two yearly in the UKNCC integrated catalogue
WEB	Currently twice a year on the web but software is being installed to update constantly

Data fields as filamentous fungi (above)

Retention: records are maintained so long as an organism remains viable. On the loss of a strain the database record is printed and stored on file before the entry is removed from the database.

Plant pathogen bacteria data fields analysis

The CABI Genetic Resources Collection (GRC) holds over 1700 plant pathogenic bacteria

Catalogue availability

The catalogue is available on CABI Website: https://www.cabi.org/services/microbial-services/culture-collection-microorganism-supply/grc/

The last printed individual catalogue was the tenth edition published in 1992. A combined catalogue of fungi was published as part of the integrated UK National Culture Collection (UKNCC). The IMI produced catalogues every 3-4 years but now makes its data available via its online catalogue and through the World Data Centre for Microorganisms Global Catalogue of Microorganisms (http://gcm.wdcm.org/).

Sample of records

CABI - Collection of Filamentous Fungi, Yeasts and Plant Pathogenic Bacteria

IMI	366616
NAME	COMAMONAS ACIDOVORANS (den Dooren de Jong) Tamaoka *et al.*, 1987
PRESERVATION	FD 06APR95 LN 07APR95 (Fr8:G8 d2)
TAXONOMY	Bacteria: Pseudomonadaceae
MYCOLOGIST	B
DATE OF ACCESSION	950216
DATE OF PROCESSING	950407
SUBSTRATE	petro-chemical wastewater
ORIGIN	India
COLLECTOR	Dharmendra
IDENTIFIED BY	G.S. Saddler

DEPOSITED BY	A. Desai (M.S. University of Baroda, India), no. D-4
HISTORY (FOR MINE)	1992 Dharmendra > A. Desai > 1995 IMI
PRESERVATION	LY (7% sucrose peptone), LN (10% glycerol)
ORGANISM TYPE	B
DATE OF INPUT	950516
DATE OF UPDATE	950907
SECURITY CODE	1
CONDITIONS	NA, 28
SUPPLY FORM	Freeze-dried

CABI - Collection of Filamentous Fungi, Yeasts and Plant Pathogenic Bacteria

IMI	349320
NAME	PSEUDOMONAS SYRINGAE pv. ERIOBOTRYAE (Takimoto) Young, Dye & Wilkie 1978
PRESERVATION	FD 29SEP76 P&V REFD 10SEP91 P&V
TAXONOMY	Bacteria:
MYCOLOGIST	B
DATE OF PROCESSING	910910
OTHER COLLECTIONS	IMI B06882; NCPPB 2801
SUBSTRATE	*Eriobotrya japonica*
ORIGIN	USA
IDENTIFIED BY	J.F. Bradbury
HISTORY (FOR MINE)	M. Lai > D.C. Hildebrand > 1975 NCPPB > 1976 IMI
NAME CHANGE	*Pseudomonas eriobotryae*
PRESERVATION	LY (7% sucrose + 7% peptone)
ORGANISM TYPE	B
DATE OF INPUT	920122
DATE OF UPDATE	960229
SECURITY CODE	1
CONDITIONS	NA, 25
RESTRICTIONS	ACDP Hazard Group 2. Containment level 2 required when handling.
SUPPLY FORM	Freeze-dried

Database information

The GRC database uses Biolomics software available from BioAware https://www.bio-aware.com/. All new entries are given security code 2 or 3 (SEC field). Data are subsequently edited by the curator of the collection.

Update:

database	constantly
catalogue	hard copy 3 yearly moving to two yearly in the UKNCC integrated catalogue
WEB	Currently twice a year on the web but software is being installed to update constantly

Data fields as filamentous fungi

Retention: records are maintained so long as an organism remains viable. On the loss of a strain the database record is printed and stored on file before the entry is removed from the database.

Identification & authentication of data

Remarks

Filamentous fungi and Yeasts
Data are entered according to the MINE format

Reference: W. Gams *et al.*, (1988) *Journal of General Microbiology* 134, 1667-1689

Bacteria
Data are entered according to the MINE format

Reference: J.A. Stalpers *et al.*, (1990) *Systematic and Applied Microbiology* 13, 92-103

General observations

1. Fields do not start with a capital letter unless grammatically necessary, e.g. 'Candida albicans', 'French cheese', 'soil', 'G.L. Hennebert'

Internal Administration

1. ACCN *Accession number*

Field

Description:	unique identification of each record in the database
Format:	ascending number; automatically given
Type:	structured, not repetitive, not numbered
Status:	mandatory
Output:	not present
Links to other fields:	none
Group of fields:	
Responsible person for modification:	database administrator
Date of modification:	none

Data

given by:	database administrator
authenticated by:	database administrator
Tests for authentication:	none
Tests for consistency:	value given by software
Responsible person for input:	database administrator
Responsible person for modification	not modified
Values:	1,2,3, ...

Notes

2. DOI Date of input

Field

Description:	date of input of a record in the database
Format:	Numerical YYMMDD 'DD' and 'DD MMM' may be optional
Status:	mandatory
Type:	structured
Output:	not present
Links to other fields:	none
Group of fields:	
Responsible person for modification:	automatic
Date of modification:	none

Data

given by:	software
authenticated by:	database administrator
Tests for authentication:	none
Tests for consistency:	none
Responsible person for input:	values given by software
Responsible person for modification:	not modified
Values:	Valid values 990125 Optional : for month: Jan, Feb, Mar, Apr, May, Jun, Jul, Aug, Sep, Oct, Nov, Dec

Notes

Examples DOI 990125; DOI 1995

3. DOU Date of update

Field

Description:	date of last update of a record
Format:	Numerical YYMMDD
Status:	mandatory when update occurs
Type:	structured
Output:	not present
Links to other fields:	none
Group of fields:	
Responsible person for modification:	database administrator
Date of modification:	none

Data

given by:	software
authenticated by:	database administrator
Tests for authentication:	none
Tests for consistency:	none
Responsible person for input:	values given by software
Responsible person for modification:	not modified
Values:	Valid values 990125

Notes

4. SEC *Security code*

Field

Description:	code that prevents or allows output or access to the record
Format:	number
Status:	mandatory if restrictions apply
Type:	structured
Output:	not present
Links to other fields:	none
Group of fields:	
Responsible person for modification:	Curator
Date of modification:	none

Data

given by:	curator
authenticated by:	database administrator
Tests for authentication:	none
Tests for consistency:	database software
Responsible person for input:	data typist
Responsible person for modification:	curator
Values:	1 = listed in catalogue
	2 = not listed in catalogue, no external access
	3 = not listed in catalogue, access by privileged staff members

Notes

5. OT *Organisms type*

Field

Description:	mark a record to belong to the filamentous fungi, yeasts or bacteria
Format:	one letter
Status:	mandatory
Type:	structured
Output:	present (implicit)
Links to other fields:	None
Group of fields:	
Responsible person for modification:	Curator
Date of modification:	none

Data

given by:	Curator
authenticated by:	taxonomist
Tests for authentication:	identification
Tests for consistency:	database software
Responsible person for input:	data typist
Responsible person for modification:	taxonomist
Values:	F = filamentous fungi
	Y = yeasts
	B = bacteria

Notes

Name

General

A list of species names is available. When both anamorph and teleomorph names are available for the same species, the teleomorph name is used always for yeasts, and in most cases for filamentous fungi. The names are yet to be updated to the "one fungus one name" convention

Author's name citations:

- full author names are used, no abbreviations
- '&' is used between authors, '*et al.*' for more than 2 authors;
- the year of publication is not included
- colons for sanctioning authors are surrounded by spaces
- initials are included when necessary to distinguish different authors with the same family name, only one letter, hyphens are used if they occur in the full name (e.g. 'M-L'). Initials are separated by dots; the last dot is followed by a space.
- no author name is mentioned of the current epithet is the same as the epithet at the next higher rank (e.g. '*Fusarium dimerum*, Penzig var. *dimerum*'), but in all other cases only the authors of the lowest epithet are mentioned.

6. NOM Species name, authors

Field

Description:	genus name and species epithet, together with the author who gave the name and/or epithet of the microorganism
Format:	genus name current rank epithet, author's name
Type:	structured
Status:	mandatory
Output:	present
Links to other fields:	none
Group of fields:	
Responsible person for modification:	database administrator
Date of modification:	none

Data

given by:	depositor
authenticated by:	curator
Tests for authentication:	identification
Tests for consistency:	database software
Responsible person for input:	data typist
Responsible person for modification:	taxonomist
Values:	Name in accordance with original publication of author, author's names are given in full

Notes

Examples: NOM *Fusarium dimerum* var. *crassum*, Wollenweber; NOM *Botrytis* sp.

7. RACE

Field

Description:	Race Name, authors
Format:	Race Name, authors
Type:	structured
Status:	not mandatory
Output:	present
Links to other fields:	None
Group of fields:	
Responsible person for modification:	database administrator
Date of modification:	none

Data

given by:	depositor
authenticated by:	depositor
Tests for authentication:	none
Tests for consistency:	database software
Responsible person for input:	data typist
Responsible person for modification:	taxonomist
Values:	

Notes

If 2 strains, the taxon only differs at the 'race' level, the strain records refer to the same species record. Example: 2, Snyder & Hansen for a strain of *Fusarium oxysporum* var. *pisi* (van Hall) Snyder & Hansen

8. TAX *Taxonomy*

Field

Description:	Taxonomy
Format:	set of names of ranks higher than genus
Type:	structured, repetitive separated by ';' (colon space), not numbered
Status:	not mandatory
Output:	
Links to other fields:	None
Group of fields:	
Responsible person for modification:	database administrator
Date of modification:	none

Data

given by:	taxonomist
authenticated by:	taxonomist
Tests for authentication:	none
Tests for consistency:	database software
Responsible person for input:	data typist
Responsible person for modification:	taxonomist
Values:	see example, subfields have to be accepted names of families, orders, classes

Notes

Example: **Basidiomycota, Agaricales, Pluteaceae**

9. TAXREM *Taxonomic remarks*

Field

Description:	Taxonomic remarks
Format:	free text
Type:	not structured
Status:	not mandatory
Output:	not present
Links to other fields:	none
Group of fields:	
Responsible person for modification:	Curator
Date of modification:	none

Data

given by:	depositor or taxonomist
authenticated by:	depositor or taxonomist
Tests for authentication:	none
Tests for consistency:	none
Responsible person for input:	data typist
Responsible person for modification:	taxonomist
Values:	free text

Notes

Example: cf. also *Penicillium simplicissimum*; anamorph. = Phoma sp.

10. SYN *Synonymy*

Field

Description:	Synonymous names
Format:	genus name current rank epithet, author's name
Type:	structured, repetitive separated by ';' (colon space), not numbered
Status:	not mandatory
Output:	present
Links to other fields:	none
Group of fields:	
Responsible person for modification:	Curator
Date of modification:	none

Data

given by:	depositor or taxonomist
authenticated by:	depositor or taxonomist
Tests for authentication:	literature
Tests for consistency:	none
Responsible person for input:	Data typist
Responsible person for modification:	Curator
Values:	

Notes

11. NAMCH *Name changes*

Field

Description:	Name changes / previous name
Format:	'received as ....'
Type:	structured, repetitive separated by ';' (semi-colon space), not numbered
Status:	not mandatory
Output:	not present
Links to other fields:	none
Group of fields:	
Responsible person for modification:	database administrator
Date of modification:	none

Data

given by:	curator
authenticated by:	curator
Tests for authentication:	none
Tests for consistency:	database software
Responsible person for input:	data typist
Responsible person for modification:	curator
Values:	see format

Notes

Example: NAMCH received as *Fomes igniarius*

12. LIT Literature

Field

Description:	Taxonomic & other literature related to a strain
Format:	authors, journal or book, volume: pages, year
Type:	structured, repetitive, not numbered
Status:	not mandatory
Output:	not present
Links to other fields:	none
Group of fields:	
Responsible person for modification:	database administrator
Date of modification:	none

Data

given by:	depositor or taxonomist
authenticated by:	depositor or taxonomist
Tests for authentication:	check the reference
Tests for consistency:	none
Responsible person for input:	data typist
Responsible person for modification:	taxonomist
Values:	free text (no checks yet)

Notes

Strain administration

13. IMI Strain number

Field

Description:	Strain number
Format:	collection acronym followed by identification
Type:	structured
Status:	mandatory
Output:	present
Links to other fields:	RT (must occur if RT = 'STR', may not occur for other record types)
Group of fields:	
Responsible person for modification:	database administrator
Date of modification:	none

Data

given by:	curator
authenticated by:	curator
Tests for authentication:	none
Tests for consistency:	database software
Responsible person for input:	data typist
Responsible person for modification:	
Values:	see examples

Notes

Example: *IMI 017454, IMI 123456, IMI 342006*

14. OCC Other Culture Collections

Field

Description:	Number in other Collections
Format:	collection acronym ' ' (space) number
Type:	structured, repetitive separated by '; ' (semi-colon space), not numbered
Status:	not mandatory
Output:	present
Links to other fields:	
Group of fields:	
Responsible person for modification:	database administrator
Date of modification:	none

Data

given by:	depositor and/or curator
authenticated by:	curator
Tests for authentication:	none
Tests for consistency:	database software
Responsible person for input:	data typist
Responsible person for modification:	curator
Values:	see example

Notes

a name of a person can be used as a collection acronym in case of well-known personal collections. Other strain designations have to be preceded by the word 'strain'.

Example: OCC CBS 5312; NRRL Y-2458; Biourge 25

15. DOA *Date of accession*

Field

Description:	Date of Accession of a strain in the collection
Format:	MMM YYYY or YYYY
Type:	structured
Status:	mandatory
Output:	present
Links to other fields:	
Group of fields:	
Responsible person for modification:	database administrator
Date of modification:	none

Data

given by:	curator
authenticated by:	curator
Tests for authentication:	not applicable
Tests for consistency:	database software
Responsible person for input:	data typist
Responsible person for modification:	not modified
Values:	see examples

Notes

Example: Nov 1968

16. Notes

Field

Description:	General remarks
Format:	free text
Type:	not structured
Status:	not mandatory
Output:	not present
Links to other fields:	none
Group of fields:	
Responsible person for modification:	Curator
Date of modification:	none

Data

given by:	curator or taxonomist
authenticated by:	curator or taxonomist
Tests for authentication:	none
Tests for consistency:	none
Responsible person for input:	data typist
Responsible person for modification:	curator or taxonomist
Values:	free text

Notes

17. PRE

Field

Description:	mode of preservation/storage
Format:	2 letter code
Type:	structured, repetitive separated by ';' (semi-colon)
Status:	not mandatory
Output:	present
Links to other fields:	
Group of fields:	
Responsible person for modification:	Curator
Date of modification:	none

Data

given by:	curator
authenticated by:	curator
Tests for authentication:	by usage
Tests for consistency:	database software
Responsible person for input:	data typist
Responsible person for modification:	curator
Values:	LY lyophilized (cryoprotectant in parentheses)
	LN liquid nitrogen (cryoprotectant in parentheses)
	MO sealed with mineral oil
	WA under water
	SG silica gel

Notes

Example: PRE LY; LN (10% glycerol); MO

18. PREREM

Field

Description:	Remarks on preservation. Indicate special care needed, results of preservation experiments, ability to survive after preservation.
Format:	free text
Type:	not structured
Status:	not mandatory
Output:	not present
Links to other fields:	none
Group of fields:	
Responsible person for modification:	database administrator
Date of modification:	none

Data

given by:	curator
authenticated by:	curator
Tests for authentication:	by usage
Tests for consistency:	none
Responsible person for input:	data typist
Responsible person for modification:	curator
Values:	free text

Notes

Example:

PREREM does not survive freeze-drying
PREREM
require DMSO for LN

19. RESTR

Field

Description:	Precautions and restrictions
Format:	free text
Type:	not structured, repetitive separated by '; ' (semi-colon space), not numbered
Status:	not mandatory
Output:	present
Links to other fields:	none
Group of fields:	
Responsible person for modification:	Curator
Date of modification:	none

Data

given by:	depositor or curator
authenticated by:	curator
Tests for authentication:	official lists of names for which precautions are necessary
Tests for consistency:	none
Responsible person for input:	data typist
Responsible person for modification:	curator
Values:	Data on the following categories can be used:- pathogenicity class (WHO standard and/or national standard) or hazard group, recombinant category - distribution and transportation restrictions and precautions - import regulations - export regulations

Notes

Example:

RESTR hazard group 2
RESTR Plant pathogen non-indigenous to the UK, requires a MAFF permit
RESTR only released to authorized customers

20. SSTR

Field

Description:	Original substratum/host on which the fungi was found
Format:	main information first to less important information
Type:	structured, not repetitive
Status:	not mandatory
Output:	present
Links to other fields:	None
Group of fields:	
Responsible person for modification:	database administrator
Date of modification:	none

Data

given by:	depositor
authenticated by:	taxonomist
Tests for authentication:	none
Tests for consistency:	visual check of list of values
Responsible person for input:	data typist
Responsible person for modification:	not modified (unless errors)
Values:	thesaurus of substrata (not yet available)

Notes

Example: SSTR wood of *Quercus robur*

21. HAB

Field

Description:	Habitat (original biotype where the strain or species was found)
Format:	free text
Type:	not structured
Status:	not mandatory
Output:	not present
Links to other fields:	none
Group of fields:	
Responsible person for modification:	database administrator
Date of modification:	none

Data

given by:	depositor
authenticated by:	taxonomist
Tests for authentication:	none
Tests for consistency:	none
Responsible person for input:	data typist
Responsible person for modification:	taxonomist
Values:	free text

Notes

	Particular of interest for the biotope of fruiting-bodies of macromycetes

Example: HAB beech forest; HAB limestone grassland

22. ORI

Field

Description:	Geographic location: location where the original material was found.
Format:	name of the country, state and other local details
Type:	structured
Status:	not mandatory (not always known)
Output:	present
Links to other fields:	None
Group of fields:	
Responsible person for modification:	database administrator
Date of modification:	none

Data

given by:	depositor
authenticated by:	depositor
Tests for authentication:	none
Tests for consistency:	database software
Responsible person for input:	data typist
Responsible person for modification:	not modified (unless errors or more information available)
Values:	thesaurus for geographical names

Notes

Examples: ORI Canada, Quebec, Montreal, La Salle Woods; ORI UK, Egham

23. COLL

Field

Description:	Collector of sample
Format:	name of collector, original number of specimen, date of collecting
Type:	structured
Status:	not mandatory
Output:	not present
Links to other fields:	None
Group of fields:	
Responsible person for modification:	database administrator
Date of modification:	none

Data

given by:	depositor
authenticated by:	depositor
Tests for authentication:	none
Tests for consistency:	none
Responsible person for input:	data typist
Responsible person for modification:	not modified (unless errors or more information available)
Values:	see format

Notes

Example: COLL F.F. Lombard, F.F.L. 1298, 18 Sep 1965

24. ISOL

Field

Description:	Isolated by
Format:	name of isolator, institute where the isolator worked, original number of isolate, isolation date.
Type:	structured
Status:	not mandatory
Output:	present
Links to other fields:	None
Group of fields:	
Responsible person for modification:	database administrator
Date of modification:	none

Data

given by:	depositor
authenticated by:	depositor
Tests for authentication:	none
Tests for consistency:	none
Responsible person for input:	data typist
Responsible person for modification:	not modified (unless errors or more information available)
Values:	For names, the same rules apply as for author names, but two names are linked by 'and'. If the number is not preceded by initials or another prefix, 'No.' is used.

Notes

Example: ISOL I.S. Nogood, I.S.N. 1548, 16 Oct 1982

25. ISOM

Field

Description:	Source & method of isolation
Format:	source, technique (medium)
Type:	structured
Status:	not mandatory
Output:	not present
Links to other fields:	RT (may only occur if RT = 'STR')
Group of fields:	
Responsible person for modification:	database administrator
Date of modification:	none

Data

given by:	depositor
authenticated by:	depositor
Tests for authentication:	none
Tests for consistency:	database software
Responsible person for input:	data typist
Responsible person for modification:	not modified (unless errors or more information available)
Values:	Source: SS single sexual spore MS mass sexual spores SA single asexual spore MA mass asexual spores MY mycelium RH rhizomorphs SCL sclerotia STR stromatic tissue AS ascoma BA tissue of basidioma CARP carpophore Methods: DP dilution plate DC damp chamber SU surface sterilization DT direct transfer EM enrichment method BAIT baiting (indicate substrate) PAST pasteurization Medium: abbreviated as in field CONDS

Notes

Example:

ISOM SS (Hagem)
ISOM BA, DT (MEA + benomyl)

26. DET Identified by

Field

Description:	name of the mycologist who identified the strain (who did a critical taxonomic study of the organism)
Format:	Name – Letter code
Type:	structured, repetitive separated by ";" (semi-colon) redet., not numbered
Status:	not mandatory
Output:	present
Links to other fields:	None
Group of fields:	
Responsible person for modification:	database administrator
Date of modification:	none

Data

given by:	depositor
authenticated by:	curator
Tests for authentication:	none
Tests for consistency:	none
Responsible person for input:	data typist
Responsible person for modification:	curator
Values:	A, B

Notes

Example: DET A

27. DEP Depositor

Field

Description:	Identification of depositor
Format:	name of depositor, institute where depositor worked
Type:	structured, not repetitive
Status:	not mandatory
Output:	present
Links to other fields:	None
Group of fields:	
Responsible person for modification:	database administrator
Date of modification:	none

Data

given by:	depositor
authenticated by:	depositor
Tests for authentication:	none
Tests for consistency:	none
Responsible person for input:	data typist
Responsible person for modification:	not modified (unless errors or more information available)
Values:	

Notes

28. HIS History (formerly HIST for MINE)

Field

Description:	history between isolation and deposit
Format:	isolator>... > year strain was sent, collection
Type:	structured
Status:	not mandatory
Output:	present
Links to other fields:	none
Group of fields:	
Responsible person for modification:	database administrator
Date of modification:	none

Data

given by:	depositor
authenticated by:	depositor
Tests for authentication:	none
Tests for consistency:	none
Responsible person for input:	data typist
Responsible person for modification:	not modified (unless errors or more information available)
Values:	'>' means 'sent to'

Notes

Example: S.I. Ahmed > 1983, CBS > 1984, IMI

Biological interactions

29. MPAR Mycoparasitism

Field

Description:	Qualification of the parasite or symptoms and host
Format:	parasite: host
Type:	structured, repetitive separated by '; ' (semi-colon space), not numbered
Status:	not mandatory
Output:	not present
Links to other fields:	none
Group of fields:	
Responsible person for modification:	Curator
Date of modification:	none

Data

given by:	depositor or taxonomist
authenticated by:	taxonomist
Tests for authentication:	none
Tests for consistency:	visual check of list of all values
Responsible person for input:	data typist
Responsible person for modification:	taxonomist
Values:	see format

Notes

Example: MPAR destructive mycoparasite: *Rhizoctonia solani*

30. PATH

Field

Description: Pathogenicity and virulence
Format: disease: susceptible group of organisms
Type: structured, repetitive separated by ';' (semi-colon), not numbered
Status: not mandatory
Output: present
Links to other fields: none
Group of fields:
Responsible person for modification: database administrator
Date of modification: none

Data

given by: depositor or taxonomist
authenticated by: taxonomist
Tests for authentication: none
Tests for consistency: visual check of list of all values
Responsible person for input: data typist
Responsible person for modification: taxonomist
Values: see format

Notes

Example:
PATH leaf spot: Hordeum; sheath rot: Triticum

31. TOX

Field

Description: Toxicity to other organisms
Format: symptom: susceptible organism <> degree of toxicity
Type: structured, repetitive separated by '; ' (semi-colon space), not numbered
Status: not mandatory
Output: not present
Links to other fields: none
Group of fields:
Responsible person for modification: database administrator
Date of modification: none

Data

given by: depositor or taxonomist
authenticated by: taxonomist
Tests for authentication: none
Tests for consistency: visual check of list of all values
Responsible person for input: data typist
Responsible person for modification: taxonomist
Values: see format

Notes

Not much data available

32. PATHREM

Field

Description:	Remarks on pathogenicity
Format:	free text
Type:	not structured
Status:	not mandatory
Output:	not present
Links to other fields:	none
Group of fields:	
Responsible person for modification:	database administrator
Date of modification:	none

Data

given by:	taxonomist
authenticated by:	taxonomist
Tests for authentication:	none
Tests for consistency:	none
Responsible person for input:	data typist
Responsible person for modification:	taxonomist
Values:	free text

Notes

Sexuality

33. SEX

Field

Description:	Sexual behaviour
Format:	One or two values separated by ', ' (comma space)
Type:	structured
Status:	not mandatory
Output:	present
Links to other fields:	none
Group of fields:	
Responsible person for modification:	database administrator
Date of modification:	none

Data

given by:	taxonomist
authenticated by:	taxonomist
Tests for authentication:	none
Tests for consistency:	database software
Responsible person for input:	data typist
Responsible person for modification:	taxonomist
Values:	A = amphithallic HO = homothallic HE = heterothallic HE1 = heterothallic unifactorial (bipolar) HE2 = heterothallic bifactorial (tetrapolar) PG = parthenogenetic (haploid)

Notes

34. SEXST Sexual State

Field

Description:	Sexual state
Format:	One or two values separated by ',' (comma space)
Type:	structured
Status:	not mandatory
Output:	present
Links to other fields:	none
Group of fields:	
Responsible person for modification:	database administrator
Date of modification:	none

Data

given by:	depositor or taxonomist
authenticated by:	taxonomist
Tests for authentication:	none
Tests for consistency:	database software
Responsible person for input:	data typist
Responsible person for modification:	taxonomist
Values:	MK = monokaryotic HOK = homokaryotic HA = haploid HEK = heterokaryotic DK = dikaryon DP = diploid MT = mating type, followed by code of mating type HO = homothallic cross = crossed mating types, followed by parental strains

Notes

Properties

35. PIGM Pigment

Field

Description:	Pigment production & autofluorescence
Format:	colour, name, (non-)diffusing, medium
Type:	structured, repetitive separated by ';' (semi-colon space), not numbered
Status:	not mandatory
Output:	not present
Links to other fields:	none
Group of fields:	
Responsible person for modification:	database administrator
Date of modification:	none

Data

given by:	depositor or taxonomist
authenticated by:	taxonomist
Tests for authentication:	microscope for colour, GCMS for substance
Tests for consistency:	none
Responsible person for input:	data typist
Responsible person for modification:	taxonomist
Values:	see format

Notes

Hardly data available

Example:
PIGM yellow < diffusible into oil

Genotype and genetics

36. GEN

Field

Description: Genotype
Format: chromosomal markers, phenotype may be added between parentheses
Type: structured, repetitive separated by ' ' (space), not numbered
Status: not mandatory
Output: not present
Links to other fields: none
Group of fields:
Responsible person for modification: database administrator
Date of modification: none

Data

given by: depositor
authenticated by: depositor
Tests for authentication: none
Tests for consistency: none
Responsible person for input: data typist
Responsible person for modification: taxonomist
Values: see format

Notes

Example:

GEN B317-9-9a Arginine 10;

37. MUT

Field

Description: Mutant
Format: '... mutant of ... <> author, date
Type: structured, not repetitive
Status: not mandatory
Output: present
Links to other fields: RT (occurs only for RT = 'STR')
Group of fields:
Responsible person for modification: database administrator
Date of modification: none

Data

given by: depositor
authenticated by: depositor
Tests for authentication: none
Tests for consistency: visual check of list of all values
Responsible person for input: data typist
Responsible person for modification: curator
Values: see format

Notes

Example:

MUT albino mutant

38. MUTMET

Field

Description:	Method of mutation
Format:	free text
Type:	not structured, not repetitive
Status:	not mandatory
Output:	not present
Links to other fields:	RT (occurs only for RT = 'STR')
Group of fields:	
Responsible person for modification:	database administrator
Date of modification:	none

Data

given by:	depositor
authenticated by:	depositor
Tests for authentication:	none
Tests for consistency:	none
Responsible person for input:	data typist
Responsible person for modification:	curator
Values:	free text

Notes

Example:

MUTMET UV, 30 min.

Growth conditions

39. CONDS

Field

Description:	Conditions for growth & maintenance on solid media
Format:	media; temperature, light
Type:	structured, repetitive (only part with media, separated by '; ')
Status:	not mandatory
Output:	present
Links to other fields:	RT (may occur for RT = 'SP' or 'STR')
Group of fields:	
Responsible person for modification:	database administrator
Date of modification:	none

Data

given by:	curator
authenticated by:	curator
Tests for authentication:	visual check of growth
Tests for consistency:	thesaurus of media
Responsible person for input:	data typist
Responsible person for modification:	curator
Values:	DL = Daylight

Notes

Example:

CONDS CMA, OA; 20-24 C, DL

40. CONDSP

Field

Description:	Conditions for fruiting or sporulation
Format:	see field CONDS
Type:	see field CONDS
Status:	not mandatory
Output:	not present
Links to other fields:	RT (may occur for RT = 'SP' or 'STR')
Group of fields:	
Responsible person for modification:	database administrator
Date of modification:	none

Data

given by:	curator
authenticated by:	curator
Tests for authentication:	visual check of sporulation
Tests for consistency:	
Responsible person for input:	data typist
Responsible person for modification:	curator
Values:	

Notes

41. CSOR

Field

Description:	Carbon sources that have been tested for growth
Format:	substance: result
Type:	structured, repetitive separated by ';' (semi-colon space), not numbered
Status:	not mandatory
Output:	not present
Links to other fields:	
Group of fields:	
Responsible person for modification:	database administrator
Date of modification:	none

Data

given by:	curator
authenticated by:	curator
Tests for authentication:	occasionally standard test kit
Tests for consistency:	database software
Responsible person for input:	data typist
Responsible person for modification:	curator
Values:	Values for result: + = growth within one week – = no growth D = growth within three weeks W = 'weak', little growth

Notes

Example:

CSOR tannic acid < (1%) I; lactic acid < (1%) I; citric acid < (1%) I; tween 80 < (1%) I;

NSOR

Field

Description:	Nitrogen sources that have been tested for growth
Format:	substance: result
Type:	structured, repetitive separated by ';' (semi-colon space), not numbered
Status:	not mandatory
Output:	not present
Links to other fields:	
Group of fields:	
Responsible person for modification:	database administrator
Date of modification:	none

Data

given by:	curator
authenticated by:	curator
Tests for authentication:	occasionally a standard test kit
Tests for consistency:	database software
Responsible person for input:	data typist
Responsible person for modification:	curator
Values:	see field CSOR

Notes

Example:
NSOR nitrite- < I; urea < I; creatine < I;

42. TOL

Field

Description:	Tolerances & sensitivities
Format:	agent, concentration: growth response
Type:	structured, repetitive separated by ';' (semi-colon space), not numbered
Status:	not mandatory
Output:	not present
Links to other fields:	none
Group of fields:	
Responsible person for modification:	Curator
Date of modification:	none

Data

given by:	curator
authenticated by:	curator
Tests for authentication:	occasionally a standard test kit
Tests for consistency:	database software
Responsible person for input:	data typist
Responsible person for modification:	curator
Values:	

Notes

Example:
TOL copper < (0.1% copper sulphate) I;
TOL TB2 resistant < not known in Europe

43. TEMPR

Field

Description:	Temperature relationships for growth
Format:	abbreviation
Type:	structured, not repetitive
Status:	not mandatory
Output:	present
Links to other fields:	none
Group of fields:	
Responsible person for modification:	database administrator
Date of modification:	none

Data

given by:	taxonomist
authenticated by:	taxonomist
Tests for authentication:	by experience
Tests for consistency:	database software
Responsible person for input:	data typist
Responsible person for modification:	taxonomist
Values:	PS = psychrophilic PT = psychrotroph ME = mesophilic TT = thermotolerant TH = thermophilic ET = extremely thermophilic

Notes

Chemistry & Enzymes

44. ENZ – PHY will change to ENZ

Field

Description:	Enzymes produced
Format:	name of enzyme, result and test where appropriate
Type:	structured, repetitive separated by ';' (semi-colon space), not numbered
Status:	not mandatory
Output:	present
Links to other fields:	none
Group of fields:	
Responsible person for modification:	database administrator
Date of modification:	none

Data

given by:	depositor or CABI Biochemistry Laboratory
authenticated by:	CABI Biochemistry Laboratory
Tests for authentication:	check literature references, or tested at CABI
Tests for consistency:	visual check of list of all values
Responsible person for input:	data typist
Responsible person for modification:	curator
Values:	see format

Notes

Example:

ENZ fatty acid hydrolysis < (tween 80) I; urease- < I; protease- < (gelatin hydrolysis) I; protease (weak) < (casein hydrolysis) I; nuclease < (DNA hydrolysis) I; cellulase- < I; amylase < (starch hydrolysis) I;

45. DEC

Field

Description:	Decomposition & deteriorating capacities
Format:	
Type:	structured, repetitive separated by ';' (semi-colon space), not numbered
Status:	not mandatory
Output:	not present
Links to other fields:	none
Group of fields:	
Responsible person for modification:	database administrator
Date of modification:	none

Data

given by:	depositor
authenticated by:	curator
Tests for authentication:	experience, or check literature reference
Tests for consistency:	visual check of list of all values
Responsible person for input:	data typist
Responsible person for modification:	curator
Values:	see format

Notes

Example:

DEC capnitrite-; gelatin; aesculin; copper sulphate < (0.1%) I;

46. MET

Field

Description:	Metabolites produced
Format:	
Type:	structured, repetitive separated by ';' (semi-colon space), not numbered
Status:	not mandatory
Output:	present
Links to other fields:	none
Group of fields:	
Responsible person for modification:	database administrator
Date of modification:	none

Data

given by:	depositor or CABI Biochemistry Laboratory
authenticated by:	CABI Biochemistry Laboratory
Tests for authentication:	check literature references, or tested at CABI
Tests for consistency:	visual check of list of all values
Responsible person for input:	data typist
Responsible person for modification:	curator
Values:	see format ; Method used and level of production (scale of 0 to 5, 0 is zero production 5 high yield) is given in brackets following the metabolite

Notes

Example:
MET aflatoxin B <D; aflatoxin G <D;
MET citric acid- (CO) <I

Practical applications

47. APLI

Field

Description:	Industrial & general applications
Format:	free text
Type:	structured, repetitive separated by ';' (semi-colon space), not numbered
Status:	not mandatory
Output:	present
Links to other fields:	none
Group of fields:	
Responsible person for modification:	database administrator
Date of modification:	none

Data

given by:	depositor
authenticated by:	curator
Tests for authentication:	none
Tests for consistency:	visual check of list of all values
Responsible person for input:	data typist
Responsible person for modification:	curator
Values:	free text

Notes

Examples:
APLI biological control agent of *Sclerotinia sclerotiorum*
APLI production of penicillin-amidase

48. TEST

Field

Description:	Application in testing and assays
Format:	free text
Type:	structured, repetitive separated by ';' (semi-colon space), not numbered
Status:	not mandatory
Output:	present
Links to other fields:	none
Group of fields:	
Responsible person for modification:	database administrator
Date of modification:	none

Data

given by:	depositor
authenticated by:	curator
Tests for authentication:	none
Tests for consistency:	visual check of list of all values
Responsible person for input:	data typist
Responsible person for modification:	curator
Values:	free text

Notes

Examples:
TEST European standard soft rot CEN/TC 38 N682; testing effectiveness of wood preservatives against attack
by soft rot fungus;
TEST assay of thiamine

49. PATENT

Field

Description:	Patents
Format:	country with patent nr, application
Type:	structured, repetitive separated by ';' (semi-colon space), not numbered
Status:	not mandatory
Output:	present
Links to other fields:	none
Group of fields:	
Responsible person for modification:	database administrator
Date of modification:	none

Data

given by:	depositor
authenticated by:	depositor
Tests for authentication:	none
Tests for consistency:	visual check of list of all values
Responsible person for input:	data typist
Responsible person for modification:	curator
Values:	see format

Notes

Example:

PATENT Canadian patent No.761.414; British Patent applied for
PATENT U.S. Pat. 4,094,739

Status

50. STAT

Field

Description:	Status, indicating type or authentic strains
Format:	abbreviation ' of ' taxonomic name
Type:	structured, repetitive separated by ';' (semi-colon space), not numbered
Status:	not mandatory
not mandatory	present
Links to other fields:	none
none	
Responsible person for modification:	database administrator
Date of modification:	none

Data

given by:	depositor
authenticated by:	curator
Tests for authentication:	check literature
Tests for consistency:	database software
Responsible person for input:	data typist
Responsible person for modification:	taxonomist
Values:	Abbreviations for strains: T = type HT = holotype LT = lectotype NT = neotype IT = isotype PT = paratype ST = syntype PTT = pathotype AUT = authentic strain OR = original strain REFS = reference strain For species, only 'T' and 'LT' may occur

Notes

Examples:
STAT T of *Penicillium abeanum* G. Smith
STAT NT of *Arthrobotrys entomopaga* Drechsler

The CABI collections agrees to follow the following recommendations from CABRI:

To keep consistency for searching and retrieving from catalogues:

- each record must contain at least the minimum dataset defined by CABRI for each type of collection
- spell checking for every field is a basic requirement
- Vocabulary has to be checked against CABRI in-house reference lists. These lists will be on the CABRI Website (http://www.cabri.org)
- Arabic numbers preferred to Roman numbers except for the case in which the latter are included in a cell line name, virus name or any other proper noun (e.g., Amdur II and ASB XIV)
- language: International English

Since it is presumed that catalogues submitted to CABRI will be maintained on a number of different platforms, it is necessary to adopt a standardized approach to certain scientific symbols. To avoid any errors due to incorrect reading of a character set, standard ASCII alternatives to symbols will be used. These alteration must be made before submitting to CABRI for indexing.

The alternatives include :

- Greek letters cannot be used, they must be fully spelled (write alpha, gamma, beta...)
- The ° symbol for temperature is to be omitted entirely. (e.g. 37C replaces 37°C).
- No subscripts or superscripts are allowed (e.g. cm3 replaces cm^3 and CO2 replaces CO_2).

Data Sets

The **Minimum Data Set** (MDS) is the set of data that is asked to describe **each product** from each collection. Each catalogue field has to be filled in (with unknown or not applicable when no information is given).

The MDS is required from new collection centres when they join CABRI.

The **Recommended Data Set** : collection centres are also asked to supply data for recommended fields; however, if data are not available, the product is still acceptable

Collection centres are free to provide data for other fields, which are part of the **Full Data Set**

A general MDS can be applied to each type of collections :

Identification	accession number, name
Origin	description, depositor, history.
Properties	
Culture conditions	medium, subculture routines...
Restrictions	
Hazard	

Minimum data set for filamentous fungi and yeasts

Here is described the minimum data set for fungi and yeast. If you don't adhere please inform the Technical Committee and your proposal will be examined.

Field name	Description
Strain number	Should contain: • collections' own accession number of the strain
Other culture collection numbers	Should contain: • accession numbers of the strain in other collections. Cannot be given if the strain is unique in the collection.
Restrictions	should contain: • Release conditions/Restrictions • country-specific shipping restrictions • restrictions imposed by the depositor etc.
Organism type	Should contain: • value «F» for filamentous fungi or • «Y» for yeasts.
Name	Should contain the full scientific name of the strain: • Genus name and species epithet • Subspecies* • Variety* • Forma* • Forma specialis* • Authors of the name *: only if applicable
Race	Should contain: • name of the race of the strain and authors of the race. Applies only to a small subset of the strains.
Status	Should contain: • nomenclature status of the strain (like «type», «neotype», etc.) Does not apply to all strains.
Misapplied names	Should contain: • names incorrectly applied to the strain (misidentifications) that may still be in use. Does not apply to all strains.

Continued

Continued.

Field name	Description
History of deposit	Should contain the history of the element from isolator to depositor. • Name of isolator • Institute of isolator • Name of depositor • Institute of depositor • Names and institutes in between
Conditions for growth	Should contain: • Culture medium • Light conditions • Temperature conditions
Form of supply	Should contain: • form in which the strain will be sent to the customer.

Minimum Data Set for bacteria and archaea

Here is described the minimum data set for bacteria. If you don't adhere please inform the Technical Committee and your proposal will be examined.

Field Name	Description
Strain number	Should contain:collections' own accession number of the strain.
Other culture collection numbers	Should contain: • accession numbers of the strain in other collections. Cannot be given if the strain is unique in the collection.
Restrictions	should contain: • Release conditions/Restrictions • country-specific shipping restrictions • restrictions imposed by the depositor etc.
Organism type	Should contain: • value «B» for bacteria or • «A» for archaea.
Name	Should contain the full scientific name of the strain: • Genus name and species epithet • Subspecies* • Pathovar* • Authors of the name *: only if applicable
Infrasubspecific names	Should contain: • variety • designation • epithet • authors and reference. Does not apply to all strains.
Status	Should contain: • nomenclature status of the strain (like «type», «neotype», etc.) Does not apply to all strains.
Other names	Should contain: • names used for the strain in the past (not synonyms) that may still be in use. Does not apply to all strains.

Continued

Continued.

Field Name	Description
History of deposit	Should contain the history of the element from depositor to isolator. • Name of depositor • Institute of depositor • Name of isolator • Institute of isolator • Names and institutes in between
Conditions for growth	Should contain: • Culture medium • Atmospheric and light conditions • Temperature conditions
Form of supply	Should contain • the form in which the strain will be sent to the customer.
Serovar	Should contain: • the serovar name and author. Required for some medically important species. Does not apply to all strains.

Appendix I

Taxonomy: An exercise to identify organisms from photomicrographs

The morphology of fungi was the basis of fungal identification until molecular techniques and automated systems to provide genomic analysis came into common use (see Chapter 3, this volume). However, to date there are no complete genomic databases representing all described species nor is it possible to distinguish these species with partial genome sequences so it is not always possible to get an identification from a DNA sequence. Microscopic examination of specimens can allow you to identify an organism. There are a variety of taxonomic keys available, not so new but useful in the process. On first glance these might look difficult to use but an an understanding of the terminology and recognition of key features will unlock these useful tools. This Appendix to *Managing Microorganisms* provides an example of how an isolate of the genus *Colletotrichum* can be identified using Sutton's *The Coelomycetes* (1980). The definitions of the terms are provided in the Glossary of this volume and also available in *The Dictionary of Fungi* (Kirk *et al.*, 2010). A set of diagrams and photomicrographs are provided with descriptions and measurements to help provide some of the information required to follow the key, several elements require observation. There are two examples presented that can be identified from the key. If the wrong species is selected key information is presented to indicate where the mistake was made. Practice will of course make perfect and familiarization with fungal morphology and observing different species using a microscope will enable useful identification skills to be developed. There are also training courses that can provide further help; CABI's training courses can be found via the website: https://www.cabi.org/products-and-services/bioscience-services/training/.

References

Kirk, P., Cannon, P., Minter, D. and Stalpers, J. (2010) *The Dictionary of the Fungi.* 10th edn. CABI, Wallingford, UK. https://www.cabi.org/bookshop/book/9781845939335/

Sutton, B.C. (1980) *The Coelomycetes. Fungi imperfecti with pycnidia, acervuli and stromata.* 696 pp. Commonwealth Mycological Institute, Kew, UK.

DOI: 10.1079/9781800622135.appi

Exercise information to aid with the identification

Colletotrichum Corda, in Sturm, Deutschl. Fl., 3 Abt. (Pilze Deutschl.) 3(12): 41, tab. 21 (1832)

Position in classification: Glomerellaceae, Incertae sedis, Sordariomycetidae, Sordariomycetes, Ascomycota, Fungi

Type species: *Colletotrichum lineola* Corda (1832)

Colletotrichum (sexual stage: Glomerella) is a genus of fungi that are obligate symbionts to plants in the form of endophytes. Many of the species in this genus are plant pathogens, although some species may express a mutualistic life-style in non-disease hosts. Colletotrichum is one of the most common and important genera of plant-pathogenic fungi. Virtually every crop grown throughout the world is susceptible to one or more species of Colletotrichum. Colletotrichum fungi cause post-harvest rots, and anthracnose spots and blights of aerial plant parts. Members of this genus cause major economic losses, especially of fruits, vegetables, and ornamentals. Colletotrichum is very damaging to important food crops, including bananas, cassava, sorghum, and pulses, grown by subsistence farmers in developing countries throughout the tropics and subtropics. In addition to their considerable economic importance, the Colletotrichum fungi are highly significant as experimental models in studies of many aspects of fungal development, infection processes, host resistance, signal transduction, and the molecular biology of plant-pathogen interactions. Dozens of laboratories around the world are studying the biology and pathology of various species of Colletotrichum.

Anamorph - Structures producing mitotic spores (or conidia) are referred to as anamorphs, and those producing meiotic spores (ascospores, basidiospores etc.) as teleomorphs. For historical reasons, the different spore-bearing forms have frequently been given separate names. As in many cases plant pathogens are most prominent as anamorphs, they are commonly referred to using their anamorph names.

The key to identify the *Colletotrichum* genus is available in Sutton *The Coelomycetes* (1980)

Description and explanation of the term 'Conidia' from *The Dictionary of the Fungi* (Kirk et al., 2010)

Conidiogenous cells and conidial development

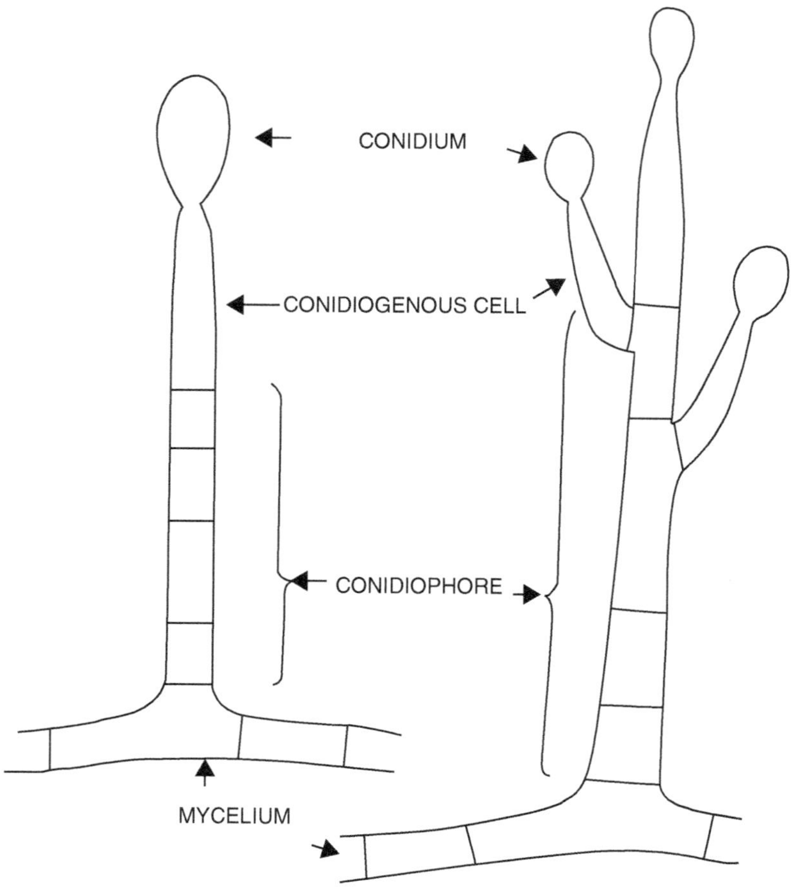

Conidia may be formed directly from more or less differentiated hyphae, or within fruit-bodies. The first recognized classification schemes for anamorphic fungi separated groups based on presence or absence of conidiomata, and on the colour and septation of conidia. More recent studies placed much emphasis on conidial development rather than conidial morphology, and features of conidial development remain very important in classification. It must be emphasized that the plasticity of fungal growth and development is at its most extreme in conidial production, and the situation in real life is often much more complicated than the pigeon-hole type of classification still widely employed for many groups might suggest.

Conidia may be formed in a variety of ways, either by the fragmentation of hyphae, and often by the subsequent differentiation of fragments (***thallic*** development) or by growth of wall material and subsequent differentiation from specialized conidiogenous cells (***blastic*** development). The two modes of development are not always easy to tell apart.

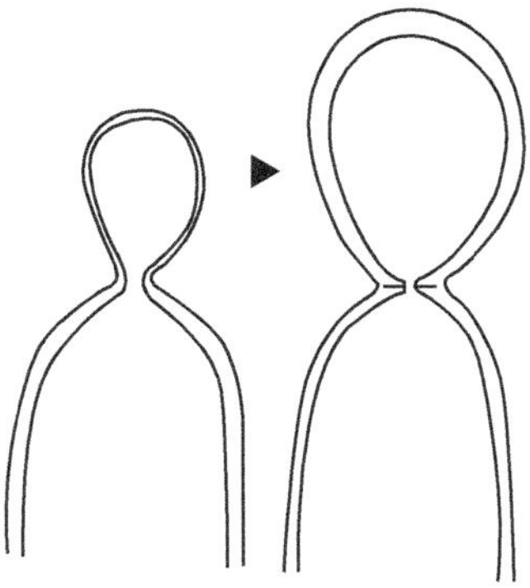

Blastic development

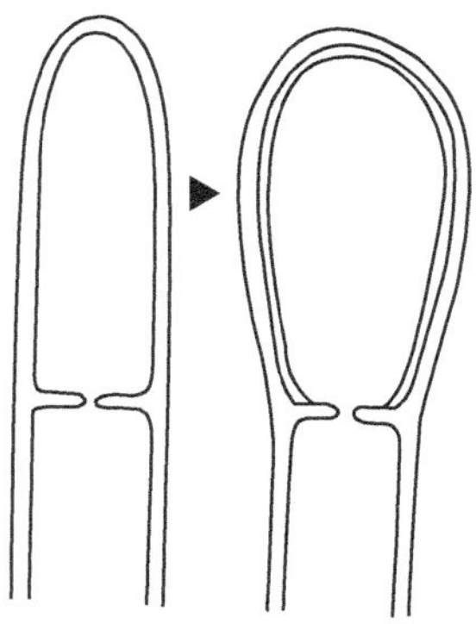

Thallic development

In fungi which produce conidia through thallic development, conidiogenous cells are effectively absent, the conidia being produced from undifferentiated vegetative hyphae. This does not imply that the production of conidia is random or uncontrolled; in many cases conidia are produced from specific portions of hyphae, either as short side branches, as modifications of alternate cells within the hypha, or indeed as hyphal lengths with specific numbers of cells. Once the conidium has been delimited, further differentiation often occurs such as wall thickening or pigmentation. Where two-ply separation septa are not formed, secession (the separation of the conidium from the hypha) occurs by the breakdown of the walls of the hyphal cell adjacent to the conidium. Loss of cytoplasm and nuclei is usually avoided by the contents of that cell becoming absorbed either into the conidium or into the supporting hyphal cell.

Most fungi producing conidia through blastic development exhibit recognizable conidiogenous cells, though in some very simple forms (e.g. *Acremonium*) these may appear to be little more than side branches of hyphae with fertile tips. In most cases, conidiogenous cells are able to produce conidia successively, either in succession from the same point or locus (***percurrent proliferation***, sometimes referred to as ***phialidic*** production) or singly from successive points (***sympodial proliferation***). In others, the benefits of producing a succession of relatively small and simply constructed conidia are overcome by those of generating large and complex structures, and a conidiogenous cell may produce only a single conidium.

Especially in cases where conidia are separated by wall breakage between pairs of one-ply septa rather than by two-ply septa, a ***scar*** may remain on the conidiogenous cell composed of the remnants of separating cell wall, showing the position of previous conidial production. In some cases in fungi employing percurrent conidial proliferation, this wall material may be relatively extended, producing a cylindrical or funnel-shaped structure known as a ***collarette***. This is thought to protect the vulnerable meristematic region as successive conidia are produced.

Where conidiogenous cells exhibit percurrent proliferation, the successive production of inner wall layers at the conidiogenous locus results in a local wall thickening, with only a narrow pore connecting the conidiogenous cell with the developing conidium. This is termed ***periclinal thickening***, and is one of the surest ways of detecting percurrent proliferation.

It is not uncommon for conidiogenous cells to exhibit a mixture of sympodial and percurrent proliferation. Frequently, this appears to be a more or less random process with one type of proliferation dominating. In a few genera the two types of proliferation are clearly controlled; in such cases the conidiogenous cells are sometimes referred to as ***polyphialidic***.

In some cases, in conidiogenous cells which exhibit percurrent proliferation, there is a small degree of proliferation (growth) of the fertile apex of the conidiogenous cell itself in between production of successive conidia. This gives the apex of the conidiogenous cell a banded appearance due to the successive breakages of the cell wall as conidia succeed and growth of the inner wall layers through the scar. This is often referred to as ***annelidic proliferation***.

The morphology of conidiogenous cells is quite variable, ranging from almost unmodified hyphal branches to clearly distinguishable cylindrical, flask-shaped or barrel-shaped cells. They are only rarely strongly pigmented, due to the difficulty of proliferating through a dark and hardened cell wall. The ***conidiogenous loci*** (i.e. the parts of the cell from which conidia are actually produced) are usually attenuated to some degree, especially where percurrent or annelidic proliferation is employed. In fungi forming conidia within fruit-bodies, the conidiogenous cells are necessarily smaller and less differentiated due to the restriction of space for development.

Conidia

Propagules which are formed without a preceding meiotic division are given the blanket term ***conidia*** (here including spermatia and resting-spores; see above). Conidia are not constrained by production within asci, and thus show an even wider range of form, especially those formed without a protective fruit-body, and those dispersed in water rather than air. Many (especially those functioning as spermatia) are small, hyaline, thin-walled and aseptate. Dispersal conidia are much more varied in form, and may be transversely and also longtudinally septate, strongly and homogenously or differentially pigmented, and variously ornamented. They are immensely varied in shape, branched or coiled, and may have cellular or gelatinous appendages from the apex, base, sides or all three of these positions. Survival conidia (or resting-spores) are usually dark and thick-walled, and may be unicellular or multicellular, in the latter case intergrading with sclerotia.

The non-specialist sometimes has difficulty distinguishing conidia from ascospores, especially where material that is not in ideal condition is examined. The mistake is often made where asci are not observed that the material must therefore be of an anamorphic fungus. However, it is rare for a specimen or culture to show conidia without conidiogenous cells, as conidial production frequently occurs over an extended period. In the absence of asci or conidiogenous cells, the most reliable way to distinguish the two types of propagule is to look for the scar where the wall between the conidium and conidiogenous cell has broken. However, many conidia are separated with double septa (see above), and the scar may be difficult to distinguish. Other pointers are that conidia tend to be more variable in size and shape than do ascospores.

Identification exercise: information to use in the key

Using the key below name the *Colletotrichum* species depicted;

Example 1: Conidia more than 12μm long not more than 4.5μm wide; colonies very variable; few sclerotia appressoria 6-20 x 4-12μm; very wide host range

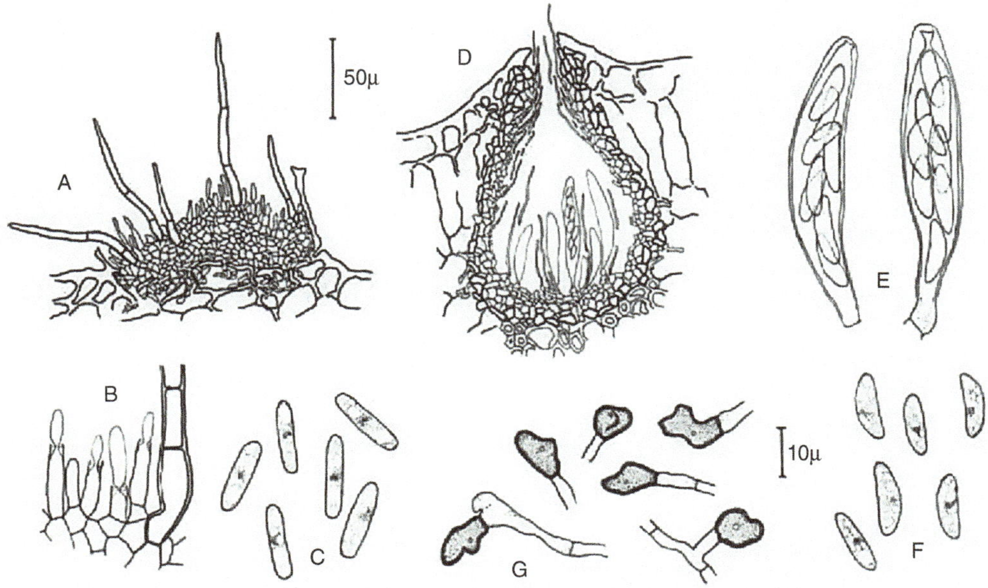

A, T.S. acervulus; B, conidiophores; C, conidia; D, T.S. perithecium; E, asci; F, ascospores; G, appressorium formation from hyphae.

Example 2: Isolated from soil

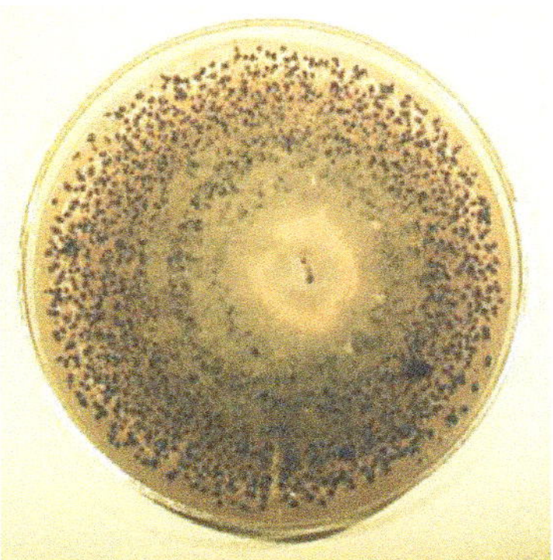

Culture with sclerotia

Setae and conidia

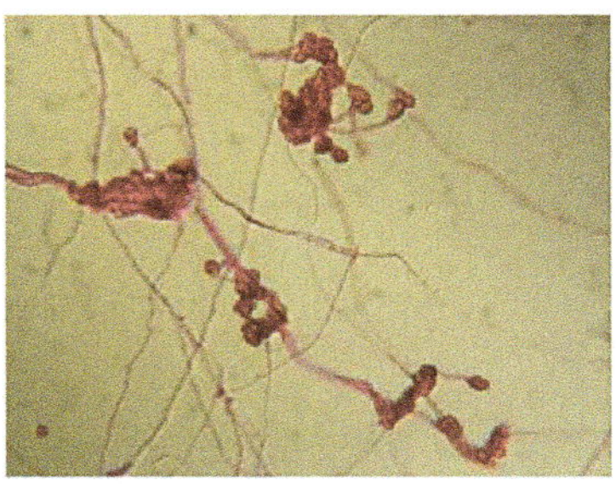

Appressoria

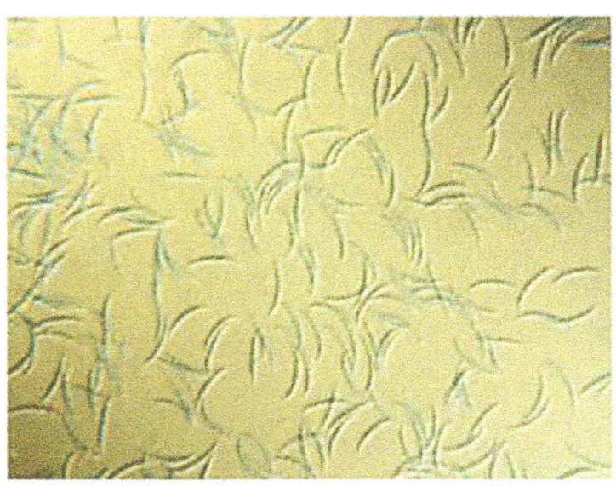

Conidia average size 2.4μm wide

Key to *Colletotrichum* anamorphs (mostly derived from Sutton, 1980).

1. Conidia straight 2
1. Conidia curved (falcate) 8
2. Conidia fusiform (narrowing towards each end) 3
2. Conidia cylindrical 4
3. Colonies not producing sclerotia; culture medium usually with pink pigmentation (especially on nitrogen-rich substrates); wide host range .... *C. acutatum*
3. Colonies producing abundant setose sclerotia; conidia slightly constricted in the mid portion; frequently on Solanaceae .... *C. coccodes*
3. Conidia sometimes slightly curved, 15-19 x 4µm; on *Brassica* spp. .... *C. higginsianum*
3. Conidia sometimes slightly curved, 14-17 x 3.5-4µm; resting-spores present; appressoria absent; on Scrophulariaceae .... *C. fuscum*
3. Conidia often slightly curved, 12-22 x 2.5-4.5µm; appressoria 6-15 x 4.5-10µm, ovate but often irregularly lobed; on Leguminosae .... *C. destructivum*
4. Conidia rarely more than 12µm long; appressoria small; spherical sclerotia present; on Leguminosae..... .... *C. lindemuthianum*
4. Conidia almost always more than 12µm long 5
5. Conidia 4.5-6.5µm wide 6
5. Conidia usually not more than 4.5µm wide 7
6. Conidia 10-15 x 4.5-6.5µm; setae present; appressoria usually scalloped (crenate); often on Orchidaceae .... *C. crassipes*
6. Conidia 12-17 x 4.5-5.5µm; sclerotia and setae absent; appressoria irregular, usually with large or deep lobes; on *Musa* .... *C. musae*
6. Conidia 14-15 x 4.5-6µm; sclerotia and setae present; appressoria clavate, elongated, entire; on Cucurbitaceae .... *C. orbiculare*
7. Colonies dense, floccose, pale chocolate brown; sclerotia absent; setae usually absent; conidia 12.5-19 x 4µm, formed from the mycelium; coffee berry disease pathogen .... *C. kahawae*
7. Colonies very variable; sclerotia occasionally present; conidia 9-24 x 3-4.5µm; appressoria 6-20 x 4-12µm; very wide host range .... *C. gloeosporioides*
8. On Gramineae 9
8. On other hosts 10
9. Apex of conidium prolonged into a narrow cellular appendage .... *C. caudatum*
9. Sclerotia absent; conidia 15.5-26.5 x 4-5µm; appressoria 12.5-14.5 x 9.5-12µm, ± entire; on *Saccharum* .... *C. falcatum*
9. Sclerotia present; conidia 18.5-27.5 x 3-4.5µm, appressoria 11.5-15 x 8.5-9µm, with an irregular margin; on *Sorghum* .... *C. sublineolum*
9. Sclerotia present; colonies vinaceous lilac; conidia 23.5-29 x 3.5-5µm; appressoria 17.5-20 x 12.5-14µm, very irregular; on *Zea* .... *C. graminicola*
10. Conidia not strongly curved (see also 3c, d) 11
10. Conidia strongly curved 12
11. Sclerotia absent; conidia straight or only slightly curved, 16-19 x 3-4.5µm; appressoria 10-13.5 x 6.5-13.5µm, very variable; on *Linum* .... *C. lini*
11. Sclerotia present; conidia moderately curved, 14.5-19.5 x 3.5-4µm; appressoria clavate or circular, 6.5-11 x 5.5-8.5µm; on *Corchorus* .... *C. corchori*
11. Sclerotia present; conidia moderately curved, 14-24 x 4-6µm; appressoria scalloped, 12-18 x 8.5-12µm; on *Hedera* .... *C. trichellum*
11. Conidia 14.5-18.5 x 5-5.5µm, widest near the base; on *Artemisia* .... *C. phyllachoroides*
12. Conidia 2-2.5µm wide; wide host range, usually saprobic .... *C. dematium*
12. Conidia 3.5-4µm wide 13
13. On *Allium* .... *C. circinans*
13. On Leguminosae .... *C. truncatum*
13. Wide host range, usually pathogenic .... *C. capsici*

Response to the species you have selected

If you find the species name in the list below you are given a hint where you may have gone wrong. If you do not find the species name you have found the correct answer. The solutions and the route to them are given on the following page.

C. acutatum – conidia not curved - fusiform (narrowing towards each end)
C. coccodes - Colonies producing abundant setose sclerotia; conidia slightly constricted in the mid portion; frequently on Solanaceae
C. higginsianum - Conidia sometimes slightly curved, 15-19 x 4μm; on Brassica spp
C. fuscum - Conidia sometimes slightly curved, 14-17 x 3.5-4μm; resting-spores present; appressoria absent; on Scrophulariaceae
C. destructivum - Conidia often slightly curved, 12-22 x 2.5-4.5μm; appressoria 6-15 x 4.5-10μm, ovate but often irregularly lobed; on Leguminosae
C. lindemuthianum - Conidia rarely more than 12μm long; appressoria small; spherical sclerotia present; on Leguminosae
C. crassipes - Conidia 10-15 x 4.5-6.5μm; setae present; appressoria usually scalloped (crenate); often on Orchidaceae
C. musae - Conidia 12-17 x 4.5-5.5μm; sclerotia and setae absent; appressoria irregular, usually with large or deep lobes; on Musa
C. orbiculare - Conidia 14-15 x 4.5-6μm; sclerotia and setae present; appressoria clavate, elongated, entire; on Cucurbitaceae
C. kahawae - Colonies dense, floccose, pale chocolate brown; sclerotia absent; setae usually absent; conidia 12.5-19 x 4μm, formed from the mycelium; coffee berry disease pathogen
C. caudatum - Apex of conidium prolonged into a narrow cellular appendage
C. falcatum - Sclerotia absent; conidia 15.5-26.5 x 4-5μm; appressoria 12.5-14.5 x 9.5-12μm, ± entire; on Saccharum
C. sublineolum - Sclerotia present; conidia 18.5-27.5 x 3-4.5μm, appressoria 11.5-15 x 8.5-9μm, with an irregular margin; on Sorghum
C. graminicola - Sclerotia present; colonies vinaceous lilac; conidia 23.5-29 x 3.5-5μm; appressoria 17.5-20 x 12.5-14μm, very irregular; on Zea
C. lini - Sclerotia absent; conidia straight or only slightly curved, 16-19 x 3-4.5μm; appressoria 10-13.5 x 6.5- 13.5μm, very variable; on Linum
C. corchori - Sclerotia present; conidia moderately curved, 14.5-19.5 x 3.5-4μm; appressoria clavate or circular, 6.5-11 x 5.5-8.5μm; on Corchorus
C. trichellum - Sclerotia present; conidia moderately curved, 14-24 x 4- 6μm; appressoria scalloped, 12-18 x 8.5-12μm; on Hedera
C. phyllachoroides - Conidia 14.5-18.5 x 5-5.5μm, widest near the base; on Artemisia
C. circinans - Conidia 3.5-4μm wide on Allium
C. truncatum - Conidia 3.5-4μm wide on Leguminosae
C. capsici - Conidia 3.5-4μm wide, wide host range, usually pathogenic

Solutions

Example 1 = *Colletotrichum gloeosporioides* (Penzig) Penzig & Sacc.

TELEOMORPH: Glomerella cingulata (Stonem.) Spaulding & von Schrenk

Key responses: 1. Conidia straight → 2. Conidia cylindrical → 4. Conidia almost always more than 12µm long → 5. Conidia usually not more than 4.5µm wide → 7. Colonies very variable; sclerotia occasionally present; conidia 9-24 x 3-4.5µm; appressoria 6-20 x 4-12µm; very wide host range

Example 2 = *Colletotrichum dematium* (Pers.) Grove; TELEOMORPH: unknown.

Key responses: 1 Conidia curved (falcate) → 8. On other hosts → 10. Conidia strongly curved → 12. Conidia 2-2.5µm wide – wide host range usually saprobic

Glossary

Absorbance

The measure of the amount of light absorbed by a suspension of bacterial cells or a solution of an organic molecule by use of a colorimeter or spectrophotometer. Absorbance may be applied in plotting the growth of bacteria in liquid cultures. It may also be used in assessing the purity and concentration of molecules (e.g. proteins) in solution.

Accreditation

Confirmation and recognition of technical competence. Procedure whereby an authority recognizes that a body or individual is competent to carry out certain tasks (ISO/IEC Guide n°2). Here, accreditation is considered to comprise the recognition that an organization uses satisfactory and pre-determined methods and is adequately organized to best answer customer satisfaction (service as well as product quality).

Acquisition

To obtain an item.

ADR Regulations

The European Agreement on the International Carriage of Dangerous Goods by Road signed in 1957 (http://www.unece.org/trans/danger/publi/adr/adr_e.html). See 2017 update at: https://unece.org/transportdangerous-goods/adr-2017-files

Aerobic

Used for an organism requiring free oxygen for growth; an aerobe.

Agar

A substance derived from some algae used as a setting agent in making culture media.

Aggressins

Proteins produced by a pathogenic microbe which inhibits the hosts immune system (mainly the white blood cells) enabling the microbe to spread.

Algae

(singular: Alga) Simple non-flowering plants or plant-like organisms that includes most of the seaweeds and many single-celled organisms.

Aliquot

A small portion or measure of a substance or microorganism. A way of measuring a substance below the scale of sensitivity

Alkalophilic

Microorganism which prefers an alkaline environment.

Alkylphenone

An alkylphenone is a small lipophilic hormone; these are used as markers in HPLC.

AMES test

A bacterial assay test developed by Bruce Ames and used to evaluate the mutagenic potential of chemical substances by screening the ability of bacteria to create mutagens or carcinogens.

Amphipathic lipids

A molecule such as a phospholipid, which contains both hydrophobic and hydrophilic parts thus causing a specific arrangement of micelles, with the hydrophobic part near the centre and the hydrophilic part at the surface.

Amplification

The creation of multiple copies of DNA or gene sequences as part of a method in molecular biology.

Antagonistic

A substance that counteracts the action of another substance, when produced by a microorganism it impedes or affects the growth of another.

Antibodies

Proteins which are created in response to the existence of a foreign body which bind to the foreign body to assist in removing it from the host organism.

Antigens

Substances which cause the host immune system to react.

Appressoria

Many parasitic fungi have specialized structures aiding penetration of their host, appressoria. These are usually produced almost directly from a hypha growing from the germinating spore, and are usually flattened and strongly pigmented.

Archaea

A group (clade) of single-celled prokaryotic organisms lacking defined nuclei, formerly called archaebacteria, which differ fundamentally from bacteria.

Artefact

A man-made object. An experimental result or structure obtained due to the incorrect manner in which the test or specimen was handled.

Asci

(singular: Ascus) A sac-like structure containing ascospores.

Ascospore

A spore produced as a result of sexual reproduction by fungi belonging to the Ascomycota;

formed by free cell formation in an ascus following fusion of nuclei and meiosis.

Aseptic technique

A method of preparing and handling materials or organisms which reduces or excludes the possibility of contamination by another substance or organism.

Atypical

An area of a culture where the growth of the organism is not typical of that organism, in that it does not exhibit the usual characteristics.

Audit

A process of tracking and evaluating the status of a test or laboratory.

Auditor

A person who is mandated to undertake the audit and report on the status of a laboratory.

Australia Group (AG)

An informal forum of countries which, through the harmonization of export controls, seeks to ensure that exports do not contribute to the development of chemical or biological weapons. Coordination of national export control measures assists the Australia Group participants to fulfil their obligations under the Chemical Weapons Convention and the Biological and Toxin Weapons Convention (BTWC) to the fullest extent possible (http://en.wikipedia.org/wiki/Biological_Weapons_Convention).

Authentic

An organism which has been examined and identified by the person who originally named it .

Authentication

The process by which biological materials are characterized up to a defined level using appropriate technology to establish a conclusive basis for accepting the material as genuine.

Autoclave

A device that uses steam (at high pressure) to sterilize equipment, waste or growth media used in microbiology.

Auxotroph

An organism which is unable to synthesize a substance that is required for its growth.

Bacteria

Unicellular prokaryotic organisms.

Bacteriophage

A virus that infects a bacterium, often shortened to phage, and usually kills it.

Ballistospore

A forcibly discharged basidiospore.

Basidiospore

A spore produced as a result of sexual reproduction by fungi belonging to the Basidiomycota; formed by a basidium.

Biodegradation

The breakdown of a substance by biological activity (usually of microbes) which often makes the substance less damaging to the environment.

Large complex molecules are broken down so that they may be more easily used or broken down further by natural processes.

Biodegradative

A substance which causes another substance to break down.

Biodeterioration

The gradual breakdown of a material or substance, usually caused by microbes. In the case of foods this may cause them to become spoilt, e.g. less palatable or with an unpleasant smell.

Bioinformatics

The use of computers to solve problems in the generation, analysis and storage of biological data.

BIOLOG

A proprietary system fused for the identification of microorganisms based upon their physiological or biochemical activities.

Biological agent

A microorganism which can cause adverse effects to human health, normally a human pathogen or toxin producer; a microorganism that presents a risk in terms of biosafety or biosecurity.

Biological materials

All materials listed in the Organisation for Economic Co-operation and Development (OECD) definition of BRCs given below.

Biological Resource Centres

An essential part of the infrastructure underpinning biotechnology. They consist of service providers and repositories of the living cells, genomes of organisms, and information relating to heredity and the functions of biological systems. BRCs contain collections of culturable organisms (e.g. microorganisms, plant, animal and human cells), replicable parts of these (e.g. genomes, plasmids, viruses, cDNAs), viable but not yet culturable organisms' cells and tissues, as well as databases containing molecular, physiological and structural information relevant to these collections and related bioinformatics. BRCs must meet the high standards of quality and expertise demanded by the international community of scientists and industry for the delivery of biological information and materials. They must provide access to biological resources on which R&D in the life sciences and the advancement of biotechnology depends.

Biomass

A quantitative evaluation of the total number of living organisms in a specific environment by weight, volume or energy.

Biopolymer

A polymer found in a living organism, such as a polysaccharide, protein or nucleic acid.

Bioremediation

The use of microorganisms to reduce the effect of substances which are damaging the environment (e.g. as pollutants).

Biorisk

The probability of harm or damage from biological materials especially in relation to human infection or toxicosis. In the laboratory setting it refers to both the level of biosafety and biosecurity.

Bios

A mixture of aneurin (thiamine, vitamin B1), biotin (vitamin B7) and other substances in yeasts which, on addition to culture media, results in better growth.

Biosafety

The use of containment principles, technologies and practices that are implemented to prevent unintentional exposure to pathogens and toxins, or their accidental release.

Biosecurity

Institutional and personal security measures and procedures designed to prevent the loss, theft, misuse, diversion or intentional release (in direct or indirect acts against humans, livestock, crops, food, water infrastructure and other economically-valuable entities) of pathogens, or parts of them, and toxin-producing organisms, as well as such toxins that are held, transferred and/or supplied.

BTWC Countries

Countries who have signed (ratified) the Biological and Toxin Weapons Convention.

Budapest Treaty

A treaty set up in 1977 which is open to those who signed the Paris Convention for the Protection of Industrial Property of 1883. The African Regional Industrial Property Organization (ARIPO), the Eurasian Patent Organization (EAPO) and the European Patent Organization (EPO) have all ratified the treaty which defines the terms for establishing a repository for the storage of microorganisms for patent purposes.

Buffer

A factor which prevents change in another substance or structure; usually used with reference to pH.

CABRI

Common Access to Biological Resources and Information.

Calibrated to standards

The process of testing the readings of laboratory equipment against a traceable standard, for example comparing measurements of temperature or pressure taken by laboratory equipment and the standard device to ensure the readings are the same (within acceptable limits) and if not, adjustments are made (the equipment is calibrated).

Capillary action

The movement of water through porous substances such as the flow of water through soil.

Carcinogenic

A substance causing or with the ability to cause cancerous growth; a carcinogen.

cDNA

RNA converted 'back' to DNA using the enzyme reverse transcriptase.

Centrifuge

An instrument that revolves around a central axis at high speed to draw liquids or to separate sediments to the bottom of a tube; used to separate components.

Certification

Confirmation of conformity to the requirements of a standard or where a third party gives written assurance that a product, process or service conforms to specified requirements or approval deriving from an authority. Issue of an authorization, recognized by the government, to perform a certain task.

Change management

The process of managing the reorganization of procedures, processes and structures in an organization with a new set of objectives in mind.

Chlamydospores

A one celled asexual spore usually thick-walled with survival as its primary function.

Chromophore

A group of atoms in a compound which absorb light, the result of which is the appearance of colour.

Chromatography

A diffusion technique in which a mobile phase (liquid or gas) is passed over a solid phase (liquid or solid) to separate the chemicals and allow their identification.

Chromogenic

A substance which produces colour.

Clavate

Club like; narrowing in the direction of the base.

Cloning

The process of reproduction in an organism which results in genetically identical copies of the original.

Coliphages

A bacteriophage that infects coliform bacteria, for example *E. coli*.

Colony

Discrete growth of a population of bacteria (or thallus of a fungus), typically a single progeny.

Concentration

The quantity of a substance 'mixed' with another.

Conidia

(singular: Conidium) Propagules which are formed without a preceding meiotic division are given the blanket term *conidia*.

Conidial

Producing the asexually derived spore.

Conidiophores

Frequently, conidiogenous cells are located singly or in clusters on specialized hyphae known as *conidiophores*. These may be largely similar in appearance to vegetative mycelium, though may be oriented vertically rather than horizontally, or strongly differentiated. They may be branched irregularly or verticillately, or have swollen apices, and are often strongly pigmented. The conidiophores themselves may be clustered together into cylindrical structures known as *synnemata* or *coremia*. Conidiophores are only well-developed in fungi producing conidia without fruit-bodies, though in these instances it is not uncommon for conidiogenous cells to be inserted singly or in small clusters on supporting cells.

Conjugated systems

The processes whereby atoms in a compound covalently bind alternately with single and multiple bonds creating an electron delocalized region.

Conservation

The process of preserving and storing organisms for later study, use or enjoyment.

Conspecificity

Belonging to the same species.

Constrict

The process of heating ampoules with a gas torch so they can be stretched to reduce the diameter and make the ampoule narrower, normally to facilitate its heat sealing later in the process.

Containment level

Microorganisms are assigned to Risk Groups (Biosafety Levels; Hazard Groups) defined for example by the WHO. The containment level numbers correlate with the risk group in which the organism falls (i.e. organisms in Risk Group 1 require Containment level, 1 and so forth).

Contaminant

The organism which has caused the culture or broth to be contaminated i.e. the presence of the organism in a culture plate or broth that is not desired.

Contaminated

The presence in a culture plate or broth of an organism that is not desired.

Corrective actions

Tasks and procedures which must be taken in order to alter procedures or protocols which caused an error or unwanted outcome to deliver the desired or required outcome.

Coryneform

Bacteria which can be recognized by the irregular, often club-shaped, cells.

Cryopreservation

The process of freezing an organism and storage at ultra-low temperature (normally below −139°C) so that it can be kept stable for future use.

Crystallography

The use of beams of energy to determine the arrangement of atoms in solids by analysing the diffraction of that beam when it passes through the solid.

Culture collection

A facility established to preserve, store and supply living representatives of organisms.

Cyanobacteria

A group of bacteria previously, and misleadingly, referred to as blue-green algae.

Dangerous Goods Regulations

Dangerous goods are liquid or solid substances and articles containing them, that have been tested and assessed against internationally-agreed criteria - a process called classification - and found to be potentially dangerous (hazardous) when carried. Dangerous goods are assigned to different classes depending on their predominant hazard. There are regulations to deal with the carriage of dangerous goods, the purpose of which is to protect everyone either directly involved (such as consignors or carriers), or who might become involved (such as members of the emergency services and public) with them. Regulations place duties upon everyone involved in the carriage of dangerous goods, to ensure that they know what they have to do to minimize the risk of incidents and guarantee an effective response. Carriage of dangerous goods by road or rail is regulated internationally by agreements and European Directives, with biennial updates of the Directives to take account of technological advances. New safety requirements are implemented by Member States via domestic regulations which directly reference the technical agreements.

Defined

The essential elements, qualities or meaning is identified, e.g. a defined medium has all the ingredients identified and quantified.

Denaturation

Structural change in proteins or nucleic acids in response to extremes in conditions, such as pH, temperature, salt concentrations or pressure, which prevent the cell from continuing to perform its original function.

Deoxyribonucleotide

A single unit of DNA consisting of a nitrogenous base, a deoxyribose sugar, and one or more phosphate groups.

Deposit

A sample of an organisms placed in a culture collection or BRC for storage and provision for confirmation of results or for further work. Biologists who collect organisms for their research and publish information on them should make their most important strains available for confirmation of those results and future use by depositing them in public service collections. This will aid collections in their major roles of: The *ex situ* conservation of organisms; Custodians of national resources; Provision of living resources to underpin the science base; Reception of deposits subject to publication; Safe, confidential and patent deposit services. To operate such functions collections must follow standard procedures to offer products and services of a consistent nature.

Dermatophyte

A fungus parasitizing keratinized tissue (hair and skin of humans or other animals) and causing dermatophytosis.

Desiccation

Drying so that a cell contains little or no residual moisture.

Deteriorate

The degeneration (worsening) of a substance or organism.

Dielectric constant

An electrical property which describes the capacity of a material to store electrostatic energy when a voltage is applied. This is the ratio between the potential for a material to store electrical energy when a source of energy is available in comparison to the amount it stores when that material is in a vacuum.

Difference spectrum

The measurement of the variations between spectra over which metabolites exist in neutral solvents as compared with those in alkaline solvents.

Diurnal

Relating to or occurring in a 24-hour period; daily. Occurring or active during the daytime rather than at night.

DNA sequencing

The process of identifying the structure and arrangement of deoxyribonucleic acid (DNA).

Dual-Use

Dual-Use items are goods, software or technology (documents, diagrams etc) and biological materials which can be used for both civil and military applications (or used maliciously by terrorists for example). They can range from raw materials or components to complete systems, e.g. aluminium alloys, bearings or lasers. They could also be items used in the production or development of military goods, e.g. machine tools, chemical manufacturing equipment and computers.

Effluent

Contaminated water or substances from a natural occurring or a man-made source or the flow of this.

Electrophoresis

The movement of dispersed particles in relation to a fluid in the presence of an electric charge.

Elution

The emergence of chemicals from a chromatography column.

Endotoxin

A toxin or poison which is a structural part of the bacteria which may be released when the organism is lysed.

Enzyme systems

A group of enzymes that interact together (in parallel or sequentially) to create a specific outcome.

Eukaryote

An organism with defined nuclei bounded by a nuclear membrane. Cell division in eukaryotes is different from that in organisms without a nucleus (prokaryotes). It involves separating the duplicated chromosomes. There are two types of division processes. In mitosis, one cell divides to produce two genetically identical cells. In meiosis, which is required in sexual reproduction, one diploid cell (having two instances of each chromosome, one from each parent) undergoes recombination of each pair of parental chromosomes, and then two stages of cell division, resulting in four haploid cells (gametes).

Ex situ genetic resources

Biological materials, capable of reproducing, that are removed from the environment in which they occur and stored in collections.

Ex situ

Removed from the environment in which it occurs, in this context, a collection of microorganisms stored in the laboratory or culture collection.

Exotoxin

A toxin which is excreted by a microorganism often to break down the cell wall of another cell.

Extracellular

A substance or process which occurs outside of the cell wall.

Extremophile

An organism that grows and reproduces in extreme environments (e.g. cold, hot, extremes of pH).

Ex-type strain

A strain of a microorganism which is derived from the physical collection (where this is not a culture) on which the name of the organism was based (the nomenclatural type).

Facultative

An organism that can live under different conditions, e.g. a facultative anaerobe can live with or without oxygen.

FAO

Food and Agriculture Organization – an agency of the UN.

Fastidious

An organism which requires a complex combination of nutrients in order to grow; meticulous in requirements.

FDA

Food and Drug Administration – an agency of the US Public Health Service.

Fermentation

The process of metabolism to produce energy to sustain growth (and often a useful by-product), under controlled conditions of light, temperature and nutrition. Typically undertaken in broth culture, often in a shaking incubator.

Filamentous

Thread-like; in microorganisms a line of adjoining cells.

Fingerprinting

Procedures used to examine microorganisms which enable the identification of a strain-specific pattern of products.

Flaming

A technique for sterilizing small pieces of laboratory equipment, for example forceps or scalpels, by dipping the implement into IMS, carefully shaking off excess and carefully passing through the flame of a Bunsen burner to burn off the IMS. The purpose of this technique is not to set alight anything that is on the bench or the person undertaking the technique.

Floccose

Cottony, byssoid – made up of delicate threads.

Flocculation

Aggregation of cells to form 'flocs', occurring during stationary phase culture; a fluffy mass of suspended particles in a liquid.

Fluorescence

The emission of light resulting from the absorption of radiation.

Fluorochrome

A dye which emits light.

Fluorogenic

Generally, an assay which will generate light through a chemical reaction if certain analytes are present.

Formulation

The end constituents of a mixture, for example the ingredients of a culture medium.

Freeze-drying

Freeze-drying (lyophilization) is the removal of water by freezing under reduced pressure and the subsequent sublimation of the ice directly to vapour (i.e. without passing through the potentially damaging ice crystal stage) at low temperature.

Fruit body

A general term for spore-bearing structures in microfungi and macrofungi.

Functional genomics

The specific application of the information from the variety of genetic projects to describe the dynamic interactions of genes.

Fungal biotechnology

The use of the biological properties of fungi to create technologies and products which can be applied in various contexts, for healthcare, environment, bioindustry and in agriculture.

Fungal taxonomy

The arrangement of organisms (fungi) into a classification.

Gene clusters

The arrangement of functionally similar genes which each encode specific proteins.

Genetic

The genome or nucleic acid component of cells containing the hereditary information.

Genome

The total inheritable genetic material of an organism.

Genomic

Pertaining to an organism's genome.

Genomics

The study of all the genes or the genome which make up a specific organism.

Germplasm

'Original cells, often stored in collections, from which an organism can be regenerated.'

Good Management Practice

GMP is a term that is recognized worldwide for the control and management of manufacturing and quality control of foods, pharmaceutical products, and medical devices.

Gradient HPLC

High-performance liquid chromatography (HPLC) is an important tool for the fast separation of complex polymers with regard to chemical composition. HPLC separations can be achieved via different mechanisms, including adsorption-desorption and precipitation-redissolution. In gradient HPLC, frequently precipitation and adsorption processes are combined.

Guideline

A non-specific rule or principle that provides direction to action or behaviour.

Haemorrhagic

An organism which causes the breakdown of cell walls and the haemorrhaging of another organism's contents.

Halophile

A microorganism which grows in a salty environment.

Halophilic

A microorganism which requires high water activity such as a salty environment to grow.

Halotolerant

A microorganism which tolerates or grows well in a salty environment.

Hazard

A substance or process that can cause injury (sometimes fatal).

Hepatotoxic

A substance that is toxic to the liver

Heterotroph

An organism using organic compounds as its primary energy source.

High fidelity

An organism which has a very specific set of characteristics adapted to a specific environment.

Histidine mutants

Organisms that have natural or induced changes with their histidine (amino acid with a pKa of 6 or 7) production or requirements.

Holotype

A specimen (or rarely an illustration) designated by the author of a scientific name of an organism as the nomenclatural type. The type serves to fix the application of a name; it is permanently linked to the name.

Hybridization

(i) The cross-fertilization of two separate species to create a hybrid or (ii) the process of creating a double-stranded nucleic acid from complementary strands of DNA.

Hydrolyses

Undergoes hydrolysis; decompose by reacting with water, a chemical reaction during which one or more water molecules are split into hydrogen and hydroxide ions.

IATA

International Air Traffic Association.

IATA Dangerous Goods Regulations

A series of regulations governing the transport of substances by air that could cause harm/problems if they are released in a public environment.

Immunosuppressant

A substance which suppresses the immune system of an organism.

In situ

Something in its original environment (in context, a microorganism in its original environment).

Incineration

The burning of waste, in a microbiological sense, to render it sterile.

Incubation

The maintenance of an organism under a set of controlled environmental parameters, for example, temperature, light or humidity.

Incubators

The apparatus used to provide incubation conditions.

Industrial alcohol

Also known as IMS or Ethyl alcohol denatured by the addition of 5% methyl alcohol; a form of methylated spirit for use in industry and the laboratory – usually free of alcohol taxes.

Industrial methylated spirit

IMS is Ethyl alcohol denatured by the addition of 5% methyl alcohol; a form of methylated spirit for use in industry and the laboratory – usually free of alcohol taxes.

Inherent risk

Is the risk associated with the nature of the substance or organism; a danger or problem related to a constituent or characteristic to a specific process or set of activities.

Injection port

A device or specific site that allows the introduction of a substance into a system or instrument.

Inoculating loop

A wire or disposable plastic device used to transfer bacteria from one source to another.

Inoculation

The act of putting an inoculum onto a fresh nutrient source using aseptic technique.

Inoculum

A defined type or volume of a microorganism which is prepared for addition to a fresh nutrient source.

Internal transcribed spacer

A sequence of DNA between genes in an organism.

Intracellular

A process or element that occurs inside a cell.

Intraspecific

A characteristic or function which occurs only in members of one species.

Inventory

System (usually numbered) used to store ampoules so they are easily retrievable; a detailed list of all items in store.

Invertase

An enzyme that causes the hydrolysis of sucrose into glucose and fructose.

Ionic strength

A measure of the concentration of ions in a solution.

ISO

International Organization for Standardization (www.iso.org); usually, but incorrectly, referred to as the International Standards Organization.

Isolate

A culture of an organism; isolation is the process of getting a microorganism into pure culture.

Leading edge

The interface between diseased and healthy tissues where a pathogenic organism is at its most active. The region to isolate from if a culture is required.

Learned societies

Professional groupings of experts in a specific field; for example, The Society for Applied Microbiology.

Legislation

(i) The act of legislating; preparation and enactment of laws; the laws enacted. (ii) Law which has been enacted by legislature or other governing body.

Lichen

A microorganism that results from the symbiotic relationship between a fungus and a green alga or less frequently a cyanobacterium.

Lichenicolous

A fungus which lives on or in a lichen.

Ligated

Nucleic acid which has been joined using DNA ligase.

Liquefaction

The process by which a solid or a gelatinous substance becomes liquid, for example, a fungus may liquefy a culture medium.

Lymphatic

Describing a system of ducts and tubules that run through a multicellular organism and carry waste matter to sites in the body where it can be processed and discarded.

Lymph system

A system of ducts and tubules that run through a multicellular organism and carry waste matter to sites in the body where it can be processed and discarded.

Lyophilization

Freeze-drying which is the removal of water by freezing under reduced pressure and the subsequent sublimation of the ice directly to vapour (i.e. without passing through the potentially damaging ice crystal stage) at low temperature.

Lysis

The deterioration and rupture of the cell wall, causing the destruction of the cell.

Maintenance

The process of ensuring that all equipment in a laboratory, through regular checks, is functioning at optimal performance.

Manifold

A piece of equipment that is part of a freeze-dryer which has multiple connection points to enable ampoules to be attached prior to a vacuum being drawn inside the ampoules.

Mass spectrometry

An analytical technique for the determination of the elemental composition of a sample or molecule. It is also used for elucidating the chemical structures of molecules, such as peptides. The principle consists of ionizing chemical compounds to generate charged molecules or molecule fragments and measurement of their mass-to-charge ratios.

Media

(singular: Medium) Artificially prepared substrates containing suitable nutrients on which microorganisms can be cultured.

Mesophilic

An organism which requires a medium temperature (usually between 10 and 40°C) for normal development.

Metabolic profiles

An impression of an organism's metabolism; for example, a series of enzymes detected; a series of metabolites resulting from analysis of a cell exudate or contents.

Metabolism

The process in an organism that maintains life, for example, respiration.

Metabolite

The by-product of an organism's metabolism.

Methylated

Containing or impregnated with methylated spirits (Industrial methylated spirit).

Methylotrophic

Organisms capable of reducing single (meth-) carbon compounds, for example, methane or methanol.

MIDI FAME Analysis

Microbial Identification Inc. (MIDI) – Fatty Acid Methyl Ester (FAME) – is a gas chromatography method to produce fatty acid profiles which can then be compared against a database of profiles from named microorganisms, for example, to provide a closest match identification.

Mobile phase

The stage at which an organism's growth rate increases.

Molecular signatures

Is a set of expressed genes that best distinguished the organism or cell from others.

Morphological

The 'form' or 'structure' of an organism ('morphology'=study of).

MRC

Microbial Resource Centre(s).

Muslin

Permeable cotton based woven cloth often used in microbiology to separate liquids from solids in a mixture.

Mutagenic

A substance which causes genetic mutations in an organism.

Mutagenicity

The ability of a substance to cause changes or mutations in a cell or organism.

Mutualistic symbiosis

The relationship between two organisms where both provide necessary nutrients or processes to benefit each other.

Mycelium

A mass of hyphae; the 'thallus' of a fungus.

Mycocins

Toxic substances produced by many yeasts that are lethal to other yeasts.

Mycologist

A scientist specializing in the study of fungi and related organisms.

Mycoplasma

A group of small (0.3–0.8 μm diam.) parasitic bacteria which have no cell wall.

Mycorrhiza

A mutualistic association between a fungus and a plant.

Mycorrhizal

A mutualistic association between a fungus and a plant.

Necrotic; necrosis

Dead, usually of a clearly delimited plant part of portion of tissue.

Necrotroph

(i) A fungus that kills tissues as it grows through them such that it is always colonizing dead tissue; (ii) an organism that obtains nutrients from dead tissues of a living host.

Nematologist

A scientist specializing in the study of nematodes.

Nephrotoxic

A substance that is poisonous to the kidneys.

NMR

Nuclear Magnetic Resonance; a type of spectroscopy using magnets and radio waves to look at the internal structure of objects.

Nomenclature

The assigning of names to taxa, for example, species or genera.

Non-conformities

Used in the quality management sense as a deviation from a specification or a standard or the non-fulfilment of specified requirements.

Non-pathogenic Protozoa

Protozoa which do not cause disease.

Non-stringent anaerobe

An organism that normally would require the absence of oxygen to grow but can grow in the presence of minimal quantities of oxygen.

Noxious fumes

Harmful or toxic vapours or gasses that should not be inhaled.

Nucleic acids

A group of complex compounds found in living cell (especially the nucleus), for example, DNA and RNA.

Nutrilites

Any organic compound necessary in small amounts for the nutrition of an organism.

Obligate

(i) A parasitic organism that can not be cultured in a laboratory. (ii) An organism which is obliged to grow on or in another organism or under specific growth conditions for its survival.

Occupational health

Health of workers in the workplace.

OECD

Organisation for Economic Co-operation and Development; an international organization helping governments tackle the economic, social and governance challenges of a globalized economy.

Oestrogenic

A substance similar to oestrogen which causes oestrus or fertility in animals.

OHSAS

Occupational Health and Safety Advisory Services of the UK (www.ohsas.org)

Oil immersion lens

A microscope lens which works with a small drop of oil between the lens and the slide to increase lens efficiency (i.e. resolution).

Oligonucleotides

Polymers made of 2-20 nucleotides.

Orchidaceous

Plants which belong to the orchid family (Orchidaceae).

Organelles

Parts of a cell which perform specific functions.

Out-compete

To exceed the progress or survival of another organism in an ecosystem.

Overarching

Encompassing or linking all that is within its scope or range.

Patent deposit

A microorganism or collection of microorganisms which is specifically designated, usually by a state, to act as a holding body for, and register the details of, unique microorganisms and the documentation which describes them.

Pathogenicity

The ability of one organism (the parasite) to induce disease in another organism (the host).

Pathologist

A scientist specializing in the study of diseases associated with plants or animals (including man).

PCR

Polymerase Chain Reaction; a technique to amplify a single or few copies of a piece of DNA across several orders of magnitude, generating millions or more copies of a particular DNA sequence. The method relies on thermal cycling, consisting of cycles of repeated heating and cooling of the reaction for DNA melting and enzymatic replication of the DNA. As PCR progresses, the DNA generated is itself used as a template for replication, setting in motion a chain reaction in which the DNA template is exponentially amplified.

Personal Protective Equipment (PPE)

An item of protective clothing or other piece of equipment that must be worn in the laboratory when undertaking certain tasks. They include but are not limited to: a coat, appropriate gloves, a mask, a hair protection net, protective glasses, etc.

Petri dish

A flat, circular, shallow transparent dish with vertical sides and an upper slightly larger dish which overlaps over the lower one when placed together; made of glass or (more frequently) plastic. Used for culturing microorganisms under sterile conditions.

Phage

A type of virus which infects prokaryotes.

Phenotypic

Relating to the expression of a specific biochemical characteristic or physical appearance resulting from the interaction between an organism's genes and its environment.

Photomicrographs

A photograph of an organism in which the image is enlarged by a microscope.

Phototroph

An organism which grows towards a light source.

Phototrophic

Describing an organism which grows towards a light source.

Phylogeny

The sequence of events in the evolution of organism.

Physiological

The biochemical 'functioning' (metabolism) of a cell.

Plant pathology

The study of the cause, nature and effects of diseases in plants.

Plasmid

A linear or circular double-stranded DNA that can replicate independently of the chromosomal DNA. They code for proteins in many bacteria.

Polymerase

An enzyme that catalyses polynucleotides of DNA or RNA to form a copy using an existing DNA or RNA template.

Preservation

The process or method of stopping or reducing metabolism to aid the long-term storage of a microorganism.

Presumptive

A group of cells which will normally develop into a certain type of tissue, although this cannot be confirmed until the culture is fully developed.

Prokaryote

An organism lacking a nuclear membrane; the bacteria and archaea.

Propagules

Structures produced at a specific stage in the life cycle of an organism, often as a result of sexual or asexual reproduction or vegetatively, with the potential to produce a new generation or substratum colonization; for example, conidia, ascospores, heterocyst.

Proprietary

Relating to the ownership of specific information or items.

Proteinaceous

Composed of protein.

Proteomics

The analysis of the expression, localization and interactions of proteomes.

Prototrophic revertant

Usually a mutant organism which has returned to its original genotype and therefore requires the same nutrients as its wild-type strain.

Prototrophy

The ability of an organism to synthesize a specific substance for nutritional purposes.

Protozoa

Predominantly unicellular, plasmodial or colonial organisms. A term used with various definitions.

Psychrophilic

An organism which requires a low temperature (usually below 10°C) for normal development.

Psychrotolerant

Able to grow below 10°C.

Public Service Collection

A Biological Resource Centre which specifically acts as a holding service from which other organizations, individuals and societies can request sample specimens or with whom they can register their own unique specimens for storage.

Pyrogenicity

Causing or caused by heat or fire.

Pyrolysed

A substance which has been heated and has broken down into smaller or simpler components.

Pyrolysis Mass Spectrometry

A process of analysing the chemical compounds (of a microorganism) by heating to break them into smaller molecules. These are separated by gas chromatography and detected by mass spectrometry.

Quarantine

(i). Quarantine is voluntary or compulsory isolation, typically to contain the spread of something considered dangerous, often but not always disease. (ii). The keeping of an organism in isolation from others, particularly in a secure, controlled location. Quarantine organisms may be harmful to the wider environment.

Recalcitrant

Difficult to preserve using current conventional methods.

Reductase

An enzyme that causes the reduction of an organic compound.

Reference Material Producer

An organization which provides source certified reference materials for use as a reference by other manufacturers and researchers. The use of certified reference materials by laboratories, particularly in the chemical and biological disciplines, is one common and practical way of establishing traceability of measurements.

Regulation

A law or administrative rule, issued by an organization, used to guide or prescribe the conduct of members of that organization.

Requested accreditation number

As part of the process for seeking accreditation, an organization is assigned a number which enables the tracking of requests for accreditation.

Respiratory sensitizers

A respiratory sensitizer is a substance which when breathed in can trigger an irreversible allergic reaction in the respiratory system. Once this sensitization reaction has taken place, further exposure to the substance, even to the tiniest trace, will produce symptoms. Sensitization does not usually take place right away, it generally happens after several months or even years of breathing in the sensitizer.

Restriction endonuclease

An enzyme, frequently occurring in bacteria and archaea, which cuts (restricts) the DNA of invading organisms, frequently viruses, at a certain point on the nucleotide chain usually as a result of a stimulus signalled by a specific nucleotide.

Rf values

It is the distance travelled by the sample or analyte divided by distance travelled by the solvent front in chromatography. The value is assigned to describe a factor or substance as separated by a specific chromatography technique.

Risk

A danger to people working in the environment.

Risk assessment

(i) The overall process of identifying all the risks to and from an activity and assessing the potential impact of each risk. (ii) The determination of the potential impact of an individual risk by measuring or otherwise assessing both the likelihood that it will occur and the impact if it should occur, and then combining the result according to an agreed rule to give a single measure of potential impact.

Saponification

The process of hydrolysing an ester under alkaline conditions to produce an alcohol and a carboxylic acid.

Saprobe

A microorganism obtaining nutrients from dead organic matter.

Saprobic

Using dead organic material as food and commonly causing decay.

Saprophyte

A plant that lives on dead or decaying matter; incorrectly used with reference to a fungus or a bacterium.

Sclerotia

(singular: Sclerotium) A compact mass of mycelium, often with a pigmented outer layer. In some instances, the resting structures may be many-celled, sometimes with more than one distinct cell layer. These are frequently overwintering organs, which germinate in the spring to produce mitotic or meiotic spores, which may or may not be contained in fruit bodies.

Screened

Tested to identify an organism or chemical.

Semi-defined

A characteristic which has only been partly defined.

Sequencing

The determination of arrangement or component structure, at the atomic level, of a gene or biopolymer.

Sequencing data

The information derived from sequencing genes or biopolymers.

Short agar slope

A Universal bottle containing nutrient agar that is set an angle of less than 45°.

Single cell protein

Typically refers to sources of mixed protein extracted from pure or mixed cultures of algae, yeasts, fungi or bacteria (grown in fermenters) used as a substitute for protein-rich foods, in human and animal feeds.

Sonification

The process of using non-audible sound to break or disintegrate cells.

Species

The basic rank in the classification of an organism.

Spectrometry

The observation and measurement of the wavelength of light as it is projected through materials.

Spore

A reproductive propagule that may be haploid or diploid.

Sporulate

The ability or process of a fungus, bacterium or alga to produce spores.

Sporulation

The ability or process of a fungus, bacterium or alga to produce spores.

Stability

The potential of an organism to remain 'unchanged' at the genetic or physiological level.

Standard Operating Procedures (SOPs)

An agreed set of actions and tasks which must be undertaken according to an agreed schedule and which is followed by all staff working in a specific environment.

Standards

A set of criteria with minimum agreed levels of action and results required.

Stationary phase

A third stage in the development of a bacterial colony, the growth rate slows as a result of nutrient depletion and accumulation of toxic products. This phase is reached as the bacteria begin to exhaust the resources that are available to them. This phase is a constant value as the rate of bacterial growth is equal to the rate of bacterial death.

Strain

A particular line of cells (microorganism) which have been cultivated from one source.

Strain drift

A change in some of progeny or characteristics of the original strain.

Stringent anaerobe

An organism that grows in the absence of and will not tolerate oxygen.

Sublimation

Transition of a substance from the solid phase directly to the vapour state without passing through the intermediate liquid phase.

Subsequent

Following on from or occurring after an initial event.

Substrate

A substance which can be used by a microorganism to grow.

Susceptible

Being at risk of disease, contamination or degeneration.

Symbiont

An organism which exists in a mutually beneficial arrangement with another organism.

Synergistic

A substance (often produced by a microorganism) that works supportively with other substances (or organisms) and facilitates a specific outcome.

Taxon

A defined unit (e.g. genus, family, species) in the classification of living organisms.

Taxonomic

The description and ordering of organisms according to presumed shared characteristics, properties and qualities.

Terpenoid lipids

Fatty acids composed of volatile aromatic compounds of 5-carbon isoprene units as found in plant essential oils (e.g. menthol, myrcene, limonene, green oil).

Terverticillate

A microorganism (usually a species of *Penicillium*) which has branches at three levels.

Thermophile

An organism that grows optimally at 40 to 50°C.

Thermophilic

An organism which requires high temperatures (optimally 40 to 50°C) for normal development.

Thermostable

The quality or characteristic of a substance or organism to resist change in its chemical or biological structure when subjected to relatively high temperatures.

Thermotolerant

An organism that is able to growth up to 50°C, but which normally has a lower optimum growth temperature.

Titre

Concentration of cells or of a substance or substances in a solution.

Tolerances

Upper and lower limits or measurements which are defined as acceptable.

Toxic metabolites

By-products of an organism's metabolism that are harmful (poisonous) to living organisms.

Toxicity

The degree to which a substance is able to damage an exposed organism. Toxicity can refer to the effect on a whole organism, such as an animal, bacterium or plant, as well as the effect on a substructure of the organism, such as a cell (cytotoxicity) or an organ (organotoxicity), for example, the liver (hepatotoxicity).

Translucent

A substance or object which permits light to pass through it but causes the light to diffuse or scatter so that the object viewed is not clear.

Type strain(s)

(i) The specific strain of a bacterial species which is chosen to define the species. (ii) A living culture derived from the nomenclatural type.

Ubiquitous

Occurring almost everywhere.

UKBRCN

United Kingdom Biological Resource Centre Network.

UKFCC

United Kingdom Federation for Culture Collections.

Ultracentrifugation

A very high-speed centrifuge, often used to separate the organelles and component molecules (according to their weight and sedimentation rate) of previously broken cells.

Ultralow temperature

Temperatures of -139°C or below, typically achieved by mechanical means of the application of cryogenic gases or liquids.

Vacuum

A space where the pressure is less than atmospheric pressure; literally an absence of matter.

Variants

A progeny or population that differs either genetically or phenotypically from its wild type or predecessor.

Vegetative

Assimilative (taking in) or growth phase before or without reproduction.

Viability

The potential of a culture or spore to produce a new culture. Cells that are living or metabolically active are described as viable.

Virulence

The degree of pathogenicity of an organism, or the relative ability of a pathogen to cause disease.

Virus

A transmissible subcellular particle, typically with DNA or RNA surrounded by a protein coat, which replicates only in living cells.

Volatile

A substance with a vaporization temperature below ambient or at relatively low atmospheric temperature that vaporize readily; in relation to microbiology these could be chemical vapours arising from culture media produced as by-products of metabolism in microorganisms.

WDCM

World Data Centre for Microorganisms.

WFCC

World Federation for Culture Collections.

WHO

World Health Organization – an agency of the UN.

Wild type

An organism as it was first isolated from its natural environment.

Xerophile

An organism which grows in an environment where (i) the sugar concentration is high, or (ii) where the humidity is very low.

Yeast

A unicellular fungus which reproduces asexually by budding.

Zones of inhibition

An area in which microorganisms are prevented from developing or certain activities do not take place as a result of some physical or chemical constraint.